Advances in High-speed Rail Technology

"Advances in High-speed Rail Technology" presents the latest and most cutting-edge theories, techniques, and methodologies in the multidisciplinary field of high-speed railways, focusing on advances and findings from China. This series includes monographs, professional books and edited volumes from dedicated conferences and workshops. All volumes are authored or edited by established experts in their fields and undergo rigorous peer review, based on the editors' preview and selection and refereeing by independent experts. The intended audience includes researchers, engineers, industrial practitioners, graduate students, and professionals. Topics of interest in "Advances in High-speed Rail Technology" include, but are not limited to: Infrastructure, Mobile Equipment, Communication & Signal, Traction Power Supply, Operation Organization, etc.

Qiuyi Li · Bin Zhu · Shijie Zhang

Continuously Welded Turnouts on High-Speed Railway Bridge

Technology and Application

SWJTUP 西南交通大學出版社

Springer

Qiuyi Li
China Railway Siyuan Survey and Design Group Co., Ltd.
Wuhan, Hubei, China

Bin Zhu
China Railway Siyuan Survey and Design Group Co., Ltd.
Wuhan, Hubei, China

Shijie Zhang
China Railway Siyuan Survey and Design Group Co., Ltd.
Wuhan, Hubei, China

ISSN 2363-5010 ISSN 2363-5029 (electronic)
Advances in High-speed Rail Technology
ISBN 978-981-97-6523-2 ISBN 978-981-97-6524-9 (eBook)
https://doi.org/10.1007/978-981-97-6524-9

Jointly published with Southwest Jiaotong University Press
The print edition is not for sale in China (Mainland). Customers from China (Mainland) please order the print book from: Southwest Jiaotong University Press.

This Springer imprint is published by the registered company Springer Nature Singapore Pte Ltd.
The registered company address is: 152 Beach Road, #21-01/04 Gateway East, Singapore 189721, Singapore

Preface

With the construction and development of high-speed railways in China, the installation of more CWR turnouts on bridge structures is becoming increasingly necessary. This is in response to environmental protection requirements and the need to adapt to terrain restrictions. To meet the operational demands of high-speed railways, it is essential that both the turnouts and bridges are constructed in an optimal relationship, possessing excellent mechanical properties. Furthermore, they must adhere to stringent technical requirements concerning safety, riding quality, passenger comfort, reliability, and regularity. The technology involved in CWR turnouts on bridges combines the technical characteristics and challenges of CWR tracks on bridges, CWR turnouts, and long-span bridges. This domain presents a significant new technical challenge in track engineering and has become a "bottleneck" constraining the progression of high-speed railway construction in China.

The technology of CWR turnouts on bridges encompasses the intricate intertwining and fusion of various fields, including CWR tracks, CWR turnouts, turnout dynamics, and the dynamics of train-track-bridge coupling. Additionally, it involves the utilization of means and techniques from diverse disciplines such as engineering tests and numerical simulations. This convergence results in a complex structural system and presents formidable technical challenges. The author team, in collaboration with domestic peers, successfully tackled a range of technical issues such as the constitutive relationships in turnout and bridge interactions, force transfer mechanisms, optimal arrangement forms, structural design parameters, and dynamic performance evaluation. They have formulated computational theories and design methodologies specific to CWR turnouts on bridges, establishing complete technical systems across various domains including theoretical research, structural design, construction and maintenance, testing and monitoring, as well as technical standards for the construction of CWR turnouts on bridges in China. These achievements offer crucial technical backing for the safe and stable operation of CWR turnouts on high-speed railway bridges in China.

This book consists of nine chapters. Chapter 1 introduces the key technologies, research ideas, and technical routes of CWR turnouts on bridges. Chapter 2 details the development of CWR turnout technologies in Germany, France, and South Korea, as well as the research and development journey of China. Chapter 3 introduces the computational model for CWR turnouts on bridges and the influence laws of design parameters. Chapter 4 outlines the design principle, design process, checking calculation indexes, and structural reinforcement measures for CWR turnouts on bridges. Chapter 5 presents the structural design approach for turnout girders. Chapter 6 introduces the dynamic simulation model for CWR turnouts on bridges, and analyzes and evaluates the dynamic indexes. Chapter 7 describes the long-term service status monitoring of CWR turnouts on Lei Bridge of the Wuhan–Guangzhou High-Speed Railway, Haihang Super Major Bridge of Shanghai–Hangzhou High-Speed Railway, and the bridge at Xuzhoudong Station of Beijing–Shanghai High-Speed Railway. Chapter 8 introduces the field tests on the mechanical properties of the CWR turnouts on bridges. Chapter 9 includes the typical design cases of CWR turnouts on high-speed railway bridges.

Our technical research on CWR turnouts on high-speed railway bridges has been supported by several projects under the Science and Technology Research and Development Program of the Ministry of Railway including "Research on Design of CWR Turnouts on Passenger Dedicated Railway Bridges and Bridge Structures", "Research on Design of CWR Turnouts with Ballastless Track and Bridge Structures on Beijing–Shanghai High-Speed Railway Bridges", and "Design and Experimental Research on CWR Turnout Group Sections on Bridges". Besides, we have received valuable support and assistance from esteemed experts in the rail transit industry, including Profs. Chen Xiufang, Xu Qingyuan, and Zeng Zhiping from Central South University, Profs. Liu Xueyi, Wang Ping, and Yang Rongshan from Southwest Jiaotong University, Profs. Gao Liang and Cai Xiaopei from Beijing Jiaotong University, Profs. Lei Xiaoyan and Feng Qingsong from East China Jiaotong University, and Researchers Jiang Cheng, Jiang Jinzhou, and Wang Shuguo from China Academy of Railway Sciences. In addition, special appreciation is extended to Mr. Wen Wangqing and Sun Li, two chief engineers from China Railway SIYUAN Survey and Design Group Co., Ltd., along with their colleagues in the track research team. Our heartfelt thanks are also extended to our colleagues for their diligent efforts in the writing and compilation process of this book. Special recognition goes to Zhang Ze, who contributed to collation, calculation, and drafting tasks, and to Liu Jie, Liu Yongcun, Lian Ziyang, and Zhao Yunzhe, who were involved in data compilation.

The publication of this book was financed by the publishing fund of China Railway SIYUAN Survey and Design Group Co., Ltd. The authors of this book hereby express their sincere gratitude to all peers and publishers who supported, helped, and showed concern regarding the publication of this book.

Any comments on inadequacies found in this textbook or suggestions will be appreciated.

Wuhan, China
July 2022

Qiuyi Li
Bin Zhu
Shijie Zhang

Contents

Chapter 1
Introduction

1.1 Background

Since the adoption of the *Mid- and Long-Term Railway Network Plan* by the State Council in 2004, China has been actively engaged in the extensive construction of high-speed railways (HSRs). In conjunction with the rapid development of HSRs and advancements in technology, there is an escalating trend of installing turnouts on bridges, super major bridges, viaducts, and even employing continuously welded rail (CWR) turnouts on bridges in certain scenarios, to accommodate environmental protection requirements and overcome topographical constraints. The emergence of CWR turnouts on bridges poses significant challenges within the realm of track engineering research. Beyond addressing the technological aspects of CWR tracks on bridges, CWR turnouts, and long-span bridges, a multitude of new technical obstacles arise from this development. During the early stage of HSR construction, China encountered an absence of systematic study on the placement of CWR turnouts on bridges, along with a scarcity of practical experience in this specific domain.

CWR turnouts on bridges represent intricate structural systems that incorporate technologies from various domains, such as CWR tracks on bridges, CWR turnouts, turnout dynamics, and the dynamics of train-track-bridge coupling. To meet the construction and operation requirements of HSRs, CWR turnouts on bridges must possess optimal compatibility with the bridges and demonstrate excellent mechanical properties. The interactions between bridges and CWR turnouts, taking into account longitudinal, vertical, and transverse coupling effects, exert a profound influence on the structural design of CWR turnouts, selection of bridge structure forms, arrangement of bearings, and their relative relationships. Despite substantial research conducted both in China and abroad on CWR tracks on bridges and CWR turnouts separately, neither the theories of CWR tracks on bridges nor those of CWR turnouts alone can adequately address the challenges faced by CWR turnouts on bridges.

Q. Li et al., *Continuously Welded Turnouts on High-Speed Railway Bridge*, Advances in High-speed Rail Technology, https://doi.org/10.1007/978-981-97-6524-9_1

Trains passing through CWR turnouts on bridges at high speeds give rise to intricate coupling dynamic interactions among the trains, turnouts, and bridges. Furthermore, an irrational bridge structure form and relative positioning of turnouts and bridges intensify their coupling effect, thereby increasing the wheel-rail dynamic response in the turnout area. In severe cases, when this intensified effect cannot be compensated by the strength margin accounted for in the design of the switch rail and point rail, the riding quality and operational safety of the trains may be compromised. Consequently, speed restrictions might be imposed on trains traversing such turnouts. Furthermore, CWR turnouts on bridges must address intricate wheel-rail interactions, deliver excellent dynamic characteristics, and fulfill technical requirements on high levels of safety, quality, comfort, reliability, and regularity.

CWR turnouts on bridges integrate the technical characteristics and challenges associated with both CWR tracks on bridges and CWR turnouts. Considered the most formidable subject in the realm of CWR track technology, it has become a prominent bottleneck that hinders China's advancements in HSR construction and has emerged as a captivating research focus in the domain of track engineering. Researching key technologies on CWR turnouts on HSR bridges holds immense practical significance. The knowledge derived from such research can guide the design, construction, and maintenance of CWR turnouts on HSR bridges, leading to an enhancement of China's technical system for CWR tracks. Additionally, these outcomes can foster progress in track technology and strengthen China's capacity for independent innovation.

1.2 Complexity Analysis

China's HSR construction, characterized by large-scale implementation, systematic development, and extensive network coverage, presents unique challenges for CWR turnouts on bridges. These turnouts are situated in diverse geographic environments and subjected to varying climatic conditions, while the bridges supporting them exhibit diverse structural forms. These substantial differences introduce exceedingly complex conditions and challenges to the construction process.

1.2.1 Environmental Conditions

China's vast territory encompasses significantly diverse geographical and climatic conditions, evident in the substantial temperature difference between the northern and southern regions. Furthermore, numerous rivers and dense water networks are scattered throughout certain areas. Given such complex environmental conditions, CWR turnouts on bridges need to possess high adaptability to effectively accommodate these variations in China.

China spans a vast range of latitudes, resulting in distinct temperature conditions and encompassing five climate types: temperate monsoon climate, subtropical monsoon climate, tropical monsoon climate, temperate continental climate, and plateau alpine climate. The northernmost point, the Mohe River located north of 53°N, lies in the cold temperate zone. Conversely, the southernmost point, Zengmu Reef located south of 4°N, experiences an equatorial climate. Most regions in China exhibit four distinct seasons, with a few exceptions. The Qinghai–Tibet Plateau, situated at an altitude of over 4,500 m a.s.l., experiences perpetual winter. The South China Sea Islands have a year-round dominance of summer, while Central Yunnan is characterized by abundant spring-like conditions throughout the four seasons.

China's expansive territory spans multiple longitudes, leading to substantial disparities in precipitation conditions and the prevalence of a monsoon climate characterized by cold winters, hot summers, dry winters, and rainy summers. The southern regions of China receive annual precipitation exceeding 1,500 mm, while the Turpan Basin experiences a significant reduction to a mere 16 mm. The Yangtze River basin receives up to 1,200 mm of annual precipitation, in contrast to the Yellow River basin, which receives over 600 mm.

The significant variation in terrain and diverse topographic conditions further amplifies the complexity and diversity of China's climate. With its complex terrain, diverse topographic types, and substantial differences in altitude, China encompasses expansive plains, widely stretching mountains, barren deserts, serene lakes, and sprawling grasslands, and is dotted with numerous deep valleys, rolling hills, and vast basins, each exhibiting distinctive climate patterns in different regions.

1.2.2 Turnout-Bridge Interaction Mechanisms

In contrast to scenarios involving CWR turnouts on subgrade and CWR tracks on bridges, the interactions between turnouts and bridges are considerably more intricate. These interactions give rise to profound mutual influences between the inner rails and stock rails, as well as between the turnouts and bridges. The action of temperature load induces a coupling effect, intertwining the expansion and contraction deformations of the bridge with the temperature influence on the turnouts. This coupling effect amplifies the complexity of longitudinal and lateral interactions, as well as the mechanisms of force transmission between the turnouts and the bridges.

1.2.3 Turnout-Bridge Structures and Parameters

The complexity associated with the turnout-bridge structures and parameters is manifested in the diverse arrangement forms of turnouts, multiple types of bridge structures, and the evident nonlinear characteristics of the relevant parameters between turnout rails and bridges. Parameters considered in bridge design comprise the span arrangements, bearing arrangements, vertical deflection limits of the girder structure in the turnout area, rotation angle at the girder end, vertical rigidity of piers, longitudinal horizontal linear rigidity at the fixed pier top, and horizontal displacement at the pier top. Parameters considered in turnout design include the primary forms and corresponding sizes of turnouts, the longitudinal resistance of fasteners, the resistance of stopper/spacing block, as well as the parameters of switch rails and stock rails. The structural form and design of the turnout girder should not only fulfill the strength and stability requirements of the bridge structure but also meet the diverse demands for laying CWR turnouts. Consequently, this leads to an elevated level of difficulty in bridge design.

1.2.4 Wheel-Rail Interactions

Turnouts are regarded as vulnerable points in track systems due to the geometric irregularities inherent in their structures. The switch rail and point rail serve to guide the rotating train wheels. However, their gradual variations in width and height introduce complexity in the geometry and creep relationships at the wheel-rail contact. The dynamic simulation analysis of the train-bridge coupling system is highly intricate, with the inclusion of turnouts further amplifying its complexity. Any significant deflection or vibration amplitude of the bridge will directly impact the regularity of the turnout track on the bridge, leading to an additional dynamic impact on the turnout bridge structure. This, in turn, can directly compromise operational safety. Due to the presence of numerous nonlinear influencing factors on the turnout structure and wheel-rail interactions, coupled with the mutual influence and interaction between wheel-rail interactions and train-bridge interactions under train load, the analysis becomes increasingly challenging as the longitudinal, transverse, and vertical three-dimensional coupling effects come into play.

1.2.5 Multi-Disciplinary Intersection

The development of CWR turnouts on bridges involves the integration and synergy of various disciplines such as track engineering, bridge engineering, rolling stock dynamics, turnout dynamics, and structural dynamics. Moreover, the analysis in this domain demands the utilization of multiple means and methods drawn from several fields, including engineering testing and numerical simulation. CWR turnouts on bridges incorporate technologies from CWR tracks on bridges, CWR turnouts, bridge engineering, and more. This intricate system presents significant technical challenges, attributed to complex interactions between turnouts and bridges, as well as between wheels and rails.

1.3 Key Technologies in CWR Turnout on Bridge

To tackle the challenges posed by highly complex CWR turnouts on bridges, the primary objective is to address key technologies, including computing theories, design parameters, turnout girder arrangements, dynamic performance, and design methods. This aims to establish a comprehensive set of computing theories and design methods that can serve as guidance for the design, construction, and operation of CWR turnouts on bridges.

1.3.1 Constitutive Relation and Force Transmission Mechanisms in Turnout-Bridge Interactions

The factors influencing the interactions between turnouts and bridges are highly intricate. The essence of researching CWR turnouts on bridges lies in understanding the interactions and force transmission mechanisms between the turnouts and bridges. This understanding helps unveil the stress and deformation rules of both the turnouts and bridges.

1.3.2 Rational Turnout-Bridge Arrangements

The turnout-bridge arrangement has a direct influence on the stress and deformation of the turnout-bridge structure, even further impacting the operational safety and comfort of trains. Analyzing the rational structural forms of the turnout girder involves considering factors such as the minimum spacing limit between the turnout expansion joints and bridge expansion joints, typical turnout-bridge arrangement patterns, and the implementation of a simply-supported girder between two long-span continuous turnout girders.

1.3.3 Rational Structural Forms and Design Parameters of Turnout Girder

The study on the structural forms of turnout girders includes the longitudinal and transverse arrangements of bearings, substructure forms, the principle of span arrangements for turnout girders, and bearing arrangements. The design parameters include the vertical deflection limits of the girder structure in the turnout area, rotation angles at the girder end, vertical stiffness of piers, longitudinal horizontal linear rigidity of piers, and horizontal displacement at the pier top.

1.3.4 Dynamic Performance Simulation and Evaluation of Train-Turnout-Bridge Coupling System

Vibration analysis of the train-turnout-bridge coupling system serves as a crucial prerequisite for the operational safety and comfort evaluation of trains, as well as the dynamic performance evaluation of turnouts and bridges. The key to scientifically evaluating the dynamic performance of CWR turnouts on bridges lies in establishing computational models and motion equations that accurately represent the train-turnout-bridge coupling system. This involves defining rational computing parameters and selecting stable and efficient numerical analysis methods capable of solving the dynamic responses within the system.

1.3.5 Design Criterion and Method of CWR Turnout on Bridge

Turnouts are acknowledged as vulnerable points within track systems. When designing CWR turnouts on bridges, it is essential to not only meet load-bearing capacity requirements, such as strength and stability but also ensure their normal operational functionality. This involves controlling the relative displacement between the turnouts and bridges to enable normal switching of the turnouts. It is critical to limit the displacement and deformation of key turnout components as failure to do so can disrupt normal switch operation, potentially jeopardizing train safety.

1.4 Research Approach of Technologies in CWR Turnout on Bridge

China differs significantly from other countries in the practices of HSR construction, encompassing various aspects such as turnout types and structures, bridge structural forms, geographical and climatic conditions for laying CWR turnouts on bridges, as well as construction and operation management. Consequently, addressing the challenges associated with implementing CWR turnouts on bridges in China necessitates the development of computing theories and design methods that are tailored to Chinese conditions through independent innovation based on the actual conditions in China, rather than simply relying on foreign models.

1.4.1 Research Objectives

(1) Explore the interactions and force transmission mechanisms between CWR turnouts and bridges and unveil the stress and deformation rules of CWR turnouts on bridges, as well as the associated pier and abutment structures.
(2) Propose the computing theories, design methods, design principles, and technical conditions specific to CWR turnouts on bridges, serving as a foundation for the design of bridges along with their piers and abutments.
(3) Present rational structural forms and design parameters for bridges that align with the stress and deformation requirements of CWR turnouts on bridges.
(4) Evaluate the operational safety and comfort related to CWR turnouts on bridges.

1.4.2 Technical Route and Research Approach

The research on CWR turnouts on bridges follows the technical route outlined below. The initial efforts are directed toward addressing challenges in collaboration among manufacturers, universities, design institutes, and contractors. Subsequent steps focus on applying the concepts and ideas of systems engineering to enable comprehensive integration and synergy across multiple disciplines, including track systems, bridges, and rolling stock, within a comprehensive system comprising trains, turnouts, and bridges. Substantial investigations are conducted following the extensive exploration of Chinese and foreign research findings and engineering applications. This involves incorporating static analysis with dynamic simulation, laboratory experiments with field tests, and combining theoretical research with practical engineering. The overall research approach is illustrated in Fig. 1.1.

1.4.3 Research Methodology

(1) Trains, turnouts, and bridges are incorporated into a comprehensive system based on the principle of systems engineering, and the longitudinal and lateral interactions between turnouts and bridges under the action of temperature load and the dynamic performance of the train-turnout-bridge coupling system are studied.

To address the dynamic interactions among HSR trains, turnouts, and bridges, in this proposed comprehensive coupling system that integrates rolling stock (EMUs), turnouts, and bridges, the dynamic characteristics of turnouts and bridge structures under high-speed train operation conditions, as well as the associated operational safety and comfort, are explored. This exploration utilizes finite element methods, incorporating rolling stock dynamics, track dynamics, turnout dynamics, and bridge dynamics, establishes connections between wheel-rail interactions and

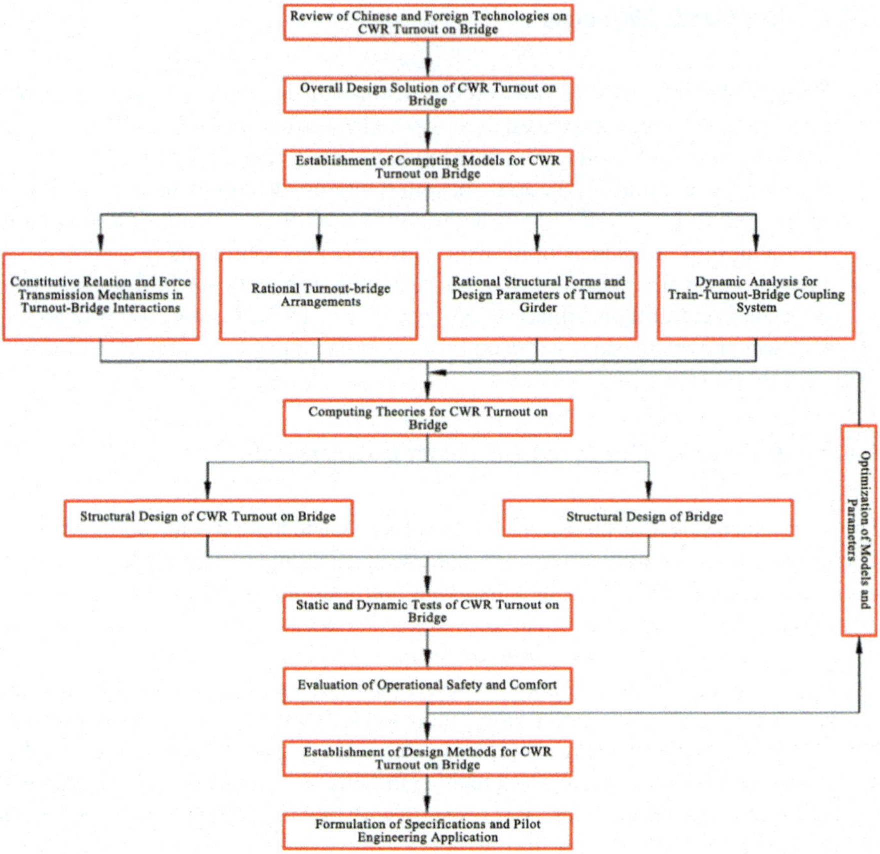

Fig. 1.1 Overall research approach

track-bridge interactions within the turnout area, and employs a numerical simulation methodology.

(2) The constitutive relation, mechanical mechanisms, and computational theories of turnouts and bridges under the train-temperature coupling effect are investigated, and the stress and deformation characteristics and rules of turnouts and bridges are analyzed, using a combination of static analysis and dynamic simulation, and applying computational models and methods based on finite element methods.

Quantitative calculations and analysis of CWR turnouts on bridges using simple mechanical mathematical models and conventional mathematical analysis methods proved challenging. However, establishing a robust train-turnout-bridge system model through the aid of modern computer simulation technology provides a distinct advantage. This approach allows for an effective simulation of the intricate constitutive relation and mechanical transmission mechanisms within the context of CWR turnouts on bridges. By doing so, it becomes possible to uncover the accurate mechanical characteristics and rules governing their behavior, leading to a scientific evaluation of the operational safety and comfort of trains.

(3) In the framework of integrating theoretical analysis with field tests, the computing theories and design methods derived are subjected to verification through static and dynamic tests conducted on actual CWR turnouts on bridges.

The interactions between turnouts and bridges constitute an exceedingly complex system that cannot be comprehensively understood solely through theoretical analysis. Therefore, continuous efforts should be devoted to enhancing the design system of CWR turnouts on bridges and advancing their technical capabilities, by verifying the feasibility and reliability of the CWR turnouts on bridges theories, and on the principle of linking theory with practice. Moreover, considering the significant influence of design parameters, such as the longitudinal resistance of turnout fasteners and ballast bed, on the stress and deformation of CWR turnouts on bridges, it is critical to define rational design parameters based on their nonlinear characteristics. These characteristics can be uncovered through actual measurements and statistical analysis conducted in laboratory experiments and field tests. Furthermore, it is imperative to carry out field tests at carefully selected sites to evaluate the mechanical properties of the CWR turnouts on bridges. This verification is essential to ensure the accuracy of the static and dynamic computational models and methods employed for CWR turnouts on bridges.

(4) The evaluation criteria and design methods for CWR turnouts on bridges are established by integrating the checking calculations of rail strength with the control of girder-rail relative displacement.

Following the design principle of dual control, which emphasizes both strength and displacement, the design requirements, checking calculation methods, and design processes for CWR turnouts on bridges are formulated. These comprehensive efforts lead to the establishment of a systematic design approach for the implementation of CWR turnouts on bridges.

Chapter 2
Overview of Technological Development of CWR Turnout on Bridge in China and Abroad

Since 2004, China has pursued a technological development route known as "introduction, digestion, absorption, and re-innovation" for HSR tracks. This strategy has involved the introduction of foreign technologies and concepts for CWR turnouts on bridges into China. Renowned foreign consulting firms have also played a crucial role in providing consulting services for the design of the Beijing–Shanghai HSR in China and the formulation of the *Provisional Specification on Design of the Beijing–Shanghai High-Speed Railway*. As part of the extensive exchanges between China and foreign countries regarding design concepts and approaches for CWR turnouts on bridges, German practices have profoundly influenced China's technological developments and advancements. This chapter aims to present the technological developments of CWR turnouts in Germany, France, and the Republic of Korea while reviewing China's research and development history in the technical field of CWR turnouts on bridges.

2.1 Overview of Technological Development of CWR Turnout on Bridge in Germany

2.1.1 Structural Characteristics

The cross-sectional view of CWR turnouts on bridges for ballastless tracks in Germany is shown in Fig. 2.1. The track structure within the turnout sections on bridges comprises the rails, turnout sleepers, track bed slabs, and deck protection layers. According to German requirements, all vulnerable mechanical equipment of turnouts should be placed on the same track bed slab (Fig. 2.2). Longitudinal bollards (Fig. 2.3) are introduced to mitigate constraints resulting from shrinkage and temperature deformation. Transverse bollards (Fig. 2.4) are positioned at the

Q. Li et al., *Continuously Welded Turnouts on High-Speed Railway Bridge*, Advances in High-speed Rail Technology, https://doi.org/10.1007/978-981-97-6524-9_2

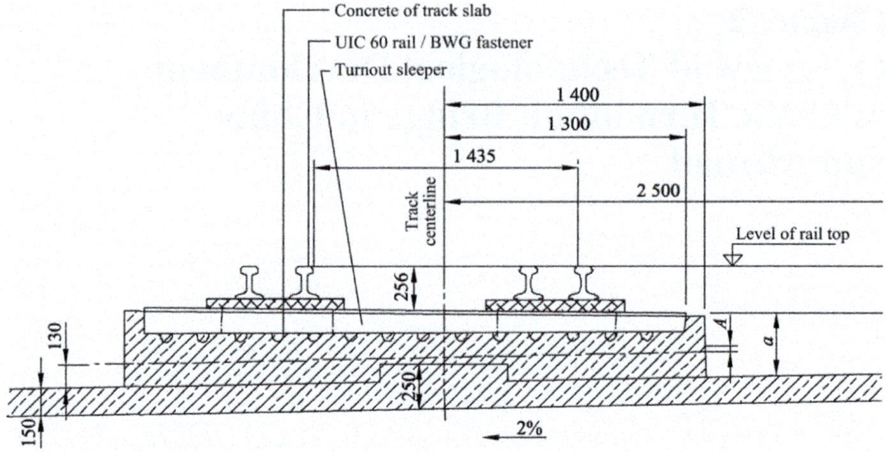

Fig. 2.1 Cross section of CWR turnout on bridge for ballastless track in Germany (Unit: mm)

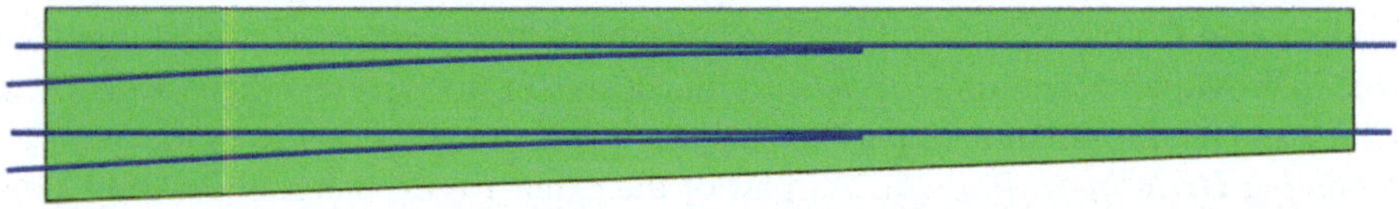

Fig. 2.2 Track bed slab within turnout section

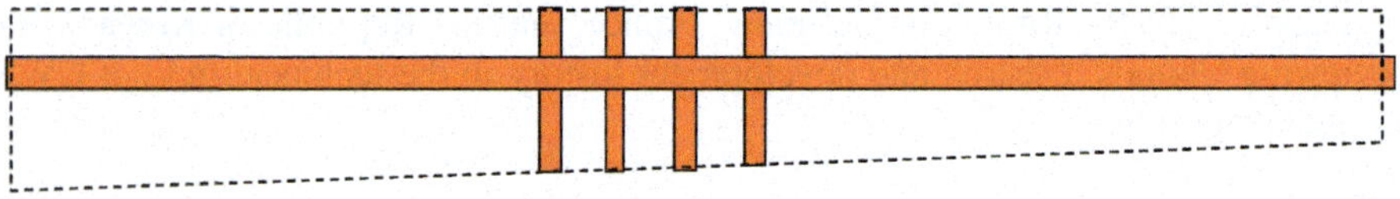

Fig. 2.3 Longitudinal and transverse bollards (unit: m)

center of the turnout slabs, and their quantity is determined according to the structural computation. Germany enforces stringent technical requirements for the design and construction of longitudinal and transverse bollards (Fig. 2.5) [1].

2.1.2 *Computational Theory*

Germany's computation methods for CWR turnouts on bridges can be classified into simple algorithms and complex algorithms. The simple algorithms, which do not consider the turnout effect, involve computing the temperature force on a single rail at both ends of the bridge according to established practices for CWR tracks on bridges. Subsequently, the total temperature force at the girder end is obtained

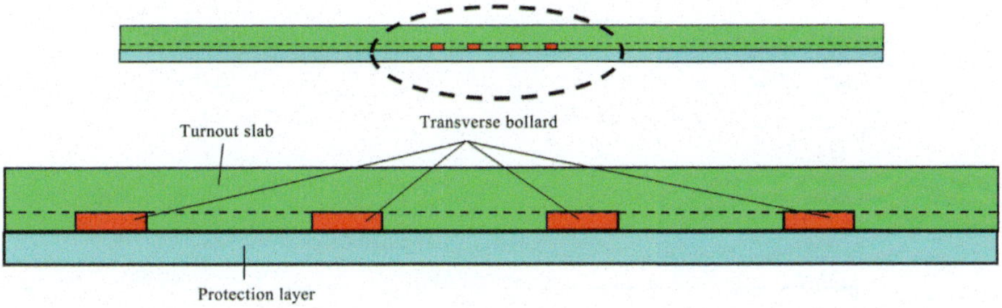

Fig. 2.4 Arrangement of transverse bollards at turnout

Fig. 2.5 Placed longitudinal and transverse bollards

by multiplying the single-rail temperature force by the number of rails at the girder end. The additional force exerted by the piers is defined as the difference between the total temperature force at the left and right girder ends. The simple algorithms, characterized by a significant relative safety factor, are commonly employed in the conceptual design phase of CWR turnouts on bridges to estimate the additional force on piers.

The complex algorithms employed for CWR turnouts on bridges in Germany utilize finite element models to calculate the interactions between track slabs and the substructure [2, 3]. These algorithms take into account various factors such as longitudinal reactions at bearings, additional rail stress, and relative displacement at switches and frog zones. The subsequent discussion primarily concentrates on the complex algorithms.

1. Computational Model

The finite element computational model utilized for CWR turnouts on bridges in Germany is shown in Fig. 2.6.

This model incorporates girder elements to represent bridges, crossovers, turnout slabs, and rails; employs elastic–plastic nonlinear spring elements to simulate unloaded fasteners, while friction spring elements to model loaded fasteners; uses elastic spring elements to reflect piers and bollards. The bridges, rails, and bollards

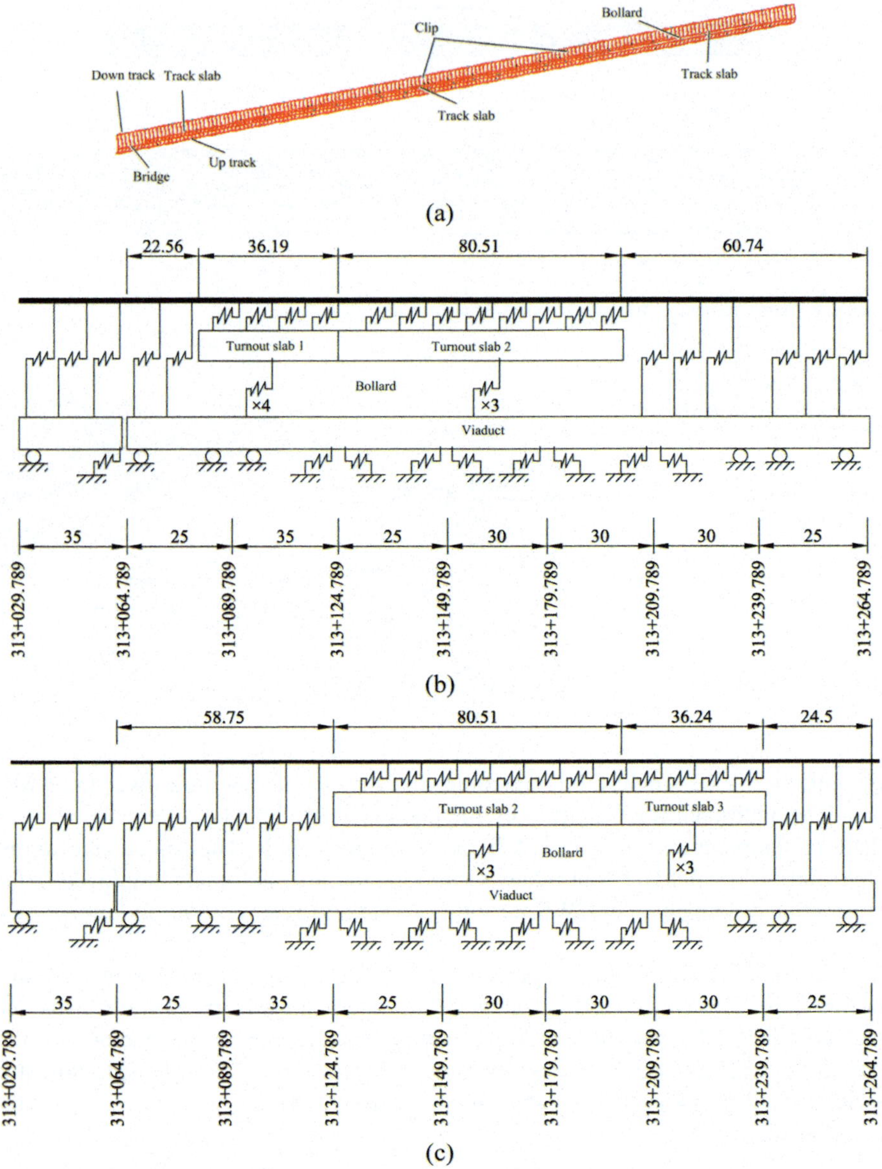

Fig. 2.6 Computational model (unit: m)

are meshed at 1 m intervals. The nonlinear computation is performed using the SOFISTIK program.

2. Computation Parameters

Fasteners for CWR turnouts on bridges in Germany can be chosen from the following three types as appropriate:

(1) Vossloh loarv300 fastener [4] (SKL 15 clip).
(2) BWG-elastic base plate and fastener (SKL 12 clip) for placement within the turnout range.
(3) BWG-elastic base plate and fasteners [5] (SKLB 15 clip) for placement at the end of long bridges.

The sliding-friction relationships of the above three types of fasteners are shown in Figs. 2.7, 2.8 and 2.9.

3. Load Cases

The analysis of track–structure interactions incorporates train-induced vertical, braking, and traction loads determined according to the UIC-71 load model for high-speed trains [6, 7] specified in the UIC Code, as shown in Fig. 2.10.

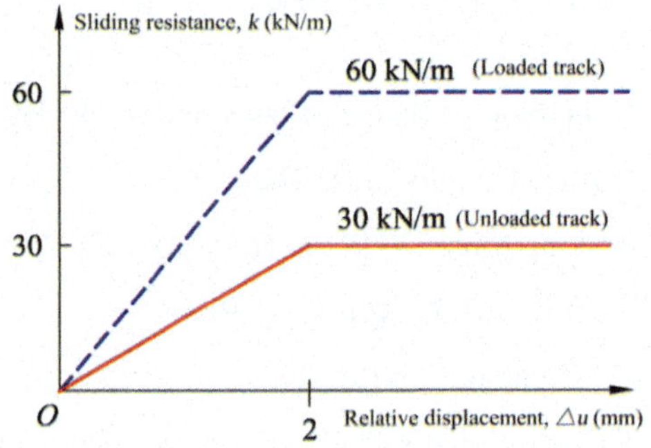

Fig. 2.7 Sliding-friction relationship of SKL 15 clip

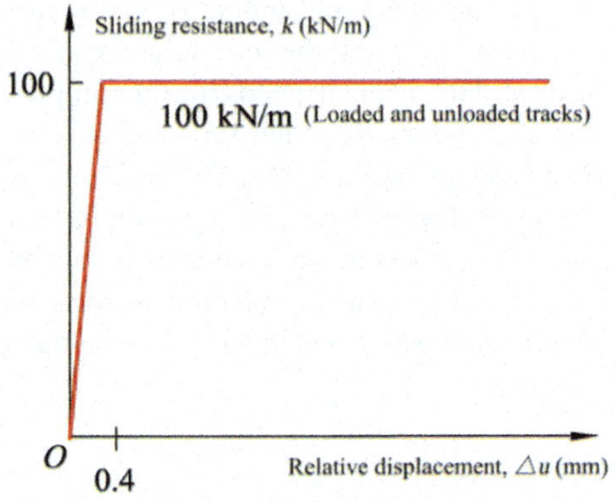

Fig. 2.8 Sliding-friction relationship of SKL12 clip

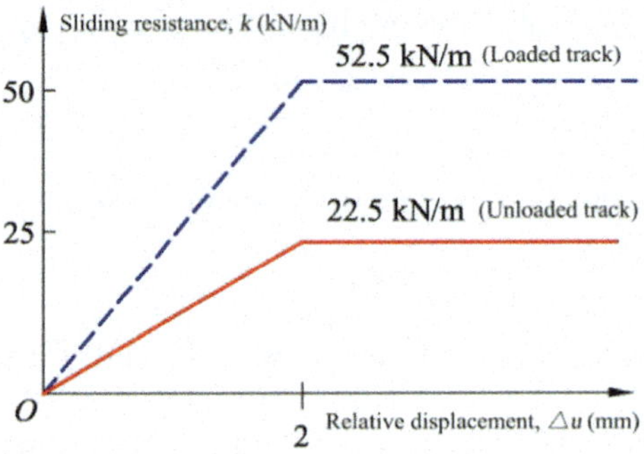

Fig. 2.9 Sliding-friction relationship of SKLB 15 clip

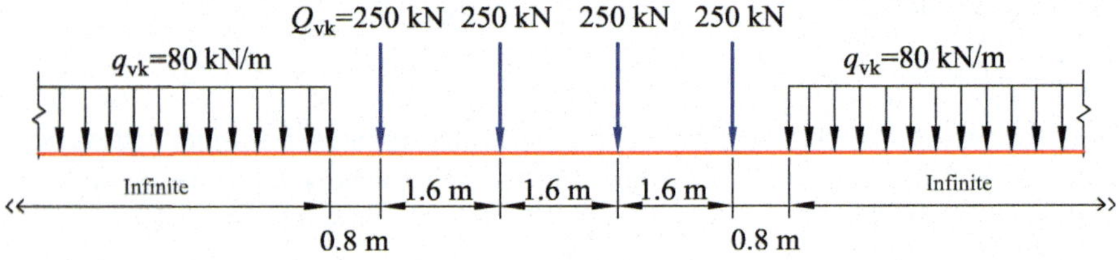

Fig. 2.10 UIC-71 load model

Temperature change of deck and track slab: ± 20°C;

Vertical train load: 60 kN/m;

Braking force: 20 kN/m ($L \leq 300$ m);

Traction: 33.3k N/m ($L \leq 30$ m).

4. Computation Results

The additional rail stress and relative displacement due to temperature changes of bridges and track slabs are shown in Figs. 2.11 and 2.12.

The computational model of CWR turnouts on bridges in Germany thoroughly takes into account the interactions among rails, fasteners, track bed slabs, and bridges. However, it needs further improvement to adequately consider the interactions between the inner rails and stock rails within the turnouts. Additionally, the successful installation of CWR turnouts on bridges in Germany relies on the utilization of high-precision and high-quality turnout products. Germany has also implemented a comprehensive set of rigorous management systems and a well-developed system of standards governing turnout production, procurement, and installation. These established systems provide valuable insights and references for China.

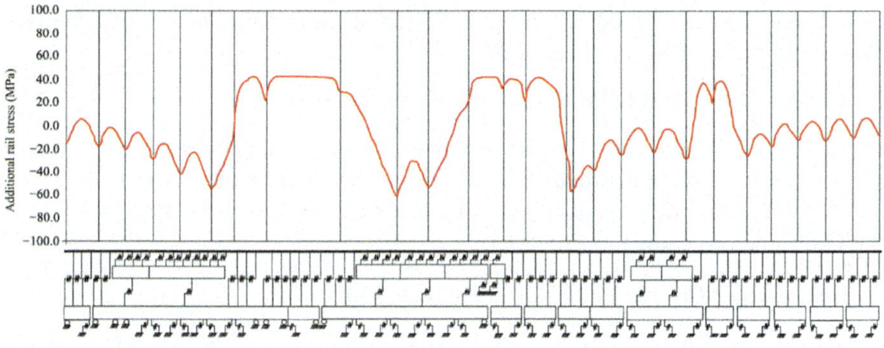

Fig. 2.11 Additional rail stress due to temperature changes

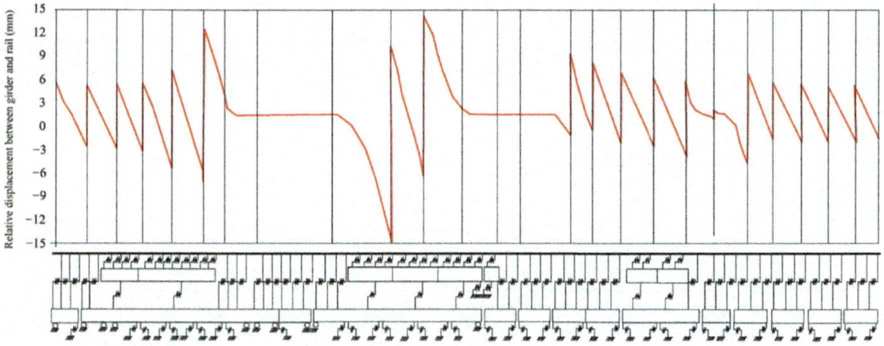

Fig. 2.12 Relative displacement between track slab/deck and rail

2.1.3 Design Approach

Based on their experience in laying CWR turnouts on bridges and research findings, Germany emphasizes stringent control over the relative displacement between the switches/frog sections within CWR turnouts and the bridges, to counter the impact of bridge expansion and contraction, vertical loads, braking force/traction from trains, and special checking calculations are necessitated to ensure the safety and comfort of train operation. The areas susceptible to relative displacement between the rail and deck are the frog and switch zones. Relative displacement limits are generally defined by turnout manufacturers in Germany.

Therefore, the basic design concepts of CWR turnout on bridges in Germany is to control the relative displacement between the rail and bridge deck in critical turnout components, including the relative displacement between the rail and fastener, between the fastener and sleeper/track slab, and between the track slab and the bridge deck.

Germany has implemented stringent regulations regarding the relative displacement limits for each component of turnouts:

1. Double-arm Movable Point Rail without Skew Joint

The maximum displacement of the base plate for point rails and sleeper/track slabs is restricted to 0.7 mm, primarily arising from the permissible deformation of 0.7 mm for the elastic dowel connecting the steel base plate and the steel cone of the BWG fastener, as shown in Fig. 2.13.

2. Single-arm Movable Point Rail with Skew Joint

The displacement between the point rail and base plate should not exceed ± 3 mm, to prevent the locking device from being obstructed by concrete, as shown in Fig. 2.14.

3. Maximum Displacement of Turnout Rail and Base Plate

The maximum displacement of the turnout rail and base plate should not exceed ± 20 mm, and that of the base plate and sleeper/track slab should not exceed ± 0.7 mm, as shown in Fig. 2.15.

4. Deformation at Switch Rail

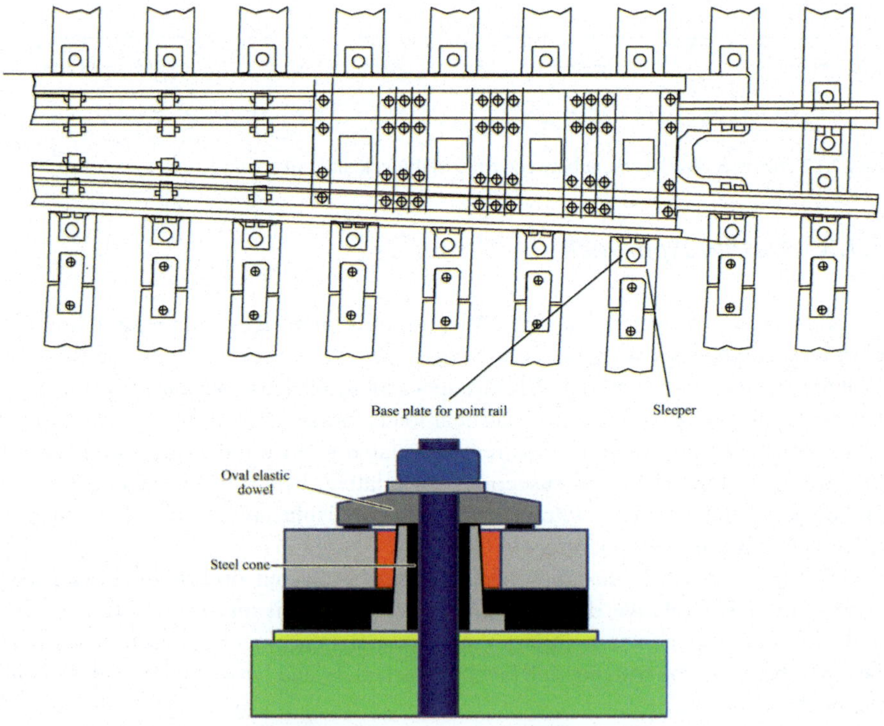

Fig. 2.13 Double-arm movable point rail without skew joint

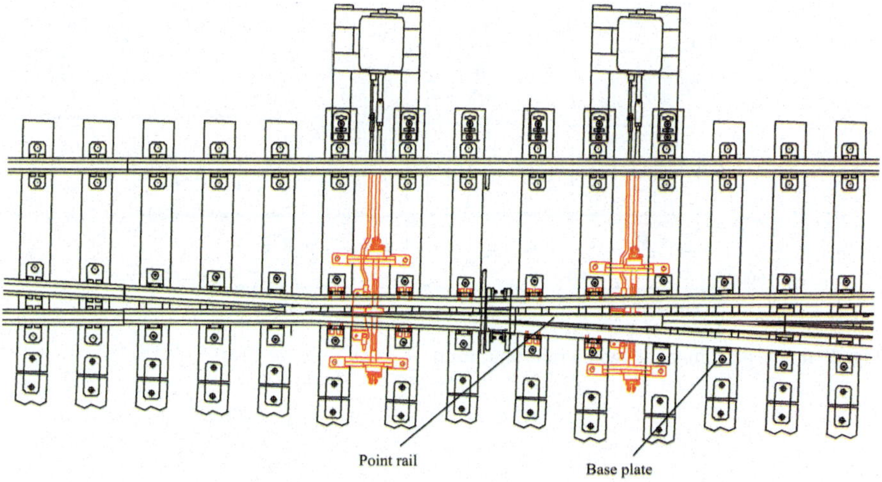

Fig. 2.14 Single-arm movable point rail with skew joint

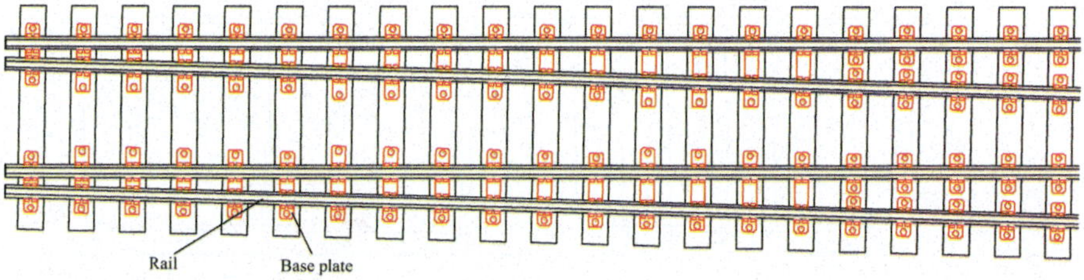

Fig. 2.15 Turnout rail and base plate

In the presence of a special arrangement, the permissible deformation at the switch rail is ± 20mm, as shown in Fig. 2.16. However, under circumstances without this special arrangement, the relative displacement of the rail and the base plate should not exceed ± 3mm.

5. Switch Zone (Frog)

The relative displacement between the turnout rail and switch base should not exceed ± 6mm, as shown in Fig. 2.17, to prevent mechanical jamming during turnout switching.

To sum up, the relative displacement limits for the components of BWG CWR turnouts on bridges in Germany are shown in Table 2.1 and Fig. 2.18.

Figure 2.19 presents the envelope for the relative displacement between the turnout girders and rails, based on the relative displacement limits for switches, frogs, and closures given in Table 2.1. To ensure that relative displacement at switches and frogs remains within permissible limits, the following checking approach is employed: First, calculate the total displacement of the stock rail under the effect of bridge

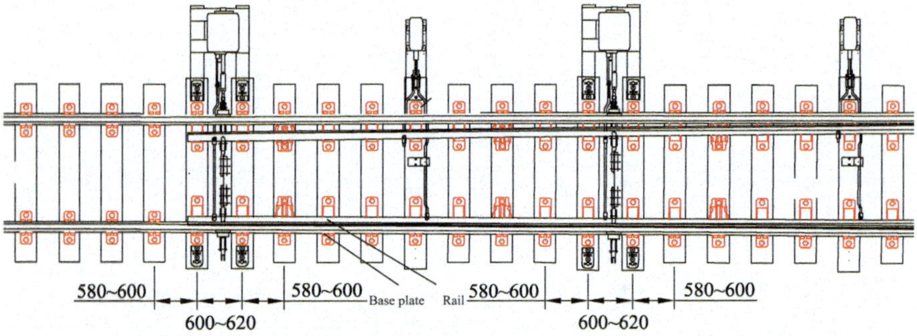

Fig. 2.16 Displacement at switch rail (unit: mm)

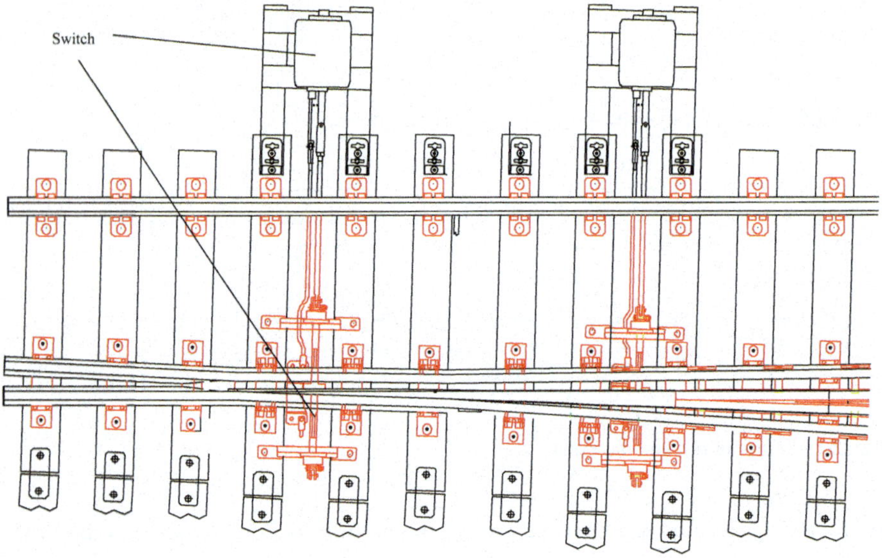

Fig. 2.17 Displacement at frog

Table 2.1 Relative displacement limits

Type	Permissible relative displacement for frog section	Permissible relative displacement for switch section (mm)	Permissible relative displacement of closure (mm)
CWR turnout on ballasted bridge	5.2 mm (single-arm and flexible)	22.2	22.2
	2.2 mm		
CWR turnout on bridge for ballastless track	3.0 mm (single-arm and flexible)	20	20
	0.7 mm (double-arm and flexible)		

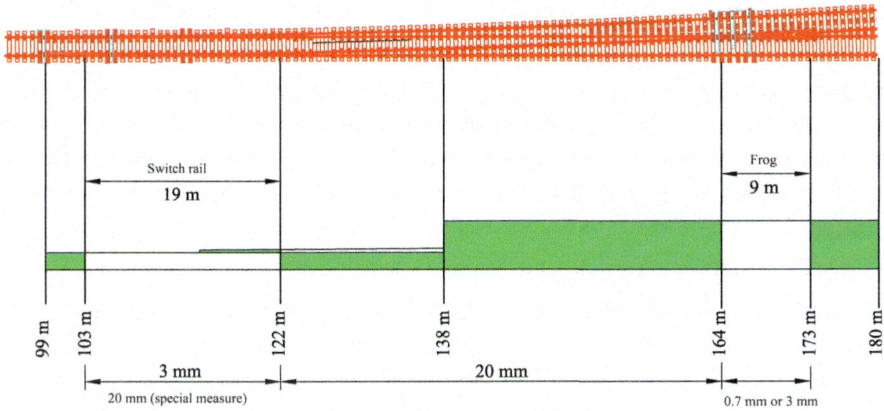

Fig. 2.18 Displacement limits of turnout components

expansion and contraction, as well as vertical loads and braking force/traction exerted by trains, using a turnout–bridge interaction program. Next, plot the calculated total displacement within the relative displacement envelope of the turnout girders and rails. Finally, compare the total displacement of the stock rail against the relative displacement envelope.

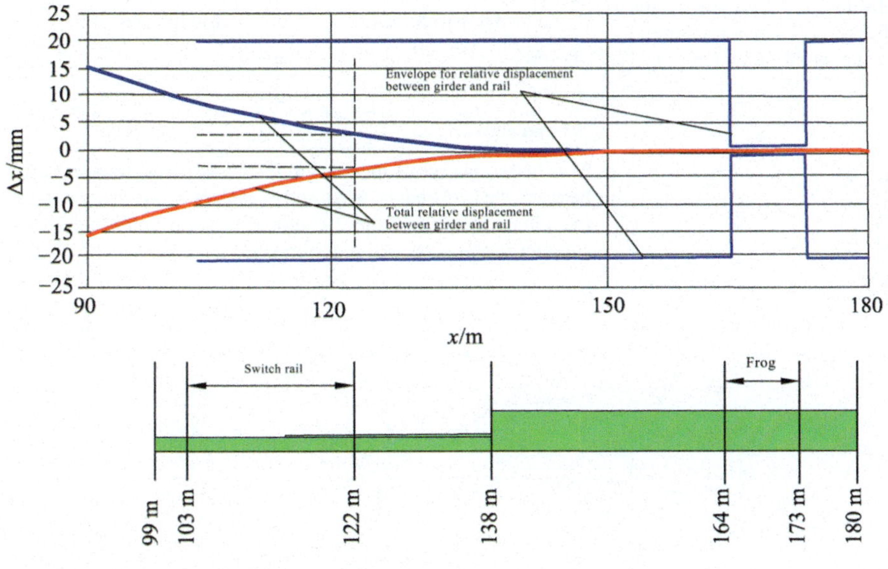

Fig. 2.19 Checking calculation of turnout rail displacement envelope

2.1.4 Design Requirements

In 2005, DB Engineering & Consulting (DB E&C) in Germany provided the following consulting recommendations concerning the design of CWR turnouts on the new Suzhou Station and Huayuan Elevated Station proposed by the Chinese design institute, as part of its engineering design consultation services for Beijing–Shanghai HSR:

1. Distance between Turnout and Bridge Expansion Joint

The German standard DS 800.0120 (7) specifies the following requirements for turnouts:

(1) Positioning turnouts on movable expansion joints on bridges is unallowed.
(2) The minimum distance between expansion joints and actual points of switch rails depends on the bridge length, as specifically shown in Table 2.2.

DB E&C has offered a different consulting recommendation regarding the arrangement of turnouts on expansion joints as specified in China's *Design Principles for CWR Tracks at Viaduct Stations of Beijing–Shanghai High-speed Railway (Interim)* [8], as shown in Fig. 2.20.

2. Longitudinal Horizontal Relative Displacement Limit

Table 2.2 Minimum distance between expansion joint and actual point of switch rail

Total bridge length (m)	Minimum distance (m)
41~60	10
61~90	20
91 and above	30

Note In the absence of specific guidelines on the minimum distance between the heel ends of frogs and expansion joints in the standard, the values given in Table 2.2 should be used unless the turnout supplier presents special requirements

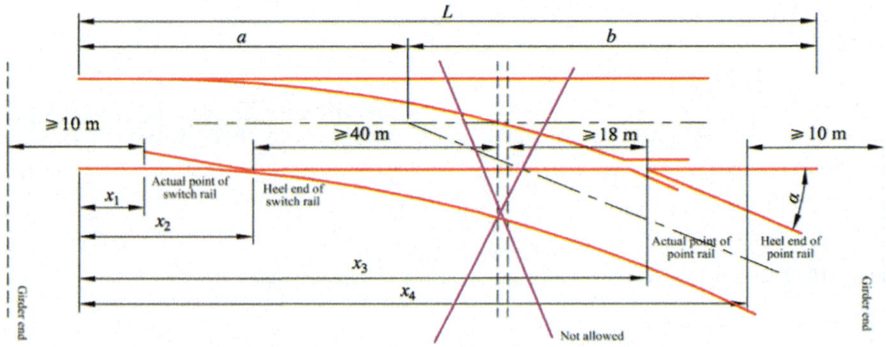

Fig. 2.20 Schematic diagram of consulting solution on turnout arrangement

In the ballasted track scenario, any relative displacement between the turnout and bridge, induced by bridge deflection or atmospheric temperature changes, typically does not impact the stability of the track bed over a short period of time. However, frequent deformation compromises the integrity of the ballast, leading to increased maintenance workload.

Relative displacement between the bridge deck and turnouts involves that between the rails and fasteners, fasteners and sleepers, as well as sleepers and bridge deck.

The German turnout supplier BWG specifies the limits for its turnout products concerning the first two displacement relations mentioned above. However, neither the German standard nor the turnout suppliers provide specific limit values for the last relative displacement relation. On the assumption that elastic deformation of ballast up to 1.5 mm is acceptable, the relative displacement between the sleepers and bridge deck at ballasted turnouts was reckoned in the Consulting Report for Beijing–Shanghai High-speed Railway, as shown in Table 2.3.

3. Turnout–bridge Interaction

The computational model for turnout–bridge interactions must incorporate the particular geometrical characteristics of turnouts, specifically focusing on the rails on the diverging routes of the turnouts.

4. Bearing Arrangement

To ensure that the maximum horizontal deformation of bridges under the action of braking force and traction remains within 30 mm, in the case of one fixed bearing at the midspan of long-span continuous girders, this single pier should possess significant horizontal rigidity. Alternatively, a cluster of fixed bearings can be positioned at the midspan of the continuous girders to distribute the longitudinal load among the respective piers. However, it is important to consider the potentially adverse effects of temperature-induced bridge deformation on the piers in this scenario.

5. Lateral Dislocation of Expansion Joint

In long-span continuous girders, bridge expansion joints can experience significant displacement due to traction/braking force and earthquakes. The longitudinal displacement of the bridges is accompanied by the expansion and contraction of the rails. Furthermore, the angle between the bridge axis and the rails on the diverging routes of the turnouts contributes to lateral displacement relative to the rail induced by the longitudinal displacement of the bridges. When this lateral displacement becomes

Table 2.3 Relative displacement between sleeper and deck at ballasted turnout Unit: mm

Type	Rail	Fastener	Sleeper	Total
General section	20	0.7	1.5	22
Switch rail	20	0.7	1.5	22
Frog, high-speed turnout (R ≥ 3,000 m)	0	0.7	1.5	2.2
Frog, high-speed turnout	3	0.7	1.5	5.2

substantial, it has adverse effects on the regularity of the track and increases the maintenance workload.

6. Longitudinal Rigidity of Pier

Considering the influence of pier foundation deformation on the longitudinal horizontal rigidity of piers, the UIC code recommends incorporating the static rigidity of the piers in the calculations of the expansion and contraction forces, and the dynamic rigidity of the piers for the calculations of the braking force and traction.

The German railway code DS 804, DIN Fachbericht, and UIC 774–3 specify the permissible values of longitudinal horizontal deformation of bridges and minimum values of horizontal rigidity.

The UIC code stipulates a permissible displacement of up to 4 mm between the rail and deck under the braking force/traction and an absolute horizontal displacement limit of up to 5 mm for the deck. For bridges with rail expansion joints at both ends, the maximum permissible absolute horizontal displacement is 30 mm.

As stipulated in DIN Fachbericht K.2.1(5), the minimum horizontal longitudinal rigidity of piers with rail expansion joints at both ends for double-track bridges should be calculated using the following formula:

$$K = \frac{20kN/m \times L}{3\,m} + 1000kN/3m$$

where K denotes the pier rigidity (kN/cm); L denotes the length of the braking load (m).

The formula is derived based on a specific scenario where a horizontal displacement of 30 mm on the pier is caused by a traction load of 1,000 kN and a braking load of 20 kN/m.

The aforementioned displacement limits are intended to prevent the loosening of ballast for ballasted tracks and indirectly restrict additional rail stress and temperature spans across the bridges.

7. Displacement Limits of Critical Turnout Components

The essence of the turnout arrangement is to control the relative displacement between the rails and the bridge deck at critical turnout components, including frogs and switches. The zones most susceptible to relative displacement between the rail and deck are found within the frog and switch zones.

Permissible relative displacement at frog: 0.7 mm; permissible relative displacement at switch: 6.0 mm.

8. Erection of Simply-supported Girder in Multi-span Continuous Girder

Expansion joints located at the largest temperature span generally experience the maximum rail force and maximum relative displacement. Introducing simply-supported girders within the multi-span continuous girders helps reduce the temperature span, subsequently lowering the rail force and displacement.

The Consulting Report recommends the following considerations for bridge arrangement:

(1) The temperature span over the bridge should not exceed 90 m.
(2) It is advisable to minimize the use of bridge arrangements with multi-span continuous girders. Whenever necessary, this type of arrangement should be alternated with at least two spans of simply-supported girders.
(3) The displacement limits for turnouts should be sourced from the turnout supplier.
(4) The bearing arrangement should be selected to minimize the temperature span.

2.2 Overview of Technological Development of CWR Turnout on Bridge in France

The design philosophy for CWR turnouts on bridges in France shares similarities with that in Germany. It is essential to avoid installing turnouts across subgrade–bridge transition sections or spanning bridge expansion joints. France imposes stricter requirements regarding the minimum distance between the turnouts and bridge expansion joints. Specifically, when the temperature span exceeds 90 m, the shortest distance between them should be 100 m. This criterion results in significantly longer girder lengths to accommodate the turnouts, posing design challenges. Furthermore, in France, the computational model for CWR turnouts on bridges primarily focuses on ballasted tracks [9].

2.2.1 Computational Model

The French consulting company SNCFI/SYSTRA conducted a study to analyze the interactions within the track structure, focusing on the turnouts 1/46 and 1/18 located at the elevated stations at DK1258 + 449 and DK1259 + 206 along Beijing–Shanghai HSR.

1. Computation Parameters

The viaducts feature 28–32m simply-supported girders, pier rigidity equal to 400 kN/cm, and a concrete thermal expansion coefficient of 1.00×10^{-5}/°C. The longitudinal resistances of the track are bilinear, as shown in Fig. 2.21.

See Table 2.4 for load cases.

2. Computation Methods

The bridge and track models are shown in Figs. 2.22, 2.23, and 2.24. These computational models incorporate rails, sleepers, bridge decks, and piers. The ballast between the bridge deck and sleepers is simulated by bilinear resistance, and the fasteners between the rails and sleepers are also simulated by bilinear resistance.

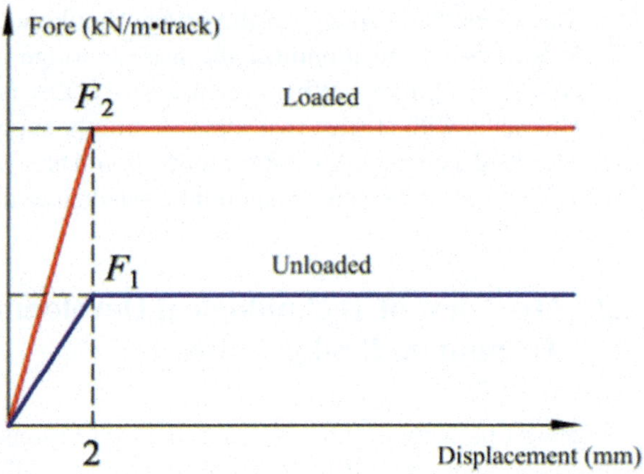

Fig. 2.21 Longitudinal resistances of track

Table 2.4 Load cases

Case	Deck (°C)	Rail (°C)
1	+20	+40
2	+20	+50
3	+20	+60

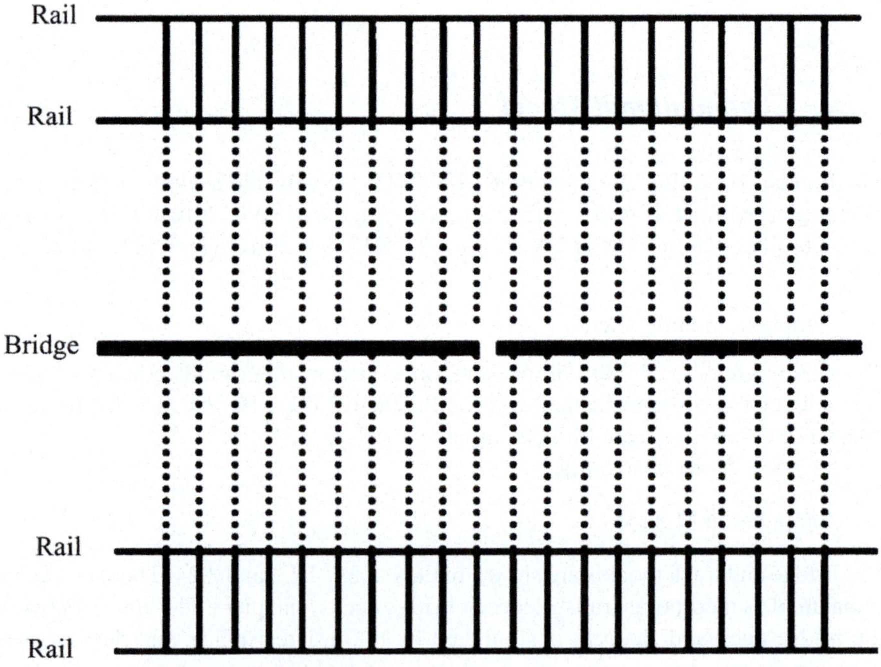

Fig. 2.22 Layout plan of rails and deck

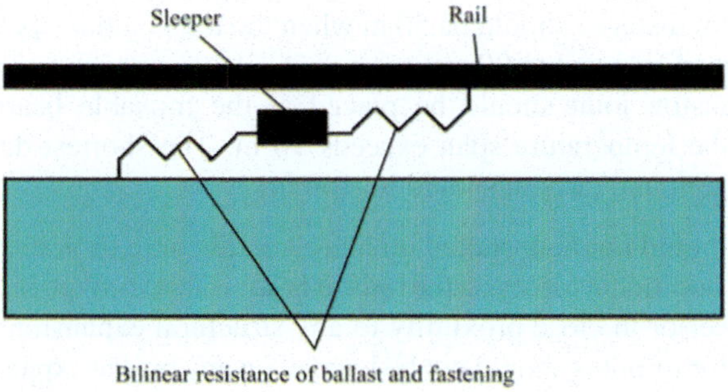

Fig. 2.23 Computational model

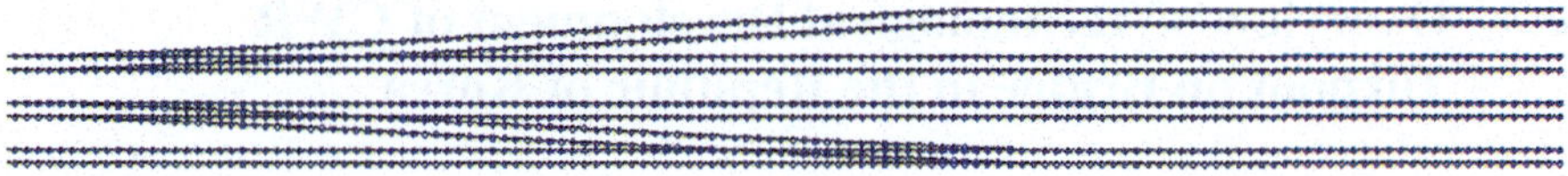

Fig. 2.24 Program-generated computational model

2.2.2 *Arrangement Requirements*

In 2004, SNCFI/SYSTRA offered the following recommendations regarding the arrangement of the turnout–bridge structure for the design of turnouts on bridges within the DK1002 + 000 ~ DK1309 + 320 section of Beijing–Shanghai HSR, as part of its consulting services:

1. French recommendations on turnout locations and bridge arrangement characteristics for this HSR
(1) Avoid excessive additional forces on the turnout rails induced by the expansion and contraction of the bridges and closure rails.
(2) Restrict the relative displacement between the turnouts and bridges resulting from temperature changes to maintain the longitudinal and lateral stability of the turnouts.
(3) Restrict the displacement of the rails and sleepers at the critical turnout components such as switches and frogs.
2. French recommendations on the setting conditions of turnouts for this HSR
(1) Turnouts on bridges should be specifically designed.
(2) Turnouts should not be partially installed on the bridges and subgrade to prevent any potential negative effects on the turnouts caused by differential settlement.
(3) The minimum distance of turnouts to any bridge expansion joint should adhere to the following criteria:

The minimum distance should be 20 m when the temperature span is less than or equal to 30 m;

The minimum distance should be 50 m when the temperature span is greater than 30 m and less than or equal to 90 m;

A rail expansion joint should be placed at the movable bearing end of the bridges when the temperature span exceeds 90 m. The shortest distance between the expansion joint and turnout should be 100 m.

(4) Turnouts should not be installed on a horizontal curve or vertical curve.
(5) To restrict additional forces on the rails, it is advisable to avoid installing turnouts directly over or in close proximity to any structural expansion joint. Switches and frogs for turnouts should not be located on any bridge expansion joint under challenging conditions.

2.3 Overview of Technological Development of CWR Turnout on Bridge in the Republic of Korea

2.3.1 Overview of CWR Turnouts on Bridges in the Republic of Korea

The Donghae Nambu Line is an HSR route connecting Busan and Pohang in the Republic of Korea, stretching roughly parallel to the east coast of the Korean Peninsula, as shown in Fig. 2.25. This line includes the following stations: Pohang Station, Ulsan Station, Changwon Station, Jinhae Station, Masan Station, and Jinju Station.

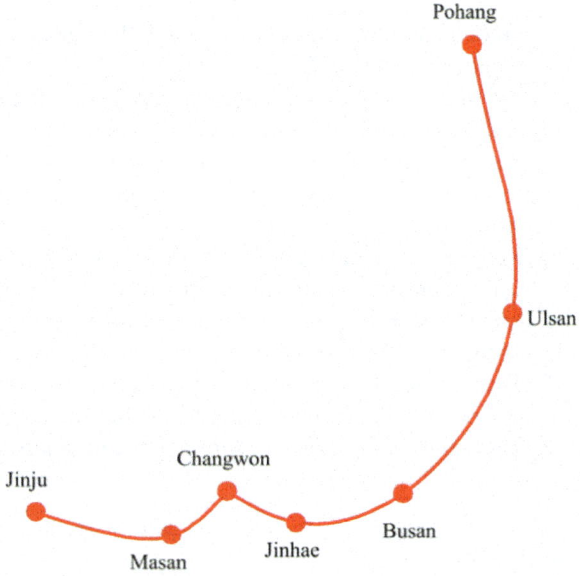

Fig. 2.25 Donghae Nambu line in republic of Korea

Table 2.5 Turnout chainage at feature points

Position	Chainage
Beginning of turnout and turnout slab	TK360+393.410
End of track bed	TK360+254.919
End of turnout slab	TK360+218.200
Frog	TK360+335.790
Heel end of frog	TK360+255.738
Toe end of frog	TK360+278.020
1st switch of switch rail	TK360+389.809
2nd switch of switch rail	TK360+384.409
3rd switch of switch rail	TK360+377.810
4th switch of switch rail	TK360+371.211
5th switch of switch rail	TK360+364.612
6th switch of switch rail	TK360+358.013
1st switch of point rail	TK360+277.422
2nd switch of point rail	TK360+272.702
3rd switch of point rail	TK360+268.002
Pull-down device	TK360+275.062

The bridge at Ulsan Station is equipped with two No. 43 single turnouts made by Voestalpine BWG, featuring UIC60 rails. These turnouts, with a total length of 138.491 m, enable trains to reach speeds of up to 170 km/h on the diverging route. Refer to Table 2.5 for specific turnout chainage and Fig. 2.26 for the layout plans.

The turnout section is constructed as a Rheda2000 ballastless track structure, incorporating embedded sleepers. The total length of the turnout slab measures 175.21 m. The slab begins with a width of 3.2 m and ends with a width of 5.9 m. The turnout slab is divided into two units, connected with an expansion joint with a width of 100 mm. The turnout slab is connected to the concrete protection layer of the bridge deck through longitudinal and transverse bollards. The bollards are positioned as shown in Fig. 2.27.

The transverse bollards are placed in the middle of each turnout slab unit. The number of bollards was determined through structural analysis. The longitudinal and transverse bollards are respectively 100 mm and 60 mm in height, as shown in

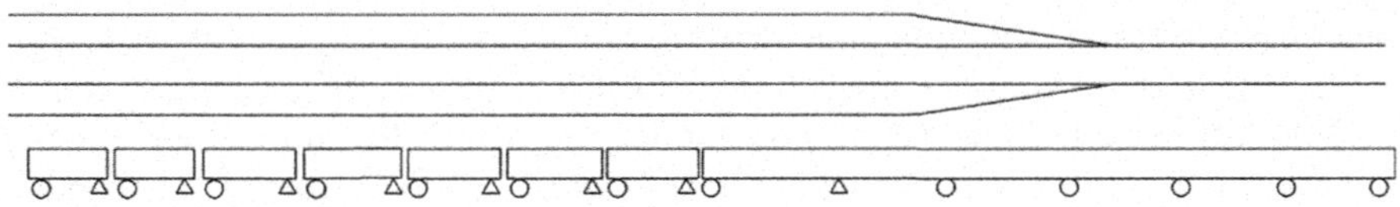

Fig. 2.26 Layout plan of turnout and bridge at ulsan station

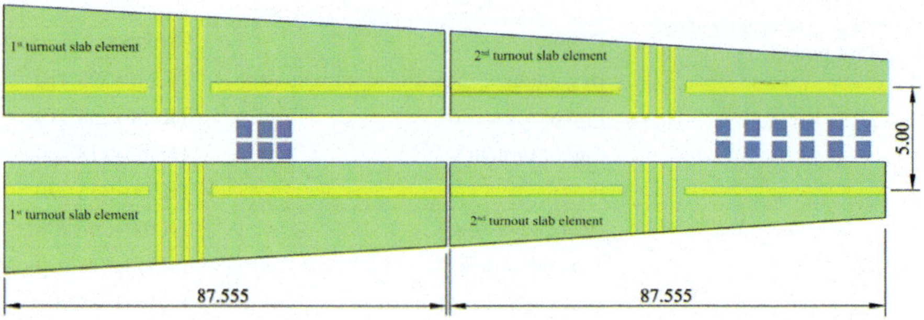

Fig. 2.27 Layout plan of longitudinal and transverse bollards (unit: m)

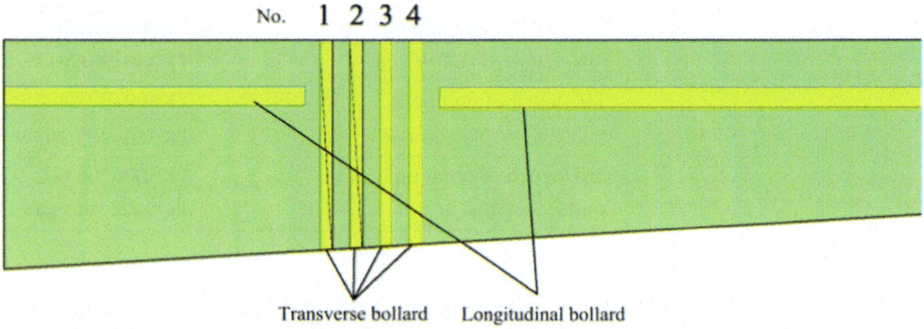

Fig. 2.28 Layout of longitudinal bollards

Fig. 2.28. The upper surface of the deck protection layer is arranged with a herring-bone drainage slope of 2%. A 4 mm thick EPDM rubber isolation layer is set between the track bed slab and the protection layer.

2.3.2 Computation of Girder-Rail Interactions for CWR Turnouts on Bridges in the Republic of Korea

KHP, a German consulting company, calculated girder-rail interactions for the CWR turnouts on the bridge at Ulsan Station. Since girder-track interactions constitute a major cause of rail fracture and track buckling, it is crucial to calculate and restrict displacement and additional stress due to such interactions, to effectively control potential adverse effects on the bridge and track and ensure their normal functioning.

In addition to calculating the additional rail stress, it is also necessary to conduct a computational analysis of the relative displacement between the bridge and rails. Since frogs and switches are most susceptible to displacement, the relative displacement analysis focused on these turnout zones to ensure the proper functioning of the turnouts in the event of relative displacement occurring under service conditions.

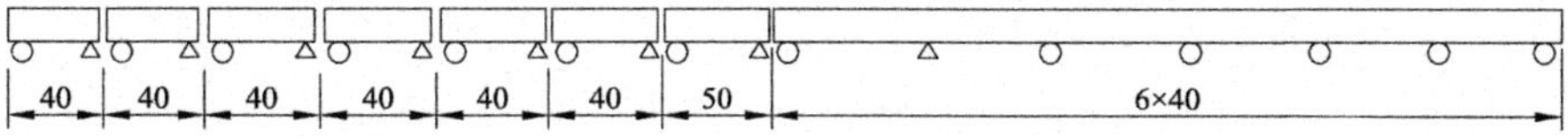

Fig. 2.29 Bridge arrangement (unit: m)

Through the analysis of track-bridge interactions, it was verified that the additional tensile and compressive stress of the CWR turnouts would not exceed a specific permissible limit under various load cases such as vertical load, temperature changes, and braking force/traction.

1. Design Parameters

This bridge is constructed in the form of 6 × 40 m continuous girders, as shown in Fig. 2.29. The cross section is shown in Fig. 2.30, the sectional parameters are shown in Table 2.6, and the longitudinal horizontal rigidity of the piers is shown in Table 2.7.

2. Computational Model

The finite element model for the bridge-turnout interactions is shown in Fig. 2.31.

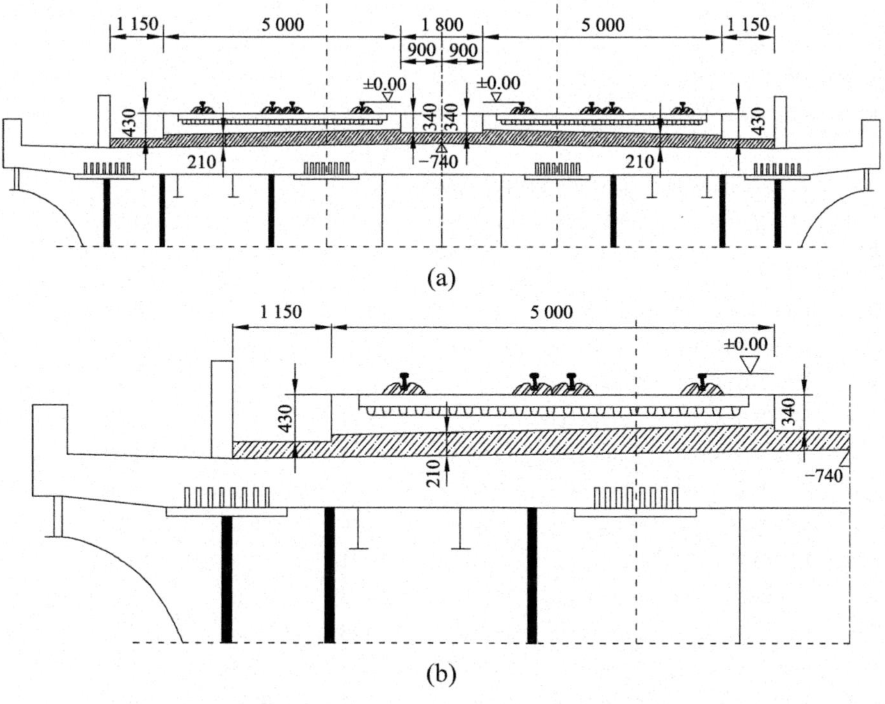

Fig. 2.30 Cross section of ballastless track in turnout section (unit: mm)

Table 2.6 Sectional parameters of turnout girders

Chainage	Girder length (m)	Type of girder	Sectional area (m^2)	Girder height (m)	Moment of inertia	
					I_y/m^4	I_z/m^4
K359+928.200	40	Simply-supported girder	24.80	1.80	40.60	332.00
K359+968.200	40	Simply-supported girder	24.80	1.80	40.60	332.00
K360+008.200	40	Simply-supported girder	24.80	1.80	40.60	332.00
K360+048.200	40	Simply-supported girder	24.80	1.80	40.60	332.00
K360+088.200	40	Simply-supported girder	24.80	1.80	40.60	332.00
K360+128.200	40	Simply-supported girder	2.04	1.80	7.00	141.00
K360+168.200	50	Simply-supported girder	2.84	2.04	7.97	116.00
K360+218.200	40	Continuous girder	2.82	2.17	8.20	89.00
K360+258.200	40		2.24	2.24	8.52	80.00
K360+298.200	40		2.77	2.27	8.39	69.00
K360+338.200	40		2.70	2.31	8.25	59.00
K360+378.200	40		2.63	2.34	8.12	50.00
K360+418.200	40		2.42	2.33	7.28	40.00

Table 2.7 Longitudinal horizontal rigidity of piers

Pier No	Chainage	Structure	Effective height (m)	Sectional area (m^2)	Moment of inertia (m^4)	Longitudinal rigidity (MN/m)
P1	K359+968.200	Cantilever	24.50	15.71	51.1	479
P2	K360+008.200	Cantilever	24.50	15.71	51.1	479
P3	K360+048.200	Cantilever	22.50	15.71	51.1	619
P4	K360+088.200	Cantilever	22.50	15.71	51.1	619
P5	K360+128.200	Cantilever	19.80	15.71	51.1	908
P6	K360+168.200	Cantilever	20.50	8.29	20.5	328
P7	K360+218.200	Cantilever	21.00	8.29	20.5	305
P8	K360+258.200	Cantilever	25.00	15.71	51.1	451
P9	K360+298.200	Cantilever	25.00	10.18	37.6	332
P10	K360+338.200	Cantilever	23.50	10.18	37.6	399
P11	K360+378.200	Cantilever	23.50	10.18	37.6	399
P12	K360+418.200	Cantilever	19.50	10.18	37.6	699

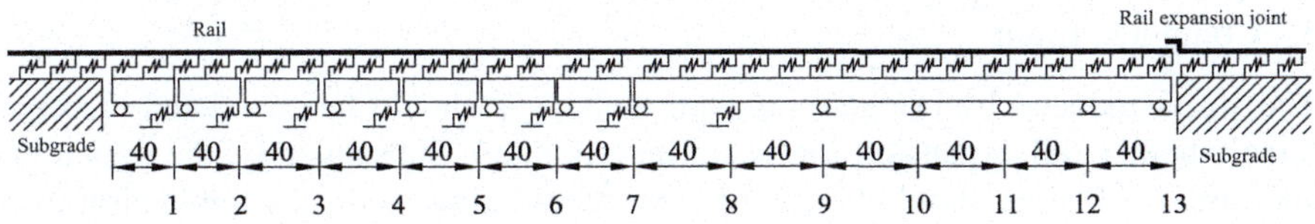

Fig. 2.31 Computational model for girder-rail interactions (unit: m)

Rail expansion joints: Due to the large temperature span of this multi-span composite bridge and the large temperature changes between the track and bridge, the relative displacement between the track and the bridge will be noticeable at both ends of the bridge, potentially causing slip of fasteners. Moreover, the additional rail stress at the expansion joint of the girder end that occurs near the tunnel portal (chainage K360 + 458.200) due to temperature changes is anticipated to exceed the permissible stress. Therefore, a pair of rail expansion joints is installed at this location.

Longitudinal dampers for bridge bearings: Within the scope of this multi-span bridge, two longitudinal dampers are placed between the top of No. 7, 9, 10, 11, and 12 piers and the bridge, facilitating longitudinal slip of the bearings. Operating as a type of viscoelastic damper, the longitudinal dampers exert acting force depending on the speed of relative motion. Specifically, at high speeds in scenarios such as braking and traction, an immediate reaction can be generated. However, in instances of slow deformation such as bridge temperature changes, the longitudinal dampers remain inactive. In the calculations, the effects of longitudinal dampers are only taken into account under braking and traction loads, while their impact is disregarded during temperature changes. Figure 2.32 shows the deformation of the longitudinal dampers versus the reactions assumed in the calculations for these two extreme cases.

The maximum reaction is 1,750 kN and the maximum permissible displacement is 12 mm.

Fig. 2.32 Deformation versus reaction of longitudinal damper

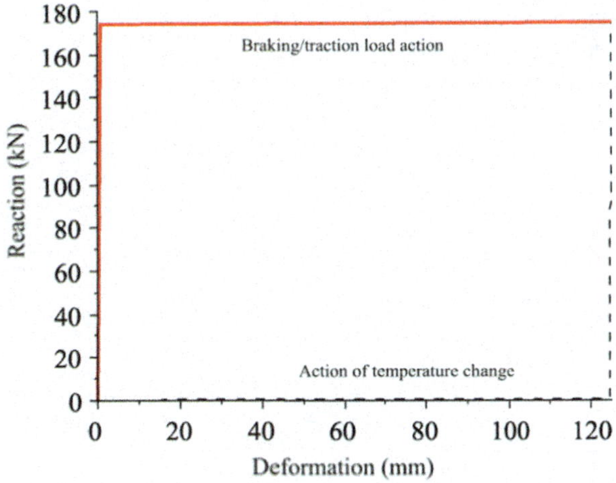

3. Computed Limit

The stress induced by bridge temperature changes should be calculated to determine whether rail expansion joints are incorporated. Subsequently, the maximum additional rail stress should be determined, considering the load combination from train braking/traction, the longitudinal force due to the deflection of the bridge, and changes in bridge load temperature. The resulting value must meet the requirement of UIC774-3 on the maximum permissible limit of additional rail stress. Apart from analyzing the additional rail stress, the relative displacement between the rail and bridge in the frog and switch zones should also be analyzed. In this regard, the requirement on the permissible relative displacement between the turnout rails and turnout slab (including the embedded sleepers) given in the technical specification of BWG turnouts must be met to ensure the normal functioning of the turnouts. See Table 2.8 for the relative displacement limits of the girders and rails.

4. Computation Results

The additional rail stress induced by bridge temperature changes should be computed, using the assumed temperature decreases and increases, as shown in Figs. 2.33 and 2.34. Since the temperature changes occur slowly on bridges, the influence of the longitudinal dampers should be ignored.

The relative displacement between the track and bridge deck is shown in Fig. 2.35.

Table 2.8 Relative displacement limits of girders and rails

Turnout position	Geometric parameters of turnout	Maximum permissible relative displacement (mm)
Frog zone	$R \leq 1200$m $R \geq 3000$m	≤ 3.70 ≤ 0.70
Switch zone	–	≤ 6.00

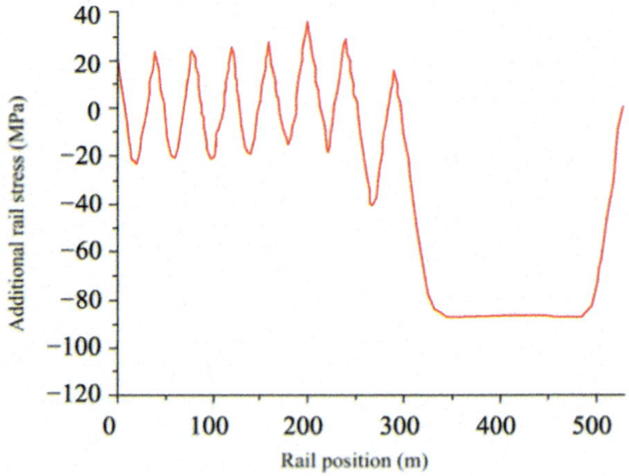

Fig. 2.33 Additional rail stress induced by bridge temperature decreases

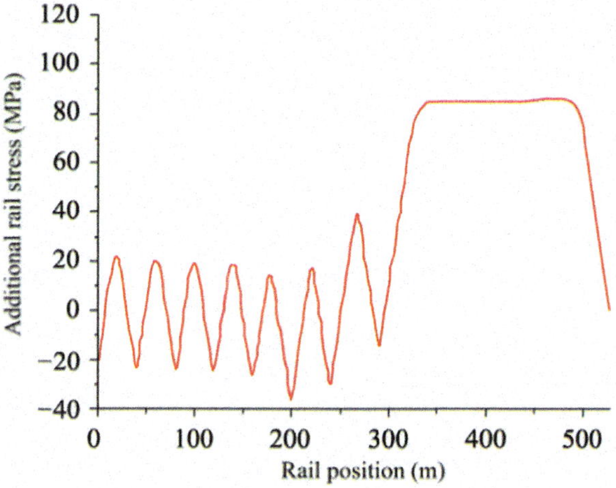

Fig. 2.34 Additional rail stress induced by bridge temperature increases

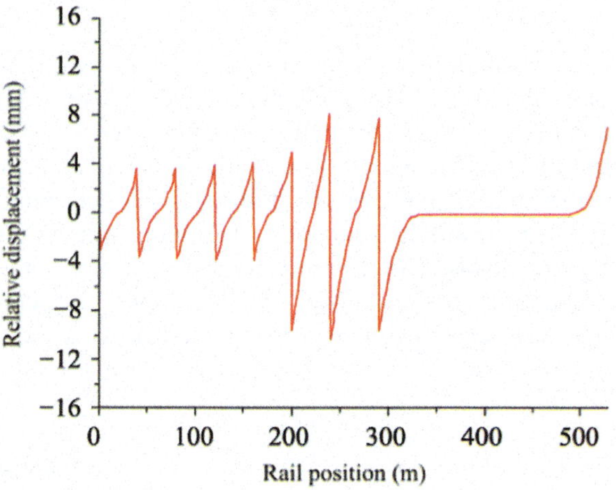

Fig. 2.35 Relative displacement induced by viaduct temperature changes

To verify the necessity of setting rail expansion joints at the girder end, calculations were also carried out for the scenario without rail expansion joints, as shown in Fig. 2.36.

The computation results revealed an additional rail stress of 151 MPa without the inclusion of rail expansion joints, far beyond the permissible stress of 92 MPa, emphasizing the necessity of incorporating rail expansion joints.

To reveal the maximum rail stress under train loads, a braking force was applied to one track while simultaneous traction in the same direction was applied to the other track, as illustrated in Fig. 2.37. Given that the rail at the girder joint experiences the most unfavorable stress in the absence of rail expansion joints, the starting point for load distribution was set at this location. Load distribution extended from the stress point to the end of the entire bridge. The vertical load was applied at the

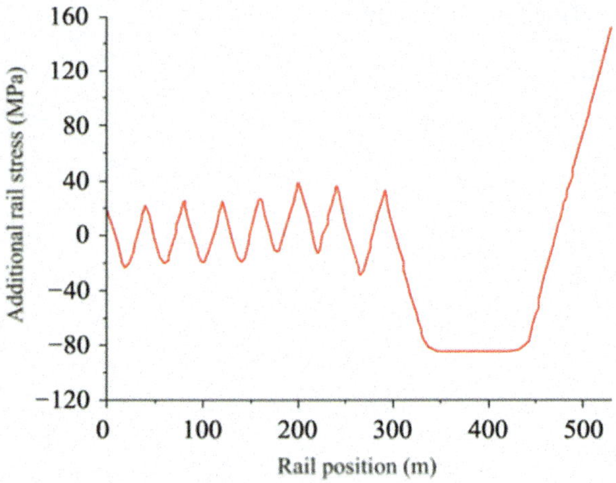

Fig. 2.36 Additional rail stress induced by temperature decrease without rail expansion Joint

same position mentioned for calculating the additional rail stress caused by bridge deflection. Considering the high speeds associated with movement under braking and traction effects, the influence of the longitudinal dampers was taken into account during the calculations. The results of these calculations are presented in Fig. 2.38.

The maximum additional rail stress was determined by combining the additional rail stress resulting from bridge deck temperature changes with the rail stress caused by train loads. The resulting combination took into account the most unfavorable additional rail stress under two conditions: temperature increase and temperature decrease. The maximum additional rail stress is shown in Fig. 2.39 and Table 2.9.

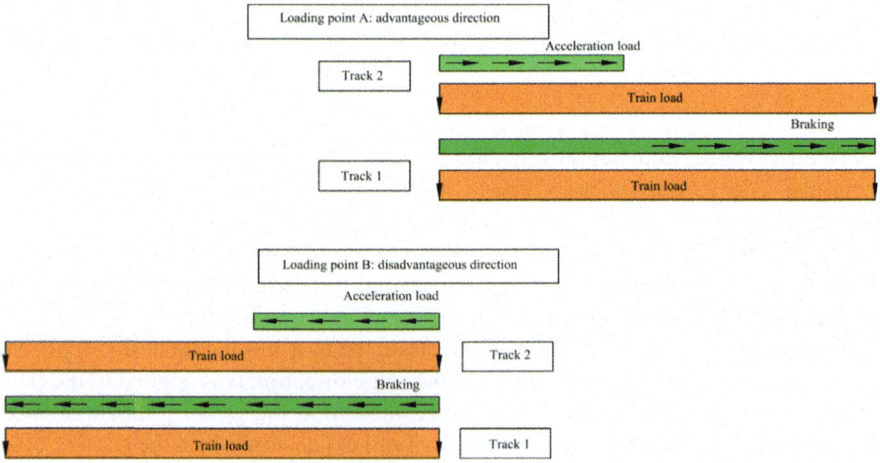

Fig. 2.37 Loading direction of braking force

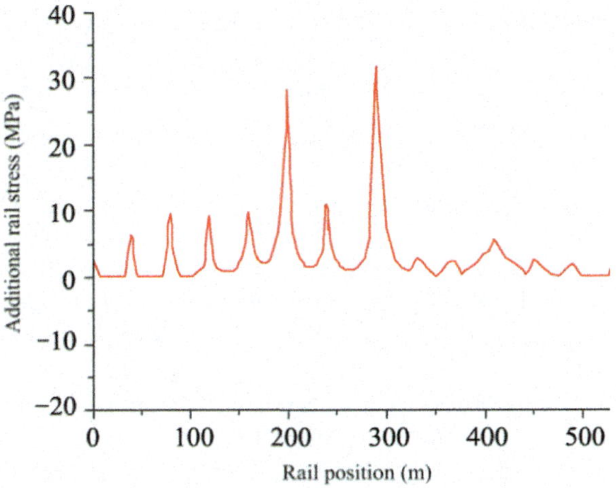

Fig. 2.38 Maximum additional rail stress caused by train traction/braking

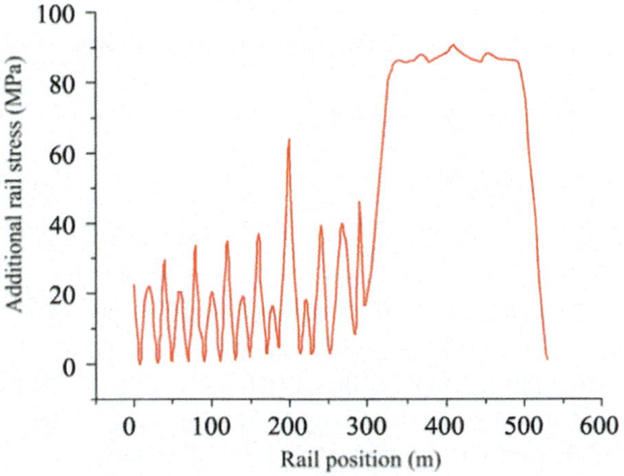

Fig. 2.39 Maximum additional rail stress envelope under load combination

Table 2.9 Maximum additional rail stress under load combination

Case	Chainage	Additional rail stress under temperature condition (MPa)	Additional rail stress under combination condition (MPa)
1	K360+128.200	36	64
2	K360+218.200	15	46
3	K360+338.200	86	91

The conclusion from the computational analysis for CWR tracks: The calculated maximum additional rail stress is 91 MPa, remaining below the permissible additional rail stress limit specified in UIC774-3, i.e. $\sigma_{rail} = 92$ MPa, as shown in Table 2.10.

Table 2.10 Checking calculation of additional forces on rails

Case	Chainage (km)	Additional rail stress (MPa)	Limit specified in UIC774-3 (MPa)	Checking conclusion
1	K360+128.200	64	≤ 92	Satisfied
2	K360+218.200	46	≤ 92	Satisfied
3	K360+338.200	91	≤ 92	Satisfied

Table 2.11 Relative displacement at frog zone on the bridge

Turnout number	Frog zone		Rail expansion joint	Frog zone	
	Chainage at starting point	Chainage at ending point	Chainage	Relative displacement at starting point	Relative displacement at end point
1#, 2#	K360+285.993	K360+263.543	K360+218.200	0.0 $\leq$ 0.7 mm	0.0 $\leq$ 0.7 mm

Table 2.12 Relative displacement at switch zone on the bridge

Turnout number	Switch zone		Rail expansion joint	Switch zone	
	Chainage at 1st switch	Chainage	Chainage	Relative displacement	Relative displacement
1#, 2#	K360+389.958	K360+357.708	K360+458.200	0.0 $\leq$ 6.0 mm	0.0 $\leq$ 6.0mm

The analysis of track-bridge nonlinear interactions confirmed compliance with the limits of 0.70 mm and 6 mm for the relative displacement respectively at the frog zone (Table 2.11) and the switch zone (Table 2.12) of the turnouts on the bridge.

To sum up, the maximum additional rail stress and girder-rail relative displacement of the turnouts on the bridge at Ulsan Station align with the requirements specified in the design specifications and technical specifications for turnouts.

2.3.3 Experience Summary of CWR Turnouts on Bridges in the Republic of Korea

The CWR turnouts on the bridge at Ulsan Station of Donghae Nambu Line in the Republic of Korea followed the technical concepts and design approaches of Germany. The computational model represented a simplified turnout structure without considering the intricate turnout–bridge interactions. However, additional measures are taken such as incorporating rail expansion joints at the ends of the long-span turnout girder and multiple longitudinal dampers at the movable bearings of the piers. These measures were intended to mitigate girder-rail interactions. This bridge offers an idea worthy of reference.

2.4 Overview of Research on Technologies of CWR Turnout on Bridge in China

2.4.1 CWR Turnouts on HSR Bridges in Taiwan

1. Overview of HSRs in Taiwan

Taiwan High Speed Rail (THSR) [10, 11] spans a total length of 345 km, connecting Taipei to Kaohsiung through 14 counties and cities. The main line is designed to achieve a speed up to 350 km/h. In terms of length, approximately 73% of the railway system comprises bridges, 18% consists of tunnels, and 9% is composed of subgrade sections. Construction for the THSR commenced in 1999, and it was inaugurated for operation from Banqiao Station to Zuoying Station on February 1, 2007. The entire line was officially opened for operation on March 2, 2007.

THSR was developed under the build-operate-transfer (BOT) mode, with construction and operation undertaken by Taiwan High Speed Rail Corporation through private investment. The track system of THSR consists of double tracks on the main line, utilizing a track gauge of 1,435 mm. The system incorporates a total of 137 turnouts and trans-sectional CWRs throughout the entire line. This system showcases various track forms including the slab ballastless track from Japan, the Rheda ballastless track from Germany, the low vibration track (LVT), the embedded track, and the ballasted track.

2. Arrangement of Viaduct Stations on THSR

THSR is designed with a total of 13 stations along the western corridor. In the initial stage of operation, 8 stations were opened to service. Among its stations, Nangang, Taipei, Banqiao, and Taoyuan are underground stations; Zuoying is a ground station; Hsinchu, Miaoli, Taichung, Changhua, Yunlin, Chiayi, Tainan, and Kaohsiung are viaduct stations. All the stations are furnished with CWR turnouts on bridges, BWG high-speed turnouts, and Rheda2000 ballastless tracks with embedded sleepers. The route and arrangement of the stations are shown in Figs. 2.40 and 2.41.

3. Design Characteristics of CWR Turnout on Bridge

To accommodate turnout systems on the viaducts, THSR is equipped with BWG high-speed turnout systems and Rheda2000 ballastless track systems that have demonstrated reliable performance. These systems are primarily installed within the station areas where turnouts are present, accounting for approximately 8.5% of the entire THSR line.

The arrangement characteristics of bridges in the turnout sections of THSR: These sections are dominated by multi-small-span continuous girders, typically spanning

Fig. 2.40 Schematic diagram of THSR route

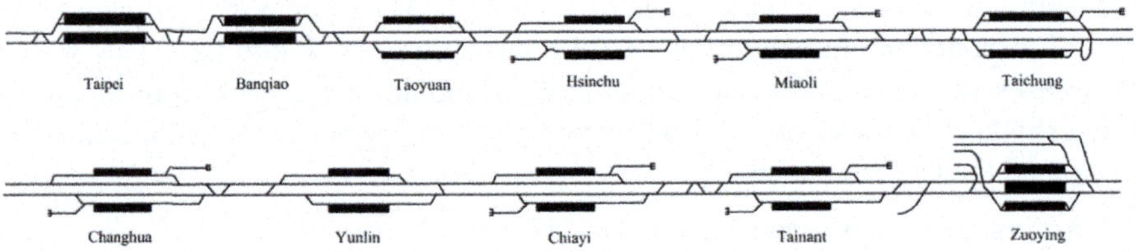

Fig. 2.41 Arrangement of THSR stations

30 or 25 m. The turnouts are positioned within these continuous girders, typically avoiding crossing the girder joint. In these turnout sections, the German Rheda2000 ballastless track technology with embedded sleepers is adopted, as depicted in Fig. 2.42.

4. Computation of CWR Turnout on Bridge

The basic model is sufficient to represent the real conditions of relatively short Rheda slabs, which are approximately 7.50 m long and typically feature 4 or 5 longitudinally placed bollards. However, for longer turnout slabs up to 80 m, which have concentrated transverse bollards located in their middle, the basic model recommended in UIC 774–3 may not accurately simulate their behavior. To account for this, a more precise model was employed in analyzing CWR turnouts on THSR bridges. This model treats the turnout slabs as separate structural units and connects them

Fig. 2.42 CWR turnout on THSR bridge

to the bridges using springs to capture the deformation of the bollards. The specific computation method is described below, using Tainan Station as an example.

(1) Computation parameters for CWR turnouts on the bridge at Tainan Station

Chainage of Tainan Station: TK312 + 734.216 ~ TK315 + 132.216; turnout numbers: W551A, W551B, W552, W553, W554, W555, W556, W557, and W558B.

The cross section of the box girder beneath the turnout at Tainan Station is shown in Fig. 2.43.

Within Tainan Station, the frog zone of the turnout is situated in close proximity to a girder joint. As a result, the relative displacement between the rails and the bridge deck/turnout slab exceeded the permissible limit of 0.7 mm. To address this issue, the turnout slab needed to be extended to cross over the girder joint. The extended portion of the turnout slab is designed as a "floating slab," meaning it maintains horizontal slips against the bridge deck. Specifically, the turnout slabs for turnouts

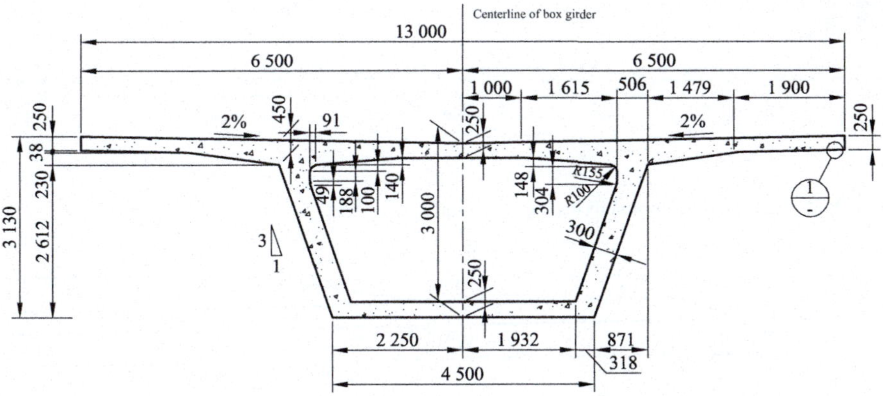

Fig. 2.43 Cross section of box girder in turnout section at Tainan station (Unit: mm)

No. W552, W553, W556, and W557 are extended by 7.5m from their tail end, as depicted in Fig. 2.44.

(2) Computational models and results for CWR turnouts on the bridge at the left throat of Tainan Station

At the left throat of Tainan Station, a total of four turnouts (W551A, W551B, W552, and W553) are designed on the main line bridge, as shown in Fig. 2.45. The single-crossover turnouts W551A and W551B are positioned on the (25+35+25+30+30+30+25) m continuous girder, using two turnout slabs. The single turnouts W552 and W553 are positioned on the 5 × 30 m continuous girder, using 3 turnout slabs.

To evaluate the interaction effects of the track-turnout slab-structure combination (e.g. bollards on turnout slabs/crossover slabs) with bearings to support longitudinal actions, a series of nonlinear computations regarding additional rail stress and displacement of the rails and bridge deck needed to be conducted. The computational models for turnouts W551 and W552 are shown in Figs. 2.46 and 2.47. These models take into account the bridge temperature effect, vertical load, longitudinal braking, and traction, and include the CWR rails, rail fasteners, crossover/turnout slabs, superstructure, and substructure.

Under the condition of a simultaneous temperature change of 20 °C for both the bridge and turnout slab, the computed additional rail stress and relative displacement between the girder and rails using the above models are shown in Figs. 2.48, 2.49, 2.50 and 2.51.

(3) Computational models and results of CWR turnouts on the bridge at the right throat of Tainan Station

On the right throat of Tainan Station, four turnouts (W556, W557, W558A, and W558B) are designed on the main line bridge, as shown in Fig. 2.52. The single turnout W556 is positioned on the 3 × 30 m continuous girder, using three turnout

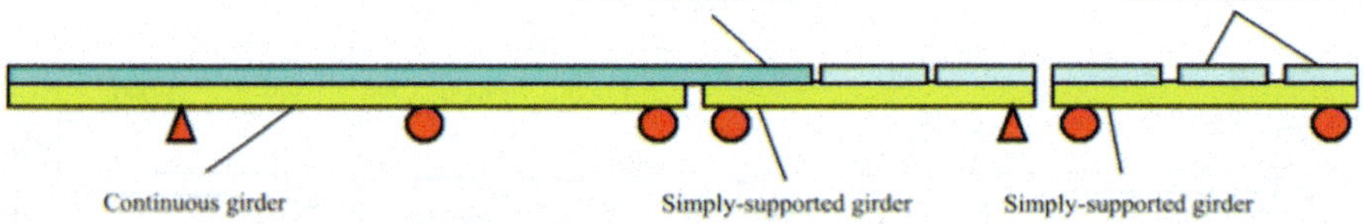

Fig. 2.44 Extended turnout slab over girder joint

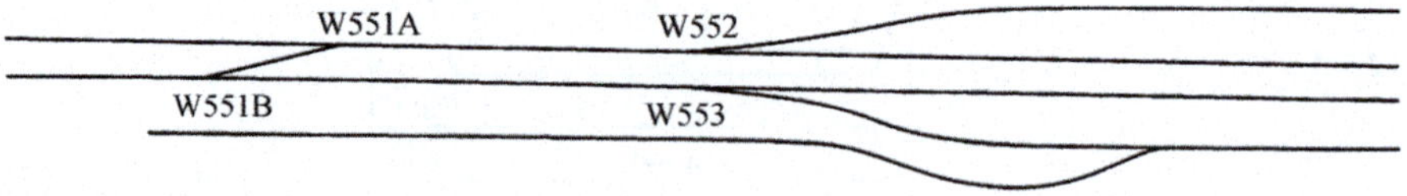

Fig. 2.45 Layout plan of left throat at Tainan station

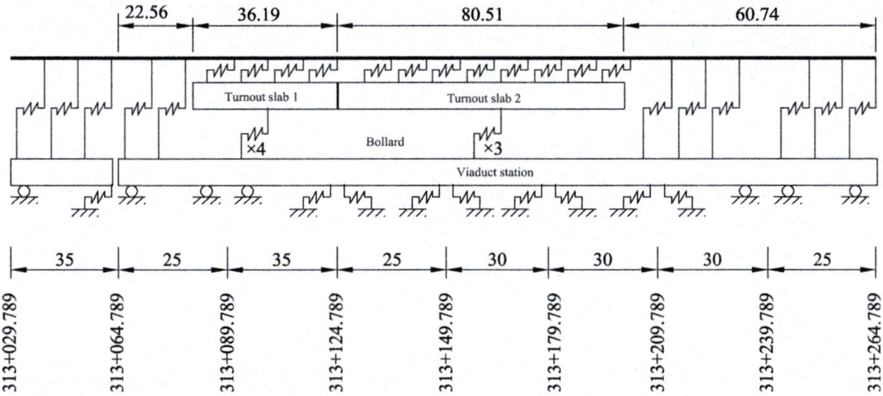

Fig. 2.46 Computational model for turnout W551 (Unit: m)

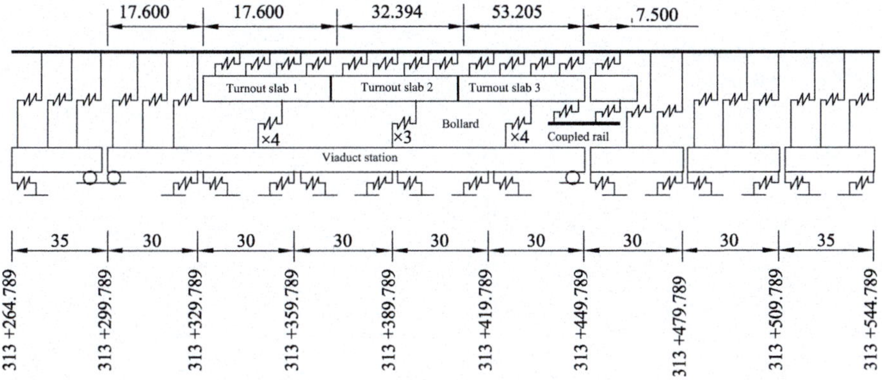

Fig. 2.47 Computational model for turnout W552

Fig. 2.48 Additional rail stress of up track

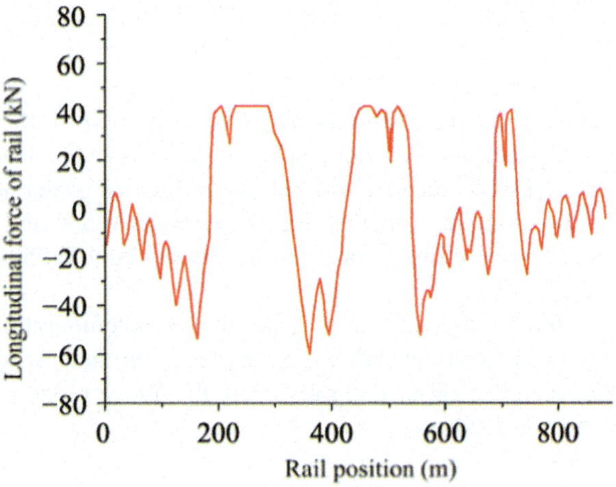

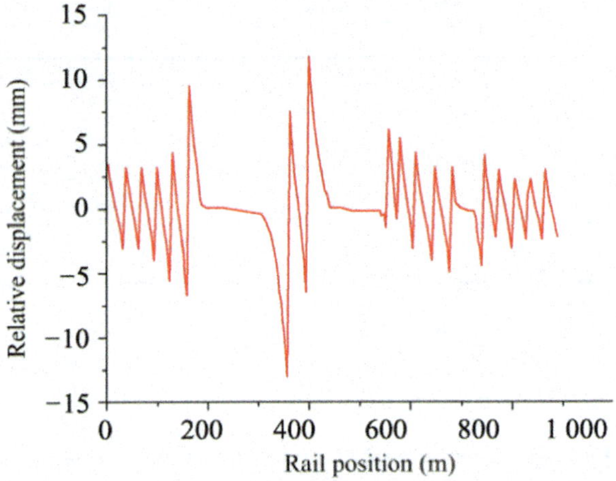

Fig. 2.49 Relative displacement between rail and track slab/deck of up track

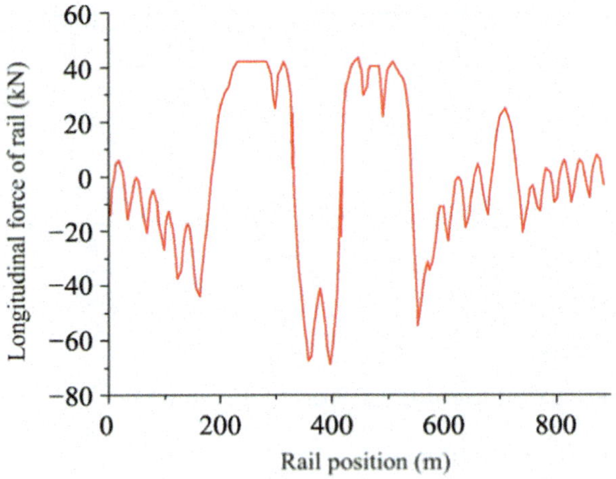

Fig. 2.50 Additional rail stress of down track

slabs. The single turnout W557 is positioned on the 5 × 30 m continuous girder, using three turnout slabs. The single-crossover turnouts W558A and W558B are positioned on the (5 × 30+25) m continuous girder, using three turnout slabs.

The computational models for turnouts W556 and W557 are shown in Figs. 2.53 and 2.54, and those for turnouts W558A and W558B are shown in Figs. 2.55 and 2.56.

Under the condition of a simultaneous temperature change of 20°C for both the bridge and turnout slab, the computed additional rail stress and relative displacement between the girder and rails using the above models are shown in Figs. 2.57, 2.58, 2.59 and 2.60.

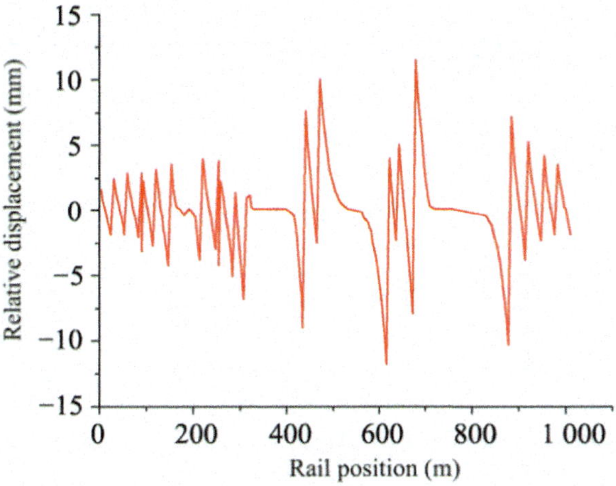

Fig. 2.51 Relative displacement between rail and track slab/deck of down track

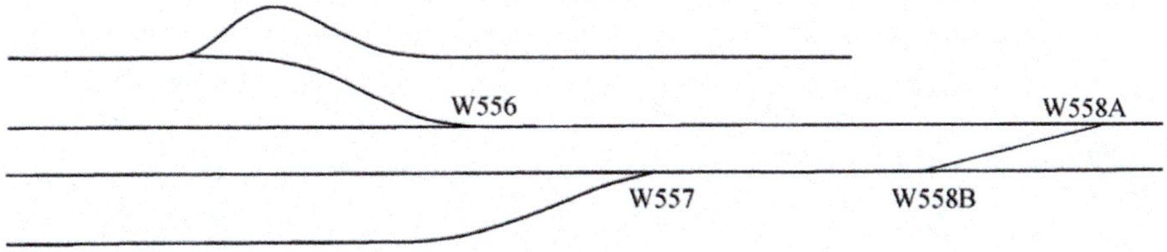

Fig. 2.52 Layout plan of right throat at Tainan station

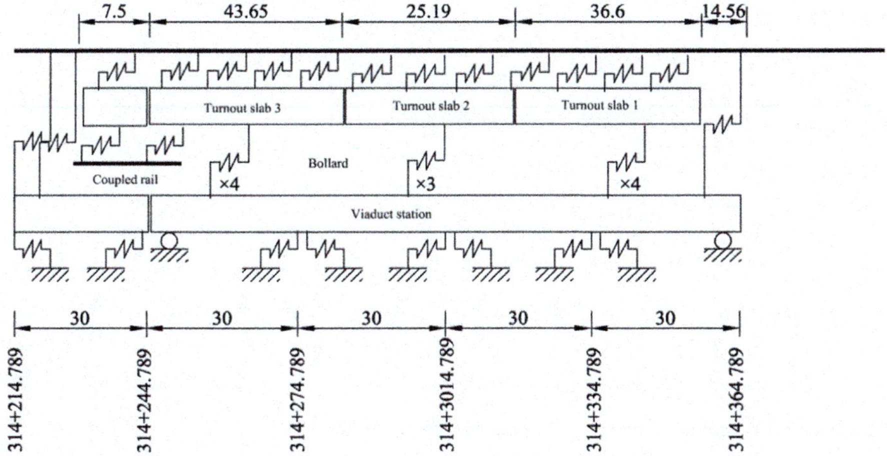

Fig. 2.53 Computational model for turnout W556 (unit: m)

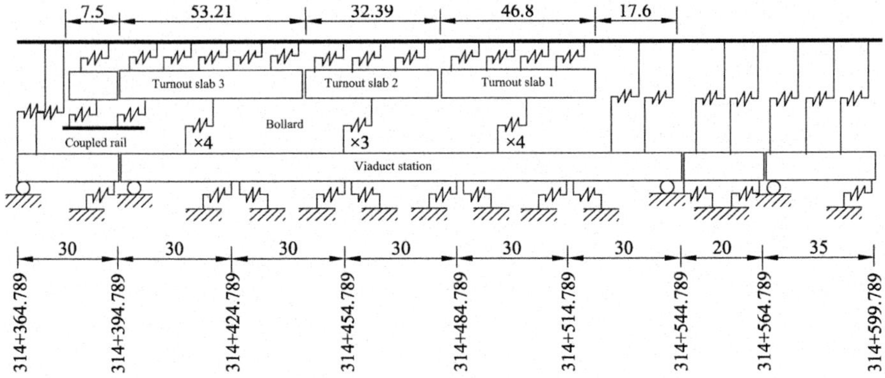

Fig. 2.54 Computational model for turnout W557 (unit: m)

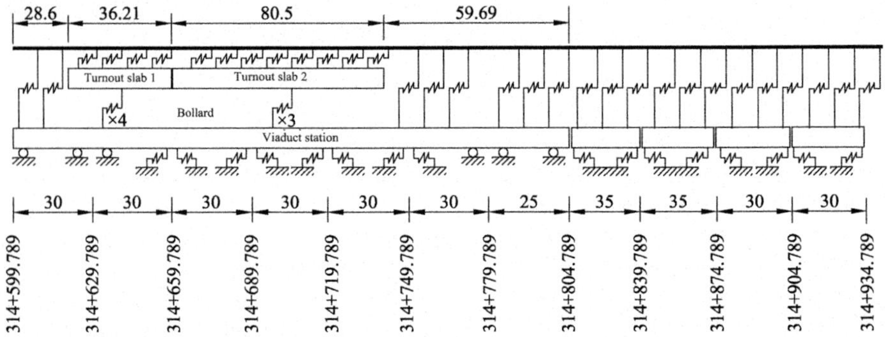

Fig. 2.55 Computational model for turnout W558A (unit: m)

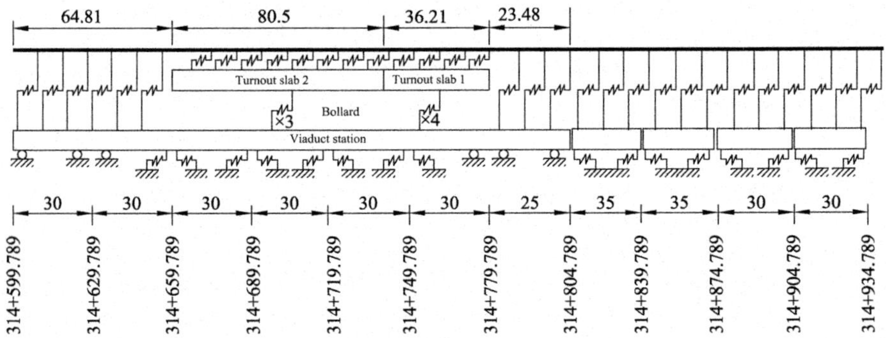

Fig. 2.56 Computational model for turnout W558B (unit: m)

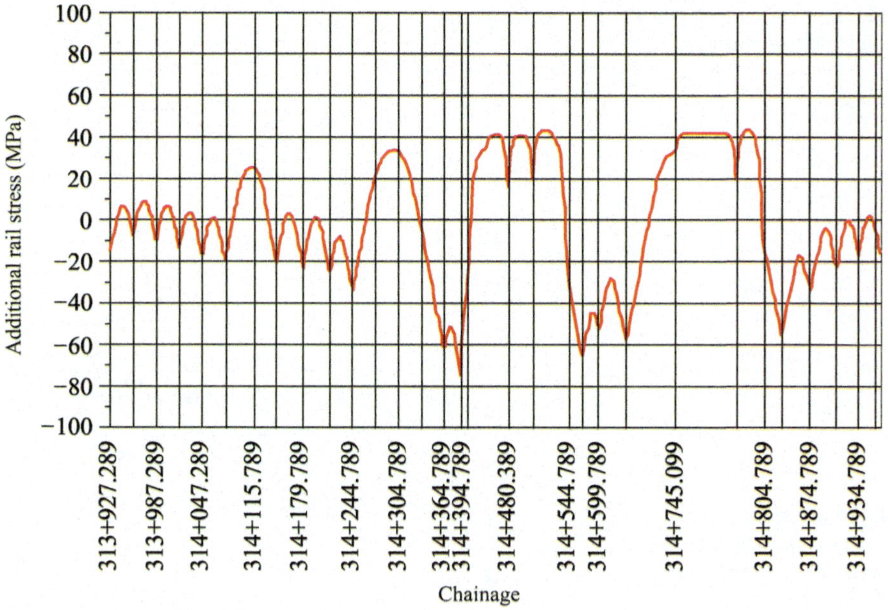

Fig. 2.57 Additional rail stress of up track

Fig. 2.58 Relative displacement between rail and track slab/deck of up track

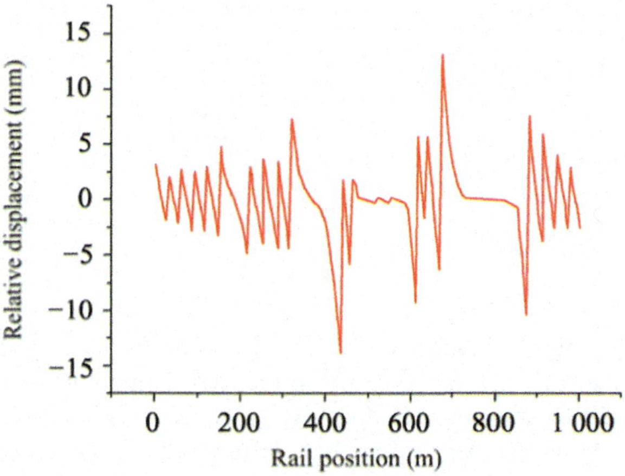

2.4.2 Overview of Research on CWR Turnout on Bridge in Chinese Mainland

The Chinese Mainland encountered challenges due to a lack of theoretical foundation and engineering practice experience in CWR turnouts on HSR bridges, primarily attributed to the late initiation of relevant research in this area. Consequently,

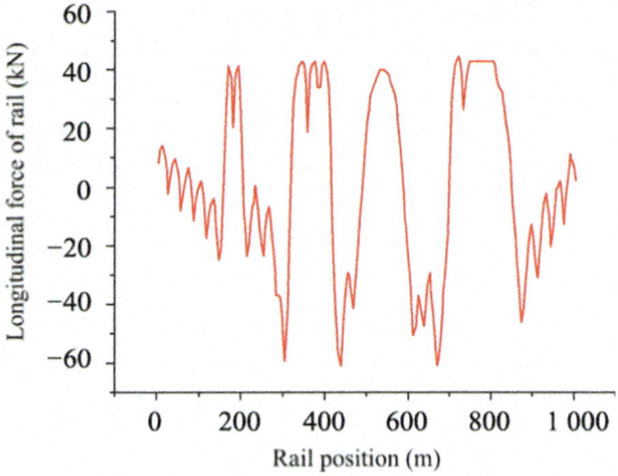

Fig. 2.59 Additional rail stress of down track

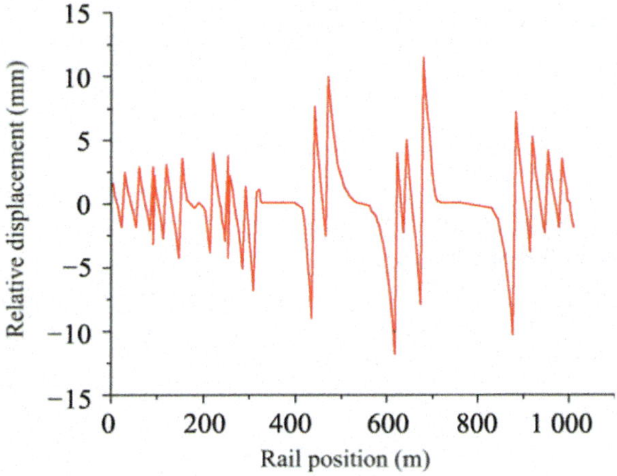

Fig. 2.60 Relative displacement between rail and track slab/deck of down track

foreign technical exchanges intensified in recent years, particularly with Germany and France, have greatly influenced China's design concepts and approaches for CWR turnouts on bridges. By learning and assimilating advanced technologies from abroad, Chinese researchers have gradually developed a comprehensive system of computational theories and design approaches specifically tailored to China's unique circumstances, as a result of significant independent innovation and careful consideration of the actual conditions within the country. The research outcomes have laid a solid technical foundation for the practical application of CWR turnout technologies on bridges.

1. Theoretical Research of CWR Turnout on Bridge

In 2004, the former Scientific Research Institute of the Ministry of Railway, which was later renamed as China Academy of Railway Sciences in 2007, developed computational models specifically for CWR turnouts on bridges. These models were designed to simulate either the plans or elevations of CWR turnouts. The former aimed to analyze the longitudinal interactions between the inner rail and stock rail within CWR turnouts, while the latter focused on analyzing the longitudinal interactions between the rails and bridges. During the calculations for CWR turnouts on bridges, the models were employed with separate perspectives for actions, considering both plans and elevations. The computation results from these perspectives were later combined. However, it should be noted that there exists a coupling effect between the plans and elevations for CWR turnouts on bridges. Using a separate approach for these perspectives led to certain deviations from the actual conditions.

In 2004, Wang Ping and Yang Rongshan et al. established models that integrated turnouts, bridges, and piers/abutments [12, 13]. Taking No.12 movable-point turnouts featuring 60kg/m rails as an example, they analyzed the stress and displacement patterns for the piers/abutments and rails associated with the installation of CWR turnouts on long continuous girders, and investigated the impact of various factors, such as the arrangement mode of CWR turnouts and the distance between the CWR turnouts and expansion joints. With the support of major science and technology projects initiated by the Chinese government, such as the National High-tech R&D Program (863 Program) and the Science and Technology Research and Development Program of the former Ministry of Railways, Southwest Jiaotong University has undertaken systematic research on CWR turnouts on bridges in collaboration with various scientific research institutes, design institutes, turnout manufacturers, and construction companies within the railway sector. Significant breakthroughs have been achieved in several key areas. They have revealed the influence of various factors on longitudinal interactions between the turnouts and bridges, such as the expansion and contraction forces on rails, bridge deflection under the influence of train loads, braking forces acting on the rail surface, occurrences of broken rails, the impact of continuous rigid frames, turnout groups, and substructure forms. Additionally, advancements have been achieved in areas such as the spatial coupling vibration of the train–turnout–bridge system and experimental studies.

In 2008, Yang Rongshan conducted in-depth research on the computational theories of longitudinal forces acting on CWR turnouts on bridges. He developed computational models and created corresponding programs [3] to facilitate these computations. Through systematic analysis of the stress and deformation patterns of CWR turnouts on bridges, he formulated design principles and approaches specific to CWR turnouts on bridges [3, 14]. The proposed computational theory for longitudinal force acting on CWR turnouts on bridges was subsequently verified through a field experimental study conducted on the CWR turnout situated on the ballasted track of a bridge on the Zhejiang–Jiangxi Railway [15].

In 2009, Ren Juanjuan conducted a theoretical analysis and experimental study to develop computational models and methods specifically focusing on the longitudinally coupled ballastless track system within the turnout section on bridges. She conducted a comprehensive computational analysis of the stress and deformation patterns of CWR turnouts, ballastless tracks, bridges, and relevant force transfer structures in the turnout section on bridges. Special attention was given to systematic mechanical computations, as well as structural design and optimization of the main structure and force transfer structure of ballastless tracks.

In 2006, Zeng Zhiping from Central South University presented a finite element model for computing the expansion and contraction forces on CWR crossovers on bridges in his doctoral dissertation titled "Study on Expansion and Contraction Forces on CWR Turnouts and Spatial Vibration of Train-Turnout-Bridge System on High-speed Railway Bridge" [16]. This model treats girders and rails as rod elements, turnout sleepers as beam elements on elastic support points, and bridge piers and abutments with fixed bearings as linear spring elements. It also accounts for longitudinal resistances of the track bed and fasteners, resistance moments of the fasteners, and resistances provided by stoppers and spacing blocks through the use of nonlinear springs. Leveraging the energy variational principle, a set of nonlinear equations was derived and solved. The model represents a detailed simulation of the turnouts and takes into account the turnout–bridge interactions. However, this model solely focuses on the connection mode using crossovers, without specifically investigating the influence of different turnout combinations on the stress and deformation of CWR turnouts on bridges.

Professor Chen Xiufang and Doctor Zeng Zhiping from Central South University, by integrating turnouts, girders, and piers/abutments into a system, established a finite element model for CWR turnouts on bridges featuring simply-supported girders and continuous girders [17, 18], and drew the following conclusions that hold guiding significance for design. (1) Under the action of temperature load on CWR turnouts on bridges featuring simply-supported girders, two peak values of rail temperature force occur simultaneously at the stopper and the girder end preceding it. (2) The maximum stress experienced by the pier beneath CWR turnouts on bridges is significantly higher compared to the scenario of CWR tracks on the same bridges. (3) The expansion and contraction of both the girder and closure rails in the same direction results in a considerable displacement of the rails within the turnout Sect. (4) The optimal position for stoppers, which contributes to advantageous stress and displacement of all turnout components, lies in the middle of the girder span. (5) Increasing the rigidity of the piers can considerably reduce the stress and displacement of the rails within the turnout section. For practical applications, the piers should be designed with appropriate rigidity, with the rigidity of the four piers preceding and following the stoppers appropriately increased. (6) Positioning rail expansion joints at the rear end of the turnouts helps to significantly reduce the longitudinal force on the fixed piers of the continuous girders. (7) With the distance between the fixed bearing of the continuous girders and the center of the rail expansion joint unchanged, positioning the rail expansion joint at the rear end of the turnouts leads to a reduction in longitudinal stress on the fixed piers of the continuous girders. Professor Gao Liang and his research team from Beijing Jiaotong University, based on the finite element method, established a computational model for longitudinal force on CWR turnouts on bridges and also drew some similar conclusions [19, 20]. Their model simplifies the rails into beam elements with the same section, and the concrete sleepers into beam elements without a specific slenderness ratio. It also accounts for longitudinal resistances of the track bed through the inclusion of longitudinal nonlinear elastic supports, resistances of the fasteners, stoppers, and spacing blocks through nonlinear springs, and the bridge span structures using solid elements. In 2005, Liu Yanfeng, Sun Daxin, Gao Liang et al. developed models to calculate the expansion and contraction force and displacement on CWR turnouts on bridges by using finite element software ANSYS. In the model, the rails and sleepers are considered as girder elements, bridges are simulated as plate-shell elements, and other closures are considered as spring elements. Taking a No. 18 turnout laid on a simply-supported girder as an example, they calculated the stress and displacement of the CWR turnout. This model offers a comprehensive simulation of the turnouts, taking into account the lateral interactions between the turnouts and bridges [21–23].

To sum up, CWR turnouts on bridges have emerged as a prominent field of research within track engineering, both in China and abroad. The finite element method has been widely utilized to develop computational methods in China and various other countries. Extensive research has been conducted on the longitudinal interactions between turnouts and bridges. While there have been in-depth investigations of scenarios involving the single turnout or crossover configuration on bridges, there is still a lack of studies focusing on CWR turnout groups on bridges.

2. Preliminary Study on CWR Turnouts on Bridges of Beijing–Shanghai HSR

In July 2004, the former Scientific Research Institute of the Ministry of Railway released the *Design Principles for CWR Tracks at Viaduct Stations of Beijing–Shanghai High-speed Railway (Interim).* This document incorporated the main design concepts of CWR turnouts on bridges that were learned and assimilated from Germany and France. The main design principles are as follows:

(1) CWR turnout girder

 (1) The structure for CWR turnout girders should be designed in compliance with the following requirements and through a technical and economic comparison:

① Corresponding technical requirements for CWR turnouts should be fulfilled.

② The vertical rigidity should be distributed as uniformly as possible.

③ Rail expansion joints are not incorporated in principle. When necessary, rail expansion joints may be included according to the computation results. The structural design with rail expansion joints should ensure the switch rail is positioned without crossing different girder spans, and the stock rail and switch rail should be positioned at least 2 m away from any girder end.

(2) CWR turnout girders may be selected from the following three types: concrete continuous girder, rigid frame with a span of 6 ~ 12 m, and simply-supported girder.
(3) Multiple fixed piers may be designed within one girder, to address specific requirements identified through checking results. It is important to assign a proper longitudinal rigidity value to these piers under this consideration.

(4) Girders beneath CWR turnouts should be positioned adjacent to the box girder, and shall be separated transversely to follow the centerline of the two main lines in the up and down directions.
(5) The viaducts should not accommodate a turnout maintenance platform.
(2) CWR turnout

The CWR turnout arrangement should meet the following requirements:

(1) CWR turnouts should be located within one girder as far as possible.
(2) The switch and frog sections should not be located on different girders To accommodate challenges posed by actual conditions, it is permissible to position them separately on different girders, under the condition that the following requirements are met:

① The actual points of the switch rails, the actual point of the point rails, and the heel ends of the point rails should be positioned at least 18 m away from any girder end.

② The heel ends of the switch rails should be positioned at least 40 m away from any girder end.

(3) Turnouts coded with a small number or single crossovers should be located within one girder.
(4) Turnouts on single-box double-track or double-box four-track girders should be placed symmetrically as far as possible.
(5) Joints by field welding, insulated rail joints by gluing, and rail joints by freezing on girders should be positioned at least 2 m away from the center of piers unless the bridges are in a rigid frame structure.

3. Overview of Scientific Research on Related Topics

From 2005 to 2012, the relevant departments of the former Ministry of Railways placed significant emphasis on research regarding CWR turnouts on bridges. Numerous projects were initiated under the Ministry's Science and Technology Research and Development Program to investigate computational theories, design methods, and engineering applications. These projects have played a crucial role in advancing the technological development of CWR turnouts on bridges.

From 2005 to 2008, China Railway SIYUAN Survey and Design Group Co., Ltd., in collaboration with Southwest Jiaotong University, conducted and completed the research project titled "Research on Design of CWR Turnouts on Passenger Dedicated Railway Bridges and Bridge Structures (Project No.: 2005G021)" as part of the aforementioned program under the former Ministry of Railways. The researchers conducted a systematic study that focused on several key aspects: the longitudinal interactions between turnouts and bridges, rational turnout–bridge arrangements, design parameters of turnout girders, lateral interactions between turnouts and bridges, dynamic simulation analysis of the turnout–bridge system, and evaluation of its dynamic characteristics. The study incorporated relevant regulations from Germany and France regarding CWR turnouts on bridges, the consulting recommendations for the Beijing–Shanghai HSR, and reference to the *Design Principles for*

CWR Tracks at Viaduct Stations of Beijing–Shanghai High-Speed Railway (Interim). The research outcomes were subsequently transformed into the *Interim Technical Conditions for the Design of CWR Turnouts on Bridges of Passenger Dedicated Lines*. These technical conditions are intended for application in the design of CWR turnouts on new high-speed railway bridges with an hourly design speed ranging from 200 to 350 km. The research results have already been successfully implemented in Hefei–Wuhan Railway, Wuhan–Guangzhou HSR, Shanghai–Nanjing Intercity Railway, and Guangzhou–Shenzhen–Hong Kong HSR, showcasing their positive effects.

From 2006 to 2011, China Railway First Survey and Design Institute Group Co., Ltd., China Academy of Railway Sciences, Beijing Jiaotong University, Southwest Jiaotong University, and the Third Railway Survey and Design Institute collaborated to conduct and successfully complete the research project titled "Research on Design of CWR Tracks on Passenger Dedicated Railway Bridges and Bridge-Turnout Systems in Turnout Sections (Project No.: 2006G011-E)" under the Science and Technology Research and Development Program of the former Ministry of Railways. Relying on the design of the CWR turnout on the bridge at Weinanbei Station of Zhengzhou-Xi'an HSR, the researchers studied the computational theories for statics and dynamics of CWR turnouts on bridges. Based on the study findings, they proposed arrangement modes and main design principles for CWR turnouts on bridges. These study results were successfully applied in the design of Weinanbei Elevated Station of the same HSR. The subsequent field tests demonstrated their compliance with the technical requirements of HSRs.

From 2006 to 2007, Southwest Jiaotong University carried out static and dynamic tests on the ballasted CWR turnout on the Meichi No. 1 Super Major Bridge at Meichi Station of Zhejiang–Jiangxi Railway, relying on the science and technology development project "Field Test of CWR Turnouts on the Bridge of Zhejiang–Jiangxi Railway" undertaken by China Railway Eryuan Engineering Group Co., Ltd. The test results preliminarily verified the correctness of the proposed computational theories for CWR turnouts on bridges. From 2007 to 2010, China Railway Eryuan Engineering Group Co., Ltd., in collaboration with Southwest Jiaotong University and China Railway Shanhaiguan Bridge Group Co., Ltd., systematically studied the design of the No. 12 CWR scissors crossover on the bridge of Hainan East Ring HSR. The research also involved the structure form of its ballastless track bed and design parameters. The study results were directly utilized to guide the design and laying of the No. 12 ballastless scissors crossover on this HSR. Subsequent field tests and operational performance confirmed the satisfactory functioning of this crossover.

From 2008 to 2010, Southwest Jiaotong University undertook and completed the research project titled "Research on Model Experiment and Dynamic Simulation Analysis of CWR Turnouts on Bridges of Passenger Dedicated Lines (Project No.: 2008G036-A)" under the Science and Technology Research and Development Program of the former Ministry of Railways. The researchers conducted a model experimental study on the expansion and contraction forces and deflection force on CWR turnouts on bridges, using 1:3 models for the first time. This study served to verify the established computational theories and methods for CWR turnouts on bridges, employing a "turnout-bridge-pier" integration pattern. Based on the theoretical analysis and model experiments, they established a systematic design theory for CWR turnouts on bridges. Through comprehensive calculations and analysis of the stress and deformation patterns of the turnouts and bridges, the researchers proposed fundamental principles and methods for the design of CWR turnouts on bridges. This significant contribution filled a gap in this field, both in China and abroad. As a result of this study, computation software for the static and dynamic design of CWR turnouts on bridges was developed. It has proven to be effective in the design and application of HSRs. The research conducted by the team provided valuable insights into the development of HSRs.

From 2008 to 2012, relying on the research project titled "Research on Design of CWR Turnouts with Ballastless Tracks Featuring Longitudinally Coupled Base Slabs within Bridges of Beijing–Shanghai High-speed Railway and Bridge Structures (Project No.: 2008G036-C)" under the Science and Technology Research and Development Program of the former Ministry of Railways, the Third Railway Survey and Design Institute, Southwest Jiaotong University, and Beijing–Shanghai High-speed Railway Co., Ltd., developed an improved computational model incorporating CWR turnouts, ballastless tracks, bridges, and piers/abutments. Using this model, they studied the longitudinal stress and deformation patterns of ballastless tracks and CWR turnouts with longitudinally coupled base slabs on bridges, along with the bridges and piers/abutments, optimized the computational theories and design methods for CWR turnouts with longitudinally coupled base slabs on bridges, and proposed the rational structural forms, design parameters, and structural design for ballastless tracks in turnout sections on bridges. The research results have been successfully utilized to offer theoretical guidance and technical support for the design of ballastless tracks and CWR turnouts with longitudinally coupled base slabs on bridges of Beijing–Shanghai HSR.

From 2008 to 2012, China Railway SIYUAN Survey and Design Group Co., Ltd., Southwest Jiaotong University, and Beijing–Shanghai High-speed Railway Co., Ltd., undertook the research project titled "Research on Design of CWR Turnouts with Ballastless Track and Bridge Structures on Beijing–Shanghai High-speed Railway Bridges (Project No.: 2008G036-D)" under the Science and Technology Research and Development Program of the former Ministry of Railways. The researchers developed a computational model integrating CWR turnouts, ballastless tracks, bridges, and piers/abutments, specific to the CWR turnouts with ballastless tracks featuring longitudinally coupled base slabs in the Xuzhou-Shanghai Section of Beijing–Shanghai HSR. They studied the longitudinal stress and deformation patterns of ballastless tracks and CWR turnouts with longitudinally coupled base slabs on bridges, along with the bridges and piers/abutments, explored the computational theories and design methods for CWR turnouts with longitudinally coupled base slabs on bridges, and proposed the rational structural forms, design parameters, and structural design for ballastless tracks with longitudinally coupled base slabs within turnout sections on bridges. The research results have been applied to the design of CWR tracks with ballastless tracks featuring longitudinally coupled base slabs on the bridges at the elevated stations of Suzhou, Wuxi, Changzhou, and Zhenjiang of Beijing–Shanghai HSR.

From 2008 to 2012, Beijing Jiaotong University, together with Beijing–Shanghai High-speed Railway Co., Ltd., Southwest Jiaotong University, East China Jiaotong University, the Third Railway Survey and Design Institute, and China Railway SIYUAN Survey and Design Group Co., Ltd., undertook the research project titled "Experimental Study of Technologies for Large Bridges, Elevated Stations, and Ballastless CWR Tracks of High-Speed Railways (Project No.: 2008G031-I)" under the Science and Technology Research and Development Program of the former Ministry of Railways. This study focused on No. 18 and No. 42 turnouts for ballastless tracks at the elevated stations, taking into account actual applications at the corresponding sites of Beijing–Shanghai HSR. The researchers systematically studied the design theory, checking and evaluation methods, lab and field experiments, and monitoring and inspection techniques for CWR turnouts on HSR bridges for ballastless tracks.

From 2008 to 2011, China Railway SIYUAN Survey and Design Group Co., Ltd. undertook and completed the research project titled "Design and Experimental Study of CWR Turnout Group Sections on Bridges (Project No.: 2008G002-A) under the Science and Technology Research and Development Program of the former Ministry of Railways. This study resulted in the development of the computational models for longitudinal force employing a "turnout-bridge-pier" integration pattern for CWR turnout groups on bridges. The researchers conducted an in-depth investigation into the longitudinal interaction patterns between turnouts and bridges within CWR turnout groups on bridges. Based on their findings, they proposed the design methods for CWR turnout groups on bridges and checked the design of the CWR turnout groups on the bridges at Guangzhounan Station and Lechangdong Station of Wuhan–Guangzhou HSR. Additionally, they established the vibration analysis models for the train–turnout–bridge coupling system, and conducted the dynamic simulation analysis for the CWR turnout groups on the bridges at Guangzhounan Station and Lechangdong Station, and evaluated the safety and comfort of EMUs passing through these turnout groups on the bridges. To verify the correctness of the proposed computational theories and design methods for longitudinal force on CWR turnout groups on bridges, static experiments were carried out at Guangzhounan Station and Lechangdong Station of Wuhan–Guangzhou HSR, focusing on longitudinal interactions for CWR turnouts on bridges, through an array of test items such as resistance of the track bed, rail temperature force in the turnout group sections, relative displacement between the rails and bridges in the turnout group sections, relative displacement between the switch rail and stock rail, and rail and bridge temperature in the turnout group sections. To verify the computational theories based on the train–turnout–bridge dynamic simulation, dynamic experiments were carried out for the CWR turnouts on Changshan Super Major Bridge at Lechangdong Station of Wuhan–Guangzhou HSR, involving the examination of numerous items such as the wheel–rail vertical (transverse) force, rail vertical (transverse) displacement, opening at the actual points of switch rails, strength of the turnout components, and bridge vibration characteristics.

4. Formulation and Application of Standards for CWR Turnout on Bridge

In 2007, the *Code for Design of Railway Continuous Welded Rail* was incorporated into the development agenda through the issuance of the *Notice on Printing and Distributing Formulation Plan of Railway Engineering Construction Standards in 2007* (TJSH [2006] No. 1112) by the former Ministry of Railways. The author team, serving as the principal development organization, undertook the responsibility for this task. During the development process, the preparation team thoroughly reviewed the characteristics of railway construction in China, as well as the technological advancements of CWR tracks both domestically and internationally. Moreover, they incorporated valuable insights gained from the laying experience of CWR tracks on various railway projects, including the Beijing–Shanghai HSR, Wuhan–Guangzhou HSR, Zhengzhou-Xi'an HSR, Guangzhou–Zhuhai Intercity Railway, Hefei–Nanjing Railway, Hefei–Wuhan Railway, Shijiazhuang–Taiyuan HSR, and Yichang–Wanzhou Railway.

After five years of dedicated efforts, the *Code for Design of Railway Continuous Welded Rail* (TB 10,015–2012) was officially promulgated and implemented in 2012. This Code marked a significant milestone as it became the first to incorporate the technologies of CWR turnouts on bridges, specifically in Sect. 7.3 titled "Design of CWR Turnouts on Bridges", among similar standards in China and abroad. TB 10,015–2012 systematically summarizes the design, construction, and operational experience of CWR turnouts on bridges from foreign countries. It also incorporates the latest research results on CWR turnout technologies on bridges from China. This standard defines the design methods and requirements for CWR turnouts on bridges and specifies crucial technical aspects such as turnout–bridge interactions, bridge structure requirements for laying CWR turnouts, turnout–bridge arrangements, checking criteria and permissible values for CWR turnouts on bridges, and additional longitudinal force on bridge piers and abutments. The standard embodies the principles of prioritizing safety, systematic optimization, and coordinated design. It further strengthens the theoretical support and verification of key technologies through scientific research and experiments. Additionally, it promotes and develops the technical system of CWR tracks on bridges within China, establishing a technical standard system with distinct Chinese characteristics. As a result, TB 10,015–2012 plays a crucial guiding role in the design of CWR turnouts on bridges and serves as essential technical support for the construction of HSRs in China.

5. Application of Technologies on CWR Turnout on Bridge

China has achieved remarkable milestones in the construction of elevated stations along its high-speed railways. Starting from the earliest elevated stations on Beijing–Shanghai HSR, such as Suzhoubei, Kunshannan, Wuxidong, and Changzhoubei, to Wuhan Station and Guangzhounan Station, which were the first to adopt the "bridge-building integration" structure in China, these stations not only serve as iconic landmarks for their respective cities but also provide passengers with a fast, convenient, and comfortable transport environment, significantly elevating the overall service quality of China's railway transportation system. Elevated stations represent a new type of HSR station in China, combining advantages such as land-saving features, improved station functions, and enhanced surrounding landscapes. The technologies of CWR turnouts on bridges have played a crucial role in the construction of these viaduct stations for HSRs.

The technologies of CWR turnouts on bridges have found extensive application in numerous HSRs across China, such as the Beijing–Shanghai, Wuhan–Guangzhou, Guangzhou–Shenzhen-Hong Kong, Shanghai–Kunming, Beijing–Xiongan, and Hangzhou–Ningbo HSR lines. These applications have demonstrated favorable results in terms of their effectiveness. One notable example is the Danyang–Kunshan Super Major Bridge on the Beijing–Shanghai HSR, which stretches 164 km, making it the world's longest bridge of its kind. This impressive structure connects four elevated stations in series: Kunshannan, Suzhoubei, Wuxidong, and Changzhoubei. All turnouts at these viaduct stations are installed on this bridge. Another remarkable project is the Guangzhounan Station on the Wuhan–Guangzhou HSR, where a total of 138 CWR turnouts on bridges are accommodated. This station holds the world record for the highest number of CWR turnouts laid on bridges. Notably, significant advancements have been made in the technologies for installing CWR turnouts on long-span bridges, even under challenging conditions. For instance, the Hangzhou–Ningbo HSR achieved a first by successfully laying CWR turnouts on a continuous girder crossing a main span of 100 m. Additionally, at Nanchangxi Station on the Shanghai-Kunming HSR, a significant milestone was reached with the successful installation of a No.18 single crossover on a 102 m long-span bridge that crosses Jiulong Avenue. So far, HSR bridges in China have been equipped with hundreds of CWR turnouts, highlighting the crucial role these specific technologies play as essential safeguards for the safe, comfortable, and stable operation of HSR systems.

References

1. Dongsheng R, Luqing Z (2006) Brief introduction to technology for high speed railway track in Germany [J]. Railw Stand Des S1:144–146
2. Juanjuan R (2009) Research on stress characteristics and structural optimization of longitudinally coupled ballastless track within CWR turnout section on bridge [D]. Southwest Jiaotong University, Chengdu
3. Rongshan Y (2008) Computational theory and experimental study on longitudinal force on CWR turnout on bridge [D]. Southwest Jiaotong University, Chengdu
4. Yang Q, Wang P (2017) Testing and analysis of rigidity and temperature changes of Vossloh300 fastener rubber pad [J]. Railw Stand Des 61(8):5–9
5. Jun L (2021) Vibration fatigue performance optimization and experimental study of fastener clip for high-speed railway [D]. Southwest Jiaotong University, Chengdu
6. Dongfang W, Mangmang G (2003) A comparison between UIC norm and the bridge design standard for the Beijing-Shanghai high-speed railway [J]. Railw Eng 7:9–12
7. Gao Y, Sheng L, Ke Z et al (2000) Study on the live load schema for bridges on high-speed railway in China [J]. Chin Railw (12):27–29
8. China Academy of Railway Sciences. Design principles for CWR track at viaduct stations of Beijing-Shanghai high-speed railway (Interim) [S]. CARS, Beijing (2004)

9. Yang S (2011) Study on rigidity homogenization of CWR turnout for slab ballastless track [D]. Southwest Jiaotong University, Chengdu
10. Yu C (2011) Investigation of Taiwan high speed rail [J]. Traffic Transp 27(2):46–47
11. Fuxiang W (2008) New records on construction of Taiwan high speed rail [J]. Chin J Undergr Space Eng 4:728–733
12. Ping W, Rongshan Y, Xueyi L (2004) Analysis on stress and deformation of CWR turnout laid on large continuous girder bridge [J]. J Transp Eng Inf 3:16–21
13. Ping W (2003) Mechanism of longitudinal force transmission of CWR turnouts with different structures [J]. J Southwest Jiaotong Univ 4:371–374
14. Rongshan Y, Xueyi L, Ping W (2010) Computational theory and experimental study on longitudinal force on CWR turnout on bridge [J]. J China Railw Soc 32(4):134–140
15. Rongshan Y, Ping W, Xueyi L (2003) Computation of temperature force on CWR turnout by finite element software [J]. Railw Eng 12:35–37
16. Zhiping Z (2006) Study on expansion and contraction force on CWR turnout on high-speed railway bridge and spatial vibration of train-turnout-bridge system [D]. Central South University, Changsha
17. Zhiping Z, Xiufang C, Guofan Z (2006) Computation of expansion and contraction and displacement of CWR turnout on continuous girder bridge [J]. J Transp Eng 1:34–38
18. Zhiping Z, Xiufang C, Guofan Z (2006) Analysis of influencing factors on temperature force and displacement of CWR turnout on simply supported girder bridge [J]. China Railw Sci 1:53–58
19. Liang G, Kai T, Cun Q et al (2009) Study on spatial mechanical characteristics of CWR turnout on bridge of passenger dedicated line [J]. China Railw Sci 30(1):29–35
20. Liang G (2012) Research and application of key technologies for CWR track of high-speed railway [M]. China Railway Publishing House, Beijing
21. Daxin S, Liang G, Yanfeng L (2007) Analysis of mechanical characteristics of CWR turnout on bridge for ballastless track [J]. J Beijing Jiaotong Univ 1:89–92
22. Yanfeng L, Liang G, Yawei F (2006) Finite element analysis on stress and deformation of CWR turnout on bridge [J]. J Beijing Jiaotong Univ 1:66–70
23. Yanfeng L, Liang G (2005) Approach to optimizing computational theory for CWR turnout on railway [J]. Railw Stand Des 4:9–12

Chapter 3
Computational Models and Methods for CWR Turnout on Bridge

Based on the theory of turnout-bridge interactions and findings of Chinese and foreign research, this chapter introduces the computational models for longitudinal force on CWR turnouts on bridges for either ballasted or ballastless track, employing a "turnout-bridge-pier" integration pattern and considering the characteristics of longitudinal interactions of CWR turnouts on bridges, along with the corresponding intelligent computation software. Based on these models and software, this chapter also presents an analysis of the longitudinal stress characteristics of CWR turnouts on bridges and an investigation into the influence patterns of various factors on the longitudinal stress of CWR turnouts on bridges, such as the span length of turnout girders, arrangement of fixed bearings, rigidity of piers, and length of intermediate tangents.

3.1 Principle for Turnout-Bridge Longitudinal Interactions

3.1.1 Stress and Deformation Characteristics of CWR Turnouts

As a special track structure, CWR turnouts exhibit the following main stress and deformation characteristics [1, 2]:

(1) Both ends of CWR turnouts are unbalanced in temperature force, as shown in Fig. 3.1. When a CWR turnout is connected to the sectional tracks by welding on both the main and diverging routes, the right end of the turnout experiences the temperature force, $4P_t$, in the non-breathing zone transmitted by four rails, while the left end only bears the temperature force, $2P_t$, in the non-breathing zone transmitted by two rails. When a CWR turnout is connected to the sectional track by welding only on the main route, the right end of the turnout experiences

Q. Li et al., *Continuously Welded Turnouts on High-Speed Railway Bridge*, Advances in High-speed Rail Technology, https://doi.org/10.1007/978-981-97-6524-9_3

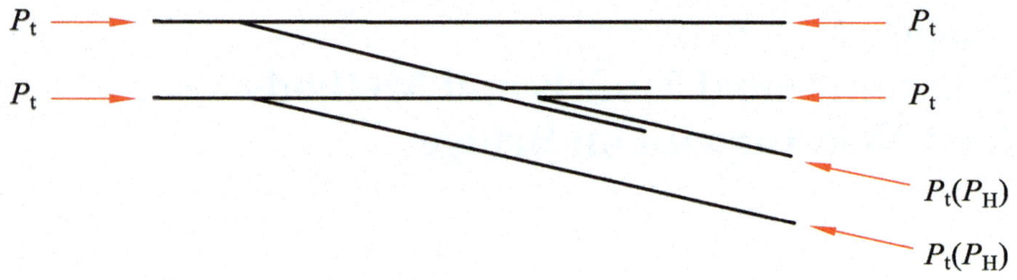

Fig. 3.1 Stress diagram of CWR turnout

$2P_t$ transmitted by two rails and the other two rails bear joint resistance, $2P_H$. Therefore, the unbalanced temperature force experienced at the left and right ends of CWR turnouts will inevitably cause the displacement of the rails and turnout sleepers toward the left end in the CWR turnouts, in the absence of adequate resistance provided by the substructure.

(2) In ballasted tracks, the turnout sleepers play a significant role in transmitting the temperature force on the CWR turnouts. Within the turnouts, the inner rails (except the straight or curved stock rails serving as the outer rails) are functionally similar to the breathing zone in CWR tracks. Consequently, they act as active forces, which can induce longitudinal displacement, deflection, and flexure of the turnout sleepers. These effects, in turn, lead to the displacement of the outer rails and the generation of additional expansion and contraction force.

(3) In CWR turnouts, force transfer devices, such as stoppers and spacing blocks, are strategically positioned between the rails on both the main and diverging routes. To restrict the expansion and contraction displacements of point rails and switch rails, a specific number of force transfer devices are typically placed at the heel end of the switch rails or the end of the movable-point rails, to facilitate the transmission of temperature force from the inner rails of the turnouts to the outer rails.

The stopper's combination blocks are intentionally designed with a specific clearance. This feature serves the purpose of transmitting significant temperature force from the inner rail to the outer rail. It occurs alongside a defined expansion and contraction displacement that takes place at the inner rail. This arrangement acts as an effective method for controlling the expansion and contraction displacement of the inner rail.

The spacing blocks function to transmit temperature force by utilizing contact friction with the rails. Once the friction is overcome, the bolt free of shear signifies the temperature force transmitted is restricted. This provides an approach to control the value of temperature force transmitted.

(4) The relationship between the longitudinal resistance of fasteners and that of the track bed in CWR turnouts is different from that in regular CWR tracks and CWR tracks on bridges. Within the lead curve of turnouts, the expansion and contraction displacement varies between the inner and outer rails, resulting in

different directions of longitudinal resistance for their respective fasteners. The total resistance of the two fasteners at the inner rail is balanced against the resistance of the turnout sleepers and the longitudinal resistance of the two fasteners at the outer rail. Consequently, the longitudinal resistance of the fasteners at the inner rail may approach its ultimate value.

3.1.2 Stress and Deformation Characteristics of Bridge Structure

Within the three typical types of substructures, namely subgrade, bridges, and tunnels, bridges are considered movable substructures. The stress and deformation characteristics of the bridges are as follows:

(1) Girders experience expansion and contraction displacement as a result of temperature changes. The expansion and contraction displacement at each girder section can be expressed as $\Delta l = \alpha \Delta T l_x$, where l_x represents the distance of the girder section to a fixed bearing, α denotes the temperature-induced linear expansion coefficient of the girder, and ΔT represents the temperature change range.
(2) In the case of a train passing through a bridge, the girder experiences deflection deformation. For example, in a deck-type simply-supported girder, the upper flange is shortened while the lower flange is elongated, resulting in rotation at each section of the girder. For the bridge with a fixed bearing at one end, the extension of its lower flange is constrained by this fixed bearing. Therefore, in the process of girder deflection, the displacement at each section is a combined effect of both translation and rotation of the girder.
(3) When braking force or other longitudinal forces are applied, the longitudinal force is transmitted to the pier through the girder. Due to the pier's longitudinal rigidity, a certain displacement occurs at the fixed bearing, thereby causing longitudinal displacement of the entire girder.

Numerous factors contribute to the stress and deformation of CWR turnouts on bridges, such as the bridge forms, arrangements of bridge bearings, connection forms of turnouts, welding procedures of turnouts, and track creeping. Each of these factors has the potential to induce significant changes in the stress and deformation experienced by CWR turnouts on bridges.

3.1.3 Turnout-Bridge Longitudinal Interactions

For CWR turnouts installed on bridges, their interactions with the bridges exhibit both similarities and differences compared to the interactions between the bridges and rails of CWR tracks on the same bridges. The girder expansion and contraction

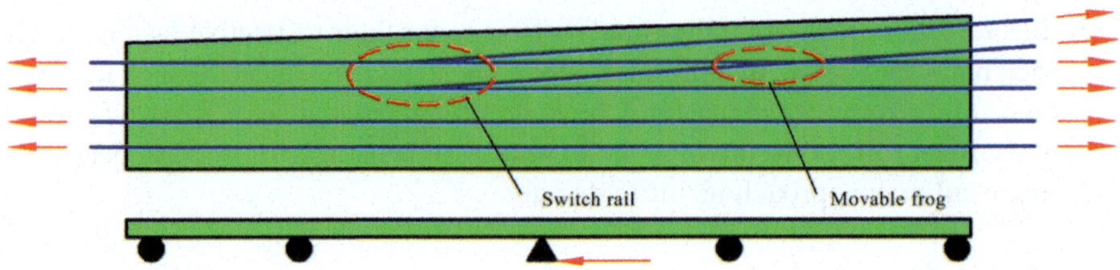

Fig. 3.2 Turnout-bridge interactions

due to temperature variations, girder deflection resulting from vertical loads, and the presence of braking forces/traction from passing trains contribute to more complex turnout-bridge interactions. Furthermore, the temperature change-induced expansion and contraction of the inner rails within the turnouts not only causes interactions between the inner rails and stock rails but also raises interactions between the turnouts and bridges through the fasteners and deck systems. Turnout-bridge interactions represent the coupled and superposed effects from both the turnouts and bridges. Differentiating between the specific effects attributed solely to the bridges from those attributed solely to the turnouts is challenging.

The principle of turnout-bridge interactions can be defined as follows: The combined effect of girder temperature changes, train loads, train braking forces/ traction, and temperature change-induced expansion and contraction of the inner rails within the turnouts leads to relative displacement between the girders and the tracks on the bridges (including the turnouts). As a result, the tracks on the bridges (including the turnouts) experience longitudinal additional forces on the rails, which react on the bridges and are further transferred to the piers and abutments through the girders and bearings. Such interactions establish a mechanical equilibrium system between the tracks on the bridges (including the turnouts) and the bridges. The principle of turnout-bridge interactions offers a theoretical basis for calculating longitudinal force and displacement of CWR turnouts on bridges. Turnout-bridge interactions are illustrated in Fig. 3.2 [3, 4].

Based on the principle of longitudinal interactions between turnouts and bridges, it is understood that when the inner rails of the turnouts undergo expansion and contraction displacement, the turnout sleepers are compelled to shift and deflect longitudinally. A portion of the acting force is transmitted to the stock rails through the fasteners, while the remaining force is initially transferred to the track bed and then to the bridges through the turnout sleepers. As a consequence, the girder surface experiences longitudinal displacement due to expansion and contraction or deflection, and the piers and abutments encounter longitudinal displacement at the pier top as a result of the forces transmitted from the turnouts, leading to longitudinal displacement of the bridges. Moreover, the girder displacement is propagated to the turnouts via the track bed, which alters the distribution of longitudinal force in the rails and subsequently influences the stress and deformation of the bridges [7]. This indicates that the turnouts, along with the bridges, their piers, and abutments, compose a coupling system of interactions and mutual influence. To elucidate the stress and deformation patterns of the turnouts and bridges, it is imperative to establish an integration model [5–7].

3.2 Turnout-Bridge-Pier Integration Computational Models

3.2.1 *Model Assumptions*

(1) The front ends of the switch rails and movable-point rails are free in expansion and contraction. The displacement at the actual point of the switch rails or that of the movable-point rails is the sum of the displacement at their heel end and the expansion and contraction displacement of their free section.
(2) Irrespective of the magnitude of the frog angle, it is assumed that the closure rails are parallel to the long rail strings.
(3) The rails are divided into finite beam elements at their support nodes and only experience longitudinal displacement. The turnout sleepers are divided into finite long beam elements at the rail support points and experience longitudinal displacement and rotation.
(4) The resistance of fasteners is nonlinearly related to the relative displacement between the rails and turnout sleepers or turnout slabs. It acts on the rail nodes and turnout sleeper nodes in a direction that opposes the displacement of the rails relative to the turnout sleepers.
(5) Relative torsion between the rails and turnout sleepers or turnout slabs is ignored.
(6) The track bed resistance of the ballasted tracks is taken as the resistance per unit length of the turnout sleepers and is nonlinearly related to the displacement of the turnout sleepers. To simplify the calculations, the track bed resistance can also be taken as a constant. The ballast resistance is uniformly distributed along the length of the turnout sleepers.
(7) The resistance of the spacing blocks shows a nonlinear relationship with the displacement of the rails, taking into account their impact on the expansion and contraction displacement of the rails. For the sake of simplifying calculations, this resistance can also be treated as a constant.
(8) The combination blocks of the stoppers are positioned at the center in laid turnouts, considering the acting force transmitted by the stoppers at the heel end of the frogs between the stock rail and closure rail. When the combination blocks are properly placed, the resistance of the stoppers is nonlinearly related to the relative displacement between the two rails.
(9) The fixed bearings of the bridges effectively eliminate the girder expansion and contraction, and the resistance of the movable bearings against expansion and contraction is ignored. The longitudinal deformation of the bearings themselves is not taken into account, and all longitudinal forces borne by the fixed

bearings are transmitted to the piers and abutments. The cantilever portion of the girders beyond the bearings is ignored in the calculations of the expansion and contraction amount. For other forms of bearings, calculations should be otherwise simplified according to their stress characteristics.

(10) Longitudinal relative displacements occur between the base slabs and the bridges, and these two components interact longitudinally through a sliding layer and a shear notch. Frictional resistances of the sliding layer are nonlinearly related to the relative displacements between them, while longitudinal acting forces of the shear notch are linearly related to the relative displacement between them.

(11) The emulsified asphalt mortar exerts a nonlinear resistance effect on the turnout slabs and base slabs, the nonlinear resistance between the turnout slabs and base slabs varies nonlinearly with their relative displacement, while the maximum nonlinear resistance is determined according to relevant test data.

(12) The base slabs interact longitudinally with the end thorns and friction plates. The longitudinal rigidity of the end thorns is linear, while the frictional resistance between the friction plates and the base slabs is nonlinear.

(13) Temperature changes in the girders are simply considered as increases or decreases to calculate the expansion and contraction force, without considering alternating temperature rise and fall. Typically, the maximum temperature difference experienced by the girders within a single day is used to calculate the expansion and contraction amount of the girders.

(14) The expansion and contraction force, deflection force, and rail breakage force within CWR turnouts on the bridges are calculated based on the maximum range of rail temperature changes. The expansion and contraction force, deflection force, rail breakage force, and braking force are calculated separately without considering the superimposed effects.

(15) The longitudinal force transmitted by the ballast section at the girder end is disregarded for ballasted bridges, and all longitudinal resistance borne by the track bed is transmitted to the piers and abutments of the bridges.

(16) The longitudinal rigidity at the top of the bridge piers and abutments is linear, including the displacement of the bearing top resulting from pier flexure, foundation inclination, foundation translation, and shear deformation of the rubber bearing under the action of longitudinal horizontal force at the bearing top. The vertical rigidity of the bridge piers and abutments as well as the foundation is equivalent to the vertical rigidity of the bridge bearing.

(17) The longitudinal constraint resistance between the turnout sleepers and the bridge as well as between the rails and the subgrade is longitudinal spring constraint.

(18) The longitudinal restraint resistance of any rail expansion joint placed at CWR turnouts on bridges is considered to be zero. The resistance of any common joint placed is taken as a constant value. In cases where the rail expansion joint is given a non-zero longitudinal resistance, it is then categorized as a common joint.

(19) When the track bed slabs in ballastless tracks are connected to the bridge using "U-shaped rebar", they are regarded as an integral for stress and deformation. When the track bed slabs and the bridge are connected with longitudinal and transverse bollards, these bollards are regarded to provide longitudinal spring restraint.

3.2.2 Computational Model for Ballasted Turnout

Figure 3.3 illustrates the "turnout-bridge-pier" integration computational model for single-crossover CWR turnouts on bridges with ballasted tracks. This model considers rails as tension–compression bar elements, with switch rails and point rails exhibiting linearly varying rigidity in parts with variable cross sections. Turnout sleepers are simplified as horizontally flexible beams with uniform sections. Fasteners, track beds, stoppers, and spacing blocks at the heel end of switch rails, as well as spacing blocks at the heel end of point rails, are all simplified as nonlinear springs. This model is suitable for analyzing both the single turnout and turnout group configurations, as well as bridges with various types of girders like simply-supported, continuous, or others. To mitigate boundary effects, a specific length of subgrade is considered at the abutments or both ends of girder spans during calculations [8–10].

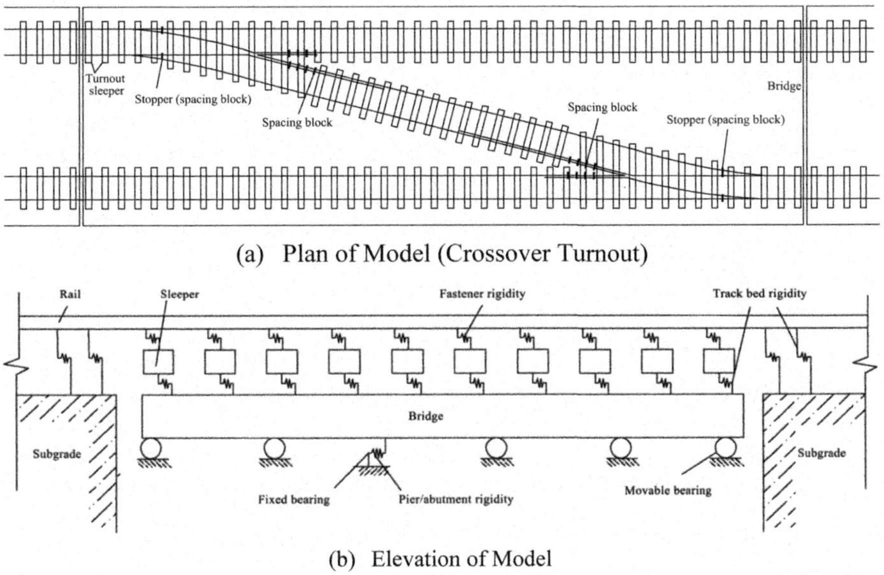

Fig. 3.3 Model of CWR turnout on ballasted bridge

3.2.3 Computational Model for Ballastless Turnout

The elevation of the "turnout-bridge-pier" integration computational model for ballastless tracks is shown in Fig. 3.4 [5, 6].

For ballastless CWR turnouts on bridges, when an intermediate cushion is placed between the turnout slabs and the bridges, relative sliding between them is permissible. The acting forces are transmitted by longitudinal and transverse bollards. In this case, the interactions between the turnout and bridge differ from those involving ballasted tracks. The CWR turnout structure on the bridges is considered a three-layer structural system comprising the turnouts, track slabs, and girders. The fasteners between the turnouts and track slabs are simulated as springs, while the track slabs and girders are connected through springs. This entire configuration is shown in Fig. 3.4a. On the other hand, when "U-shaped rebar" is used to connect the turnout slabs and bridges, relative sliding between them is not allowed. In this scenario, the CWR turnout structure on the bridges is deemed as a two-layer structural system consisting of the turnout and girder, as shown in Fig. 3.4b.

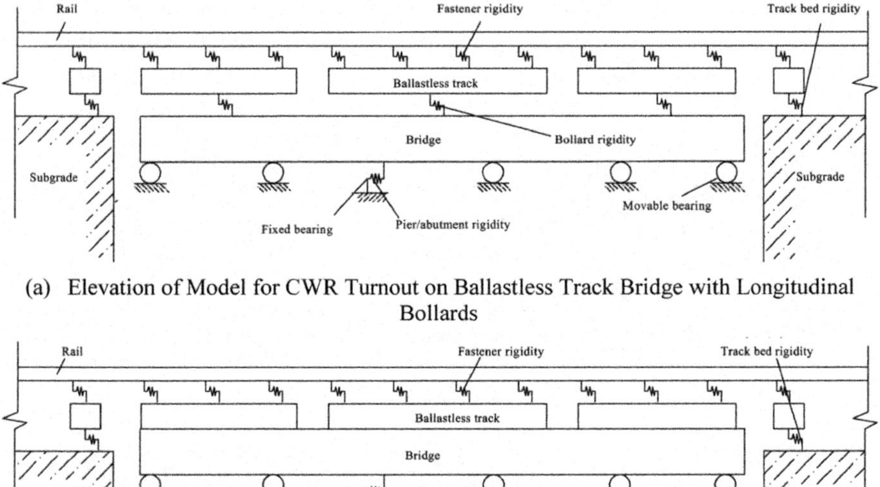

(a) Elevation of Model for CWR Turnout on Ballastless Track Bridge with Longitudinal Bollards

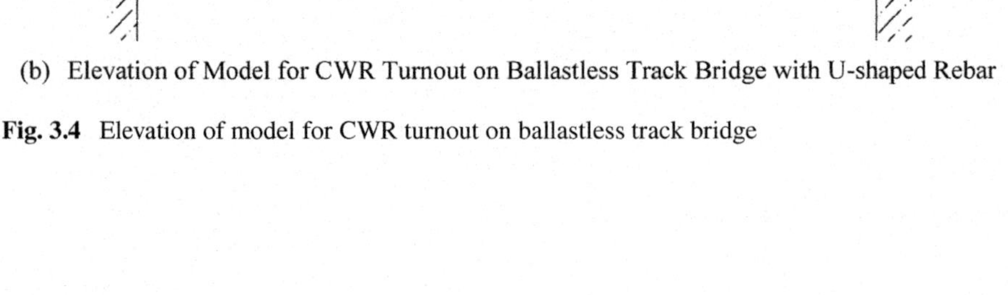

(b) Elevation of Model for CWR Turnout on Ballastless Track Bridge with U-shaped Rebar

Fig. 3.4 Elevation of model for CWR turnout on ballastless track bridge

3.2.4 Computational Model for Ballastless Track with Longitudinally Coupled Base Slabs

There are distinctive differences between CWR turnouts with ballastless tracks featuring longitudinally coupled base slabs within bridges and those with ballastless tracks featuring longitudinally coupled base slabs across bridges and adjacent subgrade. From the perspective of track structures, the ballastless tracks featuring longitudinally coupled base slabs across bridges and adjacent subgrade exhibit a simpler structural form, with uniform track conditions across girder spans throughout the entire bridge. In the case of CWR turnouts on bridges, it is imperative to consider the characteristics of longitudinal variations of rails, ballastless tracks, and bridges in the calculations, to incorporate a more complex turnout structure, with varying track conditions among girder spans.

From the perspective of interactions between adjacent tracks, the longitudinal forces on adjacent tracks in the scenario of ballastless tracks featuring longitudinally coupled base slabs across bridges and adjacent subgrade can only interact through the bridges. In the scenario of ballastless tracks featuring longitudinally coupled base slabs within bridges, the longitudinal forces on adjacent tracks may interact through the laterally connected turnout slabs or base slabs and the bridges, leading to more complex stress conditions.

From the perspective of force transfer relationships between structural layers, in the scenario of ballastless tracks featuring longitudinally coupled base slabs across bridges and adjacent subgrade, the rails and track bed slabs interact through the fastener system, the track bed slabs and base slabs interact through filled mortar, and the base slabs and bridges interact through the sliding layer and shear notch. However, in the scenario of ballastless tracks featuring longitudinally coupled base slabs within bridges, in addition to the same interlayer interactions mentioned above, the turnout slabs and base slabs also interact through anchor bars, leading to more complex interlayer interactions.

From the perspective of stress characteristics of rails, in the scenario of ballastless tracks featuring longitudinally coupled base slabs across bridges and adjacent subgrade, all the rails are situated in the non-breathing zone of the CWR tracks, and the additional expansion and contraction force on the rails is induced by bridge expansion and contraction, independent of the amplitude of temperature changes in the rails. However, in the scenario of ballastless tracks featuring longitudinally coupled base slabs within bridges, due to the structural characteristics, the turnouts accommodate breathing zones (the movable parts of the switch rails and point rails). Even if the bridges remain free of expansion and contraction, the temperature change amplitude in the rails leads to interactions between the turnouts and the substructure. Therefore, when calculating the temperature force in this scenario, the actual temperature change amplitude must be considered for the rail temperature.

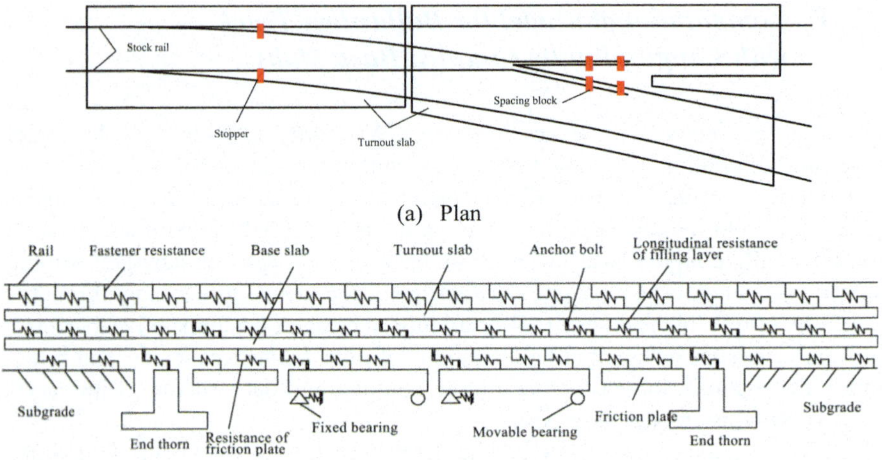

(a) Plan

(b) Elevation

Fig. 3.5 Model of CWR turnout with ballastless track featuring longitudinal coupled base slabs

To analyze the stress and deformation of CWR turnouts with ballastless tracks featuring longitudinally coupled base slabs within bridges, the turnouts, turnout slabs, base plates, girders, and piers/abutments should be considered as a unified system to establish a "turnout-slab-girder-pier" integration model [7]. This model should also incorporate the influence of key parameters such as the longitudinal resistance of fasteners, longitudinal resistance between turnout slabs and base slabs, frictional resistance of the sliding layer between the base plates and bridges, expansion and contraction rigidity of base slabs, and the longitudinal horizontal rigidity at the pier and abutment top. This model is illustrated in Fig. 3.5.

The "turnout-slab-girder-pier" integration model takes into account the interactions among various components such as rails, spacing blocks, stoppers, turnout slabs, base slabs, bridges, piers/abutments, friction plates, end thorns, longitudinal and transverse bollards on the base slabs, and shear notch between the base slabs and bridges. The resistance against longitudinal interactions between the rails and track bed slabs, turnout slabs and base slabs, base slabs and bridges, as well as base slabs and friction slabs should be considered nonlinear.

3.2.5 *Establishment of Finite Element Model and Application of Loads*

1. Finite element model for turnout

A finite element model was developed to analyze single turnouts using No.18 movable-point turnouts within a specific configuration of 3 × 32 m simply-supported turnout + 4 × 32 m continuous girder beneath girder + 3 × 32 m simply-supported girder, with 100 m subgrade at both ends, respectively, and the distance between centers of tracks of 5.0 m. The overall diagram and details of this model are shown in Figs. 3.6, 3.7 and 3.8, respectively.

Similarly, a finite element model was developed to analyze single crossovers using No.12 movable-point turnouts within a specific configuration of 3 × 32 m simply-supported girder + 4 × 32 m continuous girder beneath turnout + 3 × 32 m simply-supported girder, with 100 m subgrade at both ends, respectively, and the distance between centers of tracks of 4.3 m. The overall diagram and details of this model are shown in Figs. 3.9, 3.10 and 3.11, respectively.

The rails are simulated using beam3 elements, and the finite beam elements are divided at support nodes, only subject to longitudinal displacement along the track line. The turnout sleepers are also simulated using beam3 elements. For ballasted tracks, the turnout sleepers are represented as horizontal beam elements, assigned with nodes at the fastener positions and their both ends, and the remaining sections are divided into beam elements in equal lengths. For ballastless tracks, the turnout sleepers are generally embedded as long sleepers. In this case, the rails are assumed to be directly connected to the bridges with fasteners, without considering the influence of the turnout sleepers, monolithic track bed, and ballastless track slabs within the model.

The track bed is simulated using combin39 nonlinear spring elements. The resistance of the resistance elements under the turnout sleepers depends on the length of the turnout sleepers and the number of fasteners to ensure that the sleepers endure reasonable forces. Particularly, for ballasted tracks, the track bed resistance is assigned with its actual value. For ballastless tracks, the track bed resistance is modified to an infinite value, to simulate the fastening of the sleepers onto bridges.

The fasteners, stoppers, and spacing blocks are simulated using combine39 nonlinear spring elements. The node at the end of the stopper nonlinear spring element at the heel end of the frogs is coupled with the node at the position corresponding to the stock rail beam element, and the other end is coupled with the node at the position corresponding to the closure rail beam element. The node at the end of the

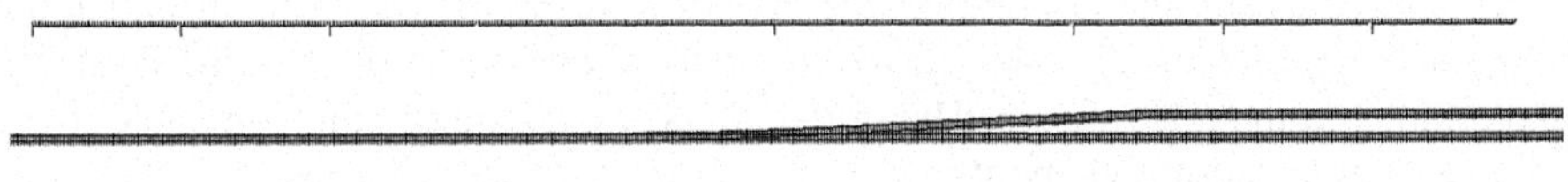

Fig. 3.6 Overall diagram of finite element model for single turnout

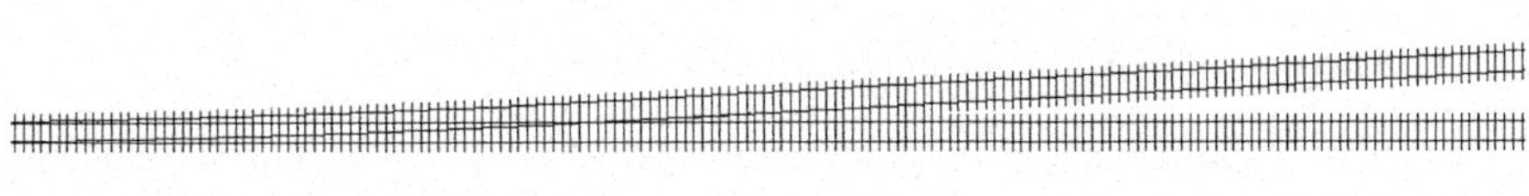

Fig. 3.7 Detail of finite element model for single turnout

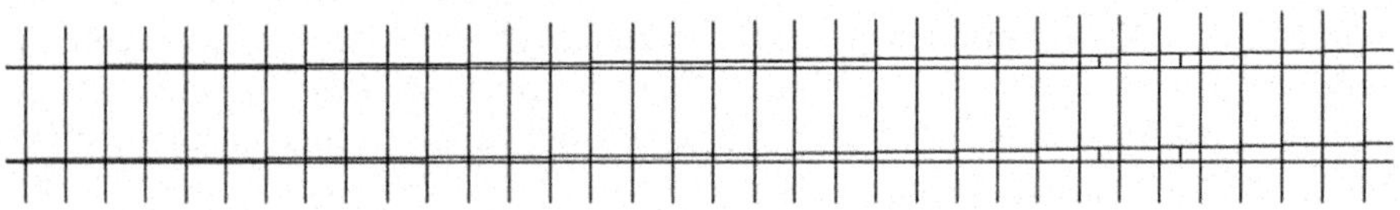

Fig. 3.8 Detail of model for switch rail and stopper

Fig. 3.9 Overall diagram of finite element model for single-crossover turnout

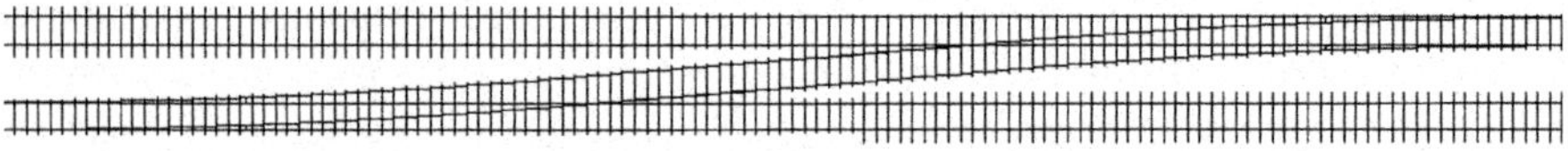

Fig. 3.10 Detail of finite element model for single-crossover turnout

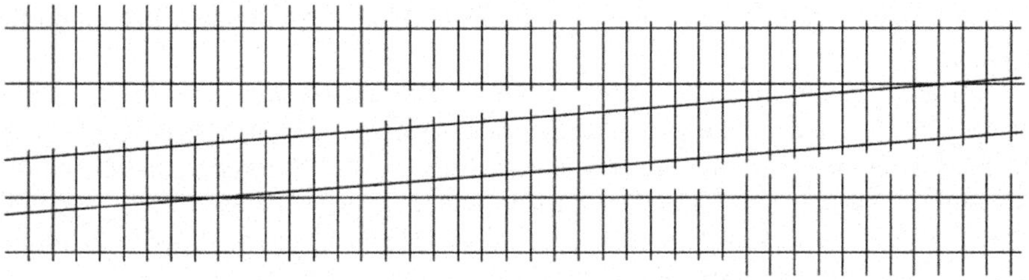

Fig. 3.11 Detail of model for frog

nonlinear spring element of the spacing blocks is coupled with the node at the position corresponding to the point rail beam element, and the other end is coupled with the node at the position corresponding to the wing rail beam element.

The bridge is simulated using beam3 elements. Rigid arms are established at the upper and lower flanges of the bridge to simulate the action of the girder surface and bottom on the neutral axis. The bridge piers and abutments are simulated using combin14 linear spring elements.

2. Application of loads in the finite element model

(1) Calculation of expansion and contraction force

For calculating the expansion and contraction force [11–13], once all bridge elements and rail elements have been selected, users can proceed to apply the corresponding temperature loads.

Fig. 3.12 Overall diagram of finite element model applied with flexural load

Fig. 3.13 Schematic diagram of braking model for the first track

Fig. 3.14 Schematic diagram of braking model for the first and second tracks

(2) Calculation of longitudinal force due to deflection [14, 15]

Once the bridge elements have been selected, users can proceed to apply loads with a vertical uniform distribution and of different magnitudes at the designated loading positions. Taking the ZK load as an example, the finite element model of the bridge applied with uniformly distributed loads is shown in Fig. 3.12.

(3) Calculation of braking force

Once the rail elements have been selected, users can proceed to apply loads with a longitudinal uniform distribution and of different magnitudes at the designated loading positions, as shown in Figs. 3.13 and 3.14.

(4) Calculation of force from rail breakage

When simulating a broken rail, the rail element at the broken rail position is deleted. Figure 3.15 gives a sketch where two rails of the track are broken at the left end of the continuous girder.

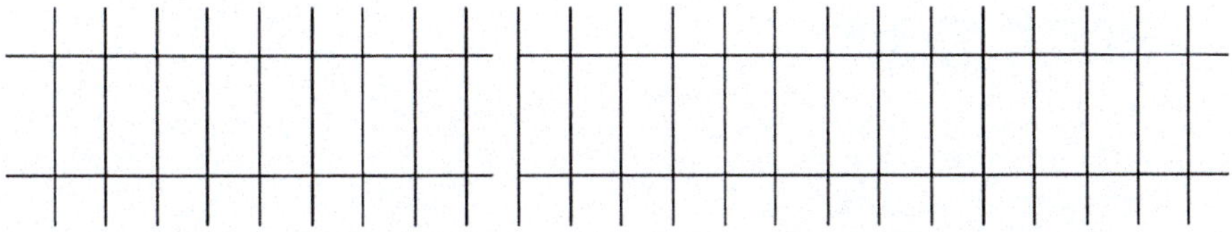

Fig. 3.15 Finite element model of broken rail at left end of continuous girder

3.3 Computation Method for CWR Turnouts on Bridges

The calculations required for CWR turnouts on bridges are known to be complex and theoretically challenging. In order to facilitate the wider adoption and implementation of CWR turnout technologies, the author's team has developed a systematic approach. This approach is centered around data-driven principles and aims to integrate the design, checking, and filing processes. Leveraging intelligence and information-based technologies, the team has developed intellectualized design software specifically tailored for CWR turnouts on railways. This software enables streamlined operations of "one-key entry," "one-key calculation," "one-key analysis," and "one-key filing" for CWR turnouts. This advanced system has effectively advanced the transformation of traditional CWR track design to an intellectualized design approach, elevating the level of intelligence in the overall design process. The intellectualized design system for CWR turnouts on bridges consists of six key modules: directory setting, data entry, calculation in scenarios, chart viewing, report generation, and storage of scenarios. This system is introduced in detail as follows.

1. Software initialization

Access to the main window interface requires entering a software password. When starting the system for the first time, users are prompted to initialize settings for the saving location of ANSYS and calculation results, as shown in Fig. 3.16:

2. Software data entry

This data entry module mainly consists of six sections of data: basic data, bridge data, turnout data, resistance data, load data, and previous parameters.

(1) Basic data entry

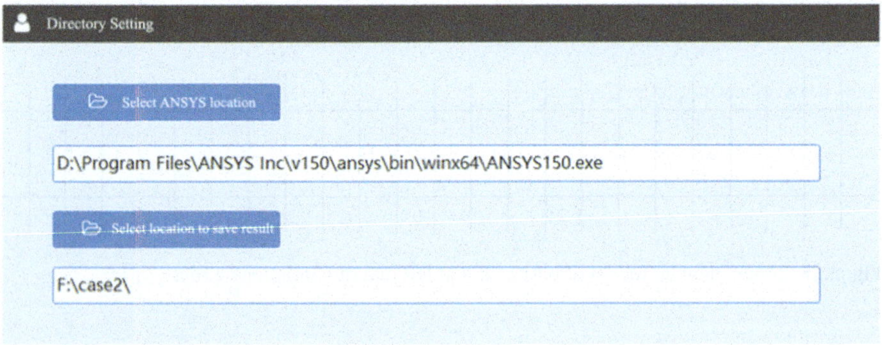

Fig. 3.16 Directory settings upon software initialization

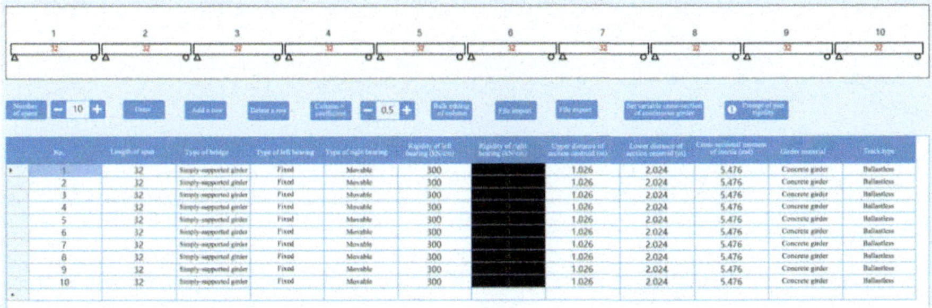

Fig. 3.17 System data entry interface

The basic data section mainly includes sleeper spacing, subgrade length at the left and right ends of the bridge, rail parameters, and sleeper parameters. This system allows users to select predefined options and make modifications to align with the specific requirements of their project.

(2) Bridge data entry

The bridge data sheet is shown in Fig. 3.17, taking 10 × 32m simply-supported girders as an example.

For each bridge span, this data section includes 12 items: number, span length, girder type, left bearing type, right bearing type, left bearing rigidity, right bearing rigidity, upper distance of section centroid, lower distance of section centroid, cross-sectional moment of inertia, girder material, and track type.

(1) Five bridge types are available for selection by double-click: simply-supported girder, continuous girder 1, continuous girder 2, rigid frame bridge, and special bridge span.
(2) The left and right bearing types are categorized into fixed bearings and movable bearings, which can be selected by double clicking the list box.
(3) Rigidity of left and right bearings: Each bridge span is equipped with either a fixed or movable bearing at the left and right ends. In the case of a fixed bearing, users need to define the longitudinal rigidity of both the pier and abutment. This setting step is omitted for a movable bearing.
(4) The upper/lower distance of the section centroid and cross-sectional moment of inertia are settable to accommodate the actual project conditions for deflection calculation.

(3) Turnout data entry

Turnout data entry is configured with five sheets: distance between centers of tracks and number of turnout, turnout details, turnout connection, girder division at turnout, and default data. The turnout data entry interface is shown in Fig. 3.18.

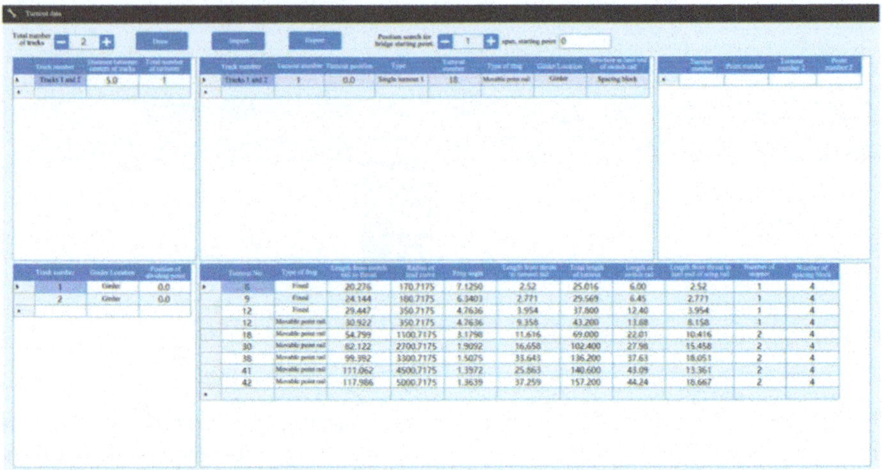

Fig. 3.18 Turnout data entry interface

The key settable items in the above figure are explained as follows:

(1) Distance between centers of tracks: Based on specific data entered, different combinations of distance between centers of tracks are supported. For example, the distance between centers of tracks 1–2 is set at 5.0 m, the distance between centers of tracks 2–3 is set at 6.5 m, and the distance between centers of tracks 3–4 is set at 5.0 m.
(2) Total number of turnouts: This item is used to count turnouts on the designated track. In this context, a single crossover is counted as one turnout. If there are no turnouts present, the value 0 is entered.
(3) Type: This system offers six types, including four types of single turnouts (up left, down left, up right, and down right) and two types of single crossovers (Z-shaped and reverse Z-shaped).
(4) Turnout number: Based on project conditions, users have the option to choose from a range of turnout numbers, including No. 8, 9, 12, 18, 30, 38, 41, and 42. These turnouts are available in both movable-point and fixed configurations and are listed in the default turnout data sheet.
(5) Type of frog: Movable-point frogs and stiff frogs are available in the system.
(6) Heel end structure of switch rail: Spacing blocks and stoppers are available in the system.

Once the above parameters have been entered, users can click on the smart drawing module of this software to convert the interface data into a schematic diagram of turnout bridge arrangement, as shown in Fig. 3.19. This function offers an increased system fault tolerance rate.

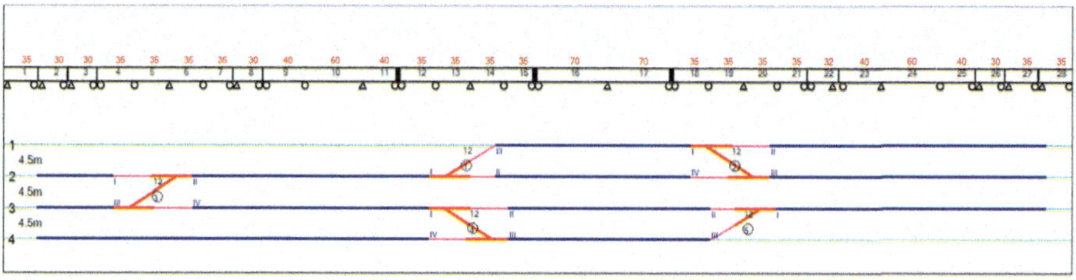

Fig. 3.19 Schematic diagram of turnout bridge arrangement of an elevated station generated by the system

(4) Load data entry

(1) Calculation of expansion and contraction force

Once the expansion and contraction force has been selected, users can proceed to enter the values for temperature increase and decrease of the rails and bridges, as shown in Fig. 3.20.

(2) Calculation of braking force and longitudinal force due to deflection

The interfaces for users to enter parameters to calculate braking force and longitudinal force due to deflection are shown in Figs. 3.21 and 3.22, respectively.

The above software interfaces present users with options to select different railway types that correspond to various loads: ZK load for high-speed railways, ZC load for intercity railways, ZKH load for mixed passenger and freight railways, ZH load for heavy-haul railways, and ZS load for suburban railways.

When calculating braking forces, users need to enter the train loading length, loading track, and braking direction.

(5) Calculation of force from rail breakage

Once the force from rail breakage has been selected, users can proceed to enter the value of rail temperature decrease and parameters of the broken rail. The number of broken rails on the designated track can be chosen from the following options: all rails broken, one broken rail (up), or one broken rail (down).

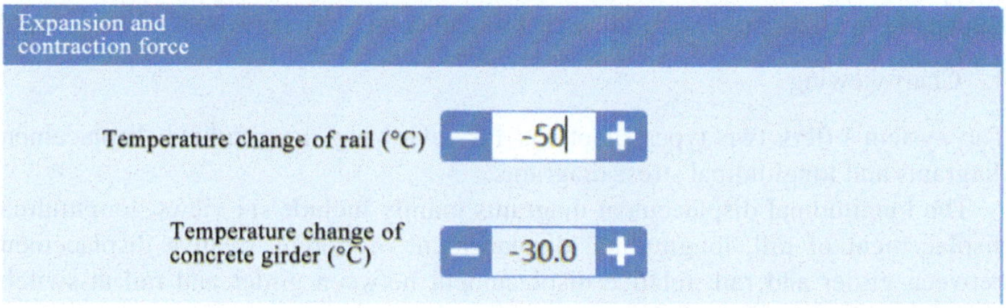

Fig. 3.20 Parameters to be entered by user for calculation of expansion and contraction force

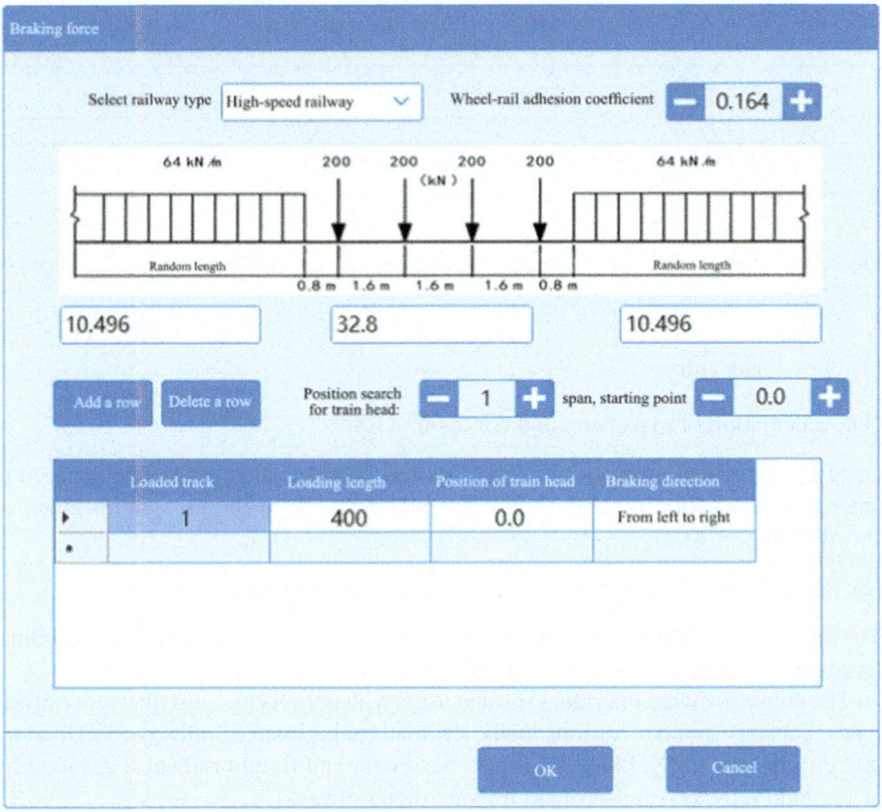

Fig. 3.21 Parameters to be entered by user for calculation of braking force

3. Calculation in scenarios

By clicking on the "Submit for Calculation" button, a callable ANSYS data file is generated. Following that, by clicking on the "Start Calculation" button, the software interface provides real-time displays on the calculation progress, as shown in Fig. 3.23.

Upon completion of the calculation, the software displays a pop-up window as shown in Fig. 3.24, presenting the actual model created by the ANSYS background.

4. Chart viewing

The system offers two types of charts for reviewing: longitudinal displacement diagrams and longitudinal stress diagrams.

The longitudinal displacement diagrams mainly include six views: longitudinal displacement of rail, longitudinal displacement of bridge, relative displacement between girder and rail, relative displacement between girder and rail at switch, displacement of actual point at switch rail relative to stock rail, and displacement at actual points of switch rail relative to wing rail, as shown in Fig. 3.25.

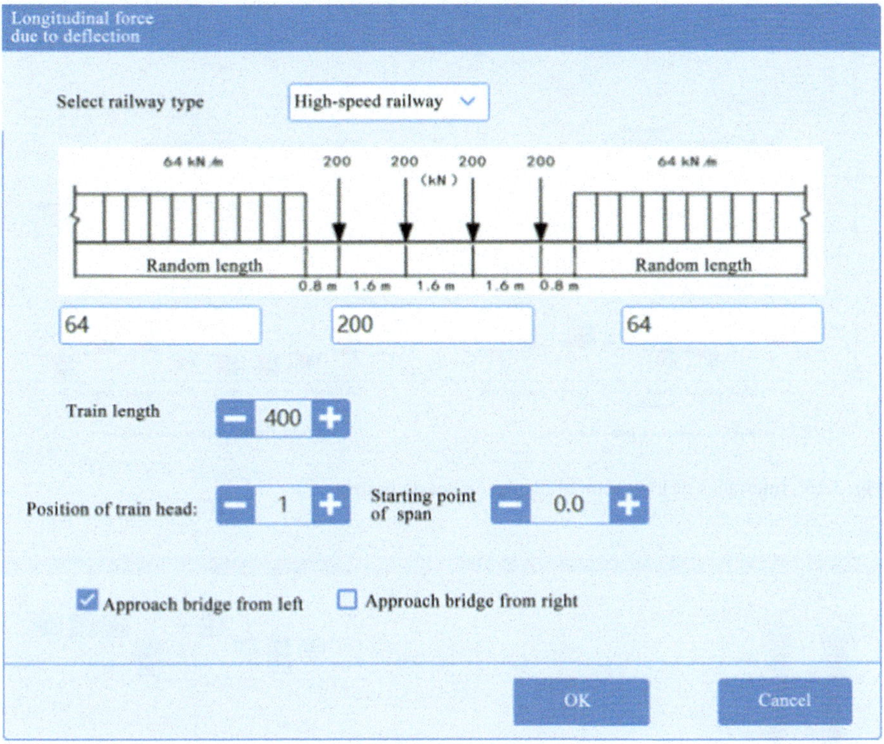

Fig. 3.22 Parameters to be entered by user for calculation of longitudinal force due to deflection

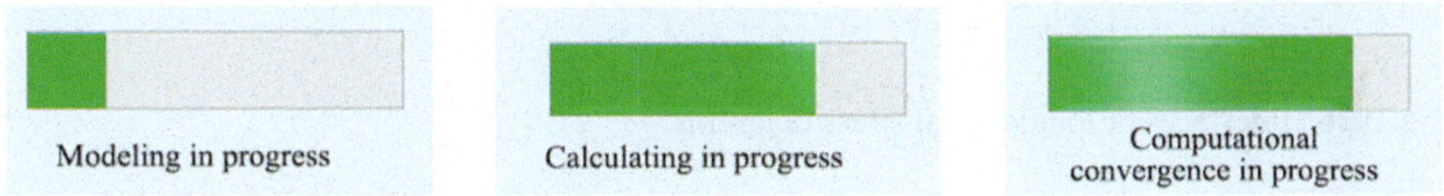

Fig. 3.23 Progress bar of calculation process

Fig. 3.24 Pop-up window after calculation completion

The longitudinal stress diagrams mainly include four views: longitudinal stress of rails, longitudinal stress of piers, stress of bolts at switches, and stress of bolts at the heel end of point rails, as shown in Fig. 3.26.

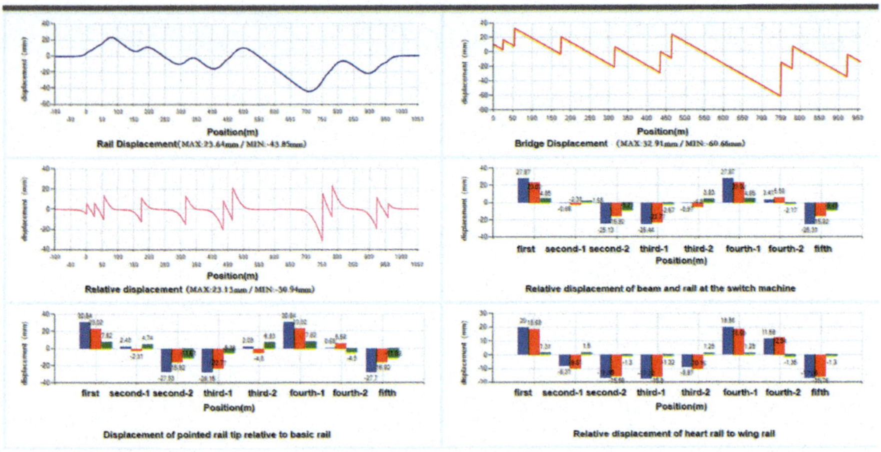

Fig. 3.25 Interfaces of longitudinal displacement diagrams

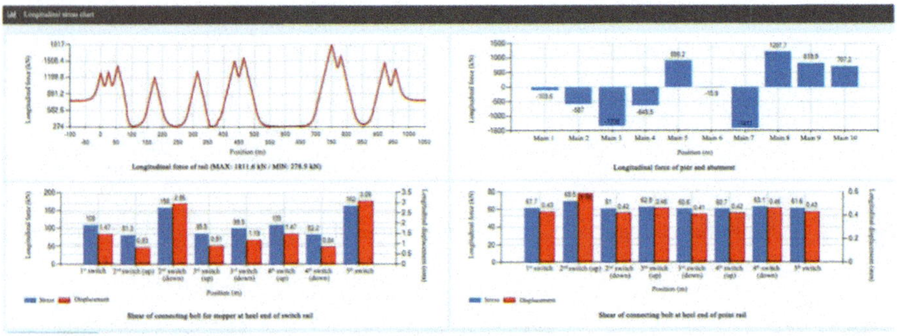

Fig. 3.26 Interfaces of longitudinal stress diagrams

5. Report generation

The report generation module enables the direct output of calculation results. It incorporates an input interface that allows users to generate reports in the MS Word format, as shown in Fig. 3.27.

For the computation of permissible rail stress, users need to predefine the rail material. Based on the predefined rail material, the software automatically extracts the corresponding computation parameters.

For the computation of temperature stress, users need to predefine the maximum and minimum rail temperatures.

For the computation of dynamic flexural stress, users need to enter the speed coefficient, eccentric load coefficient, lateral force coefficient, axle load, wheelbase, bearing rigidity, and sleeper spacing and select the rail type and wear data. The

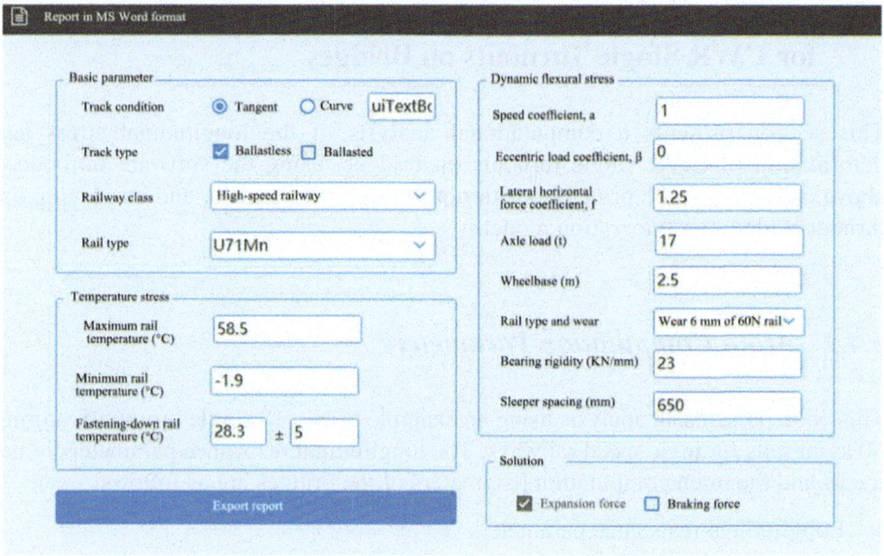

Fig. 3.27 Input interface for generation of report in MS word format

program system contains the type of rails specified in TB 10,015–2012, along with their corresponding wear data.

6. Related cases

The software program presented in this book is designed to conduct calculations for the single CWR turnout and CWR crossover configurations, as well as the CWR turnout group configuration composed of a dozen or even dozens of turnouts on bridges of different forms. Moreover, by adjusting the track or bridge structure parameters, the software can also be utilized for calculations involving CWR turnouts on subgrade or CWR tracks on bridges. The examples of ANSYS finite element models established based on relevant calculations for specific cases are shown in Figs. 3.28 and 3.29.

Fig. 3.28 Model of CWR turnout on bridge at a suburban railway elevated station

Fig. 3.29 Model of CWR turnout on bridge at a high-speed railway elevated station

3.4 Computational Analysis of Longitudinal Forces for CWR Single Turnouts on Bridges

This section presents a computational analysis of the longitudinal stress and deformation of CWR single turnouts on bridges, using the software introduced above, adhering to the principle of turnout-bridge interactions, and employing the turnout-bridge-pier integration model.

3.4.1 Main Computation Parameters

This book presents an analysis using an example of No. 18 single turnouts featuring 60 kg/m rails for high-speed railways. The longitudinal resistance parameters of the tracks and the main computation parameters of the bridges are as follows.

1. Longitudinal resistance parameters

To calculate the longitudinal interactions between the turnouts and bridges, the adoption of the constant resistance model as the longitudinal resistance model features a clear concept and simple calculations. However, when dealing with a small longitudinal relative displacement, the calculations yield significant errors. Given this, the nonlinear resistance model better aligns with the actual situation. In this section, the longitudinal resistance of the track bed, as well as the resistance of the fasteners, spacing blocks, and stoppers, are considered to be nonlinear based on measured data, and the longitudinal rigidity of the piers is considered to be linear [16–19].

(1) Resistance of sectional track bed

The ideal elastic–plastic model can be adopted for the resistance of sectional track bed, and it is expressed as follows:

$$r = \begin{cases} \frac{R}{2}u & u \le 2\,mm \\ R & u > 2\,mm \end{cases}$$

where r—longitudinal resistance of the track [kN/(m·track)];

u—longitudinal relative displacement (mm);
R—constant resistance of track bed [kN/(m·track)].

(2) Resistance of track bed in turnout section

The test values obtained from concrete turnout sleepers on Qinhuangdao–Shenyang Passenger Dedicated Line (PDL) are adopted as the longitudinal resistance of the track bed in the turnout section, as shown in Table 3.1.

Table 3.1 Longitudinal resistance of track bed in turnout section

Displacement (mm)	0	0.5	1.0	1.5	2.0
Average resistance per unit length of sleeper (kN/m)	0.327	2.272	3.541	4.267	4.628

The following expression was derived for calculation, from polynomial fitting of the data given in Table 3.1:

$$R(u) = 0.326 + 4.687 \times 10^3 u - 1.685 \times 10^6 u^2 + 2.087 \times 10^8 u^3$$

where u—relative displacement between turnout sleeper and substructure (m);

R—longitudinal resistance of turnout sleeper per unit length (kN/m). When the displacement of the turnout sleeper is greater than 2 mm, the resistance value at 2 mm should be taken.

(3) Fastener resistance

The resistance of the fasteners is considered to be nonlinear. Through polynomial fitting of the test data (Table 3.2) obtained from types II and III clip fasteners of No. 18 turnouts on Qinhuangdao–Shenyang PDL, the nonlinear resistance expression of type II fastener is derived as follows:

$$f(x) = 0.0912 + 11.278x + 3.067x^2 + 0.248x^3$$

where x—relative displacement of the rails and turnout sleepers (mm);

$f(x)$—longitudinal resistance of the fasteners (kN). When the rail displacement is greater than 2 mm, the longitudinal resistance value of the fastener at 2 mm should be taken.

(4) Resistance of spacing block

In this book, the resistance of the spacing blocks is considered to be nonlinear. Through polynomial fitting of the data (Table 3.3) from the tests conducted by China Railway Shanhaiguan Bridge Group Co., Ltd., China Academy of Railway Sciences, and Southwest Jiaotong University on the resistance of spacing blocks used at No.

Table 3.2 Measured resistance of turnout fasteners

Displacement (mm)	0	0.25	0.5	0.75	1	1.5	2
Resistance of type II fastener at a torque of 120 N·m/kN	0.0	2.0	3.8	5.1	6.4	8.9	9.8
Resistance of type II fastener at a torque of 150N·m/kN	0.0	2.8	5.2	6.9	8.2	11.2	12.3
Resistance of type III fastener (kN)	0.0	10.6	14.5	16.5	18.2	18.9	19.1

18 movable-point CWR turnouts on Qinhuangdao–Shenyang PDL, the nonlinear resistance expression for spacing blocks is derived as follows:

$$f_{\mathrm{j}}(x) = 38.027 + 68.8x - 15.392x^2 + 2.333x^3 - 0.105x^4$$

where $f_j(x)$—resistance of spacing block (kN);
x—relative displacement of point rail and wing rail (mm).

(5) Resistance of stopper

The resistance of the stoppers is considered to be nonlinear. Through polynomial fitting of the test results (Table 3.4) related to the displacement between the combination blocks of stoppers at No.18 movable-point CWR turnouts on Qinhuangdao–Shenyang PDL, the resistance expression for the stoppers is derived as follows:

$$f_x(x) = \begin{cases} 0 & x \le d \\ 224.36(x-d) - 96.36(x-d)^2 + 25.83(x-d)^3 - 3.22(x-d)^4 & x > d \end{cases}$$

where $f_x(x)$—longitudinal resistance of the stopper (kN);
x—relative displacement of the combination blocks of the stopper (mm);
d—initial clearance between the combination blocks of the stopper (mm).

2. Other Computation Parameters

The temperature variations in the rails span a range of 50°C, while those in the ballasted bridges span a range of 15°C.

Within the integral box girder across the span, the section centroid is located 1.6266 m away from the upper flange and 1.5774 m away from the lower flange, and the cross-sectional moment of inertia is 3.7044 m^4. The longitudinal rigidity of the piers and abutments on both ends is 1×10^5 kN/cm, and the rigidity of the middle pier is 1×10^3 kN/cm.

Table 3.3 Resistance of spacing block

Relative displacement of wing rail and point rail (mm)	Eccentric loading	0.26	1.05	2.79	4.12	5.53	7.00	9.00	10.55
	Center loading	0.08	1.36	2.56	4.52	6.00	7.04	8.60	10.80
	Average	0.17	1.20	2.68	4.32	5.77	7.02	8.80	10.68
Load (kN)		50	100	150	200	250	310	410	490

Table 3.4 Displacement of combination blocks of stopper

Load (kN)	50	100	150	200	250	310	410	510
Displacement of small block (mm)	0.03	0.09	0.18	0.58	1.03	1.36	1.89	2.38
Displacement of large block (mm)	0.28	0.47	0.73	1.06	1.51	2.12	3.24	4.42

3.4.2 Computation Results

CWR turnouts on bridges experience complex stress and deformation, due to various influencing factors [20, 21]. Generally, simply-supported girders on high-speed railways abroad do not accommodate CWR turnouts. This practice is rooted in the theoretical understanding that when turnouts cross joints of simply-supported girders, they introduce substantial additional expansion and contraction force and displacement, which in turn affect the stability and geometry of the turnouts. Moreover, it is known that high-speed trains generate significant dynamic impact when traversing joints of simply-supported girders. Hence, a strict prohibition exists against locating weak components like turnout switches and frogs over girder joints. However, due to design constraints related to the bridge structures and other factors, some CWR turnouts are still located on simply-supported girders in China's railways upgraded for speed increases up to 200 km/h. Nonetheless, other design principles are followed, such as ensuring switch rails and point rails do not cross girder joints.

According to the classification of longitudinal forces on CWR tracks on bridges and turnout-bridge interactions in China, the longitudinal forces on CWR turnouts on bridges are categorized into the following: expansion and contraction force, deflection force, rail breakage force, and braking force [22, 23].

This section presents a computational analysis of the longitudinal forces on CWR turnouts on bridges, using a No. 18 simple turnout laid on 9-span 32 m simply-supported girders as an example. The turnout-bridge arrangement is shown in Fig. 3.30. Following the principle that movable components of turnouts should not cross girder joints, the switch section is positioned on the girder in the 4th span. The actual points of the switch rails are located 6 m away from the fixed bearing. The frog section is situated on the girder in the 5th span. The actual points of the point rails are positioned 15.5 m away from the fixed bearing. Additionally, the girder joint at the No.4 pier is situated in the turnout closure, 10.6 m away from the heel ends of the switch rails.

For comparative calculations, the parameters of CWR tracks on bridges, including bridge span, bearing arrangement, bridge structure, and pier/abutment rigidity, are assumed to be identical to those of CWR turnouts on bridges. Likewise, the parameters of CWR turnouts on subgrade, such as structure, track bed, and fastener resistance, are assumed to be the same as those of CWR turnouts on bridges.

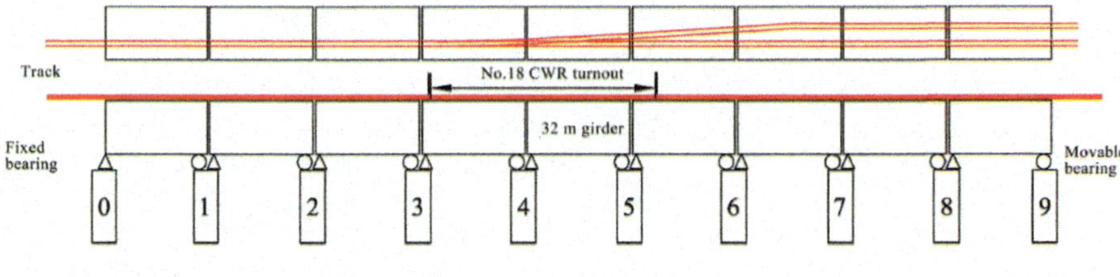

Fig. 3.30 Turnout bridge arrangement

1. Expansion and Contraction

(1) Expansion and contraction force on rail

The comparison of expansion and contraction force on rails in three scenarios of CWR turnouts on bridges, CWR tracks on bridges, and CWR turnouts on subgrade is illustrated in Fig. 3.31.

From Fig. 3.31, it is evident that the peak value of the rail temperature compressive force within CWR turnouts on bridges is identical to that within CWR turnouts on subgrade, both occurring at the heel end of the switch rails. In the case of rightward expansion and contraction of the bridge, the peak rail temperature compressive force within CWR tracks on bridges is observed at the fixed bearing of the rightmost bridge span. In sections of bridges far away from the turnout section, the distribution law of temperature stress is equivalent to that within CWR tracks on regular bridges. This indicates that the expansion and contraction of the turnout's inner rail changes the distribution of rail temperature force within CWR tracks on bridges. Furthermore, as the distance from the turnout increases, the expansion and contraction effect of the turnout diminishes. The maximum additional temperature compressive force observed within CWR turnouts on bridges is 235.3 kN, whereas, at the corresponding girder joint of CWR tracks on bridges, it is 54.7 kN. This highlights that the combined effect of girder expansion and contraction and inner rail expansion and contraction in CWR turnouts on bridges leads to significantly higher maximum additional temperature force on the rails compared to CWR tracks on regular bridges.

The maximum additional temperature compressive force at the heel end of the frogs within CWR turnouts on subgrade is 202.5 kN, which is lower than that observed within CWR turnouts on bridges. This discrepancy can be attributed to the fact that the expansion and contraction of the inner rail in turnouts on bridges occurs in the opposite direction to that of the bridge itself. Consequently, the combined

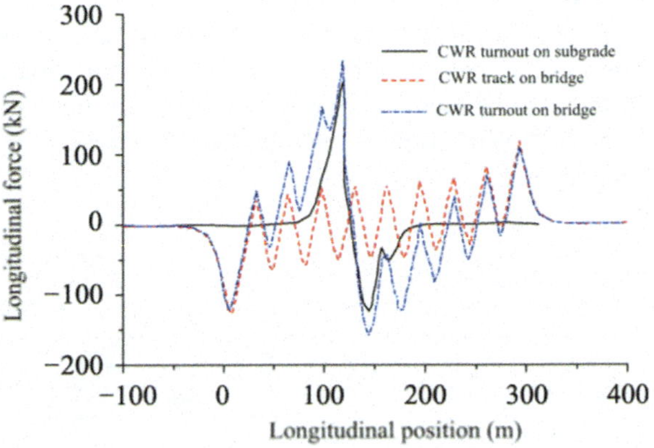

Fig. 3.31 Additional expansion and contraction force on straight stock rails

effect of these two expansions leads to the increased stress on CWR turnouts on bridges compared to CWR turnouts on subgrade.

(2) Rail displacement

The comparison of displacement experienced by rails in three scenarios of CWR turnouts on bridges, CWR tracks on bridges, and CWR turnouts on subgrade is illustrated in Fig. 3.32.

From Fig. 3.32, it is evident that the variation laws of the expansion and contraction displacement of rails within CWR turnouts on bridges are apparently different from those of CWR tracks on bridges and CWR turnouts on subgrade. The peak rail displacement for CWR turnouts on bridges is found near the heel end of the switch rails, and its absolute value exceeds that in the context of CWR turnouts on subgrade. Similar to CWR turnouts on subgrade, the straight stock rails of CWR turnouts on bridges expand and contract toward the left within the turnout section, primarily due to the leftward expansion and contraction of the inner rail. The rails adjacent to the turnout section at both ends undergo rightward expansion and contraction displacement, mainly influenced by the rightward expansion and contraction of the bridge. In contrast, the rails adjacent to CWR turnouts at both ends on the subgrade are situated in the non-breathing zone of the CWR tracks, experiencing minimal expansion and contraction displacement. On the other hand, CWR tracks on bridges exhibit their maximum expansion and contraction displacement at the movable end of the middle girder span. As the bridges expand and contract toward the right, expansion and contraction displacement on all rails occurs in the rightward direction.

Refer to Table 3.5 for a comparison of expansion and contraction displacement at the actual points of the switch rails and point rails between CWR turnouts on bridges and CWR turnouts on subgrade. It is evident that the expansion and contraction displacement of the switch rails slightly increases relative to the turnout sleepers in CWR turnouts on bridges, while the expansion and contraction displacement of the point rails decreases slightly relative to the turnout sleepers. These differences are primarily influenced by factors such as bearing arrangement, relative positioning

Fig. 3.32 Displacement of straight stock rail

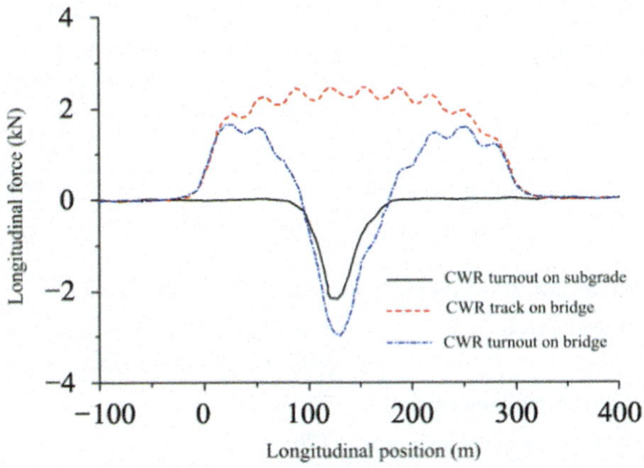

between girders and turnouts, and girder span length. In China, high-speed turnout switches are consistently installed at the end of the turnout sleepers. To ensure compliance with design requirements for the expansion and contraction displacement of the switch rails and point rails, it is essential to verify whether their expansion and contraction displacement relative to the switch meets the specified criteria. Therefore, the expansion and contraction displacement of the inner rails within the turnouts mentioned in this book mainly refers to the relative displacement between the sleepers and rails.

(3) Stress of force transfer devices

Refer to Table 3.6 for a comparison of longitudinal forces on force transfer devices, such as stoppers and spacing blocks, in CWR turnouts on bridges and CWR turnouts on subgrade. It is apparent that the stress on the stoppers slightly increases in CWR turnouts on bridges, while the stress on the spacing blocks slightly decreases. This observation is consistent with the laws observed in the longitudinal forces on the stock rails and the expansion and contraction displacement of the switch rails and point rails.

2. Deflection

(1) Calculation conditions

The ZK live load is utilized to calculate the longitudinal force due to deflection of CWR turnouts on bridges, taking into account the following two conditions in the calculations:

Condition I: From left to right, the girder spans where the main route of the turnouts is located are fully loaded (the 4, 5, and 6th spans), and the diverging routes are unloaded;

Condition II: From right to left, the girder spans where the main route of the turnouts is located are fully loaded (the 4th, 5th, and 6th spans), and the diverging

Table 3.5 Expansion and contraction displacement of inner rail in turnout

Turnout type	Expansion and contraction displacement at actual point of switch rail (mm)	Expansion and contraction displacement at actual point of point rail (mm)
CWR turnout on bridge	−19.4	−7.5
CWR turnout on Subgrade	−19.2	−7.8

Table 3.6 Stress on force transfer devices in turnout

Turnout type	Longitudinal force on stopper at heel end of switch rail (kN)	Longitudinal force on spacing block at end of wing rail (kN)
CWR turnout on bridge	130.8	156.8
CWR turnout on Subgrade	126.2	158.4

routes are unloaded. To provide a comparison, the calculation conditions for the longitudinal force due to deflection of CWR tracks on regular bridges are considered:

Condition III: From left to right, the 4, 5, and 6th spans are fully loaded, and the diverging routes are unloaded;

Condition IV: From right to left, the 4, 5, and 6th spans are fully loaded, and the diverging routes are unloaded.

(2) Distribution of longitudinal force due to deflection and displacement

The distribution of longitudinal force due to deflection and displacement on the straight stock rails of the turnouts and the comparison with CWR tracks on regular bridges are shown in Figs. 3.33 and 3.34. In the figures, the abscissa represents the distance to the left abutment, while the ordinate represents the additional longitudinal force due to deflection of the rails. Positive values indicate compressive force.

It can be seen from Fig. 3.33 that when a train is running from left to right, the maximum additional compressive force of 109.1 kN occurs at the movable bearing near the tail of CWR turnouts on bridges, which is greater than the additional compressive force of 68.2 kN at the same position for CWR tracks on bridges. Similarly, when a train is running from right to left, the maximum additional compressive force of 71.6 kN is still observed at the movable bearing near the tail of CWR turnouts on bridges, and the maximum additional compressive force of 67.7 kN occurs at the same location for CWR tracks on bridges. The maximum additional force on the rails of CWR tracks on bridges is 82.5 kN occurring at the fixed bearing of the 4th span, while the corresponding value of CWR turnouts on bridges is only 46.3 kN. It is evident that the distribution laws of additional longitudinal force due to deflection of CWR turnouts on bridges differ from those of CWR tracks on regular bridges. These distinctions are influenced by multiple factors such as the direction of train operation, bearing arrangement, relative positioning between turnouts and bridges, and

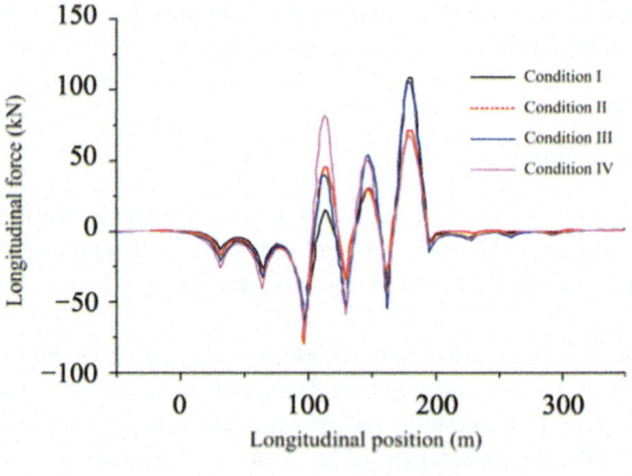

Fig. 3.33 Additional longitudinal force due to deflection of stock rail

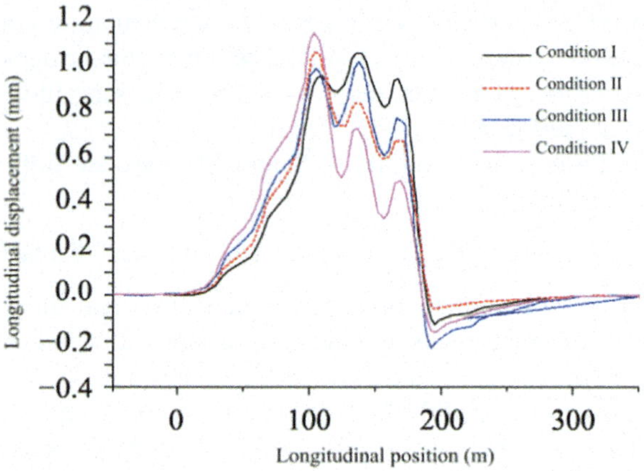

Fig. 3.34 Deflection displacement

the relative orientation between turnouts and fixed bearings on bridges. Therefore, it is crucial to perform checking calculations based on the specific circumstances.

The deflection calculation for CWR tracks on regular simply-supported girder bridges typically takes into account the load arrangement on two-span simply-supported girders. However, for CWR turnouts on bridges, the maximum additional compressive force occurring at the movable bearing near the turnout tail is 107.9 kN when using the two-span loading arrangement (from left to right) that includes the lead curve and frog. Furthermore, when considering the switch and lead curve in the two-span loading arrangement (from left to right), the maximum additional compressive force at the movable bearing located at the girder span where the lead curve is situated is 67.1 kN. Both of these values are lower than the results obtained when using the three-span loading within the turnout range. Therefore, it is apparent that distributing the load over two spans within the turnout range is preferable when calculating the longitudinal force due to deflection of CWR turnouts on bridges.

3. Rail breakage

(1) Calculation conditions

Rail breakage typically occurs at the location with the maximum longitudinal force on the rails, predominantly at the bridge end where the additional expansion and contraction force becomes evident. Moreover, the focus is solely on the breakage of one rail.

According to the distribution laws of additional expansion and contraction force illustrated in Fig. 3.31, the following four calculation conditions are considered:

Condition I: The straight stock rail, positioned at the girder end preceding the turnout (specifically, the girder joint of No.3 pier), is fractured;

Condition II: The curved stock rail, positioned at the girder end preceding the turnout (specifically, the girder joint of No.3 pier), is fractured;

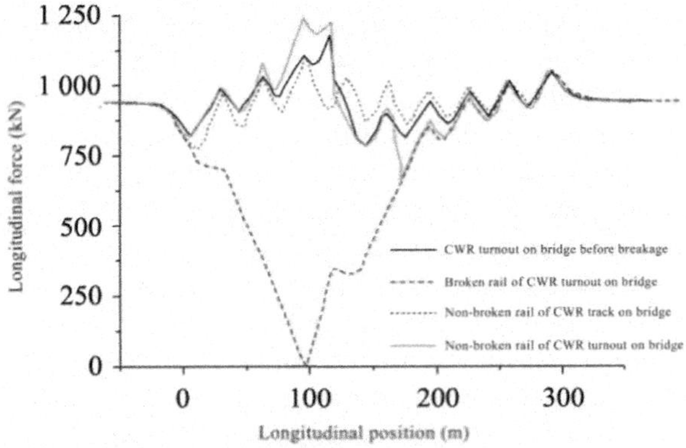

Fig. 3.35 Distribution of force from rail breakage under condition I

Condition III: The straight stock rail, positioned at the pier/abutment following the turnout (specifically, the girder joint of No.5 pier), is fractured;

Condition IV: The long point rail, positioned at the pier/abutment following the turnout (specifically, the girder joint of No.5 pier), is fractured.

(2) Force from rail breakage

Taking Condition I as an example, the distribution of longitudinal force resulting from rail breakage is shown in Fig. 3.35, where positive values represent tensile force on the rails.

Upon the fracture of the straight stock rail, the longitudinal force at the broken rail gap diminishes to zero. Additionally, the longitudinal force within the CWR turnout diminishes to a small magnitude. Nevertheless, as a result of the amplified longitudinal displacement between the turnout sleeper and the bridge, the longitudinal force on the curved stock rail at the corresponding broken rail gap increases from 1108.1 kN to 1254.8 kN.

The calculations reveal that the longitudinal displacement of the turnout sleeper at the actual point of the straight switch rail amounts to 1.8 mm, and for the curved switch rail, it measures 27.0 mm, causing a difference of 25.2 mm between the left and right sides of the turnout sleeper, which indicates the occurrence of a substantial deflection. Upon the fracture of one rail in the CWR track on a regular bridge, the additional longitudinal force on the other rail also rises from 54.8 kN to 157.1 kN due to the deflection effect of sleepers.

The current design practice for CWR tracks on bridges typically does not incorporate a strength check for the other rail at the break joint for the rail fracture scenario involving a single rail break. This is mainly because the existing simplified algorithm utilized in the design calculations only considers the breakage of a single rail, thereby lacking the ability to account for the distribution of additional longitudinal force on the remaining rail. To ensure the safety of train operation at normal speeds in case of a broken rail without timely identification, it is necessary to check the strength of the non-broken rail. This is because of the fact that broken rail gaps often exhibit large widths, which intensifies the impact on wheel-rail dynamic interactions compared to regular joint defects. Consequently, there is an increase in the dynamic actions on both rails at the broken rail gap. Since rail breakage is considered an accidental condition, it is acceptable to reduce the safety factor value appropriately during the strength checking calculations.

When the rail at the tail end of turnouts is broken, the distribution of longitudinal force on the non-broken stock rail is shown in Fig. 3.36, where positive values represent tensile force on the rails. The diagram reveals that the longitudinal force distribution on the rail preceding the turnout remains similar to that before the rail breakage. However, a significant change occurs at the right end, indicating deflection on the turnout sleeper near the broken rail gap. This deflection results in the transmission of a partial longitudinal force from the broken rail to the adjacent rail string. Typically, the inner rail in turnouts expands and contracts in a direction opposite to the bridge. However, when a rail break occurs following the turnout, the expansion and contraction direction of the broken rail changes. Consequently, the longitudinal force on the rail at the heel end of the switch rail becomes lower than its value before the rail breaks. In the case when the right end of the long point rail is broken, the influence on the longitudinal force on the stock rail at the heel end of the switch rail is relatively less significant compared to the situation when the straight stock rail following the turnout is broken. This difference primarily arises due to the long point rail's location within the breathing zone on the inner rail side of the turnout.

(3) Other calculation results

See Table 3.7 for a comparison of calculation results under these four conditions. In the context of installing the switch on the turnout sleeper, the checking calculations may only involve the relative displacement between the sleepers and rails. In the table, positive values indicate rail displacement and longitudinal force on the piers/abutments toward the right.

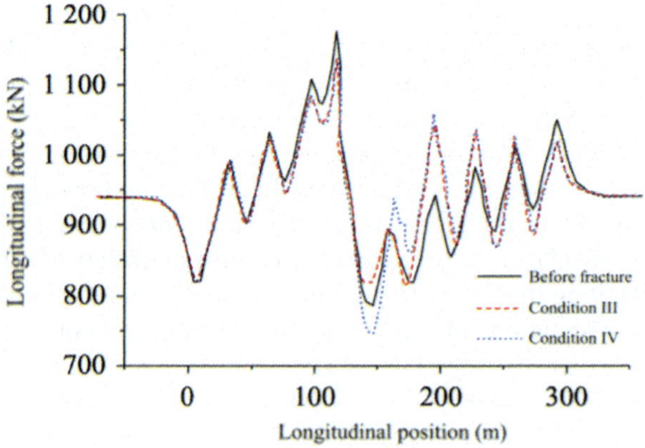

Fig. 3.36 Comparison of longitudinal force for non-broken stock rail

Table 3.7 Comparison of calculation results

Computation results	Condition I	Condition II	Condition III	Condition IV
Maximum longitudinal force on non-broken stock rail (kN)	1252.2	1254.5	1141.1	1137.4
Broken rail gap (mm)	59.8	60.7	46.3	43.6
Maximum stress on stopper at heel end of switch rail (kN)	218.2	221.7	116.2	106.5
Maximum stress on spacing block at end of wing rail (kN)	135.9	133.6	166.8	159.5
Displacement at actual point of straight switch rail relative to turnout sleeper (mm)	21.4	1.8	18.8	18.8
Displacement at actual point of curved switch rail relative to turnout sleeper (mm)	1.8	21.4	18.5	18.7
Displacement at actual point of point rail relative to turnout sleeper (mm)	6.1	5.7	7.7	4.8
Longitudinal force on no.3 pier (kN)	616.1	618.3	343.3	361.4
Longitudinal force on no.4 pier (kN)	897.4	897.6	357.3	454.0
Longitudinal force on no.5 pier (kN)	742.8	712.0	203.3	118.0

It can be seen from Table 3.7 that regardless of whether rail breakage occurs on the main route or diverging route preceding the turnout, the maximum longitudinal force on the non-broken stock rail increases, and the longitudinal displacement of the broken rail gap, switch rail, and point rail relative to the turnout sleeper and the longitudinal force on the piers and abutments are greater compared to the situation when the rail at the right end of the continuous girder following the turnout is broken. The calculations reveal that when the breakage occurs at the right end of the simply-supported girders, for both the straight/curved stock rail and long/short point rail following the turnout, the longitudinal displacement of the non-broken stock rail, broken rail gap, switch rail, and point rail relative to the turnout sleeper, as well as the longitudinal force on the piers and abutments, are roughly equivalent and smaller in magnitude compared to the situation when the rail preceding the turnout is broken. This can be attributed to the expansion and contraction displacement of the inner rail taking place in a direction opposite to that of the broken rail. To control broken rail gap angles, it is beneficial to adopt strategies that avoid the occurrence of maximum longitudinal forces on the rail preceding the turnouts when laying CWR turnouts on bridges. The purpose of this practice is to minimize the risk of rail breakage specifically preceding the expansion and contraction section of the inner rail within the turnout.

4. Braking conditions

(1) Calculation conditions

The train load is considered as the ZK live load, and the coefficient of friction on the wheel-rail tread is taken as 0.164. The braking force acts over an extended length for three girder spans (the 4, 5, and 6th spans) in the turnout section. When calculating the braking force, the following four conditions are taken into account, while excluding the expansion and contraction of bridges and the inner rails within the turnouts:

Condition I: A train passes through the turnout on the main route, and the braking force action is directed from right to left;

Condition II: A train passes through the turnout on the main route, and the braking force action is directed from left to right;

Condition III: A train passes through the turnout on a diverging route, and the braking force action is directed from right to left;

Condition IV: For comparison, the calculation conditions for braking force on CWR tracks on regular bridges are considered, in which the load also acts on the corresponding 3 girder spans, and the braking force is directed from right to left.

(2) Calculation for passage through turnout on main route

Taking Condition I as an example, the distribution of the additional braking force on the straight/curved stock rail and a comparison with the scenario of CWR tracks on regular bridges are presented in Fig. 3.37. In this figure, positive values indicate compressive force on the rails. The relative displacement between the girders and rails is illustrated in Fig. 3.38, where positive values indicate the rightward displacement of the rails relative to the bridge.

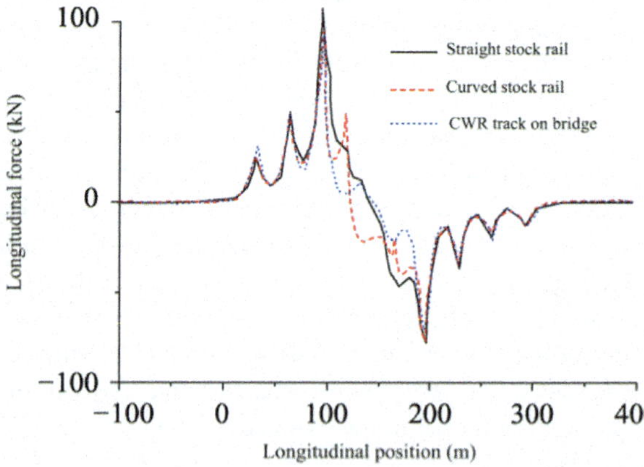

Fig. 3.37 Braking force under condition I

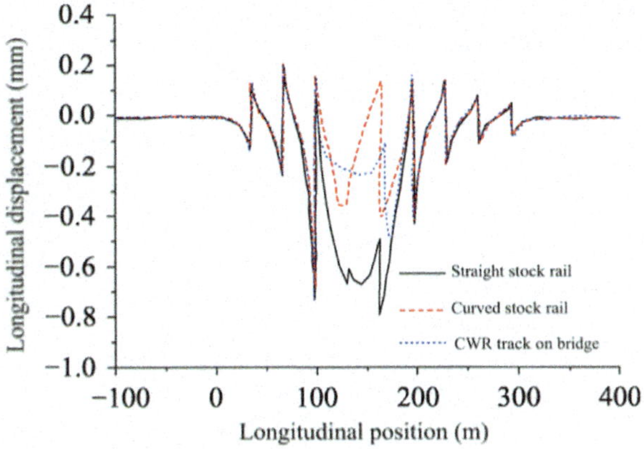

Fig. 3.38 Relative displacement between girder and rail under condition I

For CWR turnouts on bridges, the maximum additional braking force on the straight stock rail is approximately 107.1 kN and that on the curved stock rail is approximately 93.5 kN, both occurring at the left end of the girder spans. The maximum relative displacement between the straight stock rail and the bridge is approximately 0.79 mm while that between the curved stock rail and the bridge is approximately 0.62 mm. For CWR tracks on bridges, the maximum braking force is approximately 86.4 kN, which also appears at the left end of the girder spans. The maximum rapid relative displacement between the girder and rails under the braking force is approximately 0.55 mm. The main reason for the greater braking force in CWR turnouts on bridges compared to regular CWR tracks lies in the effect of the braking force on the inner rail, which acts as a breathing zone. This force drives the inner rail to expand and contract from right to left, leading to the transmission of a partial longitudinal force to the straight stock rail through the resistance provided by the fasteners and track bed. As a result, the additional braking force on the straight stock rail and its displacement relative to the bridge are greater when compared to the situation where the curved stock rail is involved. This is primarily due to the curved stock rail being partially impacted by the braking force, while the remaining force is borne by the straight switch rail and long point rail.

The calculations reveal that when subjected to the braking force, the actual point of the straight switch rail expands and contracts by approximately 3.2 mm. The curved switch rail, driven by the movement of the turnout sleeper, expands and contracts by approximately 2.0 mm. Additionally, the actual point of the point rail expands and contracts by approximately 2.4 mm. At the heel end of the switch rail, the stopper is not placed properly, and the maximum longitudinal force borne by a single spacing block at the end of the wing rail amounts to approximately 11.9 kN.

Taking Condition II as an example, subjected to the braking force from left to right, the additional braking force on the straight/curved stock rail and a comparison with the force on CWR tracks on regular bridges are presented in Fig. 3.39. In this figure, positive values indicate compressive force on the rails. The relative displacement

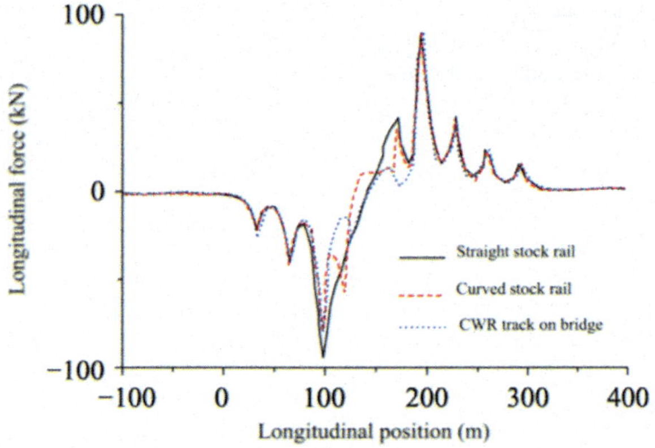

Fig. 3.39 Braking force under condition II

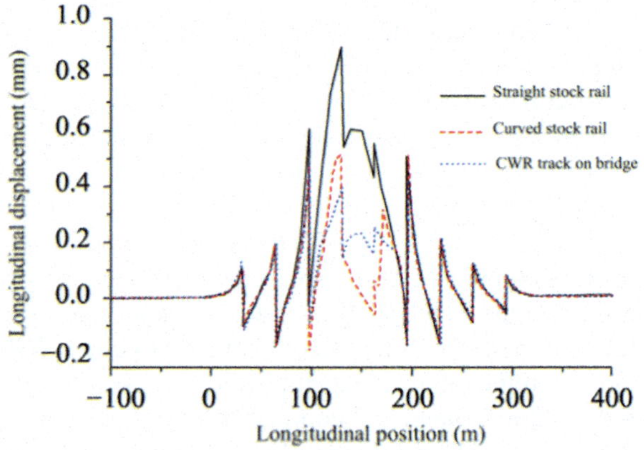

Fig. 3.40 Relative displacement between girder and rail under condition II

between the girders and rails is illustrated in Fig. 3.40, where positive values indicate the rightward displacement of the rails relative to the bridge.

The maximum additional braking compressive force occurs following the turnout at the right end of the simply-supported girder. The maximum braking force on the straight/curved stock rail is approximately 90.2 kN, slightly greater than the 82.6 kN for CWR tracks on bridges. This difference can be attributed to the significant longitudinal resistance of the track bed under the long turnout sleepers in the frog zone. The resistance weakens the longitudinal force transmitted from the inner rail to the stock rail, thus balancing the longitudinal forces among the four rails. When the braking force acts from left to right, the maximum additional compressive force for both CWR turnouts and CWR tracks on bridges is slightly less compared to the situation where the braking force acts from right to left. This distinction primarily arises from the arrangement of the bearings. The maximum relative displacement between the straight stock rail and the bridge is approximately 0.90 mm, while the maximum relative displacement between the curved stock rail and the bridge is approximately 0.50 mm. In CWR tracks on regular bridges, the maximum rapid

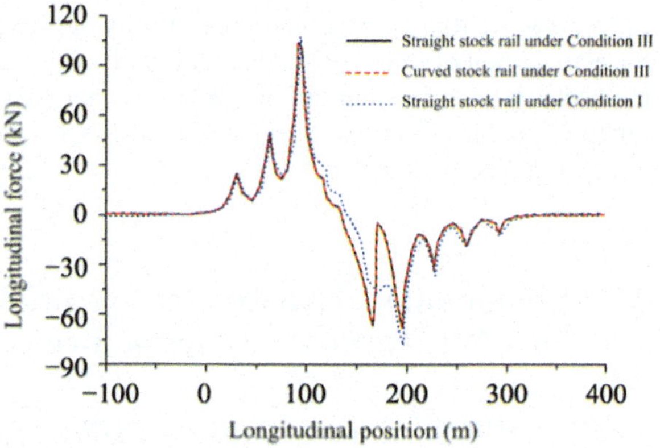

Fig. 3.41 Comparison of braking force between straight and curved stock rails

relative displacement between the girder and rails is approximately 0.41 mm. These displacements are all within limits that would cause damage to the track bed or deformation of the tracks.

The calculations reveal that when subjected to the braking force, the actual point of the straight switch rail expands and contracts by approximately 3.0 mm. The curved switch rail expands and contracts by approximately 2.0 mm. Additionally, the actual point of the point rail expands and contracts by approximately 2.4 mm. At the heel end of the switch rail, the stopper is not placed properly, and the maximum longitudinal force borne by a single spacing block at the end of the wing rail amounts to approximately 11.2 kN.

(3) Calculation for passage through turnout on diverging route

For a train passing through the turnout on a diverging route, the distribution of the additional braking force on the straight/curve stock rail and a comparison with Condition I are shown in Fig. 3.41, where positive values represent compressive force on the rails.

Within No.18 turnouts, the short point rail is connected in a skew structure at the heel end and is constrained by the long rail string at the rear end of the long point rail. The braking force on the short point rail is transmitted to the curved inner rail through the connected long point rail and the spacing block at the end of the wing rail. As a result, the additional braking compressive force generated preceding the turnout is slightly less than that generated on the straight/curved inner rail when a train passes through the turnout on the main route. However, when the turnout contains double-arm flexible point rails or fixed frogs, the symmetrical nature of the turnout between the main and diverging routes can be considered. If the bridge deck and piers form an integral structure, the distribution of additional braking force on the straight stock rail when a train passes through the turnout on the main route would be the same as that on the curved stock rail for a train passing through the turnout on the diverging route.

The calculations reveal that when subjected to the braking force, the actual point of the straight switch rail expands and contracts by approximately 1.9 mm. The curved switch rail expands and contracts by approximately 3.0 mm. Additionally, the actual point of the point rail expands and contracts by approximately 1.5 mm. At the heel end of the switch rail, the stopper is not placed properly, and the maximum longitudinal force borne by a single spacing block at the end of the wing rail amounts to approximately 4.4 kN.

3.5 Computational Analysis of Longitudinal Forces for CWR Turnout Groups on Bridges

3.5.1 CWR Turnout Groups on Bridges for Ballasted Track

1. Calculation conditions

Taking No. 18 high-speed turnout groups featuring 60 kg/m rails as an example, the analysis of the stress and deformation patterns of CWR turnout groups on bridges for ballasted tracks incorporates the following five conditions:

Condition I: opposite arrangement in opposite directions on the bridge of 3 × 32 m continuous girder + 3 × 32 m continuous girder, at the distances from the turnout centers to the left abutment of 147 m and 235.458 m, respectively, as shown in Fig. 3.42.

Condition II: opposite arrangement in the same direction on the bridge of 3 × 32 m continuous girder + 3 × 32 m continuous girder, at the distances from the turnout centers to the left abutment of 147 m and 235.458 m, respectively, as shown in Fig. 3.43.

Condition III: back-passing arrangement in opposite directions on the bridge of 3 × 32 m continuous girder + 3 × 32 m continuous girder, at the distances from the turnout centers to the left abutment of 157 m and 255.458 m, respectively, as shown in Fig. 3.44.

Condition IV: back-passing arrangement in the same direction on the bridge of 3 × 32 m continuous girder + 3 × 32 m continuous girder, at the distances from the

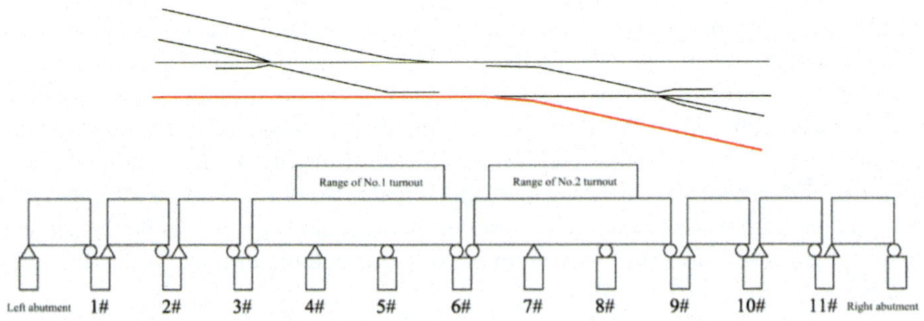

Fig. 3.42 Schematic diagram of turnout bridge arrangement under condition I

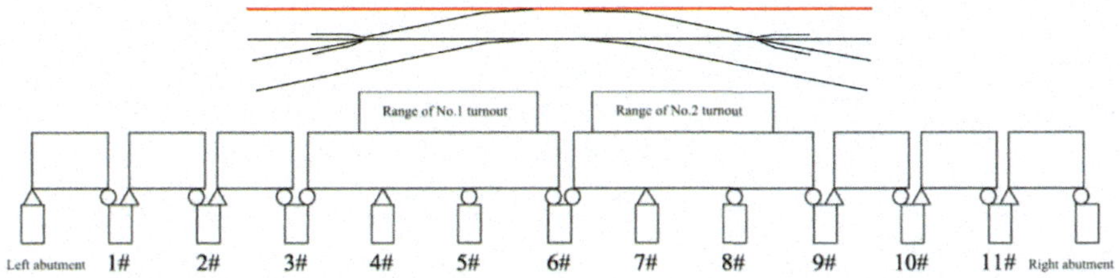

Fig. 3.43 Schematic diagram of turnout bridge arrangement under condition II

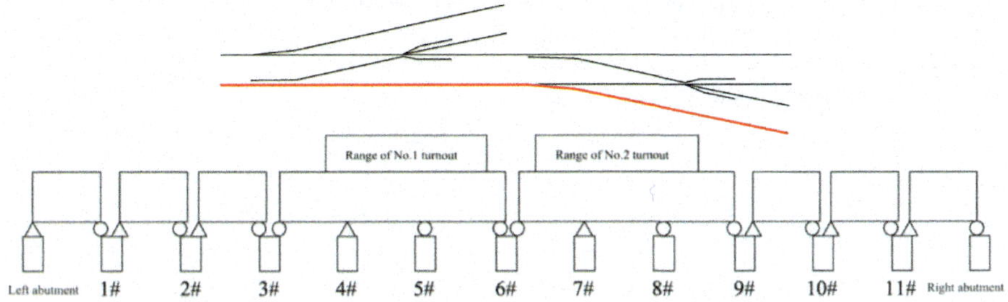

Fig. 3.44 Schematic diagram of turnout bridge arrangement under condition III

turnout centers to the left abutment of 130 m and 225.458 m, respectively, as shown in Fig. 3.45.

Condition V: single crossover on the bridge of 6 × 32 m continuous girder, at the distances from the turnout centers to the left abutment of 147 m and 235.458 m, respectively, as shown in Fig. 3.46.

2. Computation Results

The additional forces on the stock rails and the relative displacements between the girders and rails under different conditions are shown in Figs. 3.47, 3.48, 3.49, 3.50, 3.51, 3.52, 3.53, 3.54, 3.55 and 3.56, respectively.

From a comparison among the above five conditions, the calculation results are shown in Table 3.8.

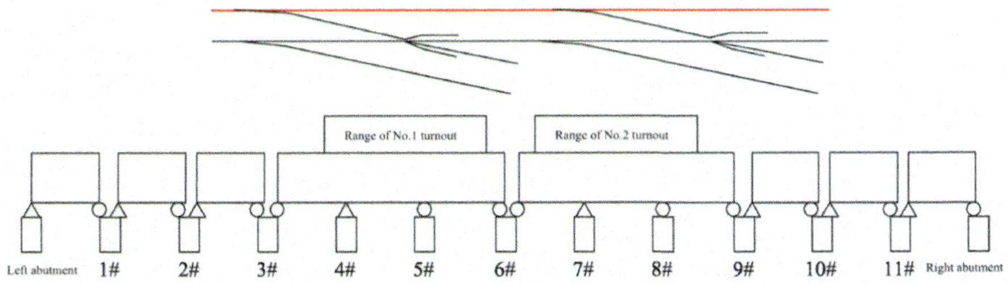

Fig. 3.45 Schematic diagram of turnout bridge arrangement under condition IV

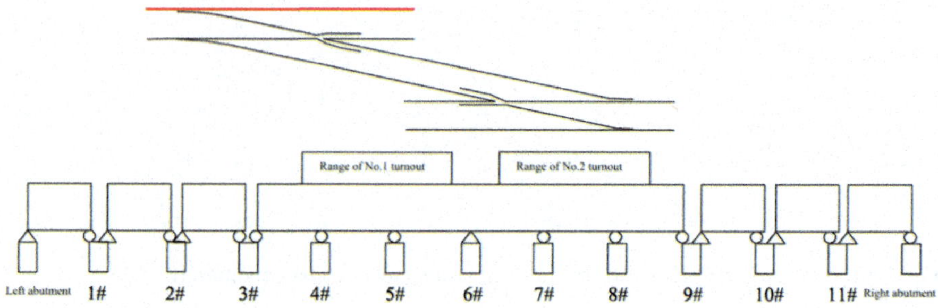

Fig. 3.46 Schematic diagram for turnout bridge arrangement under condition V

Fig. 3.47 Additional force on stock rail under condition I

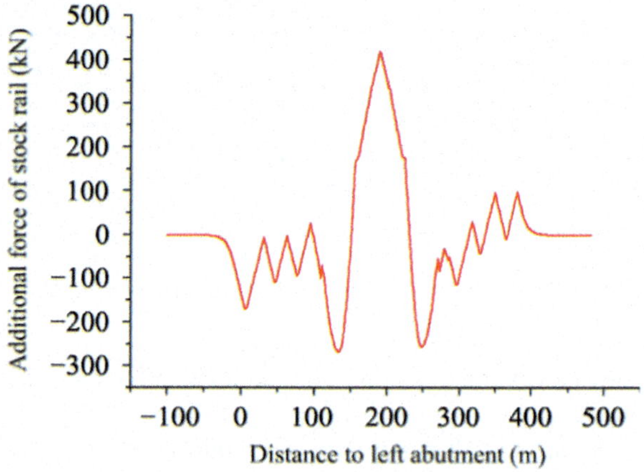

Fig. 3.48 Relative displacement between girder and rail under condition I

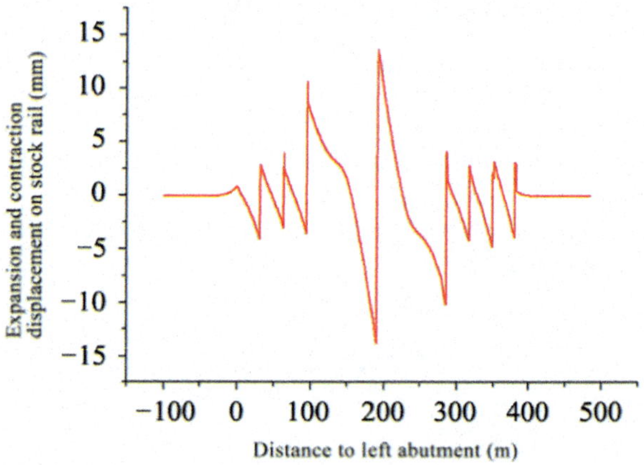

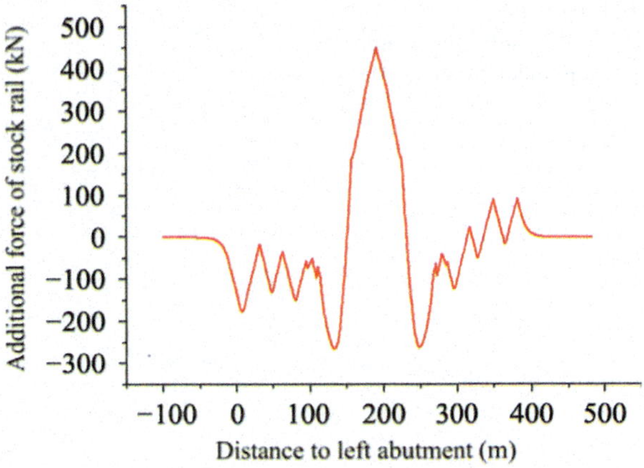

Fig. 3.49 Additional force on stock rail under condition II

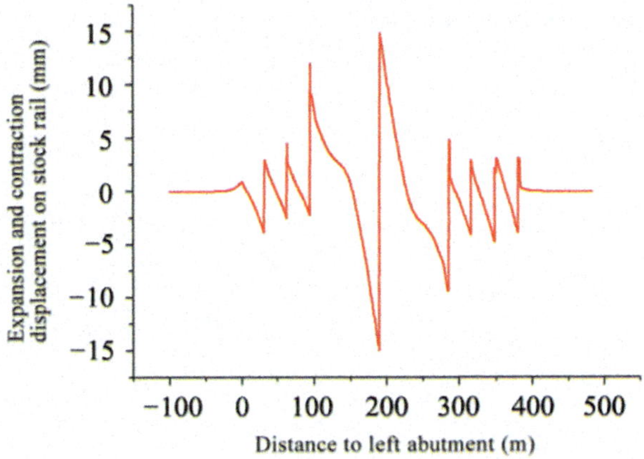

Fig. 3.50 Relative displacement between girder and rail under condition II

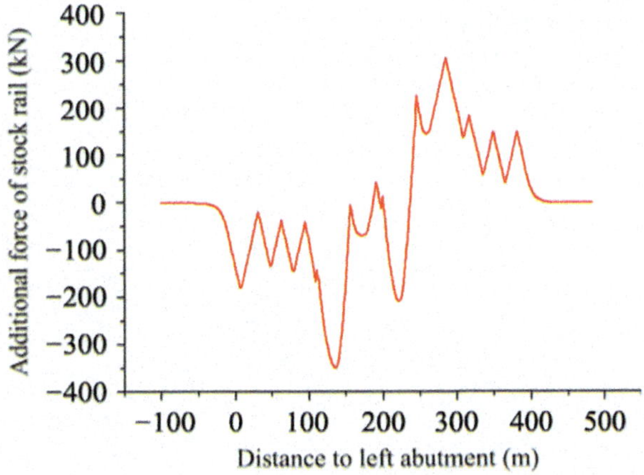

Fig. 3.51 Additional force on stock rail under condition III

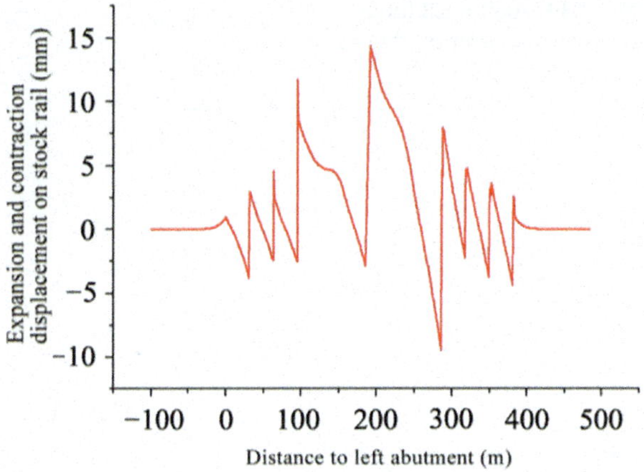

Fig. 3.52 Relative displacement between girder and rail under condition III

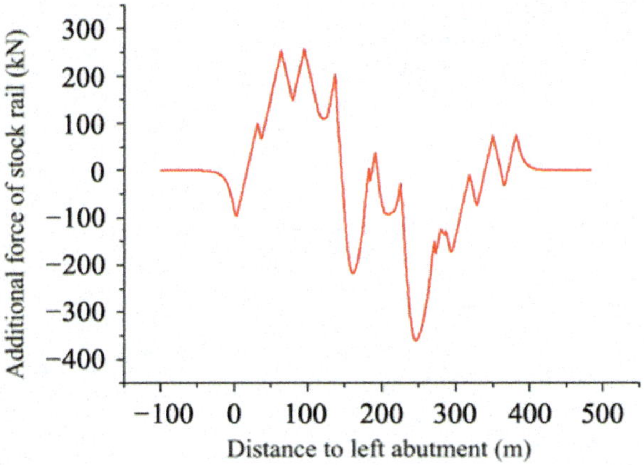

Fig. 3.53 Additional force on stock rail under condition IV

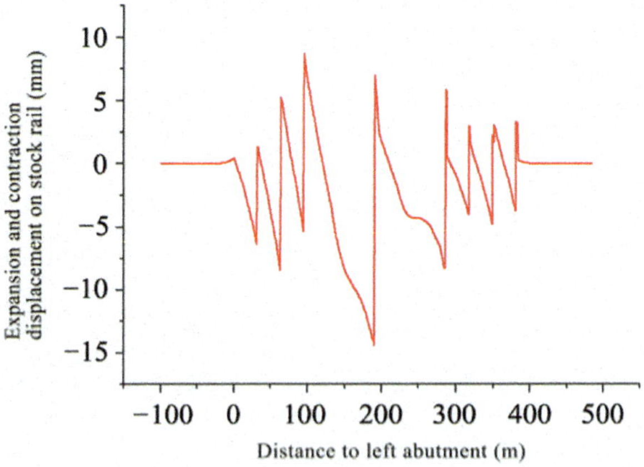

Fig. 3.54 Relative displacement between girder and rail under condition IV

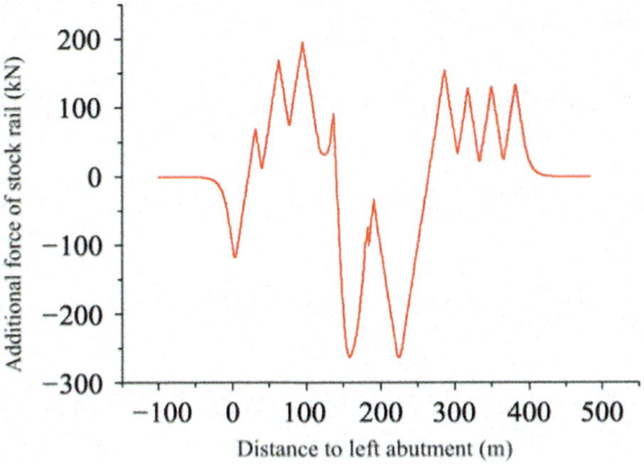

Fig. 3.55 Additional force on stock rail under condition V

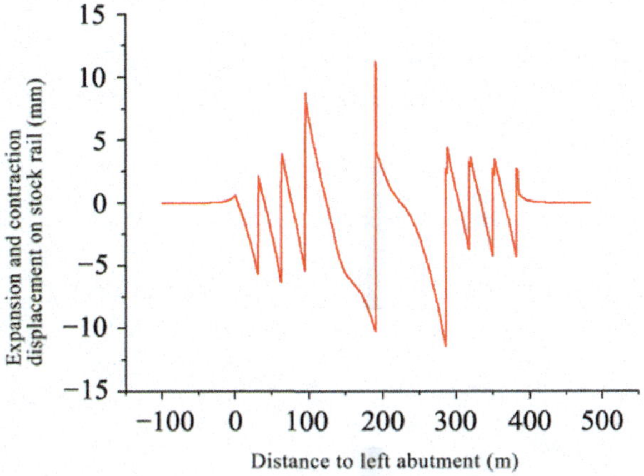

Fig. 3.56 Relative displacement between girder and rail under condition V

Table 3.8 illustrates that there are significant variations in the stress and deformation of CWR turnout groups on bridges for ballasted tracks, depending on the different connection forms. Within the turnout groups in the opposite arrangement and single crossovers, the force transfer devices at the heel end of the switch rails are not subjected to stress. However, within the turnout groups in the back-passing arrangement, the force transfer devices experience slight stress.

Table 3.8 Computation results

Calculation conditions		I	II	III	IV	V
Maximum stress on force transfer device at heel end of switch rail (kN)	No. 1 turnout	0	0	0	−12.395	0
	No. 2 turnout	0	0	23.066	0	0
Maximum stress on spacing block at end of wing rail (kN)	No. 1 turnout	162.423	161.272	149.640	−186.503	−171.269
	No. 2 turnout	−162.971	−161.803	190.957	−145.366	171.588
Displacement at actual point of straight switch rail relative to turnout sleeper (mm)	No. 1 turnout	20.897	21.224	16.786	−20.837	−17.396
	No. 2 turnout	−20.980	−21.305	−21.213	−15.457	17.749
Displacement at actual point of point rail relative to turnout sleeper (mm)	No. 1 turnout	8.168	8.071	7.331	−9.656	−8.636
	No. 2 turnout	−8.194	−8.113	9.888	−7.048	8.653
Maximum additional force on stock rail		420.5448	452.3	347.37	359	271.883
Left abutment		79.876	−116.313	−113.534	−90.383	−162.794
No. 1 pier		3.104	7.255	10.493	−38.556	−42.608
No. 2 pier		2.778	6.766	10.007	−38.698	−43.007
No. 3 pier		0	0	0	0	0
No. 4 pier		170.529	230.505	263.193	−363.799	−187.866
No. 5 pier		0	0	0	0	0
No. 6 pier		0	0	0	0	0
No. 7 pier		−163.950	−223.150	239.304	−357.993	157.959
No. 8 pier		0	0	0	0	0
No. 9 pier		−37.542	−56.163	40.945	−104.007	50.037
No. 10 pier		−18.035	−28.011	6.738	−49.894	5.477
No. 11 pier		−18.231	−28.305	6.615	−50.305	5.105
Right abutment		0	0	0	0	0

The maximum stress on the spacing blocks within turnout groups in the opposite arrangement is the smallest among all forms, measuring 162.971 kN. On the other hand, in the case of the back-passing arrangement, the maximum stress on the spacing blocks is the largest, reaching 190.957 kN. Among turnout groups connected in the same form, i.e. either opposite arrangement or back-passing arrangement, there is little difference in the maximum stress on the spacing blocks under the circumstances of opposite directions and the same direction. Furthermore, the spacing blocks within two turnouts in the opposite arrangement are stressed in opposite directions, while those within two turnouts in the back-passing arrangement are stressed in the same direction.

The maximum displacement at the actual point of the straight switch rail relative to the turnout sleeper within turnout groups in the form of single crossovers is 17.749 mm, which is the smallest among all forms. However, when comparing the opposite arrangement and back-passing arrangement, there is little difference between the maximum relative displacements of the switch rails. In both cases, the maximum relative displacements are approximately 21 mm. The maximum additional force on the stock rail in the opposite arrangement is the largest among all forms, and that within single crossovers is the smallest. Additionally, when considering the same connection form but different route contexts, the maximum additional force on the stock rail is greater in the context of the same direction compared to the context of in opposite directions.

Among all forms, the longitudinal force on the piers and abutments in the form of opposite arrangement in opposite directions is the smallest, while that on the piers and abutments in the back-passing arrangement in the same direction is the largest. The maximum longitudinal force occurs at No. 4 pier.

3.5.2 CWR Turnout Groups on Bridges for Ballastless Track

1. Calculation conditions

Taking No. 18 high-speed turnout groups featuring 60kg/m rails as an example, as shown in Figs. 3.42, 3.43, 3.44, 3.45 and 3.46, the analysis of the stress and deformation patterns of CWR turnout groups on bridges for ballastless tracks incorporates the five conditions. There are three turnout slabs under each group of turnouts under Conditions I to IV, and there are three turnout slabs in total under two groups of turnouts under Condition V, as shown in Fig. 3.57. The turnout slabs are connected to the bridge with longitudinal and transverse bollards.

2. Computation Results

The additional force on the stock rail and the relative displacement between the girders and rails under different conditions are shown in Figs. 3.58, 3.59, 3.60, 3.61, 3.62, 3.63, 3.64, 3.65, 3.66 and 3.67, respectively:

(a) Single set of turnout slab (b) Crossover turnout slab

Fig. 3.57 Schematic diagram of turnout slab arrangement

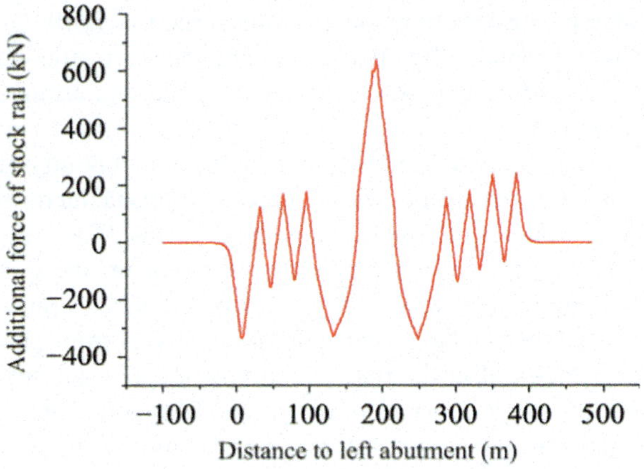

Fig. 3.58 Additional force on stock rail under condition I

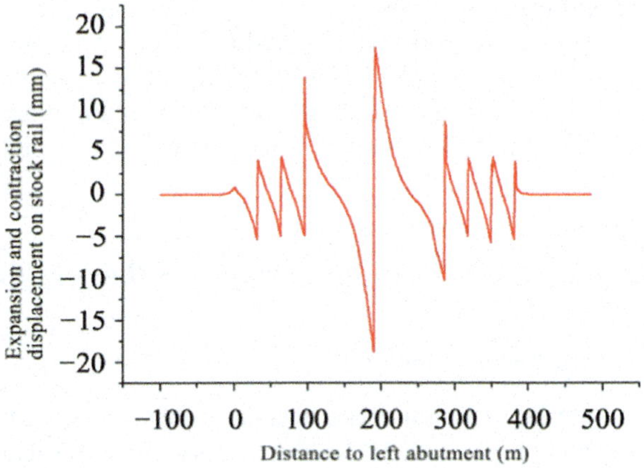

Fig. 3.59 Relative displacement between girder and rail under condition I

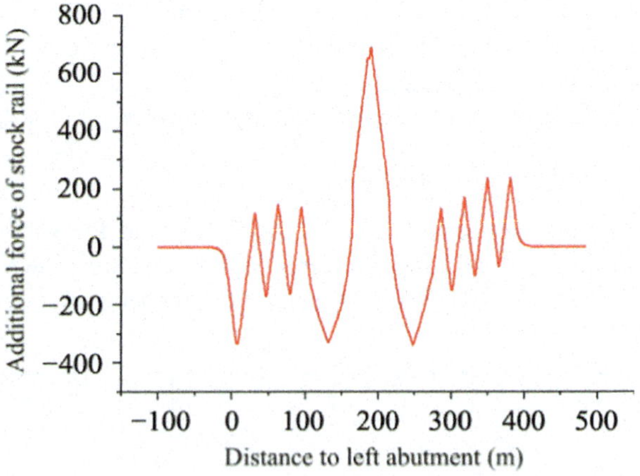

Fig. 3.60 Additional force on stock rail under condition II

Fig. 3.61 Relative displacement between girder and rail under condition II

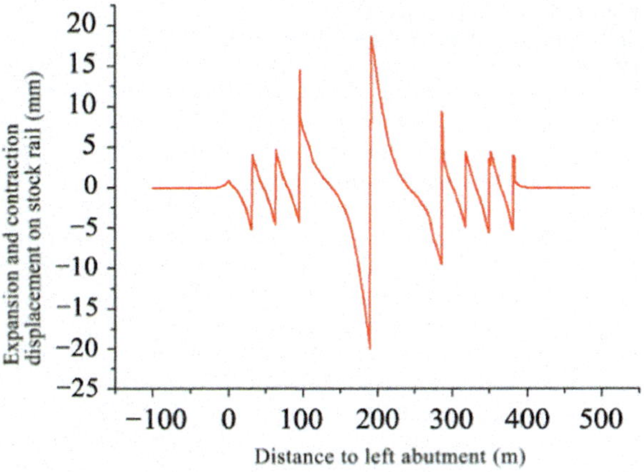

Fig. 3.62 Additional force on stock rail under condition III

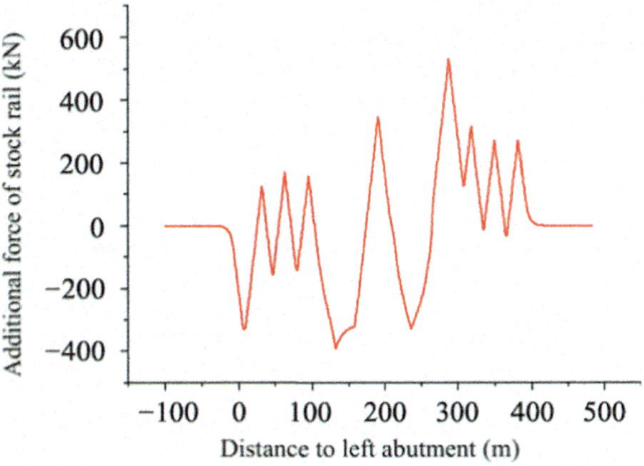

Fig. 3.63 Relative displacement between girder and rail under condition III

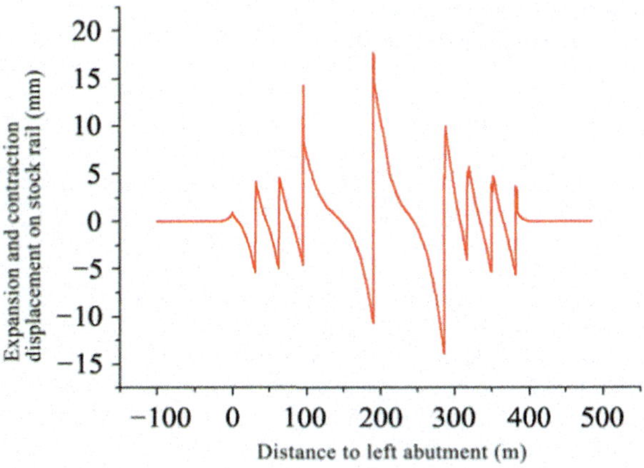

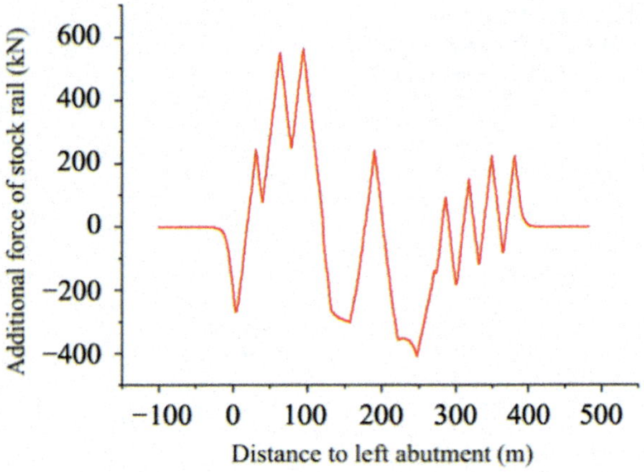

Fig. 3.64 Additional force on stock rail under condition IV

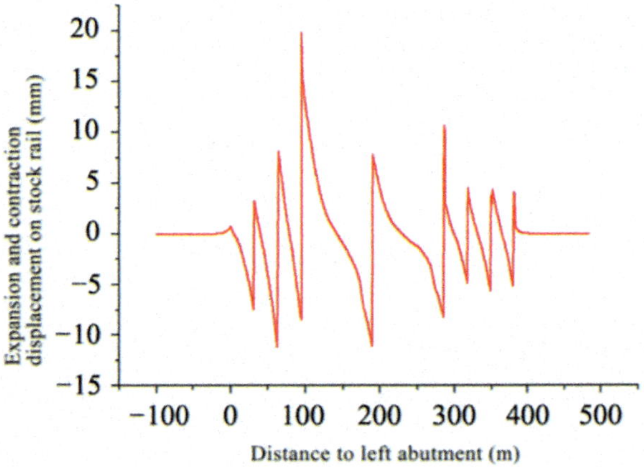

Fig. 3.65 Relative displacement between girder and rail under condition IV

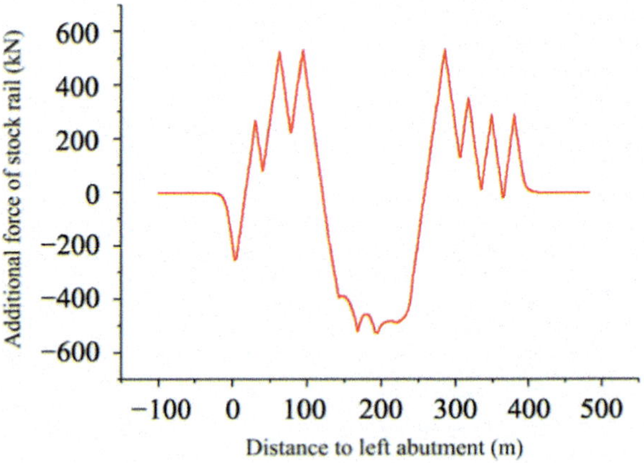

Fig. 3.66 Additional force on stock rail under condition V

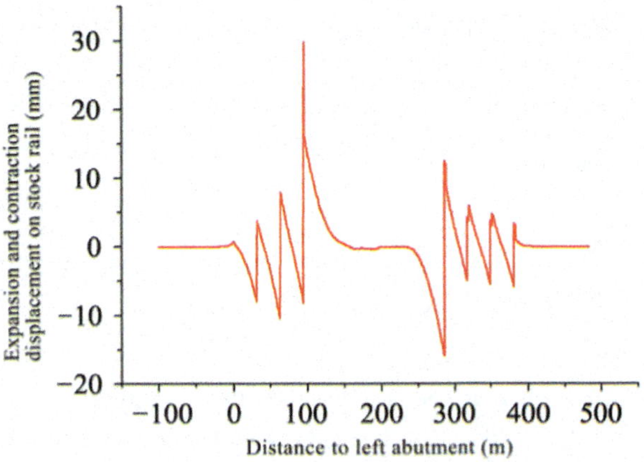

Fig. 3.67 Relative displacement between girder and rail under condition V

From a comparison among the above five conditions, the calculation results are shown in Table 3.9.

Table 3.9 reveals that among the different forms, only the force transfer devices at the heel end of the frogs in the form of single crossovers are not subjected to stress. On the other hand, the force transfer devices are most stressed in the turnout groups under the opposite arrangement.

The maximum stress on the spacing block under the back-passing arrangement is the largest among all forms, and that within single crossovers is the smallest. Under the same connection form, the maximum stress on the spacing block is greater in the context of opposite directions compared to the context of the same direction.

The maximum displacement at the actual point of the straight switch rail relative to the turnout sleeper in the single crossover form is the largest among all forms, whereas it is the smallest in the opposite arrangement. Under the same connection form, there is little difference in the relative displacement of the switch rails between the contexts of opposite directions and the same direction.

The maximum displacement at the actual point of the point rails relative to the turnout sleepers varies slightly among the five forms of connection.

The maximum additional force on the stock rails is the greatest in the opposite arrangement, while it is the smallest in the back-passing arrangement in opposite directions. Under the same connection form, the maximum additional force on the stock rail in the context of the same direction is greater compared to the context of opposite directions.

The maximum longitudinal force on the piers and abutments in the form of single crossovers is the largest among all forms, appearing on the left abutment. On the contrary, the smallest occurs in the opposite arrangement in opposite directions. The maximum longitudinal force on the piers and abutments occurs at the left abutment in the opposite arrangement, whereas it occurs at the No. 7 pier in the back-passing arrangement.

Table 3.9 Computation results

Calculation conditions		I	II	III	IV	V
Maximum stress on force transfer device at heel end of switch rail (kN)	No. 1 turnout	93.723	107.603	0	−53.862	0
	No. 2 turnout	−77.220	−87.173	61.047	0	0
Maximum stress on spacing block at end of wing rail (kN)	No. 1 turnout	136.345	135.333	131.434	−151.897	−100.844
	No. 2 turnout	−147.544	−144.147	164.923	−133.183	104.514
Displacement at actual point of straight switch rail relative to turnout sleeper (mm)	No. 1 turnout	11.086	10.822	13.104	−13.078	−12.343
	No. 2 turnout	−11.435	−11.136	12.031	−12.908	16.851
Displacement at actual point of point rail relative to turnout sleeper (mm)	No. 1 turnout	5.549	5.485	5.108	−6.720	−2.793
	No. 2 turnout	−6.001	−5.795	8.405	−4.988	5.541
Maximum additional force on stock rail		639.450	690.538	532.098	563.052	533.235
Left abutment		−222.336	−325.791	−333.629	−214.576	−443.091
No. 1 pier		−11.025	−8.418	−16.330	−76.520	−128.911
No. 2 pier		−11.951	−9.812	−17.718	−77.106	−130.183
No. 3 pier		0	0	0	0	0
No. 4 pier		219.867	281.762	248.334	−406.336	0
No. 5 pier		0	0	0	0	0
No. 6 pier		0	0	0	0	52.340
No. 7 pier		−213.811	−272.553	378.577	−407.434	0
No. 8 pier		0	0	0	0	0
No. 9 pier		−1.539	−15.937	79.640	−52.409	166.549
No. 10 pier		−14.231	−23.064	9.549	−41.595	34.648
No. 11 pier		−14.908	−24.079	9.193	−43.106	33.561
Right abutment		0	0	0	0	0

3.6 Study on Stress and Deformation Patterns of CWR Turnouts on Bridges

3.6.1 Influence of Span Length on Stress and Deformation of CWR Turnouts

1. Influence of Simply-supported Girder Span Length on CWR Turnouts

(1) Calculation conditions

Different spans may be chosen in the bridge design. This section presents a comparative analysis of CWR turnouts on simply-supported girder bridges with four spans: 24, 32, 48, and 64 m. The turnout bridge arrangement is shown in Fig. 3.68. The temperature rise range of 50 °C is considered for the CWR turnouts and 15°C for the bridges.

(2) Expansion and contract force on stock rail

Figure 3.69a illustrates the expansion and contract force on stock rails within CWR turnouts on simply-supported girder bridges with varying spans. In the figure, positive values indicate compressive force on the rails. The expansion and contract force peaks of the 24, 32, and 48 m simply-supported girders all occur at the heel end of the switch rails. On the 24 m girder, the peak of the expansion and contract force on stock rails is relatively significant. This is due to the coinciding of the girder joint with the heel end of the switch rail, resulting in the superposition of two temperature force peaks. However, the 64 m girder differs from the other three. Since this girder is relatively long, the entire turnout can be accommodated over a single girder span. Consequently, the peak of the expansion and contract force on stock rails occurs at the girder end instead of the heel end of the switch rail. This arrangement proves advantageous in maintaining the geometric state of the turnout.

The calculations show that the maximum additional compressive force on the stock rails of CWR turnouts on these four types of simply-supported girders is 243.6 kN, 235.3 kN, 239.5 kN, and 266.1 kN, respectively. As the girder span length increases, there is a slight rise in the peak of the expansion and contract force on stock rails. This minimal increase is primarily due to the bridge expansion occurring in the opposite direction of the inner rail expansion and contraction within the turnout. Furthermore, to avoid placing the switch and frog across the girder joint, the center of the turnout is positioned near the girder joint. Consequently, expansion and

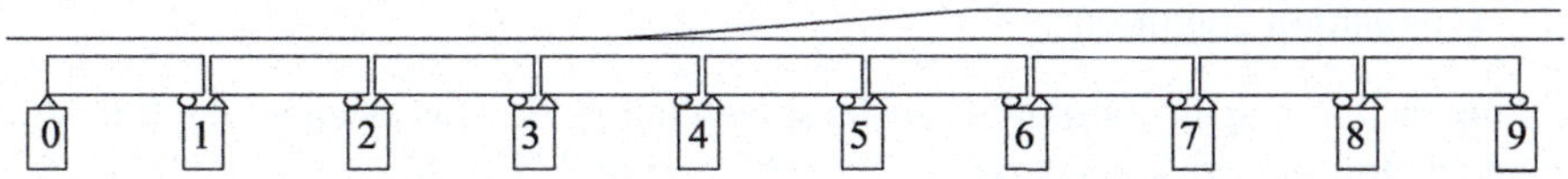

Fig. 3.68 CWR turnout across different bridge spans

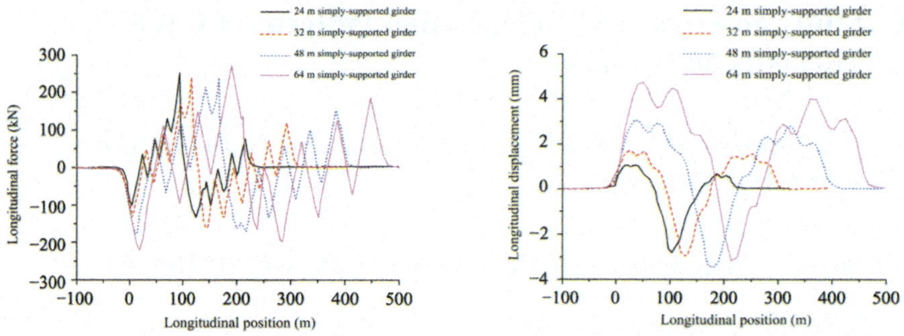

(a) Additional expansion and contraction force (b) Expansion and contraction displacement

Fig. 3.69 Additional expansion and contraction force and displacement of stock rail

contraction interactions between the girder and turnout take place at similar locations, resulting in similar expansion and contraction displacement. This is consistent with the fact that the expansion and contract force on stock rails does not increase with the rise in temperature difference experienced by the bridge within a single day.

(3) Expansion and contraction displacement of stock rail

The displacement of the stock rails within CWR turnouts on simply-supported girders with different spans is shown in Fig. 3.69b, where positive values represent rail displacement toward the right. The expansion and contraction displacement of the inner rails within the turnouts occurs toward the left, and similarly, the stock rails within the turnout section also experience leftward expansion and contraction displacement. Therefore, the maximum value shows little variations across different girder spans, indicating that the displacement is primarily attributed to the expansion and contraction of the inner rails. Furthermore, most of the resistance elements have surpassed the displacement (generally 2 mm) corresponding to the resistance limit of the track bed, resulting in no further increase in their displacement as the bridge expands and contracts, even when considering factors such as increased bridge temperature changes and elongation of the girder span. The rightward expansion and contraction displacement of the bridge preceding and following the turnout drives the expansion and contraction displacement of the rails to the right, following the same change law observed in CWR tracks on regular bridges, where the displacement increases with the lengthening of the girder span or temperature span.

2. Influence of Continuous Girder Span Length on CWR Turnouts

(1) Calculation conditions

Taking the three-span continuous girder shown in Fig. 3.70 as an example, four span length conditions are considered: (32 + 32 + 32) m, (32 + 48 + 32) m, (48 + 48 + 48) m, and (48 + 60 + 48) m (referred to as Conditions I to IV, respectively). A No.18 CWR turnout is located in the middle of the continuous girder, with equal distances from the head and tail of the turnout to both ends of the continuous girder, measuring 13.5 m, 21.5 m, 37.5 m, and 43.5m, respectively. Moreover, the fixed bearing of the continuous girder is located at the front end of the turnout, while 3-span 32 m simply-supported girders are arranged on both sides.

(2) Computation Results

See Table 3.10 for a comparison of stress and deformation calculation results for the CWR turnout and piers/abutments on the continuous girder bridge with ballasted tracks for different span lengths. In the table, positive values represent the rail displacement and longitudinal force on the piers/abutments toward the right.

Based on Table 3.10, as the continuous girder increases in span length, all the following parameters rise: the additional expansion and contraction force on the rails at the girder end, the additional tensile force on the broken rail gap and non-broken rails, and the longitudinal force on the piers on the continuous girder. Similar trends can be observed for CWR tracks on regular bridges, both due to the increased bridge temperature span. Conversely, the span length of the continuous girder has an inverse relationship with the longitudinal force on the force transfer devices within the turnouts and the displacement of the switch rails and point rails relative to the turnout sleepers. This is mainly due to the fact that the increase in the span length of continuous girders results in the head and tail of the turnout moving farther away from the girder end, causing the center of the turnout to shift rightwards from the fixed bearing of the continuous girders. As a consequence, the expansion and contraction displacement of the continuous girder acting on the turnout toward the right becomes more prominent, exerting an increasingly significant impact in offsetting the leftward expansion and contraction of the inner rails within the turnout. However, it is important to note that under such circumstances, the longitudinal force acting on the piers of the continuous girder experiences considerable amplification.

3. Calculation conclusions

From a computational analysis on the influence of girder length on the stress and deformation of CWR turnouts on bridges, using various lengths of simply-supported and continuous girders, the following conclusions can be drawn:

(1) The relative displacement between the sleepers and rails at the point rails and switch rails within CWR turnouts on simply-supported girder bridges is related to their positions on the girder. When the heel end of the switch rails and point rails is near a girder joint, the relative displacement between the stock rails of

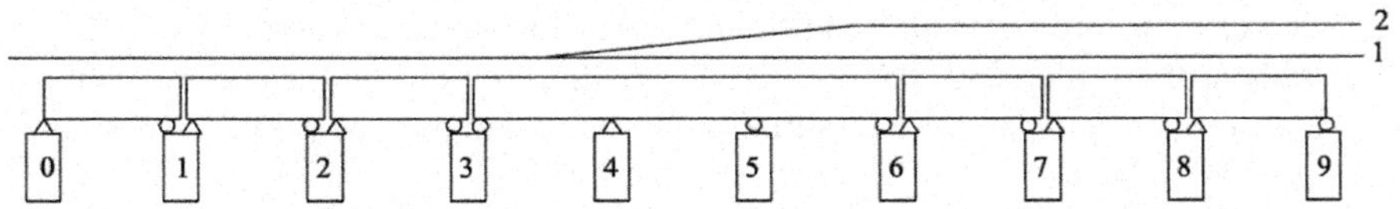

Fig. 3.70 Schematic diagram of turnout-bridge arrangement

Table 3.10 Comparison of calculation results

Calculation conditions	Condition I	Condition II	Condition III	Condition IV
Maximum additional expansion and contraction force on straight stock rail (kN)	266.7	261.9	285.2	286.2
Maximum expansion and contraction displacement of straight stock rail (mm)	−6.1	−6	−7.4	−7.4
Maximum stress on force transfer device at heel end of switch rail (kN)	39.1	27.2	8.7	8.4
Maximum stress on spacing block at end of wing rail (kN)	144.1	141.6	132.3	129.9
Displacement at actual point of straight switch rail relative to turnout sleeper (mm)	−17.7	−17.1	−16.3	−16.1
Displacement at actual point of point rail relative to turnout sleeper (mm)	−6.7	−6.5	−6	−5.9
Longitudinal force on left abutment (kN)	−294.6	−293.2	−298.7	−299
Longitudinal force on No. 1 pier (kN)	−134.7	−132.1	−144.4	−145.1
Longitudinal force on No. 2 pier (kN)	−262.8	−257	−282.9	−284.2
Longitudinal force on No. 4 pier (kN)	−1186.2	−1350.3	−1500.7	−1609.4
Longitudinal force on No. 6 pier (kN)	−259.9	−170.6	−59.3	1.6
Longitudinal force on No. 7 pier (kN)	−150.2	−115.7	−70.1	−44
Longitudinal force on No. 8 pier (kN)	−170.2	−157.4	−140	−130.2
Broken rail gap of straight stock rail at girder end preceding turnout (mm)	63.4	64.7	65.2	65.2
Maximum additional tensile force on non-broken rail (kN)	438	437.5	469.4	478.4

the CWR turnouts and the bridges increases, leading to higher stress on the force transfer devices.

(2) The longitudinal force on the piers and abutments within the CWR turnout section on simply-supported girder bridge is significantly influenced by the expansion and contraction of the inner rails within the turnouts, which is considerably greater compared to the section preceding and following the turnouts. As the bridge span increases, the stress on all the piers and abutments within the turnout section, as well as those preceding and following the turnouts, also increases.

(3) As the span length of the continuous girders increases, several parameters increase accordingly: the additional expansion and contraction force on the rails at the girder end, the additional tensile force on the broken rail gap and non-broken rails, and the longitudinal force on the piers within the continuous girders.

(4) As the span length of the continuous girders increases, the longitudinal force on the force transfer devices of the turnouts and the displacement of the switch rails and point rails relative to the turnout sleepers decrease.

3.6.2 Influence of Fixed Bearing Arrangement on Stress and Deformation of CWR Turnouts (Turnout Groups)

1. Influence of fixed bearing arrangement on simply-supported girder on stress and deformation of single turnout

(1) Calculation conditions

The influence of different bearing arrangements on the stress and deformation of CWR turnouts on bridges was studied using an example with a single No. 18 turnout laid on a 9-span 32 m simply-supported girder, as shown in Fig. 3.68. In this arrangement, the left side of each simply-supported girder span is equipped with a fixed bearing, while the right side features a movable bearing. This arrangement causes the expansion and contraction displacement of the inner rails within the turnout to occur in a direction opposite to that of the bridge. This specific bearing arrangement is referred to as Condition I.

Condition II: The turnout bridge arrangement remains consistent with Fig. 3.68, with the exception that the left side of each simply-supported girder span is equipped with a movable bearing, while the right side features a fixed bearing. As a result, the expansion and contraction displacement of the inner rails within the turnout occurs in the same direction as that of the bridge.

(2) Expansion and contract force on stock rail

A comparison of the expansion and contraction force on stock rails within CWR turnouts on bridges under these two bearing arrangements is shown in Fig. 3.71, where positive values represent compressive force on the rails.

It can be seen from the figure that longitudinal forces are distributed on stock rails within the turnout section following the exactly same law, indicating that the expansion and contraction of inner rails within the turnout section acts as the main contributing factor. Even when the bridge undergoes expansion and contraction displacements in the same direction as the inner rails, the girder-rail relative displacement within the turnout section does not vary significantly (Fig. 3.72). As a result, equivalent additional forces are transmitted from the inner rails to the stock rails through the track bed. The additional expansion and contraction force is distributed on rails of CWR tracks on regular bridges preceding and following turnouts in contrast. Specifically, under Condition I, the maximum additional compressive force occurs at the right end of the bridge, while under Condition II, the maximum additional compressive force occurs at the left abutment. Condition II is only influenced by bridge expansion due to a significant distance from the turnout section, aligning the

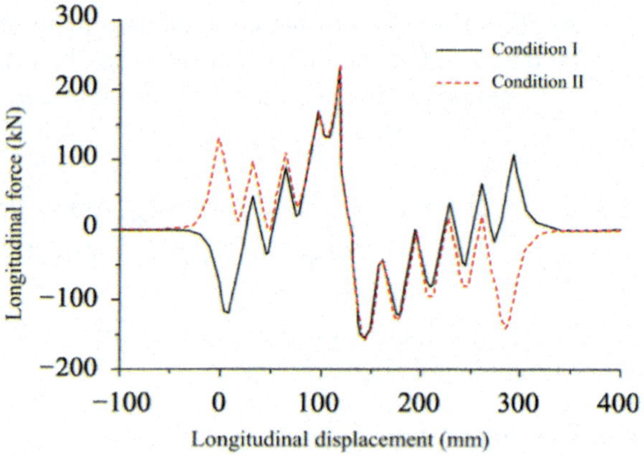

Fig. 3.71 Comparison of additional expansion and contraction force on stock rail

distribution rules of additional expansion and contraction forces on rails within CWR tracks on regular bridges.

(3) Expansion and contraction displacement between stock rail and bridge

The expansion and contraction displacement between the rails within CWR turnouts on bridges and the bridge itself under different bearing arrangements is shown in Fig. 3.73, where positive values represent rightward displacement of the rails and the bridge. The figure reveals that when the bridge's expansion and contraction displacement aligns with that of the inner rails within the turnouts in direction, the stock rails experience a leftward expansion and contraction displacement, and the maximum expansion and contraction displacement reaches 7.7 mm, which is far greater than the 3.0 mm under Condition I. This arrangement significantly contributes to noticeable longitudinal creeping of the CWR turnouts, thus negatively impacting their stability. Additionally, the deflection displacement of the rails is also leftward when subjected to train loads. Although this displacement is typically small, it can still lead

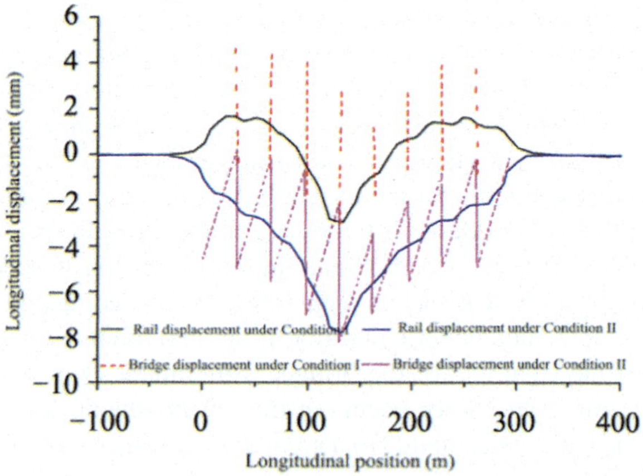

Fig. 3.72 Comparison of expansion and contraction displacement between stock rail and bridge

to creeping of the turnouts over time. When the expansion and contraction displacement and deflection displacement are combined, the turnouts become less stable in the longitudinal direction. Therefore, it is advisable to avoid bearing arrangements that result in co-directional expansion and contraction of the turnout and bridge whenever possible.

(4) Other calculation results

See Table 3.11 for a comparison of calculation results for a series of parameters for various bearing arrangements on simply-supported girders: the relative displacement between the sleepers and rails at the actual points of the switch rails and point rails, longitudinal force on the force transfer devices, and longitudinal force on the piers and abutments. In the table, positive values represent the rightward rail displacement and longitudinal force on the piers and abutments.

Based on the calculations, the positioning of the fixed bearing on the right and left sides of simply-supported girders has minimal impact on the expansion and contraction displacement of the inner rails within the turnout and the stress on the force transfer devices. This pattern aligns with the additional expansion and contraction force experienced by the stock rails. In the scenario where the straight stock rail at the girder end preceding the turnout is broken, there is a slight increase in the additional tensile force on the broken rail gap and the non-broken rail. In this context, when the fixed bearing is positioned on the right side, the right abutment bears the additional expansion and contraction force from all four rails. On the other hand, when the fixed bearing is positioned on the left side, the left abutment only experiences the additional expansion and contraction force from two rails. This results in a significant difference in stress between the two abutments. From the perspective of the longitudinal force transmitted from each girder span to the pier where the fixed bearing is located, when the fixed bearing is positioned on the right side and the expansion and contraction direction of the bridge and the inner rails within the turnout is the same, the longitudinal force borne by the pier is reduced compared to when the fixed bearing is positioned on the left side. This particular arrangement proves advantageous to the stress distribution of the piers.

2. Influence of fixed bearing arrangement on continuous girder on stress and deformation of single crossovers

The influence of different bearing arrangements on the stress and deformation of single crossovers on bridges was analyzed, taking a single crossover on a 5 × 32 m continuous girder bridge shown in Fig. 3.73 as an example. In this arrangement,

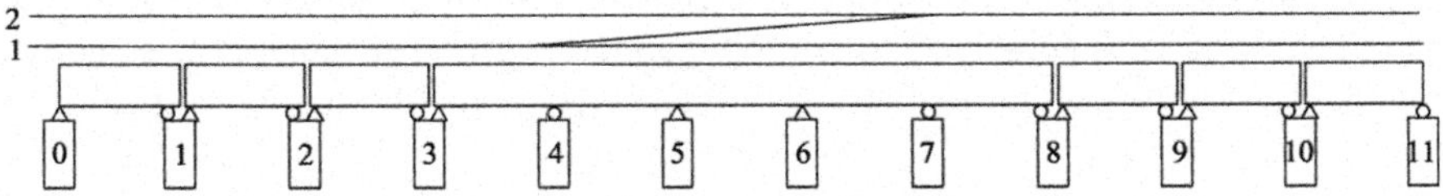

Fig. 3.73 Span arrangement of 35 × 32 m continuous girder

Table 3.11 Comparison of calculation results

Bearing arrangement of simply-supported girder	Fixed bearing on left side	Fixed bearing on right side
Maximum stress on stopper at heel end of switch rail (kN)	130.8	131.4
Maximum stress on spacing block at end of wing rail (kN)	156.8	156.6
Displacement at actual point of straight switch rail relative to turnout sleeper (mm)	−19.4	19.4
Displacement at actual point of point rail relative to turnout sleeper (mm)	−7.5	7.5
Longitudinal force on no. 0 abutment (kN)	−273	0
Longitudinal force on no. 1 pier (kN)	−83.1	67.4
Longitudinal force on no. 2 pier (kN)	−151.4	21.5
Longitudinal force on no. 3 pier (kN)	−424.7	122.8
Longitudinal force on no. 4 pier (kN)	−685.2	−414.9
Longitudinal force on no. 5 pier (kN)	−415.6	−674.6
Longitudinal force on no. 6 pier (kN)	−178.2	−383.6
Longitudinal force on no. 7 pier (kN)	−122.5	−132
Longitudinal force on no. 8 pier (kN)	−161.3	−14.6
Longitudinal force on no. 9 abutment (kN)	0	485
Broken rail gap of straight stock rail at girder end preceding turnout (mm)	59.8	61.1
Maximum additional tensile force on non-broken rail (kN)	310.9	314.9

the left side of each simply-supported girder span is equipped with a fixed bearing, while the right side features a movable bearing. The fixed bearing of the continuous girder is located at the left end (No. 5 pier) of the middle span of the continuous girder. This specific bearing arrangement is referred to as Condition I.

Condition II: The turnout bridge arrangement remains consistent with Fig. 3.68, and the bearing arrangement of the simply-supported girder remains consistent with Fig. 3.70, with the exception that the fixed bearing of the continuous girder is located at the right end of the middle span of the continuous girder (No. 6 pier).

Condition III: The turnout bridge arrangement remains consistent with Fig. 3.68, and the bearing arrangement of the simply-supported girder remains consistent with Fig. 3.73, with the exception that both left and right ends of the middle span of the

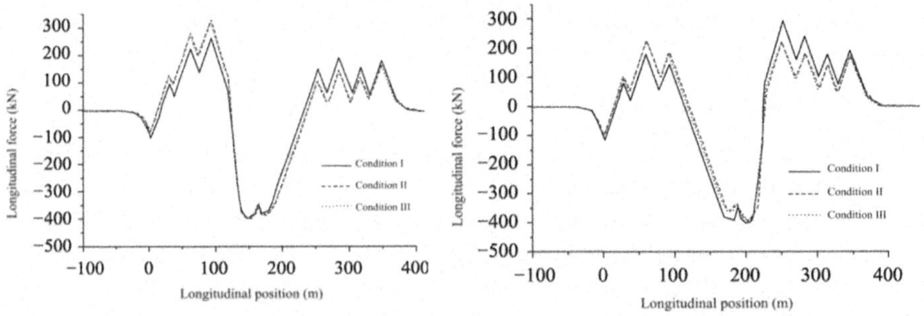

(a) No.1 Rail (b) No.2 Rail

Fig. 3.74 Additional expansion and contraction force on stock rail

continuous girder are equipped with fixed bearings (i.e. double fixed bearings or continuous rigid frame).

(1) Expansion and contraction force on stock rails

A comparison of the expansion and contraction force of the No. 1 and 2 rails within single crossovers on bridges under different bearing arrangements is shown in Fig. 3.74, where positive values represent compressive force on the rails.

As seen in the figure, the maximum expansion and contraction force on No.1 rail occurs at the left end of the continuous girder under Conditions I, II, and III. For No. 2 rail, the maximum expansion and contraction force occurs at the right end of the continuous girder under Conditions I and III, but at the left side of the continuous girder and the left end of the simply-supported girder under Condition II. This differs from CWR tracks on regular bridges, where the maximum additional expansion and contraction force typically occurs at the position with the largest temperature span of the continuous girder. The reason lies in the fact that the additional temperature force on CWR turnouts on bridges is a result of both the expansion and contraction of the inner rails within CWR turnouts and the bridge itself. In cases where the expansion and contraction of the inner rails within CWR turnouts play a major role, the maximum expansion and contraction force occurs at the end of the continuous girder, specifically the front end of the corresponding turnout.

The additional expansion and contraction force on No. 1 rail is the minimum under Condition I while it reaches the maximum under Condition II. The additional expansion and contraction force on No. 2 rail is the maximum under Condition I and it reaches the minimum under Condition II. The maximum additional expansion and contraction force on No. 1 rail under Condition II surpasses that on No. 2 rail under Condition I, which is mainly due to the different bearings utilized at both ends of the continuous girder, and the temperature span preceding No. 1 turnout under Condition II is greater than the temperature preceding No. 2 turnout under Condition I, leading to an increase in the additional expansion and contraction force. It is revealed that the fixed bearing arrangement for the continuous girder and the simply-supported girders at its both ends has a great influence on the turnout-bridge interactions. Thus, it is crucial to minimize the temperature span preceding the turnouts in practical applications.

(2) Other calculation results

See Table 3.12 for a comparison of calculation results for a series of parameters for various fixed bearing arrangements on continuous girders: the relative displacement between the sleepers and rails at the actual points of the switch rails and point rails, longitudinal force on the force transfer devices, and longitudinal force on the piers and abutments. In the table, positive values represent the rightward rail displacement and longitudinal force on the piers and abutments.

The calculations show that the longitudinal force on the pier with a fixed bearing of the continuous girder in Condition I is lower than that of the pier preceding the

Table 3.12 Comparison of calculation results

Calculation conditions		Condition I	Condition II	Condition III
Maximum stress on force transfer device at heel end of switch rail (kN)	No. 1 turnout	0	0	0
	No. 2 turnout	0	0	0
Maximum stress on spacing block at end of wing rail (kN)	No. 1 turnout	−111.086	−108.605	−108.077
	No. 2 turnout	110.45	111.622	112.232
Displacement at actual point of straight switch rail relative to turnout sleeper (mm)	No. 1 turnout	−18.291	−19.233	−19.22
	No. 2 turnout	18.958	17.779	17.982
Displacement at actual point of point rail relative to turnout sleeper (mm)	No. 1 turnout	−5.214	−5.119	−5.089
	No. 2 turnout	5.177	5.279	5.271
Longitudinal force on left abutment (kN)		−207.296	−236.017	−230.704
Longitudinal force on no. 1 pier (kN)		−56.04	−70.388	−68.041
Longitudinal force on no. 2 pier (kN)		−56.409	−70.709	−68.376
Longitudinal force on no. 3 pier (kN)		0	0	0
Longitudinal force on no. 4 pier (kN)		0	0	0
Longitudinal force on no. 5 pier (kN)		−3.699	0	−210.061
Longitudinal force on no. 6 pier (kN)		0	108.528	151.461
Longitudinal force on no. 7 pier (kN)		0	0	0
Longitudinal force on no. 8 pier (kN)		−2.815	0	146.023
Longitudinal force on no. 9 pier (kN)		52.851	28.224	32.898
Longitudinal force on no. 10 pier (kN)		13.925	11.788	12.178

turnout on the simply-supported girder. From a comparison between Condition III and Conditions I and II, it is observed that the implementation of symmetrical double fixed piers for the continuous girder within the single-crossover turnout range does not offer significant advantages over the single fixed pier option. This is due to the fact that adopting double fixed piers does not effectively mitigate the stress and deformation of CWR turnouts on bridges and actually increases the longitudinal force borne by the pier of the continuous girder. Consequently, before applying the option of double fixed piers for continuous girders, it is crucial to conduct a comprehensive comparison. This option does not always prove effective in reducing the turnout-bridge interactions under all conditions. Its effectiveness is observed only when two fixed piers serve the same function as a single fixed pier, contributing to enhancing the overall rigidity of the continuous girder. In this case, mitigating the expansion and contraction displacement in the same direction as that of the turnouts has a positive impact on reducing the turnout-bridge interactions.

Through comprehensive analysis, the fixed bearing arrangement under Condition I is the most reasonable compared with the other two.

3. Calculation Conclusions

For the design of the fixed bearing arrangement, it is advisable to avoid bearing arrangements that result in co-directional expansion and contraction of the turnout and bridge whenever possible and reduce the temperature span preceding the turnouts. Additionally, it is recommended to consider the impact on the stress of the piers and abutments.

3.6.3 Influence of Pier Rigidity on Stress and Deformation of CWR Turnouts

1. Calculation conditions

The bridge-turnout arrangement is shown in Fig. 3.68. Apart from the aforementioned computation parameters, the calculations incorporate the following three longitudinal horizontal rigidity values of piers to study their influence on the stress and deformation of CWR turnouts on bridges.

Condition I: The longitudinal horizontal rigidity per track of abutments on both sides is 1×10^7 kN/m, while that of other piers is 1×10^5 kN/m;

Condition II: The longitudinal horizontal rigidity per track of abutments on both sides is 1×10^7 kN/m, while that of other piers is 1×10^6 kN/m;

Condition III: The longitudinal rigidity per track of No. 0 and No. 9 abutments is 1×10^7 kN/m, that of No. 3, No. 4, and No. 5 piers is 1×10^6 kN/m, and that of No. 1, No. 2, No. 6, No. 7, and No. 8 piers is 1×10^5 kN/m. This implies that the rigidity of piers positioned within the CWR turnout range is 10 times that of other piers.

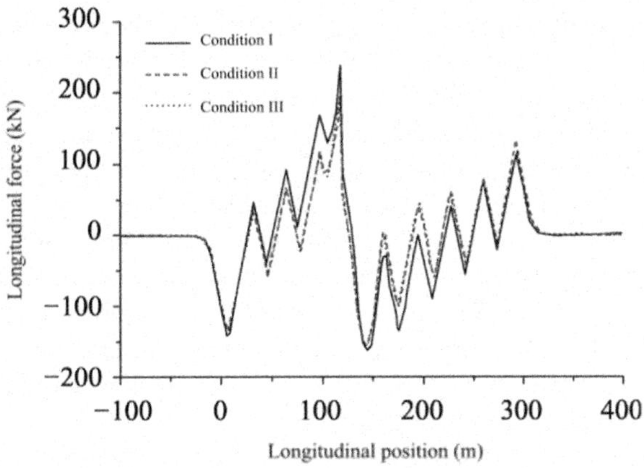

Fig. 3.75 Additional expansion and contraction force on stock rail under different pier rigidity values

2. Additional expansion and contraction force on stock rail

The comparison of the additional expansion and contraction force on stock rails of CWR turnouts on bridges with varying pier rigidity is shown in Fig. 3.75, where the longitudinal force on rails is expressed as a positive value for compression. Across the three conditions, the maximum additional expansion and contraction force on straight stock rails of turnouts appears at the same position, measuring 235.3 kN, 197.2 kN, and 195.4 kN, respectively. It is evident that with an increase in pier rigidity within the turnout range, the additional expansion and contraction force on the stock rails decreases, while alterations in pier rigidity outside the turnout range yield little influence on the additional expansion and contraction force on the CWR turnout stock rails. This trend is mainly due to increasing pier rigidity within the turnout section, which can mitigate the expansion and contraction displacement of the turnout rail (Fig. 3.76) and reduce additional forces transmitted from the inner rail to the stock rail. Additional expansion and contraction forces on rails following turnouts at girder joints of CWR tracks on regular bridges increase with escalating pier rigidity, which is consistent with the change law of CWR tracks on bridges.

3. Expansion and contraction displacement of stock rail

The comparison of the expansion and contraction displacement of stock rails within CWR turnouts on bridges with varying pier rigidity is shown in Fig. 3.76, where rail displacement to the right is expressed in a positive value. Under the three conditions, the expansion and contraction displacement of the straight stock rails at the heel end of the switch rails is −3.0 mm, −0.77 mm, and −0.77 mm, respectively, and the maximum expansion and contraction displacement preceding and following the turnout is 1.60 mm, 2.34 mm, and 2.04 mm, respectively. It can be seen that an increase in pier rigidity within the turnout section contributes to mitigating overall turnout creep, thereby maintaining the stability and geometry of the turnout. Consequently, it is recommended to enhance the longitudinal horizontal rigidity of the pier and abutment to the greatest extent possible in the design process.

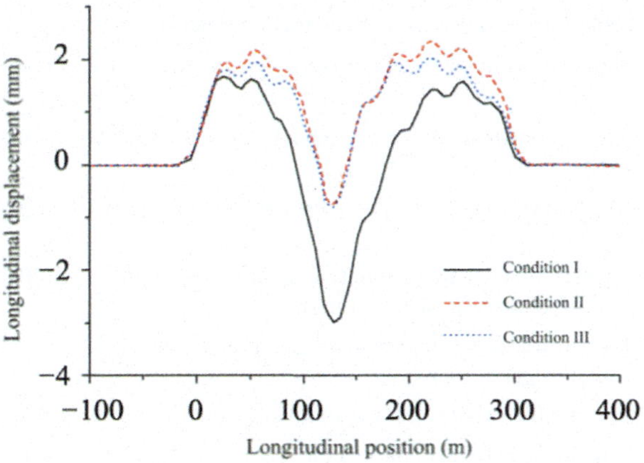

Fig. 3.76 Expansion and contraction displacement of stock rail under different pier rigidity values

4. Other calculation results

See Table 3.13 for a comparison of calculation results for a series of parameters for different longitudinal horizontal rigidity values of the piers/abutments on simply-supported girders: the relative displacement between the sleepers and rails at the actual points of the switch rails and point rails, longitudinal force on the force transfer devices, and longitudinal force on the piers and abutments. In the table, positive values represent the rightward rail displacement and longitudinal force on the piers and abutments.

As shown in Table 3.13, an increase in the rigidity of all piers/abutments of the bridge helped alleviate stress on the stopper at the heel end of the switch rails and reduced expansion and contraction displacement at the actual point of the switch rails. Furthermore, under this circumstance, the additional tensile force on the broken rail gap and non-broken rail in the event of a broken rail diminished. However, this increase in rigidity also resulted in a corresponding rise in longitudinal force on piers and abutments within the turnout section. If only the rigidity of piers within the turnout section is increased, the impact on expansion and contraction displacement of the inner rails, stress on force transfer devices, and broken rail gaps remains the same as when all piers and abutments of the bridge are made more rigid. Consequently, this leads to a more uneven distribution of longitudinal forces among the piers/abutments throughout the entire bridge. Specifically, the longitudinal force on piers within the turnout section increases, while the longitudinal force on piers preceding and following the turnout decreases. Taking economic efficiency into consideration, it is recommended to appropriately enhance the longitudinal rigidity of piers/abutments within the turnout section.

Table 3.13 Comparison of calculation results

Rigidity of pier/abutment of simply-supported girder	Condition I	Condition II	Condition III
Maximum stress on stopper at heel end of switch rail (kN)	130.8	120.0	119.2
Maximum stress on spacing block at end of wing rail (kN)	156.8	162.6	162.8
Displacement at actual point of straight switch rail relative to turnout sleeper (mm)	−19.4	−18.9	−18.9
Displacement at actual point of point rail relative to turnout sleeper (mm)	−7.5	−7.8	−7.9
Longitudinal force on no. 0 abutment (kN)	−273.0	−249.6	−260.2
Longitudinal force on no. 1 pier (kN)	−83.1	−65.7	−55.7
Longitudinal force on no. 2 pier (kN)	−151.4	−107.6	−83.9
Longitudinal force on no. 3 pier (kN)	−424.7	−441.0	−454.8
Longitudinal force on no. 4 pier (kN)	−685.2	−883.9	−892.7
Longitudinal force on no. 5 pier (kN)	−415.6	−440.8	−473.8
Longitudinal force on no. 6 pier (kN)	−178.2	−97.0	−74.7
Longitudinal force on no. 7 pier (kN)	−122.5	−83.2	−80.3
Longitudinal force on no. 8 pier (kN)	−161.3	−226.6	−145.8
Broken rail gap of straight stock rail at girder end preceding turnout (mm)	59.8	57.3	57.6
Maximum additional tensile force on non-broken rail (kN)	310.9	219.5	233.6

3.6.4 *Influence of Intermediate Tangent Length on Stress and Deformation of CWR Turnouts*

Known from the application experience of CWR turnouts, a shorter tangent between turnouts results in a more significant mutual influence between the two turnouts. In order to analyze the influence of different lengths of intermediate tangents on the longitudinal stress and deformation patterns of CWR turnouts on bridges, this section presents an analysis focusing on how different lengths of intermediate tangents affect the stress and deformation patterns of CWR turnouts. Two scenarios are considered: two turnouts on bridges in the back-passing arrangement in opposite directions, and two crossover CWR turnouts in the opposite arrangement in opposite directions (double crossovers).

1. Tangent between two turnouts in the back-passing arrangement in opposite directions

The main bridge in the middle is a 7 × 32m continuous girder, with a fixed bearing located at the No. 6 pier, and both ends are 3-span simply-supported girders, with a fixed bearing located on the left side of the girder. The turnouts on this bridge are in the back-passing arrangement in opposite directions, as shown in Fig. 3.77. Four intermediate tangents of different lengths (6.25 m, 12.5 m, 25 m, and 50 m) are laid between the two turnouts, respectively, with the midpoint of the intermediate tangent coinciding with the midpoint of the bridge. See Table 3.14 for a comparison of the calculation results. In the table, positive values for the expansion and contraction force on rails indicate compressive force and positive values for longitudinal force and displacement represent the rightward direction.

Based on the relative positioning between the turnouts and the bridge, it is observed that using a longer intermediate tangent brings the turnout closer to the end of the continuous girder. Table 3.14 shows that a longer intermediate tangent leads to increased force and larger displacements within the movable parts of the turnouts. This is primarily attributed to the linear expansion and contraction displacement of the bridge. As the positions move closer to the girder end, the expansion and contraction displacement intensifies, resulting in a more significant impact on the turnouts. Moreover, although the bridge temperature spans associated with different lengths of intermediate tangents remain unchanged, a longer intermediate tangent causes higher expansion and contraction force and increased displacement of rails at the continuous girder end due to the influence of the turnouts. The distribution law of longitudinal force on the piers and abutments indicates that a longer intermediate tangent reduces the impact of the turnouts on the fixed pier of the continuous girder but increases their impact on the simply-supported girders at both ends. However, the value change in the longitudinal force on the piers and abutments is marginal. In summary, the length of the intermediate tangents has minimal influence on both the turnouts and the bridge as a whole.

2. Intermediate tangent between two crossovers

The main bridge in the middle is an 11 × 32 m continuous girder, with a fixed bearing located at the No. 8 pier, and both ends are 3-span simply-supported girders, with a fixed bearing located on the left side of the girder. The turnout on this bridge is in the form of double crossovers, as shown in Fig. 3.78. Four intermediate tangents of different lengths (6.25 m, 12.5 m, 25 m, and 50 m) are laid between the two crossovers, respectively, with the midpoint of the intermediate tangent coinciding with the midpoint of the bridge. See Table 3.15 for a comparison of the calculation results. In the table, positive values for the expansion and contraction force on rails indicate compressive force and positive values for longitudinal force and displacement represent the rightward direction.

According to Table 3.15, focusing solely on the length changes of the intermediate tangent while keeping the bridge unchanged, a longer intermediate tangent leads to

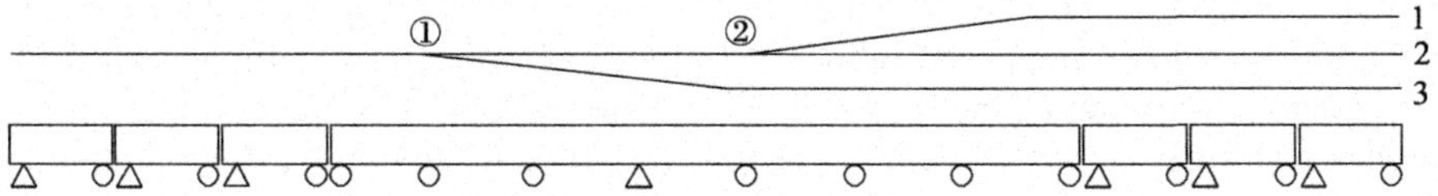

Fig. 3.77 Schematic diagram of bridge arrangement

Table 3.14 Comparison of calculation results for different intermediate tangent lengths

Length of intermediate tangent (m)		6.25	12.5	25.0	50.0
Maximum additional expansion and contraction force on No. 1 rail (kN)		422.5	426.7	437.6	466.6
Maximum displacement of No. 1 rail (mm)		−17.1	−17.3	−17.8	−18.6
Maximum stress on force transfer device at heel end of switch rail (kN)	No. 1 turnout	−16.5	−20.2	−33.0	−61.4
	No. 2 turnout	0.0	0.0	0.0	25.0
Maximum stress on spacing block at end of wing rail (kN)	No. 1 turnout	−125.3	−126.0	−126.0	−127.5
	No. 2 turnout	−132.9	−135.5	−140.5	−151.7
Displacement at actual point of straight switch rail relative to turnout sleeper (mm)	No. 1 turnout	−17.1	−17.4	−18.0	−19.4
	No. 2 turnout	−15.0	−15.1	−15.4	−16.1
Displacement at actual point of point rail relative to turnout sleeper (mm)	No. 1 turnout	−5.7	−5.8	−5.8	−5.9
	No. 2 turnout	−6.1	−6.2	−6.4	−7.1
Longitudinal force on left abutment (kN)		366.4	367.8	370.7	378.9
Longitudinal force on No. 1 pier (kN)		203.2	205.6	210.5	225.4
Longitudinal force on No. 2 pier (kN)		377.9	381.2	391.2	416.3
Longitudinal force on No. 6 pier (fixed pier of continuous girder) (kN)		3373.8	3353.6	3311.9	3177.6
Longitudinal force on No. 10 pier (kN)		−13.0	−6.1	4.5	47.7
Longitudinal force on No. 11 pier (kN)		73.6	76.3	80.3	97.7
Longitudinal force on No. 12 pier (kN)		189.0	190.2	190.7	196.9

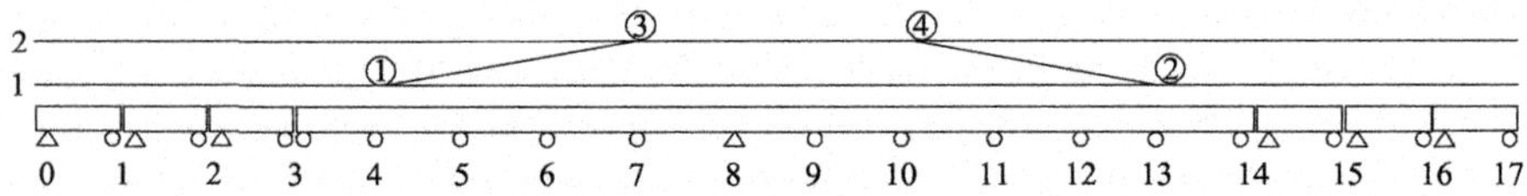

Fig. 3.78 Arrangement of bridge span with different intermediate tangents

reduced interactions between the turnouts. As the intermediate tangent increases in length, both No. ① and No. ② turnouts move closer to the end of the bridge. This proximity results in increased influence from the turnout stopper on the expansion and contraction force of the rail at the girder end, leading to higher expansion and contraction force and displacement of the No.1 rail. Additionally, the longitudinal force on the force transfer devices for No. ① and No. ② turnouts also increases. In

Table 3.15 Comparison of calculation results for different intermediate tangent lengths

Length of intermediate tangent (m)		6.25	12.5	25.0	50.0
Maximum additional expansion and contraction force on no. 1 rail (kN)		399.0	400.3	416.3	453.7
Maximum displacement at right end of no. 1 rail (mm)		20.4	20.8	21.3	22.3
Maximum stress on force transfer device at heel end of switch rail (kN)	No. ① turnout	−13.9	−15.3	−29.0	−60.4
	No. ② turnout	18.4	26.5	36.0	61.6
	No. ③ turnout	0.0	0.0	0.0	0.0
	No. ④ turnout	0.0	0.0	0.0	0.0
Maximum stress on spacing block at end of wing rail/kN	No. ① turnout	−124.8	−125.3	−125.6	−127.8
	No. ② turnout	125.9	126.1	126.5	127.8
	No. ③ turnout	124.0	124.3	124.2	130.7
	No. ④ turnout	−123.8	−123.6	−124.1	−130.7
Displacement at actual point of straight switch rail relative to turnout sleeper (mm)	No. ① turnout	−17.0	−17.1	−17.8	−19.3
	No. ② turnout	17.2	17.4	18.1	19.3
	No. ③ turnout	14.9	14.9	14.9	15.0
	No. ④ turnout	−14.9	−14.9	−14.9	15.0
Displacement at actual point of point rail relative to turnout sleeper/mm	No. ① turnout	−5.7	−5.7	−5.8	−5.9
	No. ② turnout	5.8	5.8	5.8	5.9
	No. ③ turnout	6.4	6.5	6.7	6.8
	No. ④ turnout	−6.4	−6.5	−6.7	−6.8
Longitudinal force on left abutment (kN)		709.4	710.9	714.4	722.3
Longitudinal force on no. 1 pier (kN)		366.8	367.8	375.7	388.5
Longitudinal force on no. 2 pier (kN)		684.9	687.1	699.8	722.1
Longitudinal force on no. 8 pier (fixed pier of continuous girder) (kN)		52.3	65.4	58.8	49.9
Longitudinal force on no. 14 pier (kN)		−636.5	−644.9	−653.8	−672.5

(continued)

Table 3.15 (continued)

Longitudinal force on no. 15 pier (kN)	−273.6	−279.1	−283.2	−295.5
Longitudinal force on no. 16 pier (kN)	−9.8	−13.6	−13.4	−20.5

contrast, since No. ③ and No. ④ turnouts are positioned in the middle of the bridge, the displacement of the bridge remains relatively stable across different lengths of the intermediate tangent. As a result, the variations in the intermediate tangent length have little impact on No. ③ and No. ④ turnouts. Furthermore, the change in the intermediate tangent length does not affect the overall symmetry of the turnouts, thus resulting in minimal alterations in the longitudinal force on the piers.

3. Calculation conclusions

Based on the analysis presented above, in the context of turnout groups laid on continuous girders with symmetric placement of both the fixed bearings of the bridge and the turnouts, the length of the intermediate tangent between the turnouts has a minimal impact on the longitudinal interactions of the CWR turnout system on the bridge. As a result, the selection of the intermediate tangent length should be based on the requirements specified in TB 10,015–2012 and the specific conditions of the project.

3.6.5 Influence of Bridge Connection Mode on Stress and Deformation of CWR Turnouts

1. Double crossovers

As shown in the turnout arrangement, the length of a single crossover, composed of a No.18 turnout, is approximately 140 m, and for double crossovers, which consist of No.18 turnouts, the length extends to approximately 300 m. When these double crossovers are placed on a continuous girder, regardless of the position of the fixed bearings, the temperature span of the bridge exceeds the acceptable limit. As a consequence, this leads to excessively large additional expansion and contraction force on the rails. Even if both crossovers are positioned on two connected continuous girders, the resulting bridge temperature span is still likely to exceed the acceptable limit. In light of these considerations, it is recommended in practical design to adopt a CWR arrangement on bridges where each crossover is placed on a continuous girder, with a one-span simply-supported girder between the continuous girders carrying the crossover.

Condition I: In the bridge arrangement mode shown in Fig. 3.78, the entire double crossovers are laid on a 11 × 32 m continuous girder, with the fixed bearing positioned at No. 8 pier, and the intermediate tangent between the crossovers extending for 50 m;

Condition II: In the bridge arrangement mode shown in Fig. 3.79, the left and right crossovers are respectively laid on two 5 × 32 m continuous girders, with a 32 m simply-supported girder between the two continuous girders, and the intermediate tangent between the crossovers extending for 50 m.

(1) Additional expansion and contraction force on rail

The expansion and contraction force on the stock rails within the turnouts is shown in Fig. 3.80, in which positive values represent compressive force. When the bridge arrangement is transformed from Condition I to Condition II, there is a significant decrease in the additional expansion and contraction force on the No.1 rail. Furthermore, the additional expansion and contraction force at the end of the continuous girder decreases from 453.7 to 239.9 kN. With the insertion of a one-span simply-supported girder in the middle, the 2 peaks of the expansion and contraction force on rails increase to 4. These peaks are located at the ends of the two-span continuous girder. The peak value of expansion and contraction force on the No. 2 rail at the end of the long continuous girder reduces from 359.5 to 198.2 kN. Additionally, due to the presence of the simply-supported girder in the middle, two peaks of the expansion and contraction force are formed at the end of the simply-supported girder, with the peak additional expansion and contraction force reaching 320.6 kN.

(2) Expansion and contraction displacement of rail

The expansion and contraction displacement of No.1 and No.2 rails is shown in Fig. 3.81, where positive values represent displacement toward the right. It is evident from the figure that when the bridge type is changed from Condition I to Condition II, the expansion and contraction displacement peaks of No. 1 and No. 2 rails increase from two to four. The maximum rightward expansion and contraction displacement

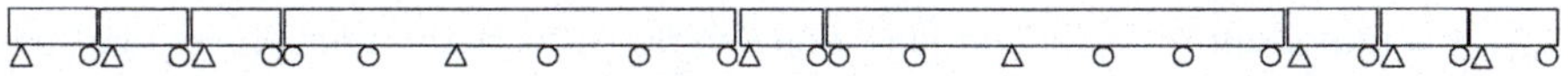

Fig. 3.79 Schematic diagram of bridge arrangement

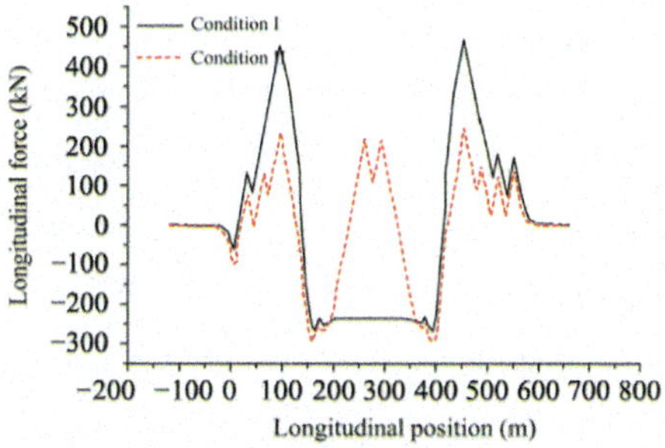

(a) No.1 Rail

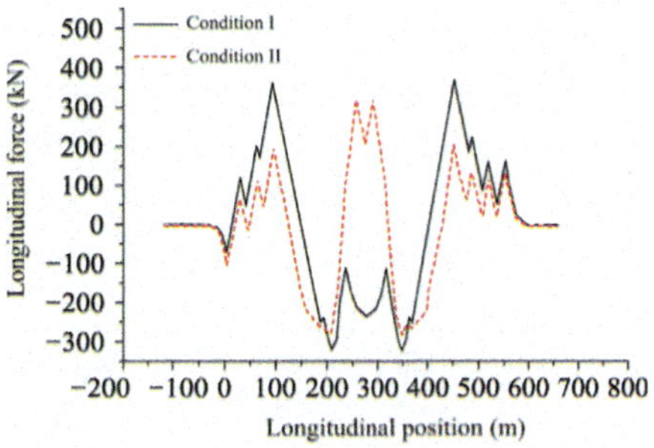

(b) No.2 Rail

Fig. 3.80 Additional expansion and contraction force on stock rail

of No. 1 rail decreases from 22.3 mm in Condition I to 10.0 mm in Condition II, while the maximum leftward expansion and contraction displacement decreases from 17.3 to 5.4 mm. Similarly, the expansion and contraction displacement of No. 2 rail is also significantly reduced due to the smaller temperature span on the bridge.

(3) Other calculation results

Other calculation results are shown in Table 3.16. Based on this table, when the bridge arrangement is transformed from Condition I to Condition II, there is a significant reduction in the expansion and contraction force and displacement of the stock rails within the turnout. However, other indices show minor changes. The insertion of a simply-supported girder in the middle of the continuous girders leads to a decrease in the expansion and contraction displacement of the bridge under No. ① and No. ② turnouts, while the expansion and contraction displacement under No. ③ and No. ④ turnouts increases. Consequently, the expansion and contraction displacement of the movable parts in No. ① and No. ② turnouts decreases, while it increases in No. ③ and No. ④ turnouts. The weakened girder-rail interactions resulting from the reduced temperature span lead to smaller longitudinal forces on the piers in Condition II compared to Condition I.

It can be seen from the above calculations that when double crossovers are laid on bridges, a substantial force is transferred from the diverging route of the turnout to the main route, resulting in significant displacements of the movable components, particularly on the simply-supported girder. This configuration poses a disadvantage in maintaining the stability of the turnout. In instances where they are placed on continuous girders, the fixed bearings should be positioned as symmetrically as possible on the bridge, to align with the longitudinally symmetrical structure of the turnout.

To prevent the bridge from experiencing excessive temperature spans, it is recommended to position the two crossovers on two separate continuous girders. Additionally, it is important to include at least one simply-supported girder across one span between the continuous girders.

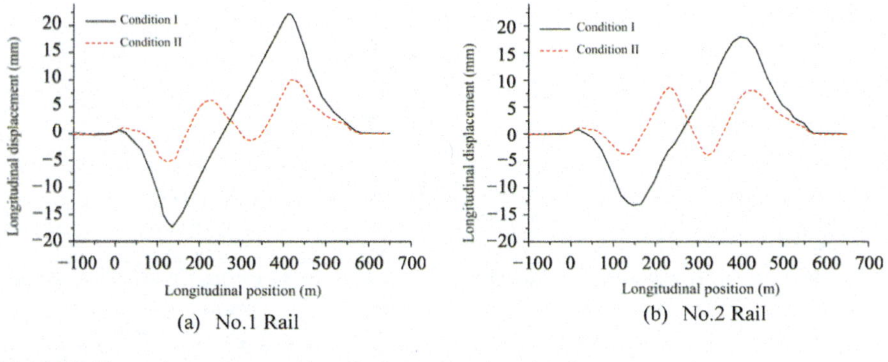

Fig. 3.81 Expansion and contraction displacement of stock rail

Table 3.16 Other calculation results

Calculation conditions		Condition I	Condition II
Maximum additional expansion and contraction force on no. 1 rail (kN)		453.7	239.9
Maximum displacement at right end of no. 1 rail (mm)		22.3	10
Maximum stress on force transfer device at heel end of switch rail (kN)	No. ① turnout	−60.4	−13.7
	No. ② turnout	61.6	14.1
	No. ③ turnout	0	39.6
	No. ④ turnout	0	43.1
Maximum stress on spacing block at end of wing rail (kN)	No. ① turnout	−127.8	−124.5
	No. ② turnout	127.8	124.9
	No. ③ turnout	130.7	124.9
	No. ④ turnout	−130.7	−125.9
Displacement at actual point of straight switch rail relative to turnout sleeper (mm)	No. ① turnout	−19.3	−17
	No. ② turnout	19.3	17
	No. ③ turnout	15	18.4
	No. ④ turnout	15	−18.6
Displacement at actual point of point rail relative to turnout sleeper (mm)	No. ① turnout	−5.9	−5.7
	No. ② turnout	5.9	5.7
	No. ③ turnout	6.8	5.8
	No. ④turnout	−6.8	−5.8
Longitudinal force on left abutment (kN)		722.3	606.1
Longitudinal force on no. 1 pier (kN)		388.5	204.6
Longitudinal force on no. 2 pier (kN)		722.1	373.6
Longitudinal force on fixed pier within turnout group (kN)		–	201.3
		49.9	3.1
		–	−170.2

(continued)

Table 3.16 (continued)

Calculation conditions	Condition I	Condition II
Longitudinal force on no. 14 pier (kN)	−672.5	−318
Longitudinal force on no. 15 pier (kN)	−295.5	−102.2
Longitudinal force on no. 16 pier (kN)	−20.5	−68

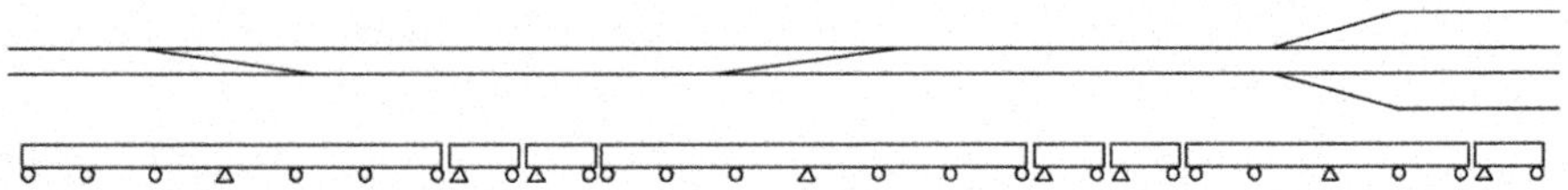

Fig. 3.82 Schematic diagram for turnout bridge arrangement at typical station

2. Typical stations

(1) Calculation conditions

Condition I: In the bridge arrangement shown in Fig. 3.82, the CWR turnout group on the station bridge consists of double crossovers and turnouts on the station tracks. The two crossovers are laid on separate 6 × 32 m continuous girders, with two-span simply-supported girders in the middle. The turnouts on the station tracks are positioned in parallel on the same 6 × 32 m continuous girder. Two-span simply-supported girders are inserted between the continuous girders for both the crossover turnout and the turnouts on the station tracks. The fixed bearing of each continuous girder is located at the intermediate pier, and the center of the turnout coincides with that of the continuous girder.

Condition II: addition of a one-span simply-supported girder between continuous girders in Condition I.

Condition III: The simply-supported girder between the crossover and station track in Condition I is replaced with a continuous girder. This continuous girder is equivalent to a 12-span 32 m structure, with a fixed bearing positioned at its intermediate pier.

Condition IV: The simply-supported girder between the crossover and station track in Condition II is replaced with a continuous girder. This continuous girder is equivalent to a 11-span 32 m continuous girder structure, with a fixed bearing positioned at its 7th pier (the same location as in Condition III).

(2) Expansion and contraction force on rail

The expansion and contraction force on the stock rails within the turnout is shown in Fig. 3.83, where positive values represent compressive force. When comparing with Condition III, the maximum additional expansion and contraction force on No. 1 rail changes from 712.2 to 429.8 kN in Condition I. Under Condition I, the insertion of two-span simply-supported girders between the continuous girder of the turnout on the station track and the continuous girder of the crossover turnout leads to a peak value (333.8 kN) of additional expansion and contraction force at the left end of

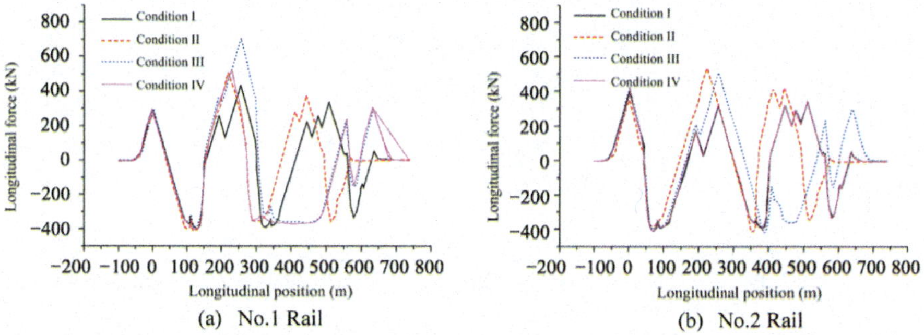

Fig. 3.83 Additional expansion and contraction force on stock rail

the continuous girder of the turnout on the station track. Conversely, the maximum additional expansion and contraction force on No. 2 rail changes from 519.9 kN to 403.1 kN, with a peak value of 346.0 kN at the simply-supported girder. The same pattern is followed under Condition II and Condition IV, with only varying values. By observing the diagrams of the additional expansion and contraction force under different conditions, it is evident that the expansion and contraction force is significantly reduced when simply-supported girders are inserted between continuous girders. Additionally, the additional expansion and contraction force is lower when using two-span simply-supported girders compared to one-span girders.

(3) Expansion and contraction displacement of rail

The expansion and contraction displacement of No.1 and No.2 rails is shown in Fig. 3.84, where positive values indicate rightward displacement. It is evident from the figure that when transitioning from Conditions III and IV to Conditions I and II in terms of bridge arrangement, there is a considerable reduction in the maximum leftward expansion and contraction displacement of No. 1 and No. 2 rails. Specifically, the maximum leftward expansion and contraction displacement of No. 1 rail decreases from 32.6 mm in Condition III to 10.1 mm in Condition I, and from 34.8 mm in Condition IV to 9.0 mm in Condition II. Similarly, the expansion and contraction displacement of No. 2 rail to the left is significantly reduced due to the smaller bridge temperature span. Conversely, the maximum amplitude of rightward expansion and contraction displacement of both No. 1 and No. 2 rails slightly increases. In summary, the insertion of simply-supported girders between continuous girders can effectively reduce the expansion and contraction displacement of rails within a certain range.

(4) Other calculation results

Other calculation results are shown in Table 3.17. Based on this table, due to the decreased temperature span, the girder-rail interactions are weakened. As a result,

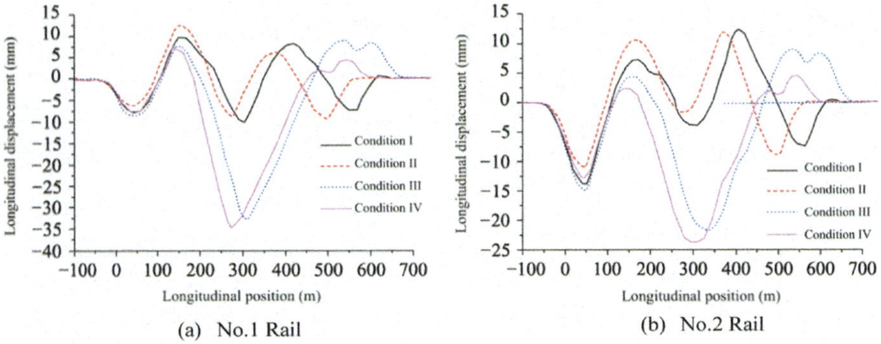

Fig. 3.84 Expansion and contraction displacement of stock rail

the longitudinal forces on the piers under Conditions I and II are smaller when compared to Conditions III and IV, respectively.

From the above calculations, it can be seen that when the common double crossovers and turnouts on station tracks at typical stations are laid on bridges, there are also challenges such as a substantial force transmitted from the diverging route of the turnout to the main route, resulting in significant displacements of the movable components, particularly on the simply-supported girder. This configuration poses a disadvantage in maintaining the stability of the turnout.

3. Calculation conclusions

Through the computational analysis and comparison on the insertion of simply-supported girders between two crossovers and between the crossover on bridges at typical stations and the station track, conclusions are drawn as follows:

(1) To prevent the bridge from experiencing excessive temperature spans, it is recommended to position the two crossovers on two separate continuous girders. Additionally, it is important to include at least one simply-supported girder across one span between the continuous girders.
(2) To prevent the bridge from experiencing excessive temperature spans, it is recommended to position two single crossovers and station tracks on bridges at typical stations respectively on three continuous girders. Additionally, it is important to include at least one simply-supported girder across one span between the continuous girders.

Table 3.17 Comparison of other calculation results

Calculation conditions		Condition I	Condition II	Condition III	Condition IV
Maximum stress on force transfer device at heel end of switch rail (kN)	No. ① turnout	0	0	0	0
	No. ② turnout	0	0	−34.461	−67.615
	No. ③ turnout	0	0	−27.878	0
	No. ④ turnout	0	0	0	0
	No. ⑤ turnout	0	0	0	0
	No. ⑥ turnout	0	0	−27.869	0
Maximum stress on spacing block at end of wing rail (kN)	No. ① turnout	111.680	110.308	114.063	117.263
	No. ② turnout	−109.503	−119.346	−118.246	−118.841
	No. ③ turnout	−148.658	−141.252	−201.536	−180.857
	No. ④ turnout	−107.095	−105.565	−107.568	−115.113
	No. ⑤ turnout	112.211	118.686	147.642	151.203
	No. ⑥ turnout	−149.315	−141.447	−201.520	−180.873
Displacement at actual point of straight switch rail relative to turnout sleeper (mm)	No. ① turnout	16.591	17.363	18.342	22.725
	No. ② turnout	−19.193	−20.824	−24.685	−25.692
	No. ③ turnout	−17.552	−17.842	−21.054	−19.215
	No. ④ turnout	−18.688	−17.987	−19.155	−18.614
	No. ⑤ turnout	17.807	19.367	0.253	0.394
	No. ⑥ turnout	−17.710	−18.583	−21.053	−19.220
Displacement at actual point of point rail relative to turnout sleeper (mm)	No. ① turnout	5.223	5.137	5.363	5.607
	No. ② turnout	−5.167	−5.642	−5.707	−5.743
	No. ③ turnout	−7.282	−6.789	−3.424	−9.117

(continued)

Table 3.17 (continued)

Calculation conditions		Condition I	Condition II	Condition III	Condition IV
	No. ④ turnout	−5.034	−4.934	−5.061	−5.426
	No. ⑤ turnout	5.308	5.596	2.299	2.799
	No. ⑥ turnout	−7.303	−6.804	−3.424	−9.118
Longitudinal force on left abutment (kN)		0	0	0	0
Longitudinal force on no. 1 pier (kN)		0	0	0	0
Longitudinal force on no. 2 pier (kN)		0	0	0	0
Longitudinal force on no. 3 pier (kN)		−127.772	38.857	−194.193	−54.210
Longitudinal force on no. 4 pier (kN)		0	0	0	0
Longitudinal force on no. 5 pier (kN)		0	0	0	0
Longitudinal force on no. 6 pier (kN)		0	0	0	0
Longitudinal force on no. 7 pier (kN)		26.016	−102.912	−53.621	−111.302
Longitudinal force on no. 8 pier (kN)		−107.117	0	−113.402	0
Longitudinal force on no. 9 pier (kN)		0	0	0	0
Longitudinal force on no. 10 pier (kN)		0	51.493	0	0
Longitudinal force on no. 11 pier (kN)		26.747	0	0	0
Longitudinal force on no. 12 pier (kN)		0	0	0	0
Longitudinal force on no. 13 pier (kN)		0	−23.894	0	−317.602
Longitudinal force on no. 14 pier (kN)		29.552	0	−194.490	0
Longitudinal force on no. 15 pier (kN)		−32.966	0	0	0

References

1. Rongshan Y (2008) Computational theory and experimental study on longitudinal force on CWR turnout on bridge [D]. Southwest Jiaotong University, Chengdu
2. Liang G, Kai T, Cun Q et al (2009) Study on spatial mechanical characteristics of CWR turnout on bridge of passenger dedicated line [J]. China Railw Sci 30(1):29–35
3. Daxin S, Liang G, Yanfeng L (2007) Analysis of mechanical characteristics of CWR turnout for ballastless track on bridge [J]. J Beijing Jiaotong Univ 31(1):89–92
4. Jingmang X, Rong C (2011) Computation and research on longitudinal force on CWR turnout on bridge [J]. Railw Eng 3:93–95
5. Ping W (2003) Mechanism of longitudinal force transmission of CWR turnouts with different structures [J]. J Southwest Jiaotong Univ 38(4):371–374
6. Ping W (2002) Research on the influence of CWR turnout groups on rail displacement and longitudinal force [J]. J China Railw Soc 24(2):74–78
7. Juanjuan R (2009) Research on stress characteristics and structural optimization of longitudinally coupled ballastless track within CWR turnout section on bridge [D]. Southwest Jiaotong University, Chengdu
8. Ping W, Rongshan Y, Xueyi L (2004) Analysis on stress and deformation of CWR turnout laid on large continuous girder bridge [J]. J Transp Eng Inf 3:16–21

9. Ping W, Xueyi L (2007) Computational theory and design approach of CWR turnout [M]. Southwest Jiaotong University Press, Chengdu
10. Ping W, Rong C, Rongshan Y et al (2011) Design theory of CWR turnout on bridge [M]. Southwest Jiaotong University Press, Chengdu
11. Zhang L, Liu Y, Yang R (2010) Experiments on expansion and contraction force and simulation analysis using models of CWR turnouts on bridges of passenger dedicated lines [J]. Railw Engg 2010(10):135–138
12. Zhiping Z, Xiufang C, Guofan Z (2006) Computation of expansion and contraction and displacement of CWR turnout on continuous girder bridge [J]. J Transp Eng 1:34–38
13. Zhiping Z (2006) Study on expansion and contraction force on CWR turnout on high-speed railway bridge and spatial vibration of train-turnout-bridge system [D]. Central South University, Changsha
14. Ye Q, Tang J, Yu L, et al. (2009) Analysis of longitudinal force due to deflection and displacement characteristics of CWR turnout on bridge for ballastless track [J]. Sci Technol West China 7–9
15. Liu Y, Zhang L, Yang R (2011) Model experimental study on longitudinal force due to deflection of CWR turnout on bridge for ballastless track [J]. Railw Eng (1):93–96
16. Wang P, Huang S (2001) Research on computational theory for nonlinearity of CWR turnout with movable point frog [J]. China Railw Sci 22(1):89–96
17. Qiuyi L (2010) Study on computational models and methods of CWR turnout on bridge for passenger dedicated line [J]. Railw Stand Des 2:64–66
18. Qiuyi L, Li S, Yanli Y (2008) Research on design approach of CWR turnout on bridge of passenger dedicated line [J]. J Railw Eng Soc 12:50–53
19. Yanfeng L, Liang G, Yawei F (2006) Finite element analysis on stress and deformation of CWR turnout on bridge [J]. J Beijing Jiaotong Univ 30(1):66–70
20. Song Y (2007) Finite element analysis of CWR turnouts under special conditions [D]. Central South University, Changsha
21. Chuanyong Z (2007) Development of research and analysis system for temperature force on CWR turnout group based on ANSYS [D]. Central South University, Changsha
22. Huang Z (2009) Research on design technology of CWR turnout groups for Hefei-Nanjing railway [J]. J Railw Eng Soc (6):54–57
23. Ping W, Xueyi L (2003) Analysis of influencing factors on stress and deformation of CWR turnout [J]. China Railw Sci 24(2):61–69

Chapter 4
Design Methodology of CWR Turnout on Bridge

CWR turnouts on bridges represent a complex structural system consisting of turnouts, bridges, and foundations. This system is subjected to various loads, including temperature fluctuations of rails and girders, train loads, braking forces, and rail shrinkage upon breakage [1–3]. Based on the design approaches of CWR tracks on bridges and CWR turnouts [4, 5], this book presents the design principles, design process, and checking calculation indexes for CWR turnouts on bridges and explores the rational structural forms for turnout girders, the positional relationships between turnouts and bridges, bearing arrangements, and linear rigidity limits of piers, taking into account the structural characteristics of CWR turnouts on bridges. Moreover, it establishes design methods and a design system specifically tailored for CWR turnouts on HSR bridges, based on the principle of turnout–bridge interactions and guided by the idea of dual control of turnout components: bearing capacity and displacement.

4.1 Content and Methods of Checking Calculation for CWR Turnout on Bridge

The design concept distinguishing CWR turnouts on bridges from general CWR tracks lies in the dual control of turnout components: bearing capacity and displacement. While ensuring compliance with structural safety requirements, encompassing aspects such as strength and stability, additional attention needs to be directed towards the normal operational condition of turnouts. This entails constraining the rail displacement of critical components within the turnout section to guarantee the normal functioning of the switches, as well as to ensure optimal operational smoothness and riding quality for trains.

Q. Li et al., *Continuously Welded Turnouts on High-Speed Railway Bridge*, Advances in High-speed Rail Technology, https://doi.org/10.1007/978-981-97-6524-9_4

4.1.1 Basic Requirements for Design of CWR Turnout on Bridge

CWR turnouts on bridges shall be designed to meet the following requirements:

(1) The actual points of switch rails and point rails should be controlled against displacement to prevent the switches from jamming due to their excessive displacement.
(2) The longitudinal tension on long rails should be controlled to ensure the rail strength.
(3) Various additional forces on CWR turnouts on bridges should be considered, and the longitudinal compressive force value shall be controlled for long rails, to prevent track buckling within CWR turnouts on bridges, particularly in the sections preceding the turnouts.
(4) Forces exerted on the force transfer devices within the turnouts should be controlled to ensure their strength.
(5) The broken rail gap value upon fracture of rails should be controlled to ensure the operational safety of trains.
(6) The longitudinal horizontal forces acting on bridge piers/abutments should be controlled to ensure the safe operation of the bridges.
(7) The bridge superstructures should be designed to prioritize high rigidity, structural integrity, and stability, and meet the requirements of turnout stability and regularity, as well as operational safety and riding comfort of trains.
(8) The bridges and track structures shall be designed using a systematic approach. The bridge deck shall be designed to adapt to different track structural forms, incorporating designated positions for mounting various facilities and connecting devices and high-performing waterproofing and drainage facilities.
(9) In the turnout section of railway main lines, the bridge girder shall be designed with a continuous structure and an equal span arrangement. The maximum span should not exceed 48 m; otherwise, a separate design is necessary. Two adjacent units of continuous girders should be separated by one or more spans of simply-supported girders. Similarly, in the turnout section of station tracks, the bridge girder should also be designed with a continuous structure.
(10) The turnouts on railway main lines shall not be placed across any girder joint, and the distance from the beginning and end of the installed turnouts to any girder joint shall not be less than 18 m. Similarly, on station tracks, turnouts should also not be positioned across any girder joint. To cross girder joints under challenging conditions, the turnouts shall be placed to maintain the minimum distance from the actual points and heel ends of the switch rails and point rails to the girder joint aligned with the requirements for the structural safety of turnouts and bridges and ensure the normal functioning of switches.
(11) The design stress-free rail temperature for CWR turnouts on bridges shall be consistent with that for the adjacent sectional CWR tracks at both ends, to facilitate operation management. CWR turnouts on bridges shall be fastened within ± 3 °C of the design stress-free rail temperature. Additionally, the

fastening-down rail temperature difference between adjacent unit rail links shall not exceed 5 °C.

4.1.2 Checking Calculation Items for Longitudinal Forces on CWR Turnout on Bridge

CWR turnouts on bridges shall be designed on the premise of calculations for additional forces and displacements of rails within them. The turnout–bridge interactions include expansion and contraction force, deflection force, traction/braking force, and rail breakage force [6, 7]. These calculations shall incorporate the interactions between closure rails and stock rails within the turnouts, as well as those between the turnouts and bridges. To perform accurate computations, the "turnout–bridge-pier" integration computational model and method outlined in Chap. 3 of this book can be utilized. The calculations for longitudinal forces and displacements of CWR turnouts on bridges involve the following items:

(1) Maximum additional longitudinal force on stock rail
(2) Maximum shear of force transfer device at the heel end of frog
(3) Maximum shear of force transfer device of frog
(4) Maximum displacement at actual point of switch rail
(5) Maximum displacement at actual point of point rail
(6) Displacement of stock rail within turnout
(7) Longitudinal additional forces on pier.

4.1.3 Checking Calculation Items for Design of CWR Turnout on Bridge

CWR turnouts on bridges shall be designed to accommodate the permissible temperature rises and drops required for trans-sectional CWR tracks and to ensure all connecting parts within the turnouts are secure, durable, and reliable. Additionally, the displacements of the switch rails and point rails within the turnouts shall meet the requirements for the normal functioning of the turnout structures and switches. Moreover, the requirements for relative displacement limits between switches and frogs within the turnouts and the bridges shall be met.

The checking calculation items for CWR turnouts on bridges include the following:

1. Displacement of rails within turnout section

For CWR turnouts on bridges, the checking calculation shall include not only the displacements of the switch rails and point rails but also the displacements of the switches and frogs relative to the bridges, as well as the displacement of the stock rails relative to the bridges at the switch zone.

(1) Displacement of Switch Rail and Point Rail Relative to Stock Rail and Wing Rail

Checking calculations for the displacements of the switch rails and point rails relative to the stock rails and wing rails aims to ensure that the expansion and contraction displacement at each traction point of the switch rails and point rails remains within the permissible values for switches. Excessive displacements of the switch rails and point rails relative to the stock rails and wing rails may cause defects, such as jamming during the switching of CWR turnouts, lateral bending of the switch rails, and creeping of the point rails onto bollards. To ensure that the turnout structures and switches meet the normal functional requirements, it is necessary to conduct checking calculations on the displacements at the actual points of the switch rails and movable point rails within CWR turnouts. On ballasted tracks, the relative displacement between the actual points of the switch rails and the sleepers shall be checked in the context of the switches mounted on the sleepers. Additionally, the relative displacement between the actual points of the switch rails and the bridges shall be checked in the context of the switches mounted on the bridges. On ballastless tracks, the relative displacement between the actual points of the switch rails and the bridges shall be checked.

The relative displacement at the actual points of the switch rails may be checked using Formula (4.1):

$$\Delta l_{\max} \leq \left[\Delta l'\right] \tag{4.1}$$

where $\triangle l_{\max}$—relative displacement at the actual point of switch rail (mm);

$[\triangle l']$—permissible relative displacement at the actual point of switch rail (mm).

The relative displacement at the actual points of the movable point rails may be checked using Formula (4.2):

$$\Delta l_{0\max} \leq [\Delta l''] \tag{4.2}$$

where $\triangle l_{0\max}$—relative displacement at the actual point of movable point rail (mm);

$[\triangle l'']$—permissible relative displacement at the actual point of point rail (mm).

The turnout products used on China's high-speed railways include three series: independently developed turnouts for passenger dedicated lines, CZ turnouts, and CN turnouts. Different series of turnouts are equipped with distinct switches, corresponding to various permissible values for expansion and contraction displacements of the switch rails and point rails. The permissible expansion and contraction displacements of the switch rails and movable point rails typically vary depending on the turnout structures and switch performance. Refer to Table 4.1 for the specific permissible expansion and contraction displacements of the switch rails and movable point rails within these three series of turnouts.

Table 4.1 Permissible expansion and contraction displacements of switch rails relative to stock rails and movable point rails relative to wing rails

Turnout type	Permissible expansion and contraction displacement of switch rail (mm)	Permissible expansion and contraction displacement of point rail (mm)	Remarks	
			Locking device	Structure at heel end of switch rail
Passenger dedicated line series	± 40	± 20	Hook-shaped external locking at multiple traction points and with multiple switches	Stopper, spacing block, or without force transfer device
CZ series	+ 45	± 30	External locking at the crank of first traction point	No force transfer device
CN series	± 40	± 20	Self-adjusting external locking at multiple points and with multiple switches	Stopper

(2) Displacements of Switch and Frog

Based on the laying experience and research findings of CWR turnouts on bridges abroad, relative displacements between the switch/frog sections and the bridges shall be strictly checked to ensure the safety and comfort of train operation under the action of bridge expansion and contraction, vertical loads of trains, and braking/traction force of trains. Frog and switch zones have been identified as the areas prone to relative displacement between the rails and deck. The specific limits for relative displacement typically depend on the structural performance of the turnouts and are provided by the turnout manufacturer upon supply. Due to the utilization of both independently developed series and those incorporating foreign technologies such as CZ and CN series on China's high-speed railways, it becomes challenging to establish accurate displacement limits for these turnouts. Therefore, it is advisable to refer to the relative displacement limits applied in Germany (Table 2.1). For the purpose of checking calculations, the displacement envelope shown in Fig. 2.19 provides an effective method.

(3) Relative Displacement between Stock Rail and Bridge at Switch

To ensure the normal functioning of the switches, it is crucial to strictly limit the relative displacement between the turnouts and bridges, particularly the relative displacement between the stock rails and bridge deck at the switch zone. Considering the specific structural characteristics of switches in the turnouts developed independently for high-speed railways in China, the joint research team focusing on key technologies for turnouts on passenger dedicated lines has proposed a limit of not exceeding 5 mm for the relative displacement between the stock rails and the bridges in the switch zone. This limitation plays a crucial role in aligning the throw rod, operating rod, and locking rod of the switches, thereby preventing any potential jamming during the switching process. When designing CWR turnouts on bridges, it is imperative to enhance the checking calculations concerning the relative displacement between the stock rails and bridges at the switch zone. Moreover, it is essential to formulate effective control measures to ensure the normal functioning of the switches even under extreme conditions.

2. Strength of Rails within Turnout Section

Within CWR turnouts on bridges, the rails experience a range of stresses, including temperature stress, dynamic flexural stress caused by train loads, additional stress, and traction (braking) stress. To maintain the safe utilization of rails within the turnout section, it is crucial to ensure that the sum of all these operating stresses remains below the permissible stress designated for the rails.

The checking calculation formula of rail strength is given below

$$\sigma_{\mathrm{d}} + \sigma_{\mathrm{t}} + \sigma_{\mathrm{f}} + \sigma_{\mathrm{z}} \leq [\sigma] \tag{4.3}$$

where σ_{d}—dynamic flexural stress of rail (MPa);

σ_{t}—temperature stress of rail (MPa);

σ_{f}—maximum additional stress of rail, taken as the maximum value of the additional expansion and contraction stress and additional deflection stress of rail (MPa);

σ_{z}—traction (braking) stress of rail (MPa);

$[\sigma]$—permissible stress of rail (MPa).

3. Stability of CWR Turnout on Bridge for Ballasted Track

In light of the longitudinal additional compressive force on rails in the turnout section on bridges, CWR turnouts on bridges with ballasted tracks are subject to checking calculation, focusing on the stability of the CWR tracks within the turnout section, to prevent track buckling. The checking calculations involve superimposing the maximum temperature compressive stress endured by the turnout at the highest rail temperature with the maximum additional temperature stress of the stock rails. It is vital to concentrate on the switches within CWR turnouts as the critical zone for stability checking, due to the significant additional forces exerted on this comparatively weaker structure.

The stability of CWR turnouts on bridges with ballasted tracks may be checked using Formula (4.4):

$$2(\max P_t + \max \Delta P \leq [P]) \tag{4.4}$$

where $\max P_{\mathrm{t}}$—maximum temperature compressive force endured by the stock rails within the CWR turnout when the rail temperature rises (kN);

$\max \Delta P$—maximum additional compressive force on the stock rails at the stopper or spacing block at the heel end of the frog (kN);

[P]—permissible temperature compressive force (kN).

4. Bolt Strength for Stopper and Spacing Block

Stoppers and spacing blocks act as the key connecting parts within CWR turnouts. They experience significant longitudinal forces when subjected to substantial temperature fluctuations. To ensure the safe operation of CWR turnouts, it is essential to conduct checking calculations on the strength of their connecting bolts.

(1) Shear strength of connecting bolt of stopper

The combination blocks of stoppers at the heel end of the switch rails are fastened to the rails with bolts. During significant fluctuations in rail temperature, the combination blocks come into contact, resulting in shear on the stopper bolts. When the rail temperature reaches its maximum/minimum value, the shear borne by the bolts reaches its extremum as well.

The shear strength of connecting bolts for the stoppers (spacing blocks) may be checked using Formula (4.5):

$$\tau = \frac{4T}{\pi d^2} \leq [\tau] \tag{4.5}$$

where τ—shear strength of bolt (MPa);

T—maximum shear endured by the connecting bolt of stopper (spacing block) (N);

d—bolt diameter (mm);

$[\tau]$—permissible shear stress of bolt (MPa).

(2) Shear strength of bolt connecting movable point rail and long wing rail

The movable point rails and the long wing rails within high-speed turnouts are connected using long spacing blocks. The spacing block bolts connecting them are subjected to shear under the action of temperature force, so it is necessary to conduct a checking calculation on the shear strength of these bolts.

The shear strength of connecting bolts at the heel ends of the point rails may be checked using Formula (4.6):

$$\tau = \frac{4T'}{\pi d^2} \leq [\tau] \tag{4.6}$$

where T'—maximum shear endured by the connecting bolt at the heel end of point rail (N);

d—bolt diameter (mm);

$[\tau]$—permissible shear stress of bolt (MPa).

5. Broken rail gap

In the event of any defects in rail materials or insufficient welding strength at rail joints, rail breakage may occur during low-temperature seasons. Any excessively widened broken rail gap poses a risk to train operations. Consequently, the checking calculations on broken rail gaps of CWR turnouts on bridges are necessitated in the design process.

The broken rail gap shall be checked using Formula (4.7):

$$\lambda \leq [\lambda] \tag{4.7}$$

where λ—calculated value for broken rail gap (mm);

$[\lambda]$—permissible value for broken rail gap, taken as 70 mm in general conditions and 90 mm under difficult conditions (mm).

4.2 Design Stress-Free Rail Temperature for Turnouts

The stress-free rail temperature is considered an important parameter in the design of CWR turnouts on bridges, due to its direct association with the stress and deformation of the CWR turnouts. Moreover, the design stress-free rail temperature for CWR turnouts on bridges is subject to checking calculations that incorporate factors such as the strength and stability of rails within the turnout section, the strength of turnout components, and the displacement of switch rails and point rails.

The design stress-free rail temperature may be checked using Formula (4.8):

$$T_e = \frac{T_{\max} + T_{\min}}{2} + \frac{[\Delta T_d] - [\Delta T_u]}{2} \pm \Delta T_k \tag{4.8}$$

where T_e—design stress-free rail temperature (°C);

T_{max}—local maximum rail temperature over the years (°C);

T_{min}—local minimum rail temperature over the years (°C);

$[\Delta T_d]$—permissible temperature drop (°C);

$[\Delta T_u]$—permissible temperature rise (°C);

ΔT_k—corrected value of design stress-free rail temperature, taken in a range of 0~5 °C.

4.2.1 Principle of Load Combination for Checking Calculation of Bridge Piers and Abutments

For bridges laid with CWR turnouts, their pier and abutment structures are subject to checking calculations that incorporate longitudinal additional forces on these turnouts. The load combinations for these calculations involve the scenarios of main forces, as well as main forces combined with special forces. The selection of longitudinal force combinations depends on the number of tracks. See Table 4.2 for the load combinations.

Table 4.2 Longitudinal force combination for checking calculation of bridge piers and abutments

Girder type	S/N	Combination of longitudinal forces
Double-track bridge (crossover)	1	Expansion and contraction force or deflection force on double tracks
	2	Braking flexural force acting on the crossover or braking on one track and starting on the other track
	3	Train braking or traction on one track + expansion and contraction force or deflection force on the other track + longitudinal additional forces such as wind force
	4	Force from breakage of one rail + expansion and contraction force on the other rail on one track + expansion and contraction force or deflection force acting on the pier and abutment top by two rails on the other track
Three-track bridge	1	Expansion and contraction force forces or deflection force on three tracks
	2	Braking flexural force acting on the crossover or braking on one track and starting on another track + expansion and contraction force or deflection force on the third track
	3	Braking on one track + expansion and contraction force or deflection force on the other two tracks
	4	Force from breakage of one rail + expansion and contraction force on the other rail on one track + expansion and contraction force or deflection force on the other two tracks
Four-track bridge	1	Expansion and contraction force or deflection force on four tracks
	2	Braking flexural force acting on the crossover
	3	Braking on one track + starting on another track + expansion and contraction force or deflection force on the other two tracks
	4	Rail breakage force on one rail + expansion and contraction force on the other rail on one track + expansion and contraction force or deflection force on the other three tracks

4.3 Design Process of CWR Turnout on Bridge

Based on the above analysis, the design process of CWR turnouts on bridges is summarized into the following four steps, as shown in Fig. 4.1:

(1) Turnout and bridge arrangements;
(2) Calculations of rails forces and displacements;

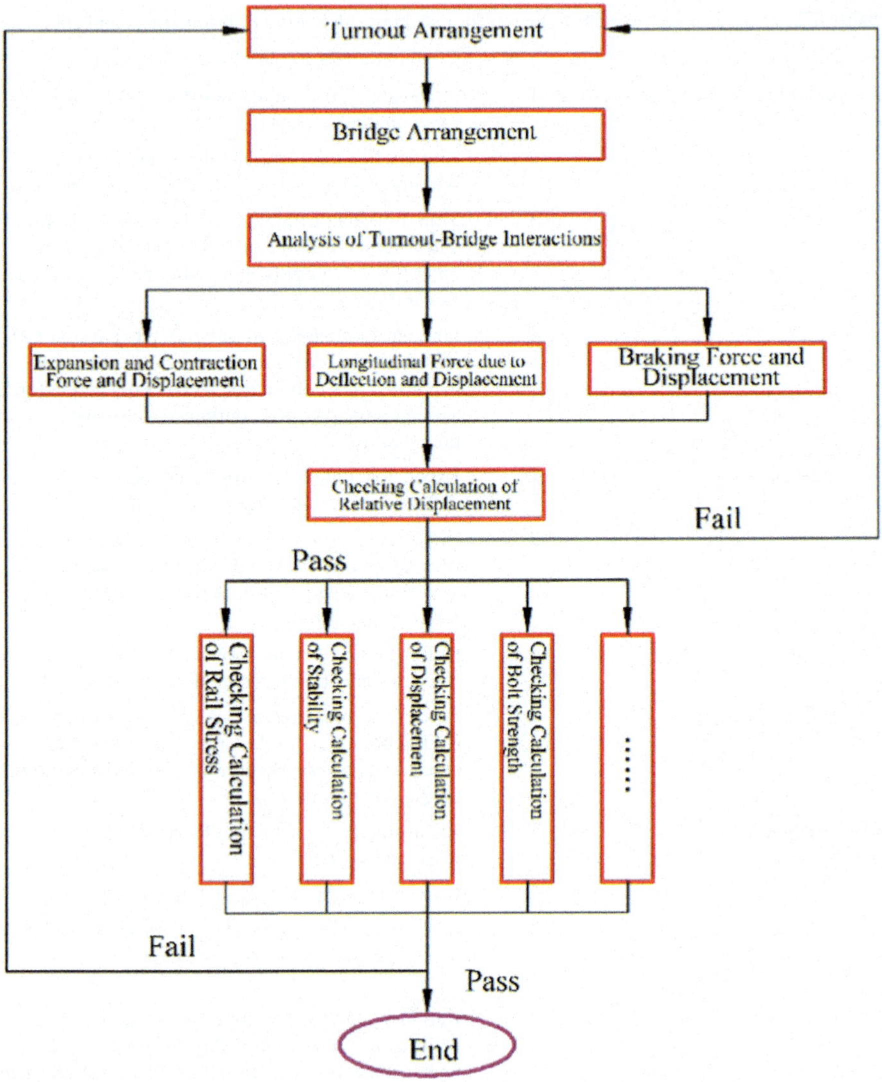

Fig. 4.1 Design process of CWR turnout on bridge

(3) Checking calculations of relative displacement between girders and rails at frog and switch zones;
(4) Relevant checking calculations associated with turnout and bridge.

4.4 Study on Rational Turnout and Bridge Arrangements

To ensure the operational safety and riding comfort of high-speed trains on high-speed railways, turnouts on bridges should generally align with those on the subgrade in geometric dimensions, geometric alignment, technical parameters, and processing technology. This principle also reflects the basic experience gained from foreign countries regarding CWR turnouts on bridges. Maintaining the normal geometry and operational condition of turnouts on bridges under the influence of temperature variations, vertical train loads, and braking/traction necessitates appropriate adjustments and modifications in the arrangement and design of bridges within the turnout section. Adhering to foreign experience, a meticulous focus on the turnout–bridge arrangement is essential during the initial stage of high-speed railway development. This approach can significantly reduce inconveniences during subsequent construction and future operation, without a significant increase in overall investment.

The design of CWR turnouts on bridges requires close cooperation among three disciplines: station/yard, bridge, and track. The arrangement of turnouts should be initially designed according to the station functions within the station/yard discipline. Following that, the span arrangements and turnout girders should be designed within the bridge discipline. The subsequent checking calculations should be conducted regarding CWR turnouts on bridges within the track discipline, leading to recommendations for adjustments in turnout and bridge arrangements, as well as additional forces acting on piers and abutments within the bridge discipline. Checking calculations should be conducted again within the bridge and station/yard discipline regarding the adjustments proposed by the track discipline until all calculation results are accepted. A finalized design scheme should be established upon the approval of all three disciplines involved.

4.4.1 Study on Rational Structural Forms for Turnout Girders

CWR turnouts on bridges should be designed on the premise of bridge structures offering an integral and continuous substructure for the turnouts. The study on rational structural forms of turnout girders is one of the key research topics in the field of CWR turnouts on bridges, involving structural selection and span arrangements for turnout girders. The adaptability of simply-supported girders, rigid frames, and continuous girders to the laying of CWR turnouts is studied below, adhering to the fundamental law governing the distribution of girder-rail displacements.

1. Simply-supported Girder

Simply-supported girders, being one of the most prevalent girder types in the railway sector, are known for their mature technology and convenient construction. Figures 4.2, 4.3 and 4.4 illustrate the arrangement, expansion and contraction forces, and girder-rail displacements of 20–32 m simply-supported girders. The distribution rule of girder-rail relative displacements caused by the expansion and contraction of these girders is outlined below. The maximum girder-rail displacement occurs at the ends of girder spans, while there is a zero displacement point at the midspan. Under the action of temperature force, girder-rail displacements can escalate to 3.45 mm. Superimposing the relative expansion and contraction displacement with those arising from deflection and braking amplifies the overall girder-rail relative displacement. In the span of 32 m, only one zero displacement point is present at the midspan, with both ends experiencing substantial displacements. This displacement pattern renders simply-supported girders ill-equipped to meet the necessary turnout-girder relative displacement criteria for high-speed railway applications. In China, there are ongoing debates regarding whether simply-supported girders are suitable for accommodating high-speed CWR turnouts. One viewpoint in these discussions favors implementing such applications on high-speed railways, citing the successful precedent of No.12 turnouts on simply-supported girders for specific operational trunk railways with a speed of 120 km/h in China. This book's standpoint is rooted in the observation that larger numbered high-speed turnouts are employed on high-speed railways, accompanied by elevated demands for their regularity and geometry on these railways. In scenarios of simply-supported girders, high-speed turnouts are inevitably laid across girder joints. Any expansion/contraction and lateral dislocation of these joints, as well as rotation at girder ends, can potentially alter the geometry of the turnouts. Any irregularity in high-speed turnouts exceeding permissible limits can trigger substantial dynamic reactions from trains, potentially leading to unforeseen consequences in terms of operational safety. In addition, there is a lack of experience in laying high-speed turnouts on simply-supported girders in China and abroad. The consulting recommendations put forward by Germany and France suggest installing high-speed turnouts on girders extending over entire spans. To sum up, turnouts should not be laid on simply-supported girders on the main line of high-speed railways, and potential applications of turnouts on simply-supported girders could be explored for arrival and departure tracks to facilitate future trials.

2. Multi-span Rigid Frame

Multi-span rigid frames have emerged in the field of bridge engineering in China as a structural form tailored to accommodate the characteristics of CWR turnouts on bridges. This structural configuration is distinguished by small temperature spans,

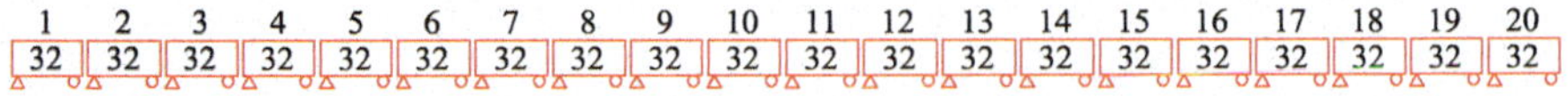

Fig. 4.2 Simply-supported girder

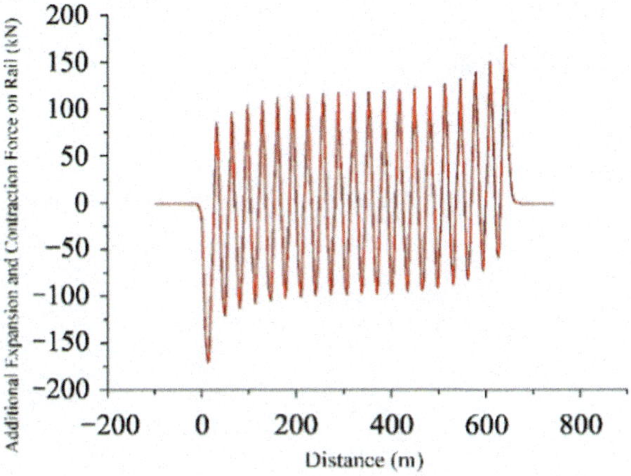

Fig. 4.3 Additional expansion and contraction force on rail

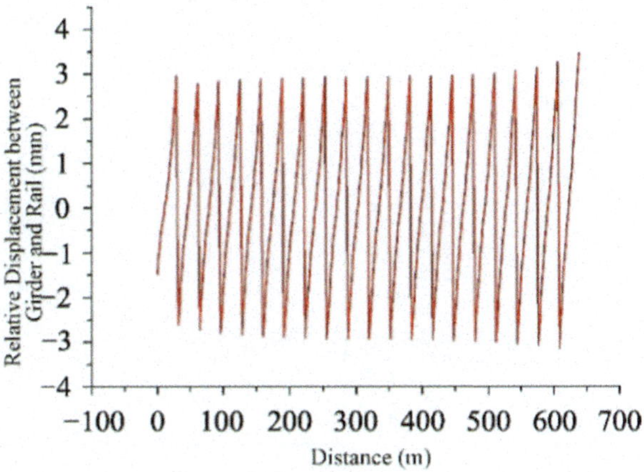

Fig. 4.4 Relative displacement between girder and rail

high rigidity, relatively modest flexural and braking displacements under the effect of girder-rail expansion and contraction displacement and vertical loads, as well as the ease of managing relative displacements between the turnouts and girders beneath. Figures 4.5, 4.6 and 4.7 depict the calculated additional expansion and contraction force and girder-rail relative displacements within 20–20 m rigid frames. Nevertheless, the application of multi-span rigid frames has been restricted primarily due to their limited spanning capability. Moreover, the landscape effect of this bridge design is compromised in urban settings owing to its inherent discordance with the surrounding environments. Additionally, the presence of numerous piers in this design poses challenges in controlling differential settlement across the piers and abutments.

Fig. 4.5 Multi-span rigid frame girder

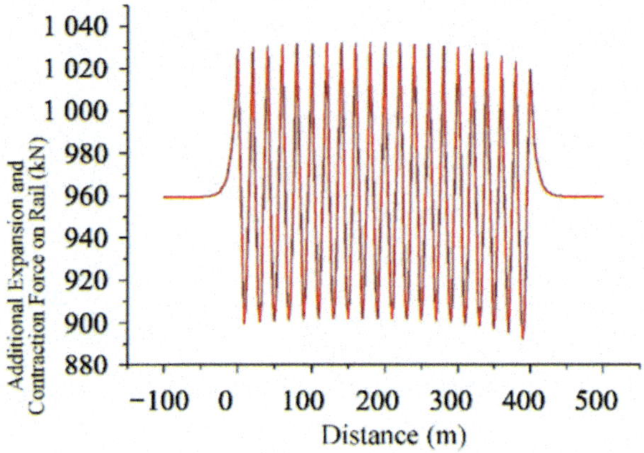

Fig. 4.6 Additional expansion and contraction force on rail

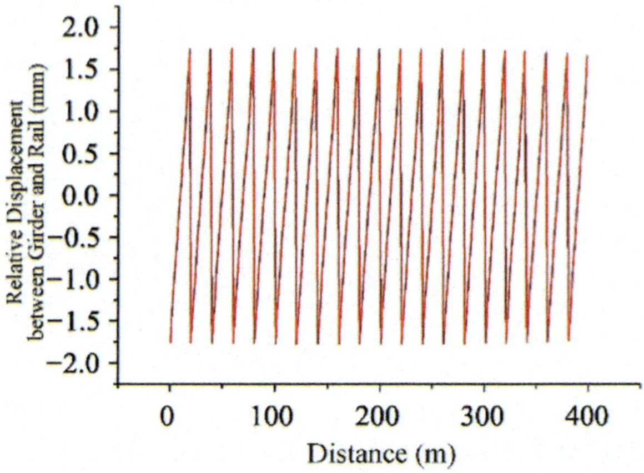

Fig. 4.7 Relative displacement between girder and rail

3. Continuous Girder

Continuous girders are believed to have difficulty meeting the demands of laying CWR turnouts when solely considering their characteristics: large temperature spans leading to substantial additional forces on rails, and significant girder-rail relative displacements. Nonetheless, a profound examination of the relative displacement pattern governing long-span continuous girders can demonstrate the limitations of this viewpoint.

To deeply reveal the change law of girder-rail relative displacements within continuous girders with varying unit lengths and temperature spans, expansion and contraction forces and relative displacements of several continuous girders with different unit lengths were calculated. The specific span arrangements of these continuous girders are as follows: (32 + 48 + 32) m, (40 + 64 + 40) m, (48 + 80 + 48) m, (60 + 100 + 60) m, (80 + 120 + 80) m, (36 + 60 + 60 + 36) m, (50 + 80 + 80 + 50) m, (64 + 4 × 116 + 64) m, (60 + 5 × 100 + 60) m, and (7 × 32) m. See Table 4.3 for the calculation results of expansion and contraction forces and girder-rail relative displacements in the aforementioned continuous girders.

The conclusion drawn from Table 4.3 is as follows: The curve representing girder-rail relative displacements indicates that the maximum relative displacement occurred at the ends of the continuous girders, with a zero point appearing in the middle. As the unit length and temperature span increased, the zero relative displacement point expanded into a segment, growing in length based on the unit length and temperature span. This consistent pattern was observed across all continuous girders with three, four, and more spans, highlighting this inherent characteristic of continuous girders irrespective of the computational models and methodologies employed.

The presence of a zero girder-rail relative displacement segment in the middle of continuous girders offers advantageous conditions for placing CWR turnouts, setting this girder type apart from others. Leveraging this feature in design allows for precise control of relative displacements in critical turnout areas within permissible limits by optimal placement of turnouts within this zero girder-rail relative displacement segment.

Multi-span 32 m continuous girders have been prevalently applied beneath turnouts so far in China, showcasing several benefits such as girder heights consistent

Table 4.3 Expansion and contraction forces and girder-rail displacements of continuous girders with different unit lengths

Continuous girder (m)	Temperature span (m)	Unit length of continuous girder (m)	Maximum longitudinal compressive force due to expansion and contraction at girder end (kN)	Maximum longitudinal tension due to expansion and contraction in the middle of girder (kN)
(32+48+32)	80	112	123.59	−231.604
(40+64+40)	104	144	159.975	−238.713
(48+80+48)	128	176	194.608	−239.273
(60+100+60)	160	220	239.177	−239.317
(80+120+80)	200	280	295.355	−239.32
(36+60+60+36)	128	192	190.768	239.306
(50+80+80+50)	162	260	253.796	239.319
(64+4 × 116+64)	328	592	488.588	239.32
(60+5 × 100+60)	364	620	521.542	239.32
(7 × 32)	128	224	260	243

with simply-supported girders at both ends, aesthetic appearance, integration with the surrounding urban environment, small flexural deformation, and reliable track alignment.

4.4.2 *Study on Spacing Limits Between Turnout and Girder Joint*

Controlling girder-rail relative displacements in critical turnout components specifically switches and frogs, is essential in the design of CWR turnouts on bridges. Adhering to the distribution law of girder-rail relative displacements in long-span continuous girders, wherein the maximum relative displacement occurs at the ends with a zero point in the middle, it is crucial to maintain an adequate distance between turnouts and girder joints to ensure that girder-rail relative displacements fall within the specified limits [8–10]. Given the prominent use of No.18 turnouts on high-speed railway bridges in China, the following study emphasizes the rational spacings of switch rails and point rails within No.18 turnouts with girder joints on bridges featuring continuous girders. As outlined in the Design Principles for CWR Track at Viaduct Stations of Beijing–Shanghai High-speed Railway, the document specifies the minimum distances from switch rails and point rails within turnouts to girder joints on high-speed railways. Refer to Table 4.4 and Fig. 4.8 for the key parameters and arrangement requirements for high-speed turnouts.

The table and figure above specify the requirements for a minimum distance of 18 m between switch rails within No.18 high-speed turnouts and girder joints on high-speed railways, as well as a minimum distance of 18 m between the heel end of turnouts and girder joints. These established values are widely employed in the design of CWR turnouts on bridges in China. their rationality and compliance with requirements are verified below using the turnout–bridge interaction program.

1. Single No.18 High-speed Turnout on Bridge

The minimum girder length for No. 18 turnouts can be derived from the distances from their beginning and end to the girder joints, i.e. $(69 + 18 + 18 - 2.585) = 102.415$ m. There are two bridge arrangement options for the minimum girder length:

Table 4.4 Main parameters of high-speed turnout

Turnout no	Total length (L/m)	Front end length (a/m)	Rear end length (b/m)	Coordinate at actual point of switch rail (x_1/m)	Coordinate at heel end of switch rail (x_2/m)	Coordinate at actual point of point rail (x_3/m)	Coordinate at heel end of point rail (x_4/m)
18	69.000	31.729	37.271	2.585	24.592	55.325	69.000
43	164.400	56.892	107.508	3.220	44.400	113.612	143.612
58	220.800	75.794	145.006	3.220	59.400	150.759	180.759

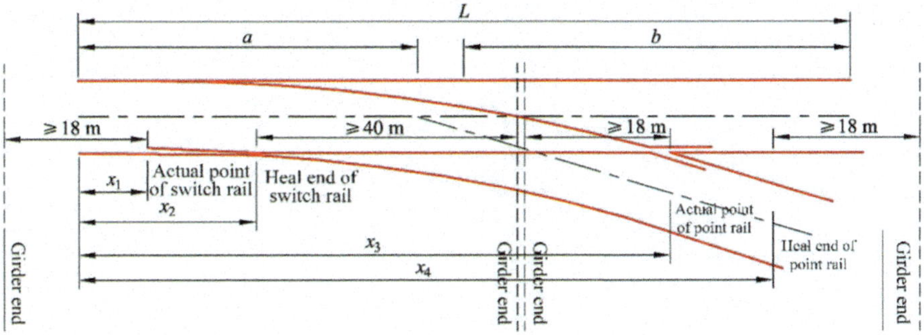

Fig. 4.8 Arrangement requirements of high-speed turnout

Option I: (32 + 48 + 32) m continuous girder with a girder length of 112 m and a turnout-girder arrangement shown in Fig. 4.9.

Option II: (4 × 32) m continuous girder with a girder length of 128 m and a turnout-girder arrangement shown in Fig. 4.10.

(1) Feasibility demonstration for Option I

The expansion and contraction forces on rails and girder-rail relative displacements due to expansion and contraction, deflection, and braking respectively were calculated using the turnout–bridge interaction program, for the case of a No.18 turnout placed on a (32 + 48 + 32) m continuous girder (calculation parameters: ballasted track, pier linear rigidity of 200 kN/cm, maximum rail temperature variation range of 55 °C, daily girder temperature variation range of 15 °C). The calculation results are shown in Figs. 4.11, 4.12, 4.13 and 4.14.

The relative displacements between the rails and bridge shown in Figs. 4.12, 4.13 and 4.14 were used to infer the relative displacements due to expansion and contraction, deflection, and braking respectively within the full length of the CWR turnout, as shown in Figs. 4.15, 4.16 and 4.17. The total relative displacement between the turnout and bridge (total turnout-girder displacement = relative displacement due

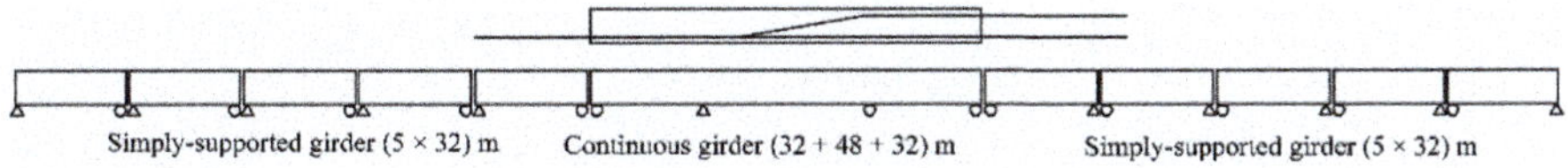

Fig. 4.9 Arrangement of turnout and girder in option I

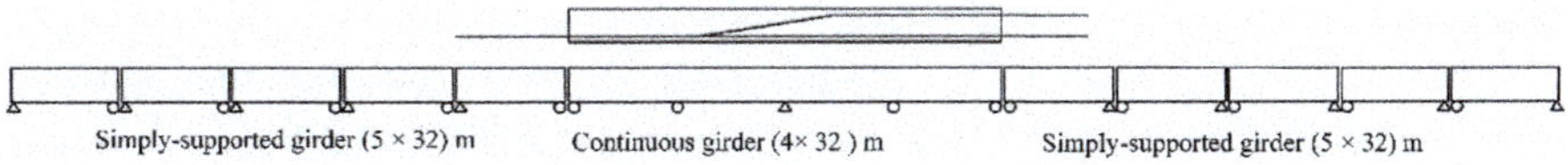

Fig. 4.10 Arrangement of turnout and girder in option II

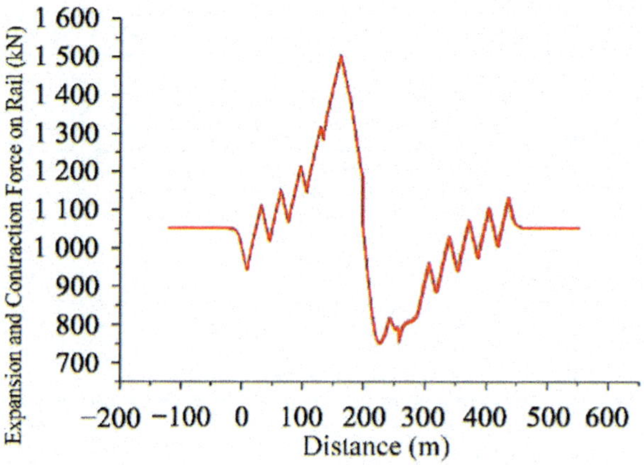

Fig. 4.11 Expansion and contraction force on rail

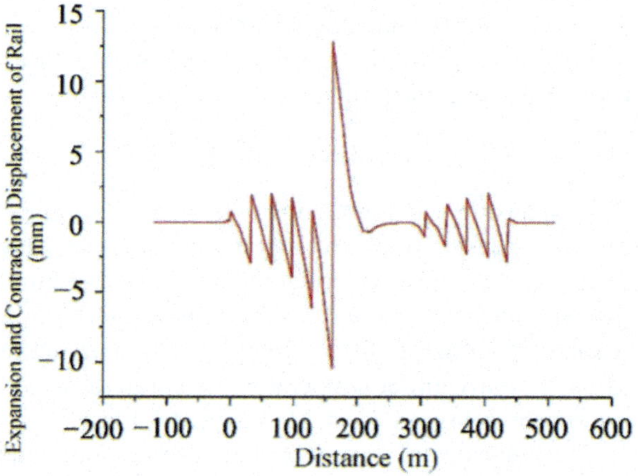

Fig. 4.12 Relative girder-rail expansion and contraction displacement

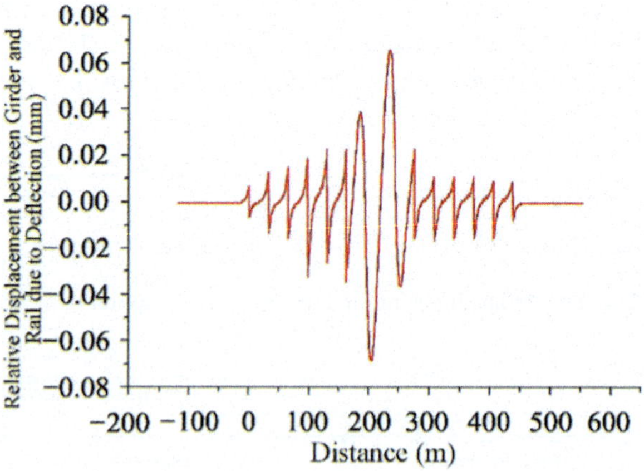

Fig. 4.13 Relative displacement between girder and rail due to deflection

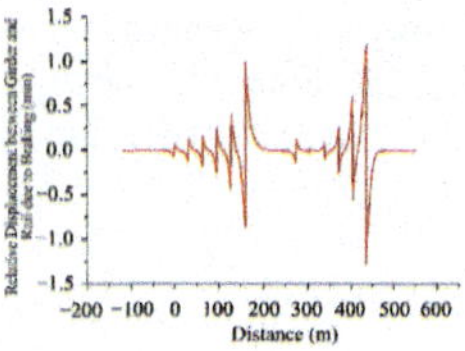

Fig. 4.14 Relative displacement between girder and rail due to braking

to expansion and contraction + relative displacement due to deflection + relative displacement due to braking) is shown in Fig. 4.18.

As depicted in Fig. 4.18, the relative displacement at the frog amounts to 0.72 mm, while at the switch, it measures 6.6 mm. Germany's consulting recommendations for the design of Beijing–Shanghai High-speed Railway in China stipulate turnout-girder displacement limits of 5.2 mm at frogs and 22 mm at switches. Upon

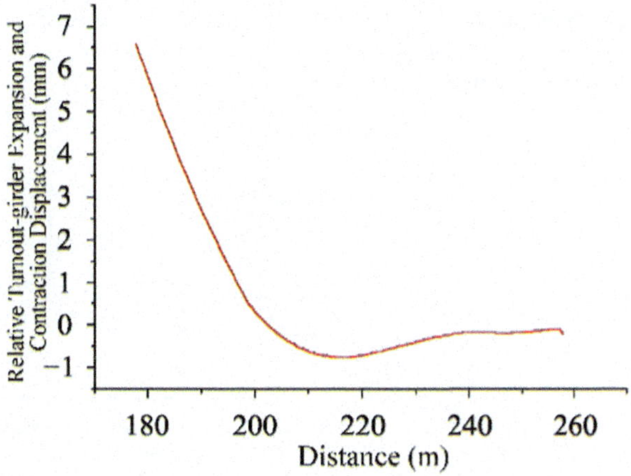

Fig. 4.15 Relative turnout-girder expansion and contraction displacement

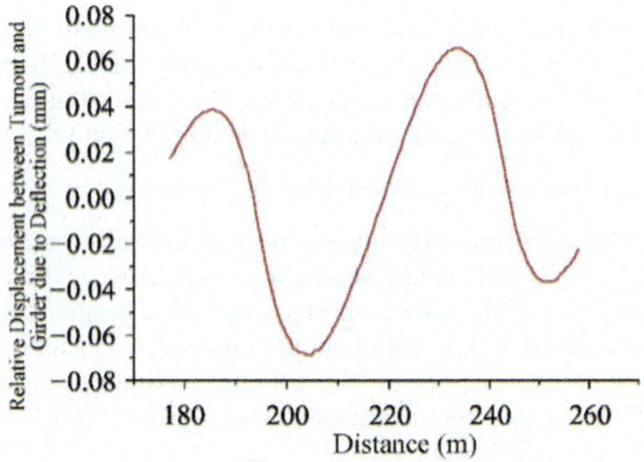

Fig. 4.16 Relative displacement between turnout and girder due to deflection

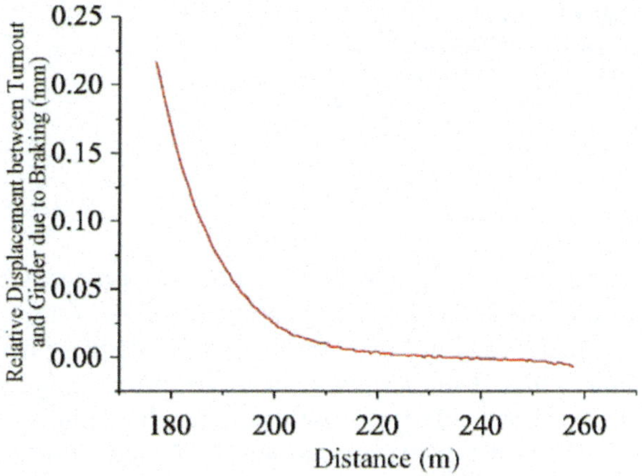

Fig. 4.17 Relative displacement between turnout and girder due to braking

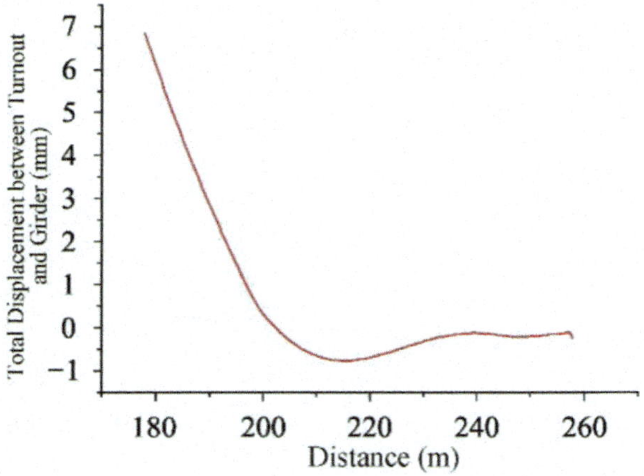

Fig. 4.18 Total displacement between turnout and girder

assessment based on these standards, the turnout–bridge relative displacement satisfactorily adheres to the criteria for the No. 18 high-speed turnout placed on the (32 + 48 + 32) m continuous girder. The demonstration results prove the feasibility of the bridge and turnout arrangements in Option I.

(2) Feasibility demonstration for Option II

The expansion and contraction forces on rails and girder-rail relative displacements due to expansion and contraction, deflection, and braking respectively were calculated using the turnout–bridge interaction program, for the case of a No.18 turnout placed on a (4 × 32) m continuous girder (calculation parameters: ballasted track, pier linear rigidity of 200 kN/cm, maximum rail temperature variation range of 55°C, daily girder temperature variation of 15°C). The calculation results are shown in Figs. 4.19, 4.20, 4.21 and 4.22.

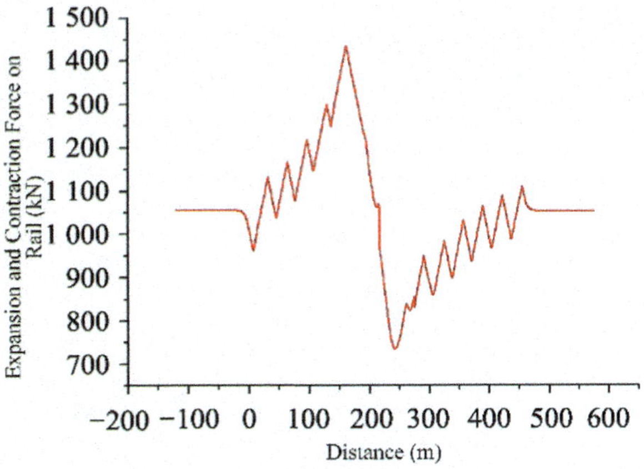

Fig. 4.19 Expansion and contraction force on rail

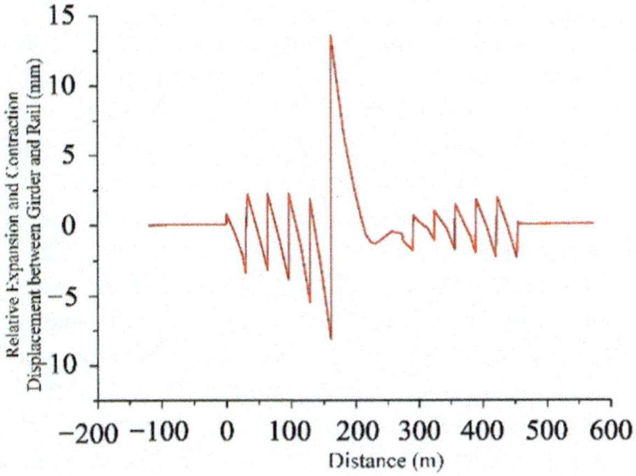

Fig. 4.20 Relative girder-rail expansion and contraction displacement

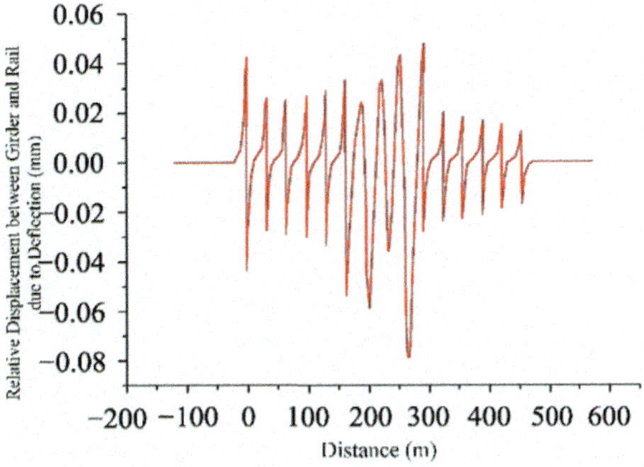

Fig. 4.21 Relative displacement between girder and rail due to deflection

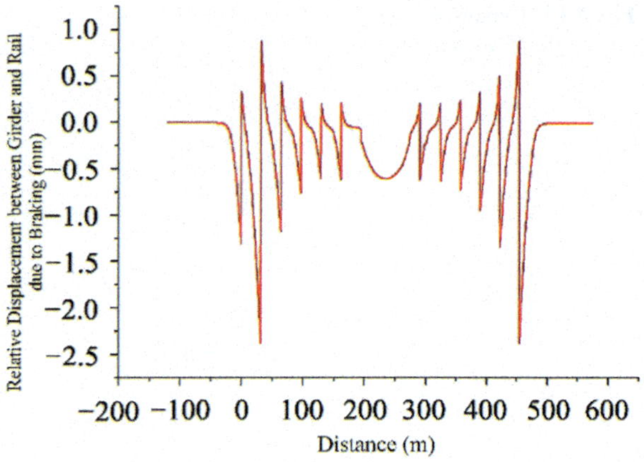

Fig. 4.22 Relative displacement between girder and rail due to braking

The relative displacements between the rails and bridge shown in Figs. 4.20, 4.21 and 4.22 were used to infer the relative displacements due to expansion and contraction, deflection, and braking respectively within the full length of the CWR turnout, as shown in Figs. 4.23, 4.24 and 4.25. The total relative displacement between the turnout and bridge (total turnout-girder displacement = relative displacement due to expansion and contraction + relative displacement due to deflection + relative displacement due to braking) is shown in Fig. 4.26.

As depicted in Fig. 4.26, the relative displacement at the switch amounts to 2.9 mm, while at the frog, it measures 1.0 mm. Germany's consulting recommendations for the design of Beijing–Shanghai High-speed Railway in China stipulate turnout-girder displacement limits of 5.2 mm at frogs and 22 mm at switches. Upon assessment based on these standards, the turnout–bridge relative displacement satisfactorily adheres to the criteria for the No. 18 high-speed turnout placed on the (4

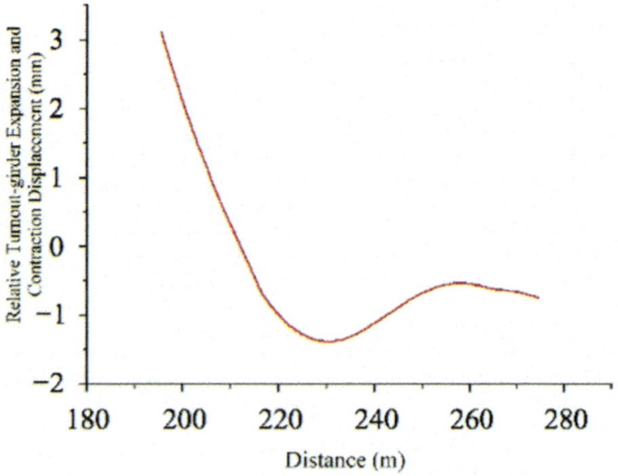

Fig. 4.23 Relative turnout-girder expansion and contraction displacement

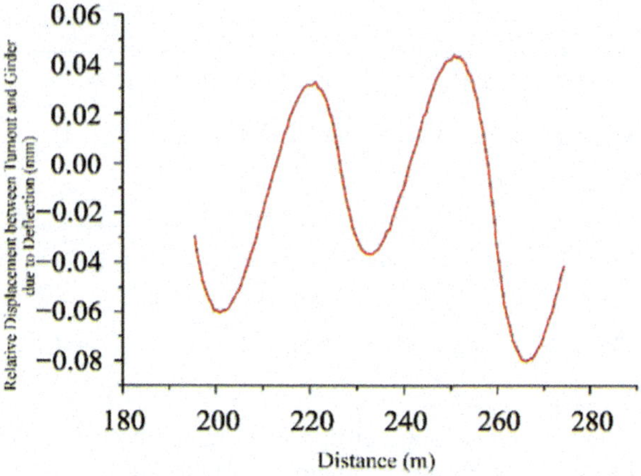

Fig. 4.24 Relative displacement between turnout and girder due to deflection

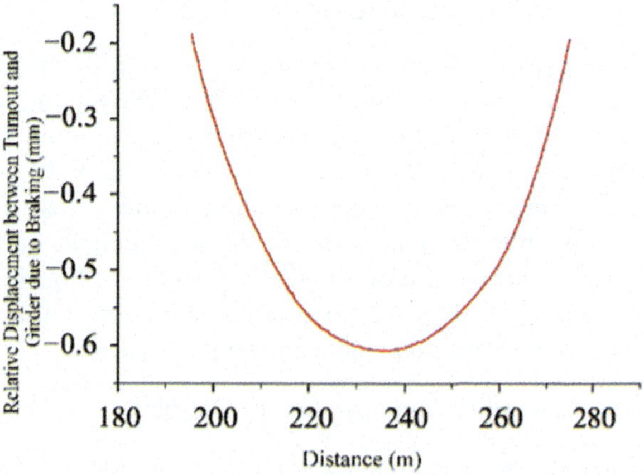

Fig. 4.25 Relative displacement between turnout and girder due to braking

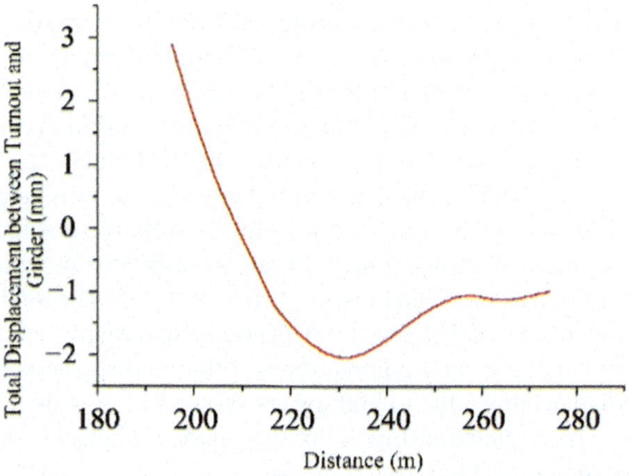

Fig. 4.26 Total displacement between turnout and girder

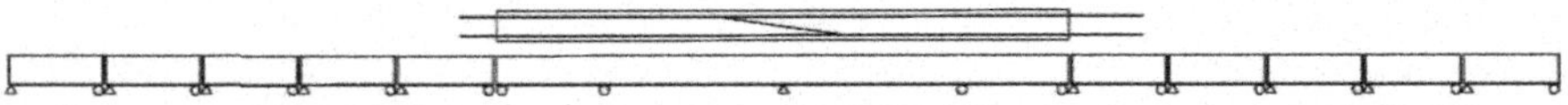

Fig. 4.27 Arrangement of turnout and girder in option I

Fig. 4.28 Arrangement of turnout and girder in option II

× 32) m continuous girder. The demonstration results prove the feasibility of the bridge and turnout arrangements in Option II.

2. No.18 Single Crossover on Bridge

The total length of No. 18 crossovers is 153.458 m, corresponding to a distance between the centers of tracks of 5.0 m. The minimum length of turnout girders can be inferred from the distance from the beginning and end of No. 18 high-speed turnouts to girder joints, i.e. 153.458 m + 2 × 18 m = 189.458 m. There are two bridge arrangement options for the minimum girder length:

Option I: (36 + 60 + 60 + 36) m continuous girder with a girder length of 192 m and a turnout-girder arrangement shown in Fig. 4.27.

Option II: (6 × 32) m continuous girder with a girder length of 192 m and a turnout-girder arrangement shown in Fig. 4.28.

(1) Feasibility demonstration of Option I

The expansion and contraction forces on rails and girder-rail relative displacements due to expansion and contraction, deflection, and braking respectively were calculated using the turnout–bridge interaction program, for the case of a No.18 single crossover placed on a (36+60+60+36) m continuous girder (calculation parameters: ballasted track, pier linear rigidity of 200 kN/cm, maximum rail temperature variation range of 55 °C, daily girder temperature variation range of 15 °C). The calculation results are shown in Figs. 4.29, 4.30, 4.31 and 4.32.

The relative displacements between the rails and bridge shown in Figs. 4.30, 4.31 and 4.32 were used to infer the relative displacements due to expansion and contraction, deflection, and braking respectively within the full length of the CWR turnout, as shown in Figs. 4.33, 4.34 and 4.35. The total relative displacement between the turnout and bridge (total turnout-girder displacement = relative displacement due to expansion and contraction + relative displacement due to deflection + relative displacement due to braking) is shown in Fig. 4.36.

As depicted in Fig. 4.36, the relative displacement at the switch amounts to 2.9 mm, while at the frog, it measures 0.6 mm. Germany's consulting recommendations

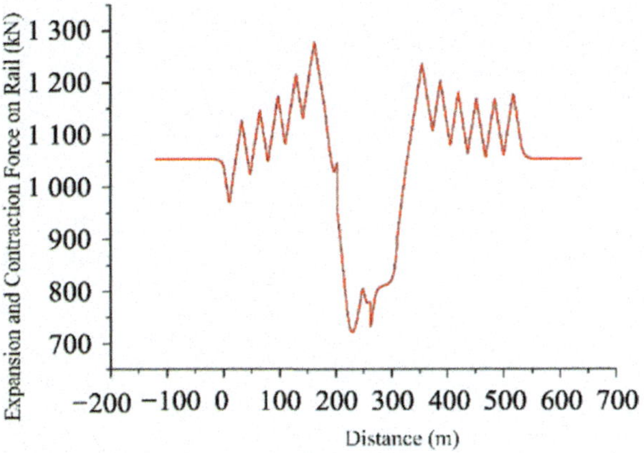

Fig. 4.29 Expansion and contraction force on rail

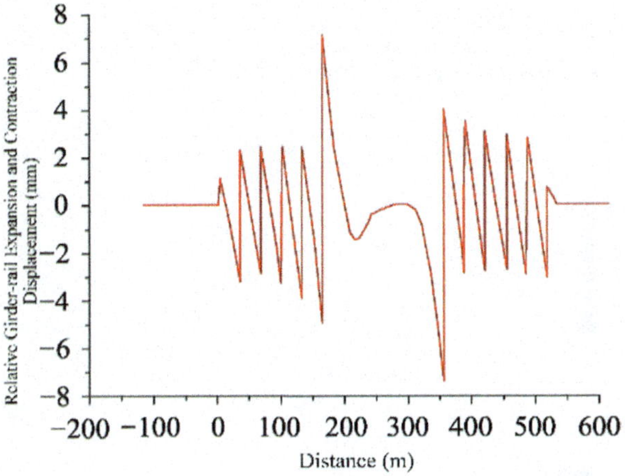

Fig. 4.30 Relative girder-rail expansion and contraction displacement

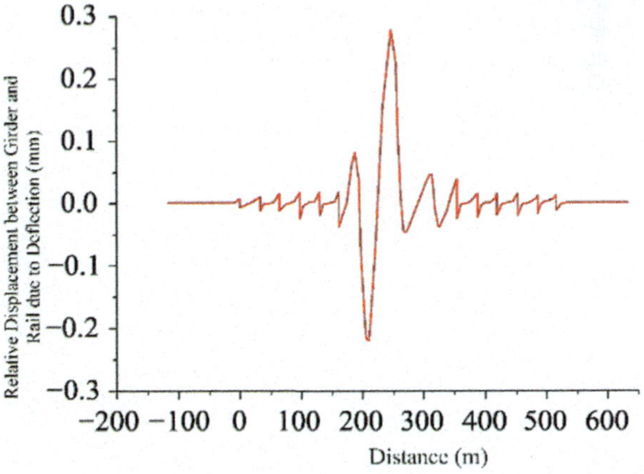

Fig. 4.31 Relative displacement between girder and rail due to deflection

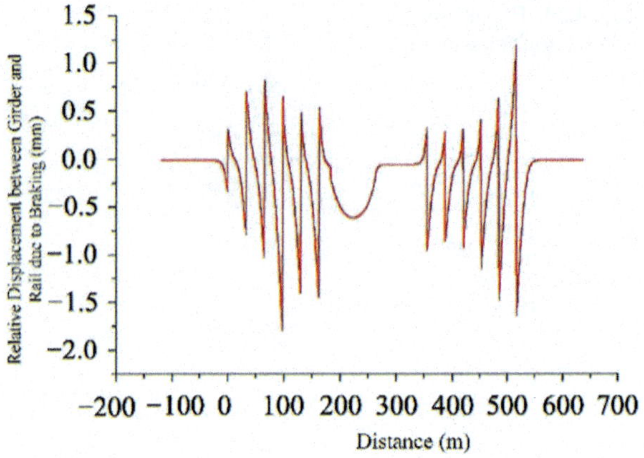

Fig. 4.32 Relative displacement between girder and rail due to braking

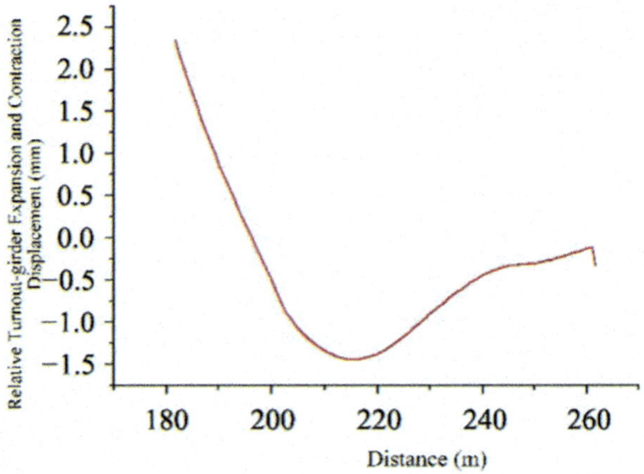

Fig. 4.33 Relative turnout-girder expansion and contraction displacement

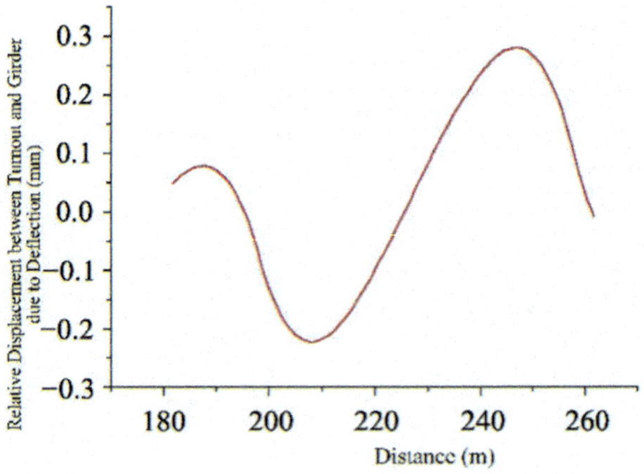

Fig. 4.34 Relative displacement between turnout and girder due to deflection

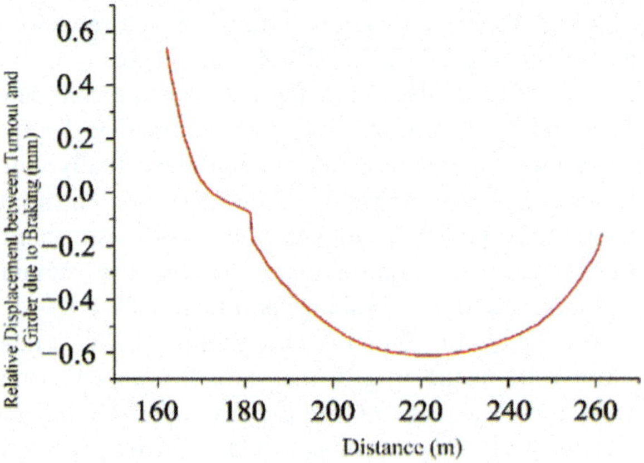

Fig. 4.35 Relative displacement between turnout and girder due to braking

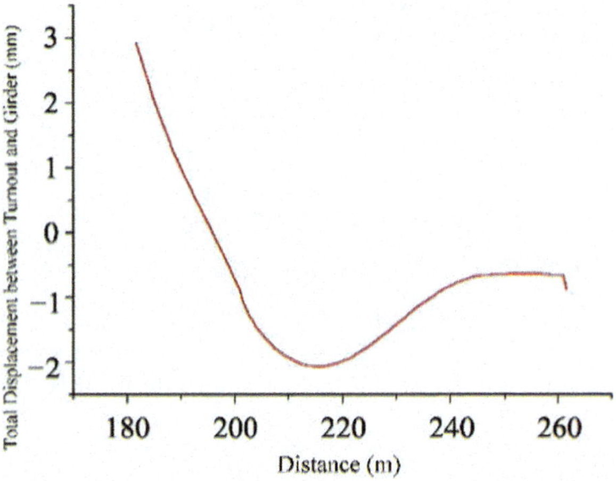

Fig. 4.36 Total displacement between turnout and girder

for the design of Beijing–Shanghai High-speed Railway in China stipulate turnout-girder displacement limits of 5.2 mm at frogs and 22 mm at switches. Upon assessment based on these standards, the turnout–bridge relative displacement satisfactorily adheres to the criteria for the No. 18 single crossover placed on the (36+60+60+36) m continuous girder. The demonstration results prove the feasibility of the bridge and turnout arrangements in Option I.

(2) Feasibility demonstration for Option II

The expansion and contraction forces on rails and girder-rail relative displacements due to expansion and contraction, deflection, and braking respectively were calculated using the turnout–bridge interaction program, for the case of a No.18 single crossover placed on a (6 × 32) m continuous girder (calculation parameters: ballasted track, pier linear rigidity of 200 kN/cm, maximum rail temperature variation range of 55 °C, daily girder temperature variation range of 15 °C). The calculation results are shown in Figs. 4.37, 4.38, 4.39 and 4.40.

The relative displacements between the rails and bridge shown in Figs. 4.38, 4.39 and 4.40 were used to infer the relative displacements due to expansion and contraction, deflection, and braking respectively within the full length of the CWR turnout, as shown in Figs. 4.41, 4.42 and 4.43. The total relative displacement between the turnout and bridge (total turnout-girder displacement = relative displacement due to expansion and contraction + relative displacement due to deflection + relative displacement due to braking) is shown in Fig. 4.44.

As depicted in Fig. 4.44, the relative displacement at the switch amounts to 2.2 mm, while at the frog, it measures 0.8 mm. Germany's consulting recommendations for the design of Beijing–Shanghai High-speed Railway in China stipulate turnout-girder displacement limits of 5.2 mm at frogs and 22 mm at switches. Upon

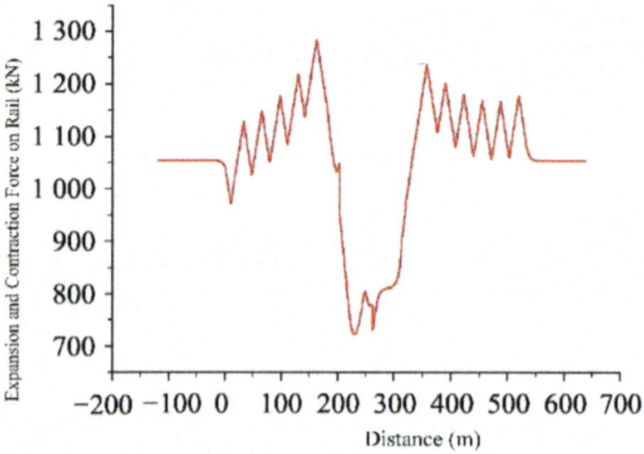

Fig. 4.37 Expansion and contraction force on rail

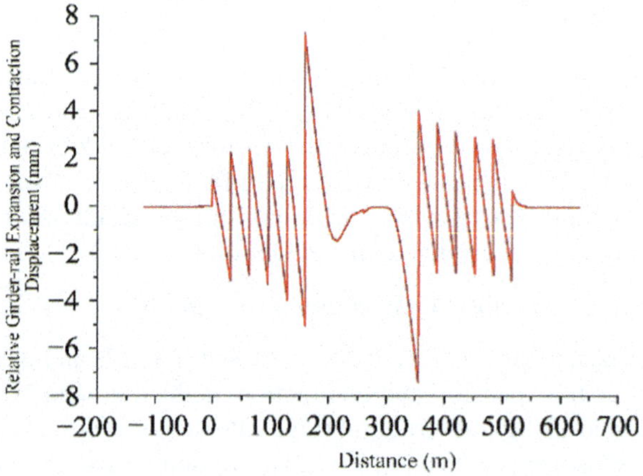

Fig. 4.38 Relative girder-rail expansion and contraction displacement

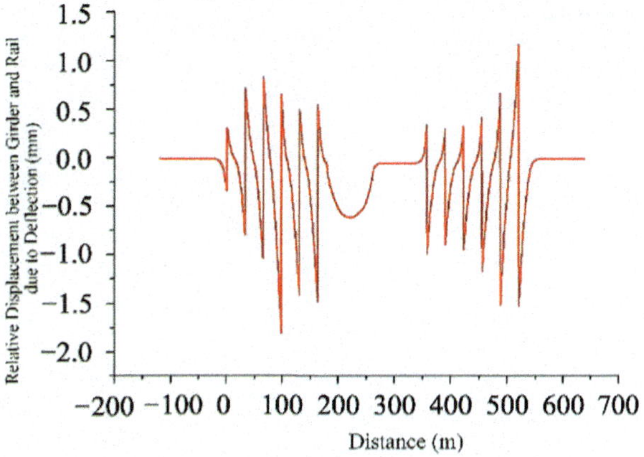

Fig. 4.39 Relative displacement between girder and rail due to deflection

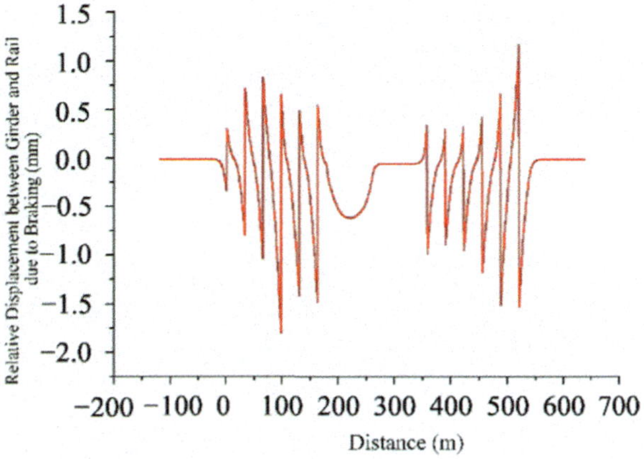

Fig. 4.40 Relative displacement between girder and rail due to braking

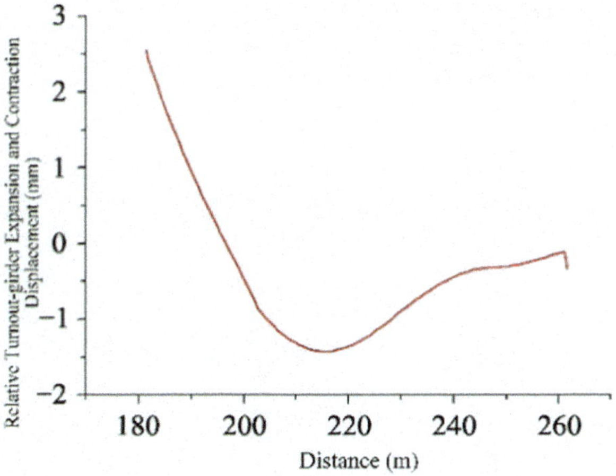

Fig. 4.41 Relative turnout-girder expansion and contraction displacement

Fig. 4.42 Relative displacement between turnout and girder due to deflection

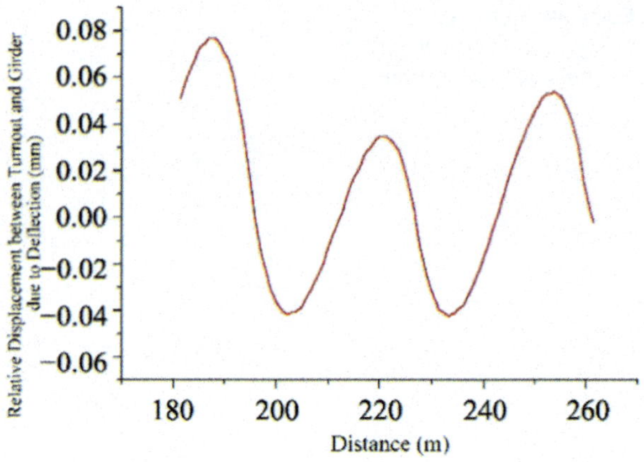

Fig. 4.43 Relative displacement between turnout and girder due to braking

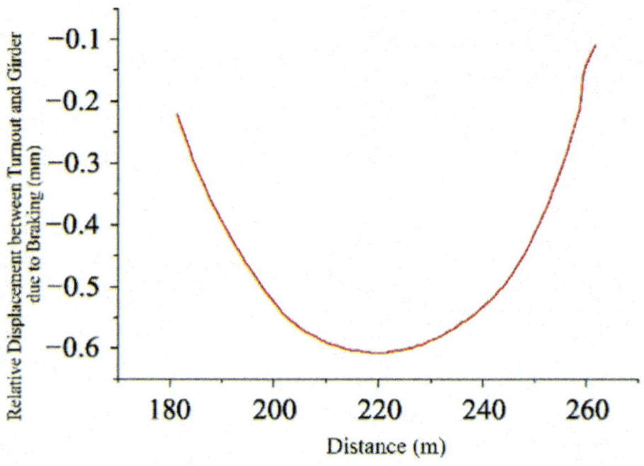

Fig. 4.44 Total displacement between turnout and girder

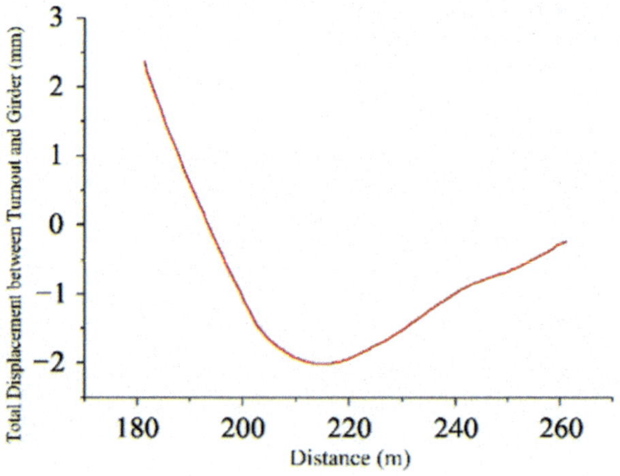

assessment based on these standards, the turnout–bridge relative displacement satisfactorily adheres to the criteria for the No. 18 single crossover placed on the (6 × 32) m continuous girder. The demonstration results prove the feasibility of the bridge and turnout arrangements in Option II.

From the above computational analysis results, it is evident that maintaining a minimum distance of 18 m from the beginning of No. 18 high-speed turnouts placed on passenger-dedicated lines to girder joints, as well as a minimum distance of 18 m from the end of the turnout to girder joints, can ensure that turnout–bridge relative displacements fall within the specified limits.

4.4.3 Study on Bearing Arrangement of Continuous Girder Beneath Turnout

1. Bearing Positioning

When continuous girders beneath CWR turnouts are furnished with only one fixed bearing, the bearing positioning has a great influence on the stress and deformation of the turnouts. This section takes a (40 + 48 + 40) m continuous girder as an example to analyze the influence of fixed bearing positioning. There are two arrangements for the single No. 18 high-speed turnout configuration on the (40 + 48 + 40) m continuous girder:

(1) Expansion and contraction displacement towards the same direction for turnout closure rails and continuous girder

The positioning of the fixed bearing of the continuous girder at the right end of the 48 m span and the placement of the turnout in the middle of the continuous girder bridge, with the actual points of the switch rails pointing to the left end, indicate that expansion and contraction displacements of the turnout closure rails occur in the same direction as those at the left end of the fixed bearing of the continuous girder. The fixed bearing arrangement is illustrated in Fig. 4.45. Calculations were conducted using the other parameters without any modifications. Rail expansion joints were not integrated into this arrangement.

(2) Expansion and contraction displacement towards the opposite directions for turnout closure rails and continuous girder

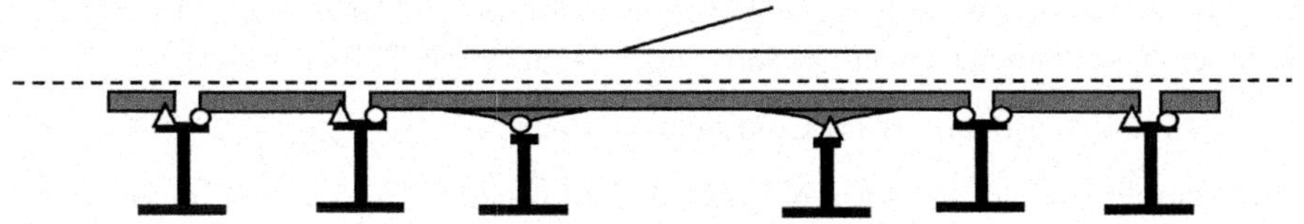

Fig. 4.45 Bearing arrangement I of continuous girder

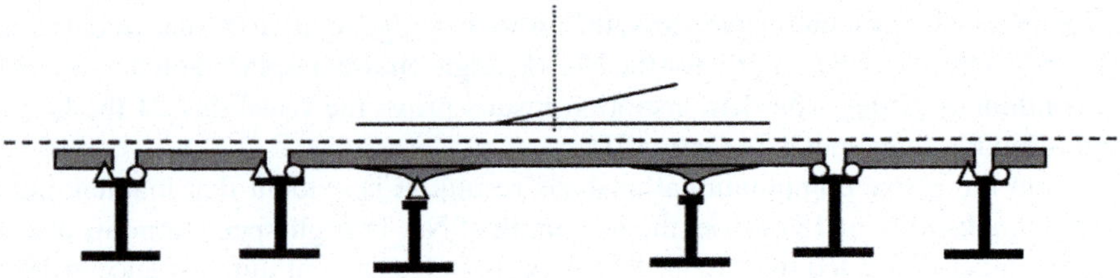

Fig. 4.46 Bearing arrangement II of continuous girder

Fig. 4.47 Rail force in arrangement option I

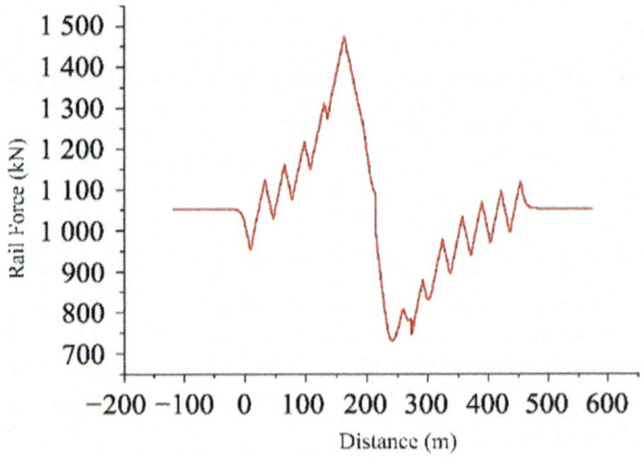

The positioning of the fixed bearing of the continuous girder at the left end of the 48 m span, while keeping the other girder spans and turnout positions unchanged, indicates that expansion and contraction displacements of the turnout closure rails occur in an opposite direction to those of the continuous girder, as shown in Fig. 4.46.

The calculation results of the longitudinal forces on rails in the two arrangements are shown in Figs. 4.47 and 4.48.

See Table 4.5 for the comparison of calculation results of rail forces, pier/abutment forces, and displacements at switch rails and point rails within the turnout section in the two arrangements.

As seen from Table 4.5, the longitudinal forces on the fixed pier under Bearing Arrangement I are less than those under Bearing Arrangement II, indicating favorable stress on the pier. The additional forces on rails at turnout stoppers, the displacements at the actual point of switch rails, and the displacements at the actual point of point rails under Bearing Arrangement II are less than those under Bearing Arrangement I, indicating advantages in stress and deformation of the CWR turnouts.

2. Double Fixed Bearings for Continuous Girder

An example was taken with a No. 30 turnout placed on a (5 × 32) m continuous girder at the junction between a connecting line and the main line for a branching off from the main line, where there are two rails at one end of the turnout girder and four

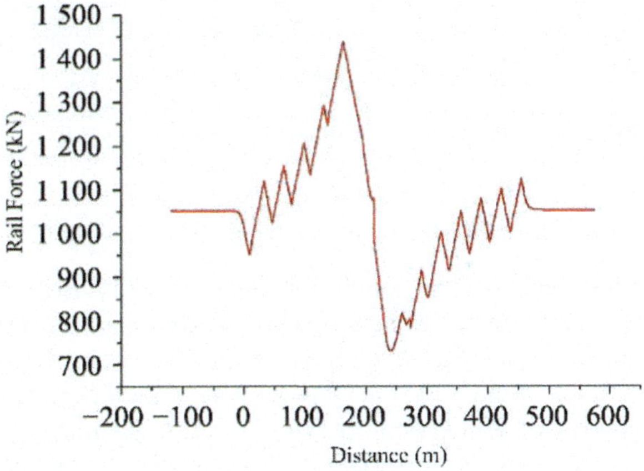

Fig. 4.48 Rail force in arrangement option II

Table 4.5 Comparison of results

Item	Bearing arrangement I	Bearing arrangement II
Additional force on rail at turnout stopper (kN)	33	19
Longitudinal force on fixed pier under continuous girder (kN)	612	852
Displacement at actual point of switch rail (mm)	28.3	26.1
Displacement at actual point of point rail (mm)	19.8	17.6

rails at the other end. In the case of only one fixed bearing, significant forces will act on the fixed pier of the continuous girder, leading to demanding rigidity requirements for this fixed pier. A comparative analysis was performed to explore the feasibility of setting double fixed bearings for the continuous girder, as elaborated below.

The total length of the No. 30 turnout is 102.4 m. The minimum length of the turnout girder can be inferred from the specified minimum distance from the beginning and end of No. 30 high-speed turnouts to girder joints, i.e. 102.4 m+2 × 18 m=138.4 m. There are two fixed bearing arrangements:

Option I: The continuous girder is provided with one fixed bearing, and the arrangement of the girder under the turnout is shown in Fig. 4.49.

Option II: The continuous girder is provided with two fixed bearings, and the arrangement of the girder under the turnout is shown in Fig. 4.50.

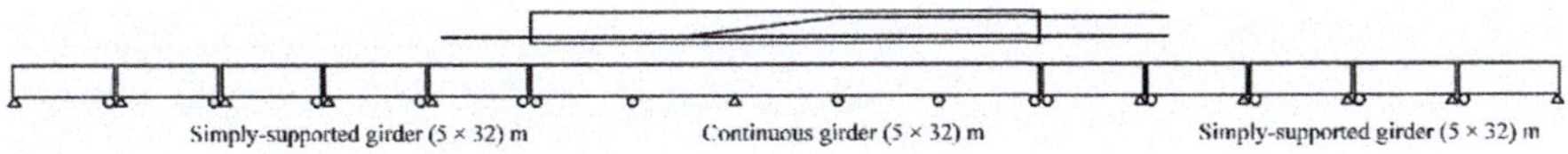

Fig. 4.49 Arrangement of turnout and girder in option I

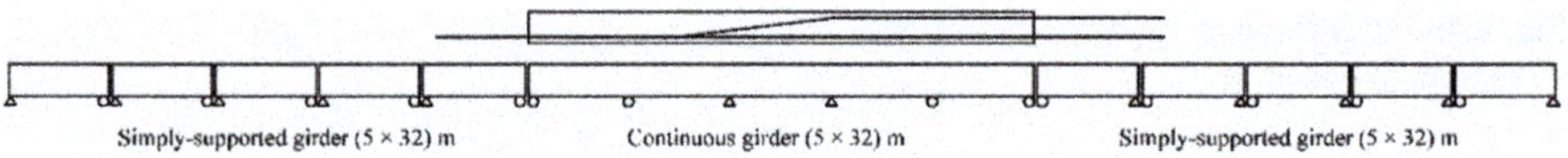

Fig. 4.50 Arrangement of turnout and girder in option II

(1) Computational analysis of CWR turnout on bridge in Option I

The expansion and contraction forces on rails, girder-rail relative displacements due to expansion and contraction, and additional forces on the pier and abutment were calculated using the turnout–bridge interaction program, for the case of a CWR turnout on the bridge in Option I (calculation parameters: ballasted track; pier linear rigidity of 200 kN/cm; maximum rail temperature variation range of 55 °C; daily girder temperature variation range of 15 °C). The calculation results are shown in Figs. 4.51, 4.52 and 4.53.

Fig. 4.51 Expansion and contraction force on rail

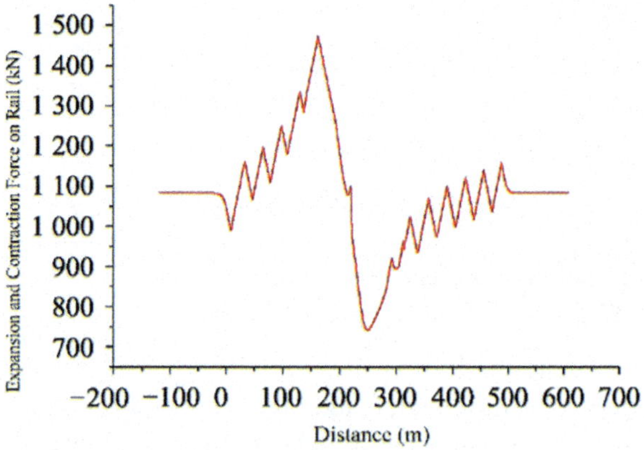

Fig. 4.52 Relative girder-rail expansion and contraction displacement

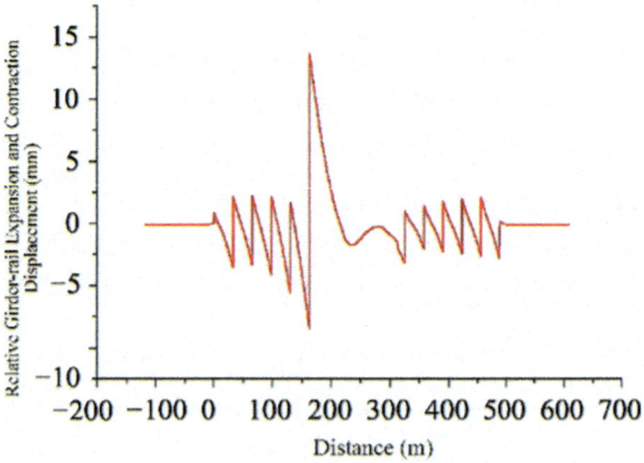

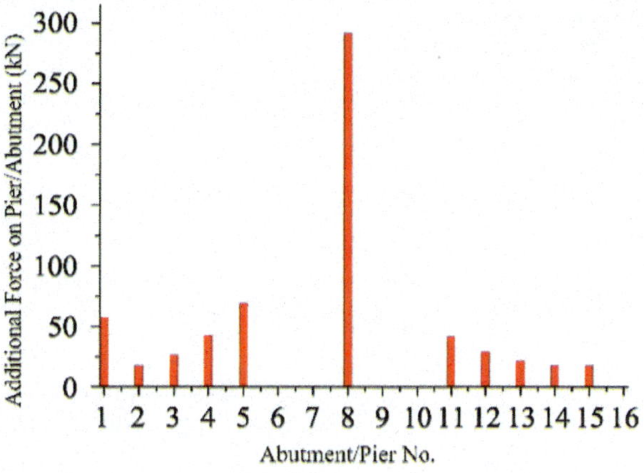

Fig. 4.53 Additional force on pier/abutment

(2) Computational analysis of CWR turnout on bridge in Option II

The expansion and contraction forces on rails, girder-rail relative displacements due to expansion and contraction, and additional forces on the pier and abutment were calculated using the turnout–bridge interaction program, for the case of a CWR turnout on the bridge in Option II (calculation parameters: ballasted track; pier linear rigidity of 200 kN/cm; maximum rail temperature variation range of 55 °C; daily girder temperature variation range of 15 °C). The calculation results are shown in Figs. 4.54, 4.55 and 4.56.

(3) Comparative analysis

Refer to Table 4.6 for the comparative analysis of calculation results between Options I and II.

As shown in the table above, in the case of supporting turnout girders with double fixed bearings, the additional forces carried by a single fixed bearing in Option I are

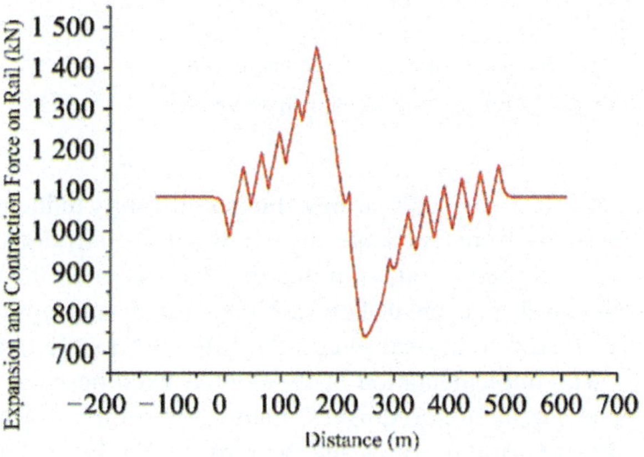

Fig. 4.54 Expansion and contraction force on rail

Fig. 4.55 Relative girder-rail expansion and contraction displacement

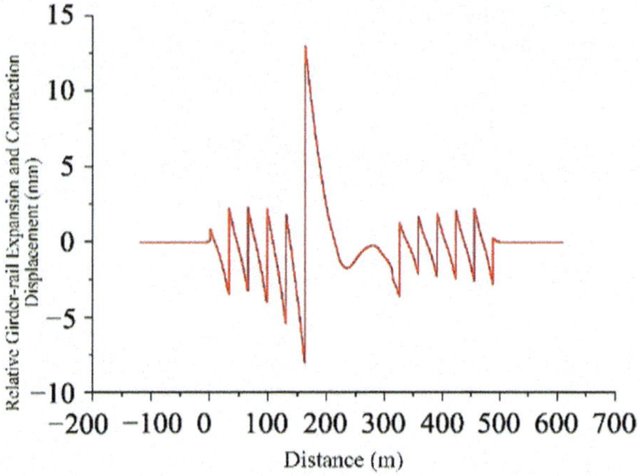

Fig. 4.56 Additional Force on Pier/Abutment

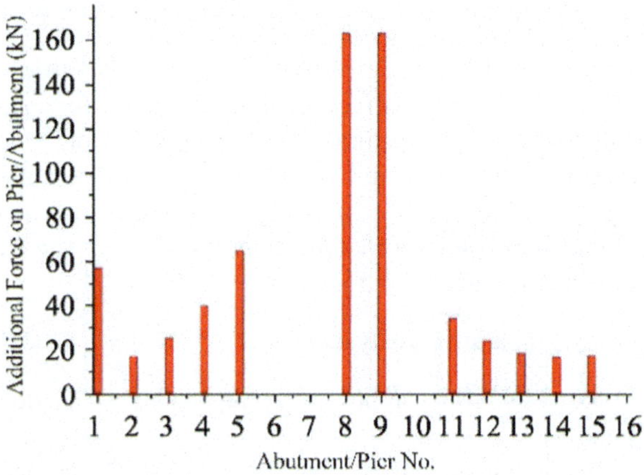

Table 4.6 Comparative analysis between options I and II

Item	Option I	Option II
Expansion and contraction force on rail (kN)	384	363
Additional forces on pier/abutment (kN/rail)	291	163

shared by two fixed bearings, thus significantly reducing additional forces on the pier/abutment from 291 kN/rail to 163kN/rail. Introducing an additional fixed bearing proves advantageous in mitigating the stress on the fixed bearing and minimizing relative displacement between the turnout and girder. When turnout girders bear substantial additional forces, the option of double fixed bearings is feasible.

The implementation of the double fixed bearings option for continuous girders entails certain prerequisites, including specific restrictions for the rigidity of piers carrying fixed bearings and the girder span. Integrating double fixed bearings leads to a statically indeterminate structure, where temperature-induced expansion and contraction in the continuous girder generates forces acting on the two fixed bearings. This resultant force rises with increases in pier rigidity and girder span. Consequently, it becomes imperative to conduct checking calculations to ascertain that the stress on the bearings remains within the permissible limit based on pier rigidity and girder span, before introducing this option.

4.4.4 Study on Inserting Simply-Supported Girder Between Multiple Units of Continuous Girders Beneath Turnouts

Double crossovers are commonly placed between the throat area and home/starting signals at high-speed railway stations to support station functions. To ensure compliance with the spacing limits between turnouts and girder joints as above-mentioned, two units of (6 × 32) m continuous girders are generally erected to support the double crossovers. In scenarios involving Option I, where the temperature span extends to 192 m, substantial additional forces are exerted on the rails and result in significant girder-rail displacement, proving disadvantageous for the stress and deformation of both the turnout and the bridge. This section discusses the feasibility of inserting a simply-supported girder between two units of continuous girders from the point of view of turnout–bridge interactions.

Option I: without inserting simply-supported girder, as shown in Fig. 4.57.

Option II: inserting a span of 32 m simply-supported girder, as shown in Fig. 4.58.

Option III: insert two spans of 32 m simply-supported girder, as shown in Fig. 4.59.

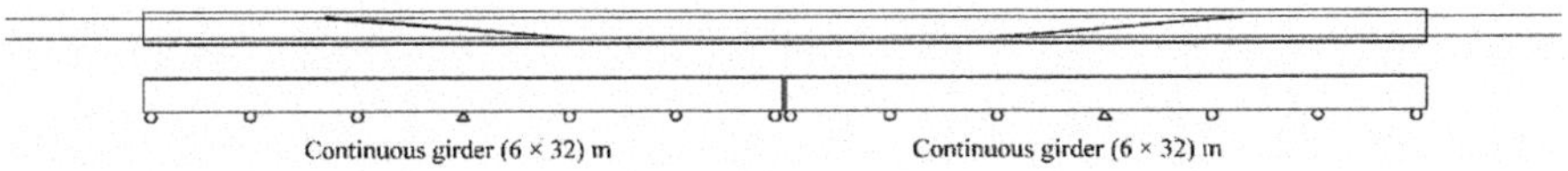

Fig. 4.57 Arrangement option I of bridge and turnout

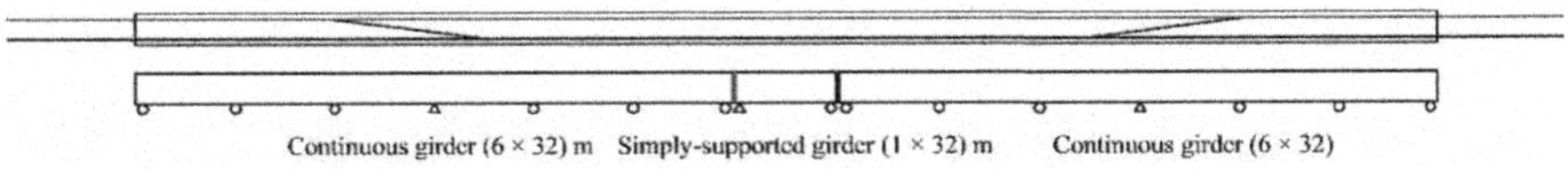

Fig. 4.58 Arrangement option II of bridge and turnout

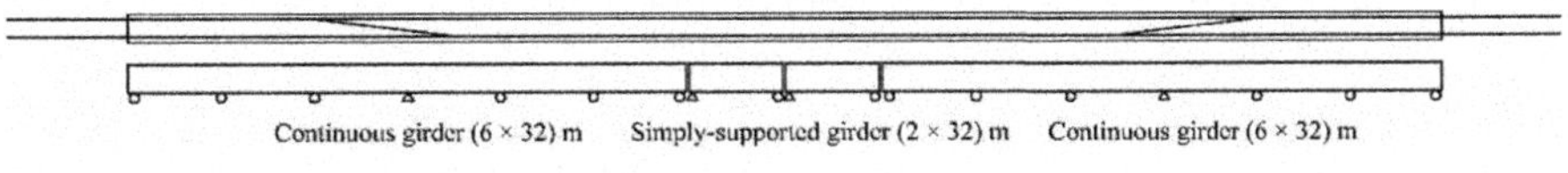

Fig. 4.59 Arrangement option III of bridge and turnout

1. Computational analysis of CWR turnout on bridge in Option I

The expansion and contraction forces on rails, girder-rail relative displacements due to expansion and contraction, and additional forces on the pier and abutment were calculated using the turnout–bridge interaction program, for the case of a CWR turnout on the bridge in Option I (calculation parameters: ballasted track; pier linear rigidity of 200 kN/cm; maximum rail temperature variation range of 55 °C; daily girder temperature variation range of 15 °C). The calculation results are shown in Figs. 4.60, 4.61 and 4.62.

2. Computational analysis of CWR turnout on bridge in Option II

The expansion and contraction forces on rails, girder-rail relative displacements due to expansion and contraction, and additional forces on the pier and abutment were calculated using the turnout–bridge interaction program, for the case of a CWR turnout on the bridge in Option II (calculation parameters: ballasted track; pier linear rigidity of 200 kN/cm; maximum rail temperature variation range of 55 °C; daily

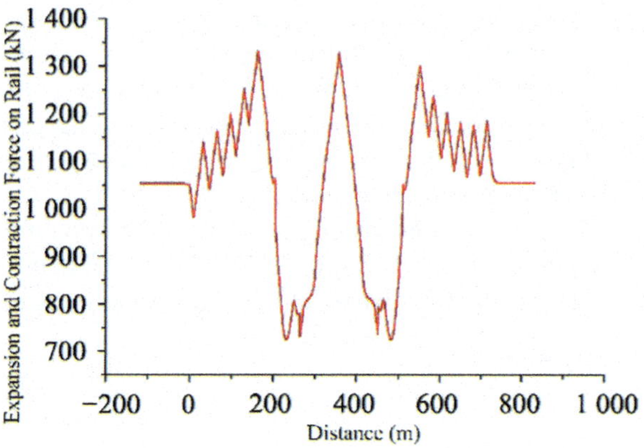

Fig. 4.60 Expansion and contraction force on rail

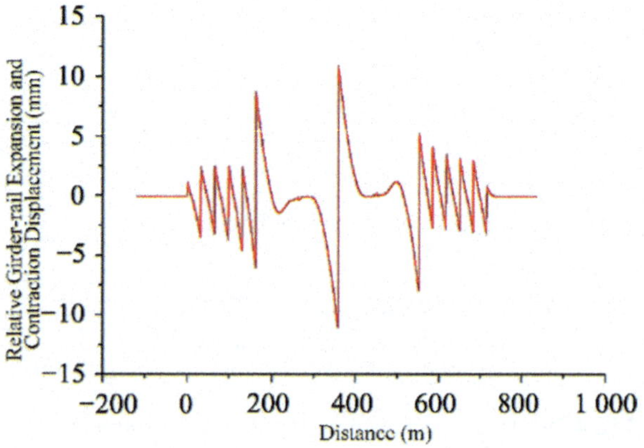

Fig. 4.61 Relative girder-rail expansion and contraction displacement

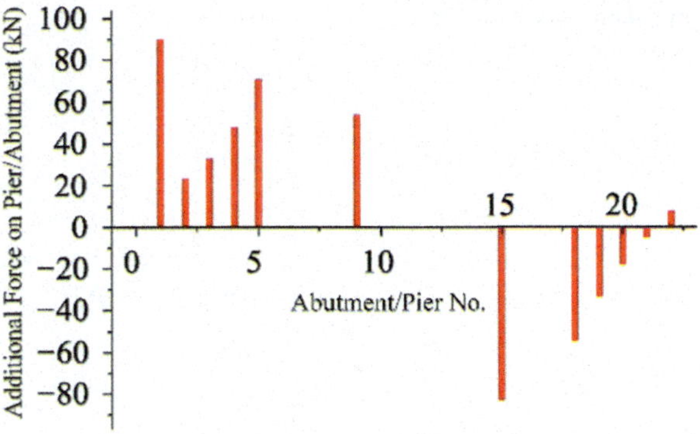

Fig. 4.62 Additional force on pier/abutment

girder temperature variation range of 15 °C). The calculation results are shown in Figs. 4.63, 4.64 and 4.65.

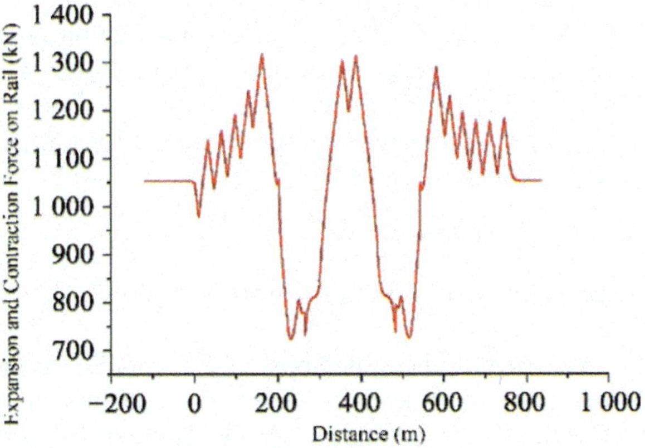

Fig. 4.63 Expansion and contraction force on rail

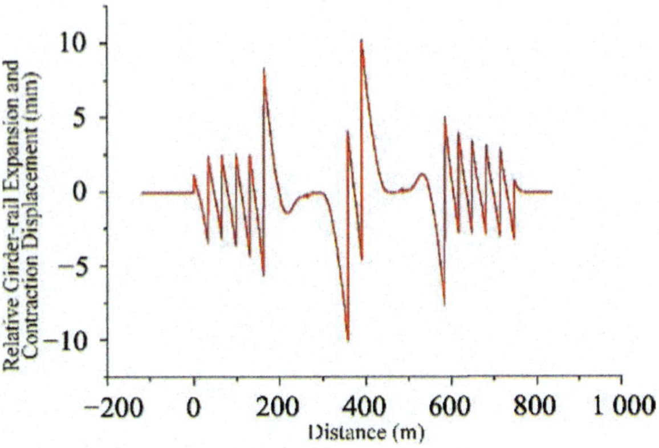

Fig. 4.64 Relative girder-rail expansion and contraction displacement

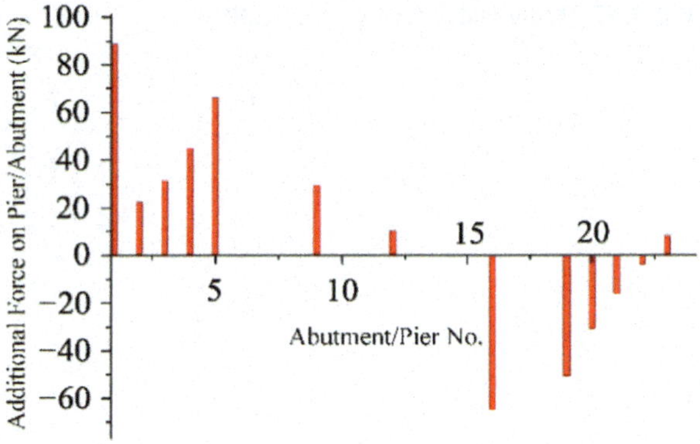

Fig. 4.65 Additional force on pier/abutment

3. Computational analysis of CWR turnout on bridge in Option III

The expansion and contraction forces on rails, girder-rail relative displacements due to expansion and contraction, and additional forces on the pier and abutment were calculated using the turnout–bridge interaction program, for the case of a CWR turnout on the bridge in Option III (calculation parameters: ballasted track; pier linear rigidity of 200 kN/cm; maximum rail temperature variation range of 55 °C; daily girder temperature variation range of 15 °C). The calculation results are shown in Figs. 4.66, 4.67 and 4.68.

4. Comparative analysis

Refer to Table 4.7 for the comparative analysis of calculation results among Options I, II, and III.

As shown in the table above, the escalation in the number of spans of simply-supported girders leads to a notable reduction in additional forces on the piers/abutments of the continuous girders, concurrently diminishing the additional forces on the rails. However, integrating simply-supported girders elongates the spacing

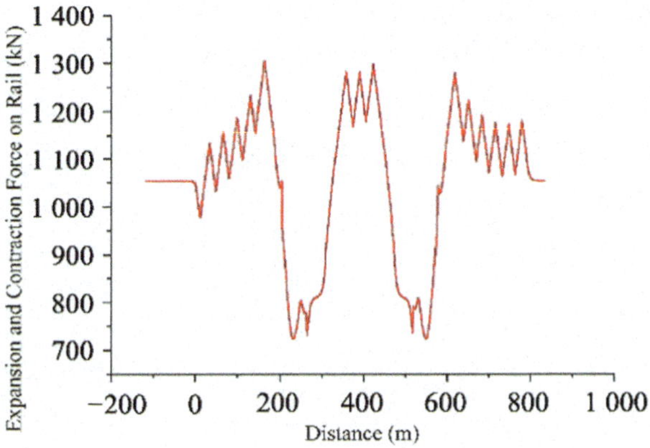

Fig. 4.66 Expansion and contraction force on rail

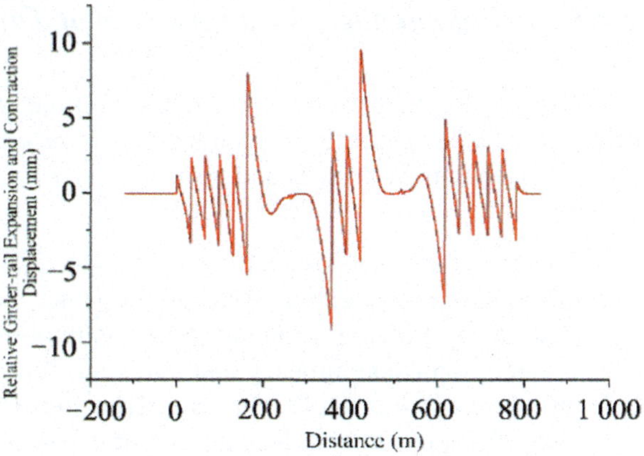

Fig. 4.67 Relative girder-rail expansion and contraction displacement

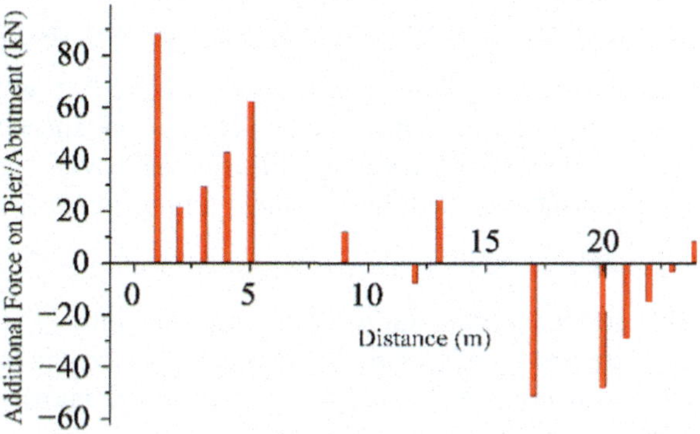

Fig. 4.68 Additional force on pier/abutment

Table 4.7 Comparative analysis of calculation results among options I, II, and III

Item	Option I	Option II	Option III
Expansion and contraction force on rail (kN)	279	260	245
Additional forces on pier/abutment of continuous girder (kN/rail)	53 82	29 63	12 24

between single crossovers. Thus, this approach proves effective in mitigating the expansion and contraction forces on rails and additional forces on the piers/abutments of continuous girders, depending on the provision of essential conditions within the station arrangement.

4.4.5 Recommended Arrangements of Turnout and Bridge

According to the analysis of the stress and deformation patterns of CWR turnouts on bridges, from the perspective of longitudinal interactions, the principles for reasonable arrangements of bridges and turnouts on high-speed railways are proposed as follows:

(1) CWR turnouts on the main line of high-speed railways should be placed within the range of one girder as far as possible, and the girders should form a continuous structure along the entire length of the turnouts.
(2) Under difficult conditions, girder joints are allowed at the turnout closure within turnouts on station tracks. However, a specific distance shall be maintained from the girder joints to both the heel end of the switch rails and the toe of frogs.
(3) To reduce girder-rail relative displacement and the additional forces on rails, the introduction of simply-supported girders is recommended for one span or more between two continuous girders beneath the double crossovers area.

Adhering to the aforementioned principles, this book presents typical turnout and bridge arrangements based on No.18 high-speed turnouts. These arrangements, proven effective in various engineering endeavors, serve as valuable references for similar projects. The typical turnout arrangements are shown as follows:

1. Single Turnout

This arrangement features a turnout placed on a (4 × 32) m continuous girder. The front and rear ends of the turnout are located 29 m away from the girder joints. A fixed bearing is located on the pier at the midspan. The specific turnout-girder arrangement is shown in Fig. 4.69.

2. Single Crossover

This arrangement features a crossover (distance between centers of tracks: 5 m) on a (6 × 32) m continuous girder. The front and rear ends of the turnout are located 19.2 m away from the girder joints. A fixed bearing is located on the pier at the midspan, and the bridge and crossover are symmetrical along their centerlines. The specific turnout-girder arrangement is shown in Fig. 4.70.

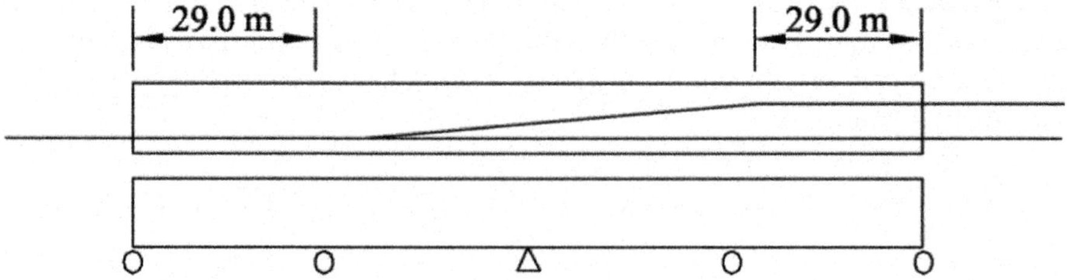

Fig. 4.69 Turnout-girder arrangement of single turnout

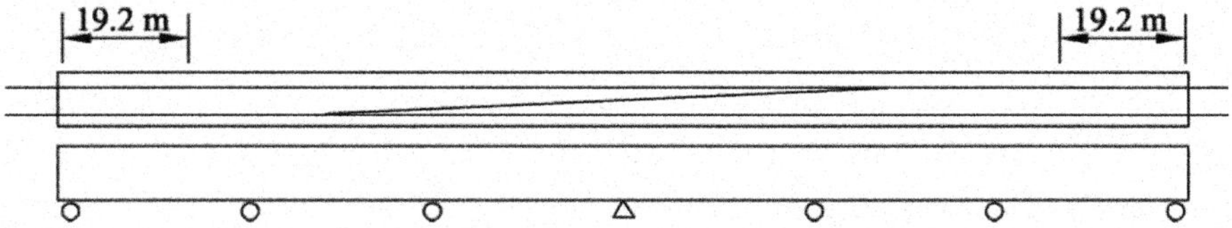

Fig. 4.70 Arrangement of single crossover turnout girder

3. Double Crossovers

According to the analysis in Sect. 4.3.4, inserting a simply-supported girder between two crossovers is beneficial to reducing the expansion and contraction forces on the stock rails.

When the required topographical conditions are met, introducing two simply-supported girders can provide more advantageous conditions. The arrangements of CWR turnouts on the bridge in the form of double crossovers are recommended as follows:

Option I: The bridge span arrangement consists of (6 × 32) m continuous girder + 32 m simply-supported girder + (6 × 32) m continuous girder, and the turnout-girder arrangement is shown in Fig. 4.71.

Option II: The bridge span arrangement consists of (6 × 32) m continuous girder + 2–32 m simply-supported girder + (6 × 32) m continuous girder, as shown in Fig. 4.72.

4. Typical Arrangements at Stations

The bridge span arrangement at overtaking stations features (6 × 32) m continuous girder + 2–32 m simply-supported girder + (6 × 32) m continuous girder + 2–32 m simply-supported girder + (4 × 32) m continuous girder, as shown in Fig. 4.73.

The bridge span arrangement at intermediate stations features (6 × 32) m continuous girder + 2–32 m simply-supported girder + (6 × 32) m continuous girder + 2–32 m simply-supported girder + (4 × 32) m continuous girder + (3 × 32) m continuous girder, as shown in Fig. 4.74.

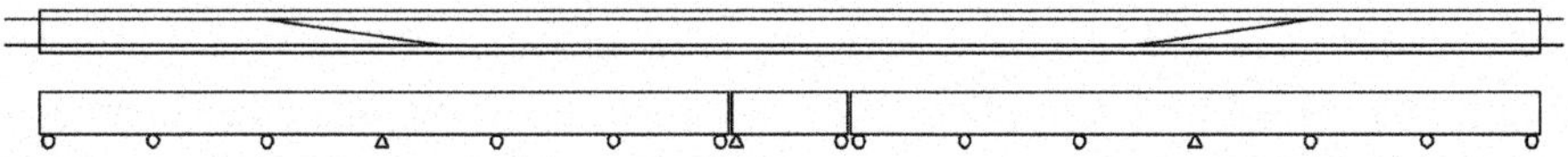

Fig. 4.71 Option I of double crossovers

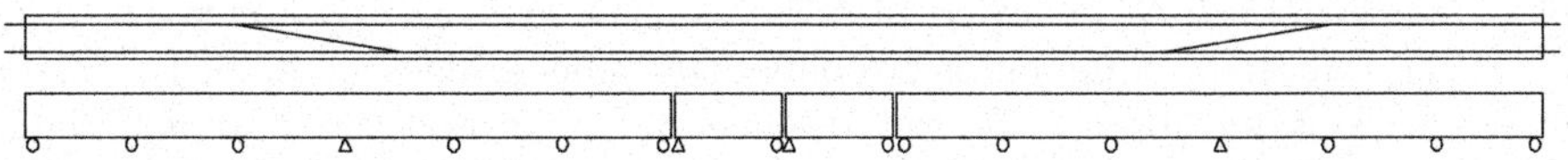

Fig. 4.72 Option II of double crossovers

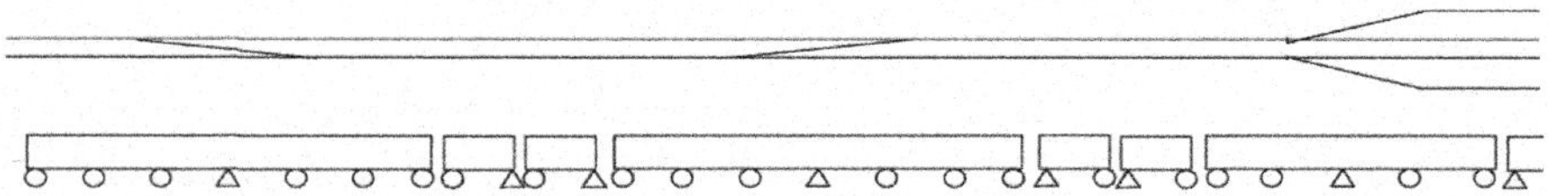

Fig. 4.73 Arrangement of bridge and turnout at overtaking station

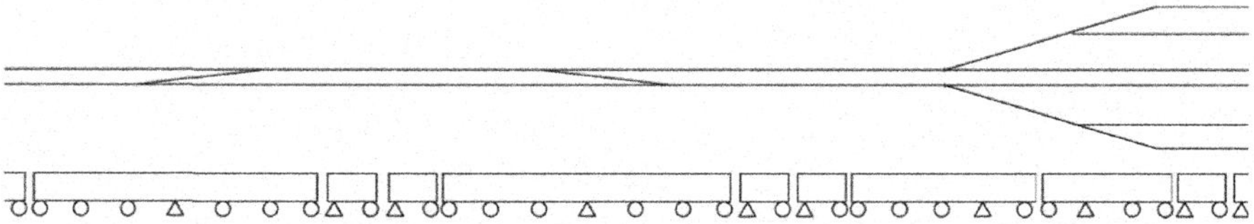

Fig. 4.74 Arrangement of bridge and turnout at intermediate station

4.5 Influence of Heel End Structure of Switch Rail Within Turnout

CWR turnouts on bridges play a crucial role in constructing trans-section CWR tracks. Generally, one end of their inner side rails bears huge temperature force, while the other end is approximately free. To ensure the switching performance through the controlled displacement of the switch rails or movable point rails, corresponding force transfer devices are placed at the heel ends of the switch rails and point rails, such as spacing blocks and stoppers, to transmit the temperature force on the inner rail side to the stock rails, in tandem with other components such as turnout sleepers, thus resulting in the additional temperature force on the stock rails. It can be known accordingly there are extremely complex relationships involving force bearing, force transfer, and displacement among rail strings within CWR turnouts.

The influence of force transfer devices at the heel end of switch rails on the turnout–bridge interactions is one of the focal points of this book.

4.5.1 Calculation Conditions

The form of force transfer devices at the heel end of switch rails directly depends on the permissible expansion and contraction displacement of switch rails under the control of the locking device within the switch zone of CWR turnouts. There are typically three structural options at the heel end of switch rails: stopper, spacing block, and no force transfer device. In this section, the stress and deformation patterns of the single turnout configuration on continuous girder bridges are studied by taking three scenarios: the absence of a force transfer device, a single stopper, and a single spacing block at the heel end of switch rails based on the other calculation parameters without any modifications.

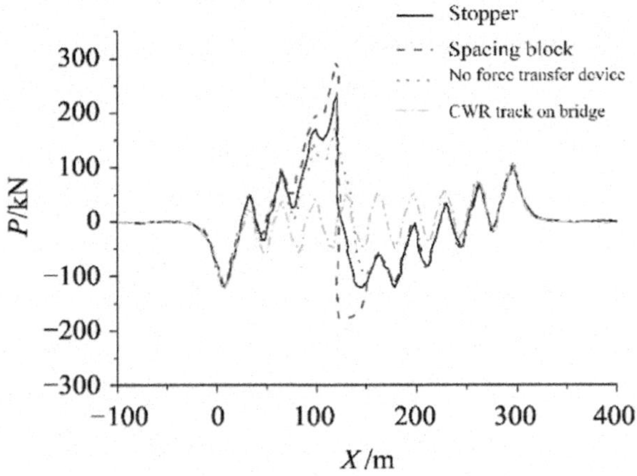

Fig. 4.75 Additional expansion and contraction force on stock rail

4.5.2 Expansion and Contraction Force on Stock Rail

The additional expansion and contraction force on the stock rails of CWR turnouts with different force transfer structures at the heel end of the switch rails on simply-supported girder bridges is shown in Fig. 4.75. As shown in this figure, across the 3 scenarios, the maximum additional longitudinal compressive force due to expansion and contraction on the stock rails appears at the heel end of the frogs, with these scenarios involving a spacing block, a stopper, and no force transfer device following a descending trend. The distribution of additional expansion and contraction force preceding and following the turnout on simply-supported girders showcases a similar pattern, contrasting with the behavior observed within the turnout section. This distinction highlights that the selection of force transfer structures at the heel end of switch rails significantly impacts the stress and deformation tendencies within the turnout section, whereas their influence on CWR tracks preceding and following the turnout on the bridge is relatively minor.

4.5.3 Expansion and Contraction Displacement of Stock Rail

The expansion and contraction displacement of the stock rails within CWR turnouts with different force transfer structures at the heel end of the switch rails on simply-supported girder bridges is shown in Fig. 4.76, where the rail displacement to the right is expressed with a positive value. As shown in this figure, the distribution of expansion and contraction displacement of the stock rails across the three scenarios showcases similar trends as follows. The maximum leftward expansion and contraction displacement occurs in the middle of the turnout, while the expansion and contraction displacement occurs rightward on the rails preceding and following the turnout. Across the spacing block, stopper, and no force transfer device scenarios, there is a

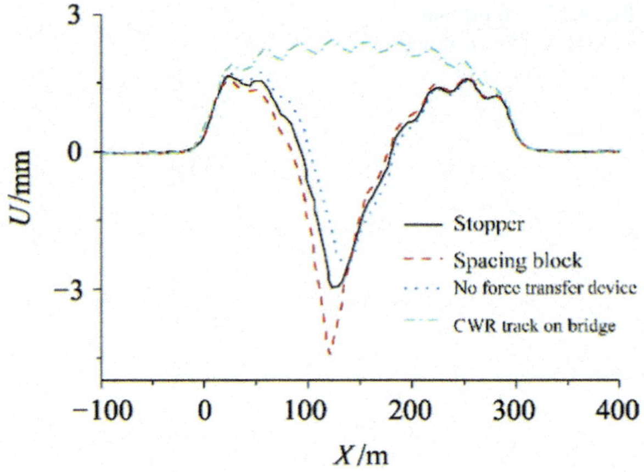

Fig. 4.76 Expansion and contraction displacement of stock rail

descending trend in the expansion and contraction displacement of the stock rails. Notably, the maximum expansion and contraction displacement is about −4.5 mm in the spacing block scenario, greater than the stopper scenario of −2.9 mm. This suggests that incorporating a spacing block at the heel end of the switch rails leads to altered turnout–bridge interactions and an increase in the overall displacement of turnouts on bridges.

4.5.4 Other Calculation Results

Table 4.8 shows the calculation results for different force transfer structures at the heel end of the switch rails on simply-supported girder bridges, including the relative displacement between the sleepers and rails at the actual points of the switch rails and point rails within CWR turnouts, longitudinal force of the force transfer devices, longitudinal force on the piers and abutments, and their comparison with those corresponding to same structures in the case of subgrade. In the table, positive values represent the rightward rail displacement and longitudinal force on the piers and abutments.

As shown in Table 4.8, when CWR turnouts are located on bridges featuring simply-supported girders, the additional expansion and contraction force on the stock rails at the heel end of the switch rails, the stress on the force transfer device, and the expansion and contraction displacement of the inner rails relative to the turnout sleepers are all greater compared to CWR turnouts on subgrade, irrespective of the type of the force transfer structure used at the heel end of the switch rails. In both bridge and subgrade settings, CWR turnouts exhibit similar trends in the stress and deformation effects from the force transfer device at the heel end of the switch rails as follows. Across the three scenarios of force transfer structure at the heel end of the switch rails, a spacing block contributes to maximizing additional compressive force

Table 4.8 Comparison of calculation results

Calculation conditions	CWR turnout on bridge			CWR turnout on subgrade		
	Stopper	Spacing block	No force transfer device	Stopper	Spacing block	No force transfer device
Maximum additional longitudinal compressive force due to expansion and contraction on straight stock rail (kN)	231.9	331.4	167.0	202.5	321.5	140.1
Maximum expansion and contraction displacement of straight stock rail (mm)	−2.9	−4.5	−2.4	−2.3	−3.9	−1.9
Maximum stress on force transfer device at heel end of switch rail (kN)	130.8	236.6	0	126.2	235.6	0
Maximum stress on spacing block at end of wing rail (kN)	156.8	167.0	150.0	158.4	191.5	153.0
Displacement at actual point of straight switch rail relative to turnout sleeper (mm)	−19.4	−14.8	−23.9	−19.2	−14.0	−22.0
Displacement at actual point of point rail relative to turnout sleeper (mm)	−7.5	−7.9	−7.8	−7.8	−6.9	−7.5
Longitudinal force on left abutment (kN)	−273.0	−280.7	−266.7	–	–	–
Longitudinal force on no. 1 pier (kN)	−83.1	−101.2	−71.4	–	–	–
Longitudinal force on no. 2 pier (kN)	−151.4	−193.4	−122.9	–	–	–
Longitudinal force on no. 3 pier (kN)	−424.7	−498.9	−346.6	–	–	–
Longitudinal force on no. 4 pier (kN)	−685.2	−556.4	−696.7	–	–	–
Longitudinal force on no. 5 pier (kN)	−415.6	−408.4	−488.1	–	–	–
Longitudinal force on no. 6 pier (kN)	−178.2	−172.4	−203.0	–	–	–
Longitudinal force on no. 7 pier (kN)	−122.5	−120.3	−132.3	–	–	–
Longitudinal force on no. 8 pier (kN)	−161.3	−160.3	−164.7	–	–	–

on the stock rails, stress on the force transfer device, and expansion and contraction displacement of the stock rails, while minimizing expansion and contraction displacement at the actual end of the switch rails and point rails. Opting for the absence of a force transfer device at the heel end of the switch rails contributes to maximizing additional compressive force and expansion and contraction displacement at the stock rails, as well as expansion and contraction displacement at the actual end of the switch rails and point rails.

The longitudinal forces on the piers within the expansion and contraction range of the turnout inner rails and following the turnout section exhibit a descending order in the force transfer device, stopper, and spacing block scenarios. In contrast, the stress on the piers and abutments preceding the turnout section shows an opposite order. Specifically, the greater the effect from the force transfer device at the heel end of the switch rails, the greater the acting force transmitted from the inner rails to the stock rails preceding the turnout section. This results in a stress increment on the piers and abutments preceding the turnout section and a stress decrement on those following the turnout section. In scenarios without a force transfer device, the longitudinal forces transmitted from the inner rails to the track preceding the turnout section become the minimum, as these forces are mostly borne by the bridge. Therefore, the longitudinal forces on the piers within the range of the inner rails become the maximum.

4.6 Study on the Limit of Minimum Linear Rigidity for Pier of Turnout Girder

From the above analysis, it is evident that the longitudinal horizontal linear rigidity of piers is a very important parameter in the design of CWR turnouts on bridges, given its significant impact on turnout stress and deformation. An excessively small longitudinal horizontal linear rigidity of piers may lead to excessive additional forces and displacements on the track, potentially causing failures. Conversely, an excessively large longitudinal horizontal linear rigidity of piers can result in a larger section size. This influence not only leads to material wastage but also has adverse effects on the clearance under the bridge and the structural appearance. Hence, it is vital to establish a reasonable limit on the design value of the minimum longitudinal horizontal linear rigidity of piers to ensure track safety on bridges and the structural safety, while taking into account the project cost and landscape effect.

According to the stress characteristics of CWR tracks and turnouts on bridges, the longitudinal horizontal linear rigidity of the piers and abutments for bridges carrying CWR turnouts depends on three factors as follows:

(1) Strength and stability of the track structure on bridges;
(2) Girder-rail relative displacement under braking force;
(3) Displacement of the critical areas within turnouts (switches and frogs).

Therefore, to guarantee the strength and stability of the track structure on bridges and comply with the girder-rail relative displacement limit and the displacement limit at critical areas within the turnout section, it is essential to set limits for the substructure rigidity of turnout girders in various types.

In the structure system of CWR turnouts on bridges, various turnout and bridge arrangements correspond to different minimum limits for the longitudinal horizontal linear rigidity of the piers. These limits should be derived through an analysis of specific conditions. This section establishes the limits of the minimum longitudinal horizontal linear rigidity of the piers by examining the relationship of turnout stress and deformation with the longitudinal horizontal linear rigidity of piers in three typical arrangements of CWR turnouts and bridges with ballasted tracks to facilitate the design process.

4.6.1 Single no.18 Turnout on Continuous Girder

Calculations were conducted to examine the influence of varying linear rigidity on the stress and deformation of a single No.18 turnout on a (4 × 32) m continuous girder, using the longitudinal linear rigidity of the fixed pier under the continuous girder (double-track, the same below) of 400 kN/cm, 600 kN/cm, 800 kN/cm, 1,000 kN/cm, 2,000 kN/cm, 4,000 kN/cm, 8,000 kN/cm, 12,000 kN/cm, and 20,000 kN/cm respectively, along with the other parameters without any modification. Figure 4.77 shows the relationship of the additional expansion and contraction forces, deflection forces, and braking forces on the CWR turnout with the linear rigidity of the pier and abutment. Figure 4.78 shows the relationship between the displacement of critical components within the CWR turnout section and the linear rigidity of the pier/abutment.

As shown in Fig. 4.77, the expansion and contraction forces on rails decrease with an increase in the longitudinal horizontal linear rigidity of the fixed pier under the continuous girder, which indicates that enhancing this rigidity results in a decrease in

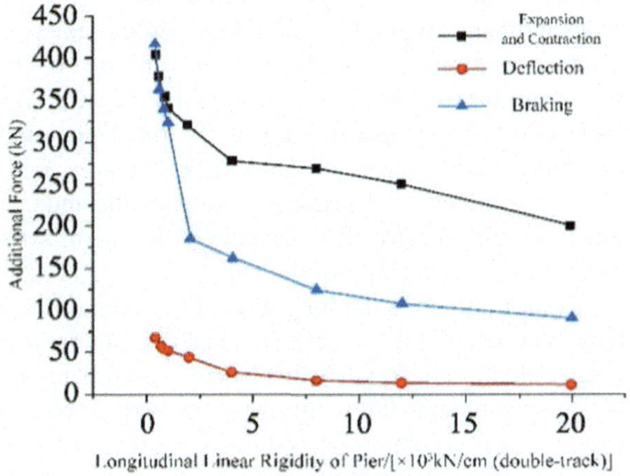

Fig. 4.77 Relationship between additional force and pier rigidity

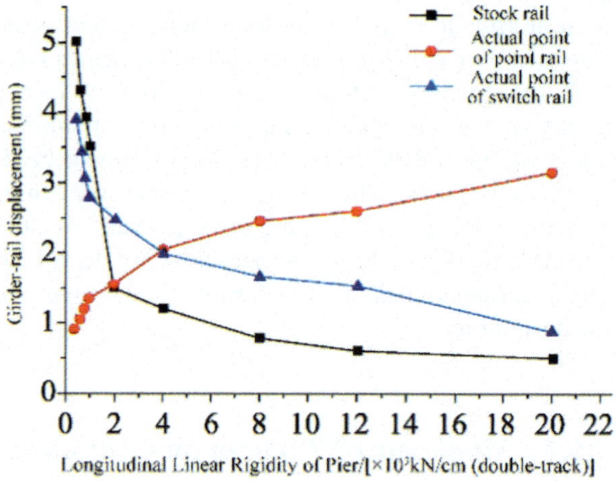

Fig. 4.78 Relationship between girder-rail relative displacement and pier rigidity

the expansion and contraction forces on rails. Similarly, the maximum longitudinal force due to deflection of the rails decreases with an increase in the longitudinal horizontal linear rigidity of the fixed pier under the continuous girder. However, the longitudinal force due to deflection of the rails is not the main factor influencing this rigidity. Moreover, the braking force on the rails decreases with an increase in the longitudinal horizontal linear rigidity of the fixed pier under the continuous girder. For rigidity values below 2,000 kN/cm·(double-track), the additional braking force on the rails exhibits a linear relationship with pier rigidity, with a notable decline in the additional forces on the rails corresponding to escalating rigidity. Conversely, for rigidity values surpassing 2,000 kN/cm, the maximum braking force on the rails tends to stabilize, implying a reduced impact of rigidity enhancement on the additional forces on the rail. These calculations reveal that the braking force is a pivotal factor influencing the longitudinal horizontal linear rigidity of the piers.

As shown in Fig. 4.78, with an increase in the longitudinal horizontal linear rigidity of the fixed pier under the continuous girder, the maximum girder-rail quick relative displacement decreases. For rigidity values greater than 2,000 kN/cm, this displacement tends to stabilize, suggesting its effect on the longitudinal horizontal linear rigidity of the piers also as a main factor. With an increase in the longitudinal horizontal linear rigidity of the fixed pier under the continuous girder, the maximum displacement of the switch decreases, while the maximum displacement of the frog increases. These calculation results underscore the importance of controlling the change in frog displacement within permissible limits when determining the minimum linear rigidity of the pier.

To sum up, when the longitudinal horizontal linear rigidity of the pier is greater than 1000 kN/cm, the overall stress of the rails remains below the permissible level, the girder-rail rapid relative displacement is kept under 4 mm, and both the switch and frog displacements fall within the permissible values. Therefore, it is recommended to adopt a minimum horizontal linear rigidity of 1,000 kN/cm·(double-track) for the pier supporting the continuous girder that carries a single turnout.

4.6.2 Single Crossover on Continuous Girder

Calculations were conducted to examine the influence of varying linear rigidity on the stress and deformation of a single crossover (distance between centers of tracks: 5m) on a (6 × 32) m continuous girder in the settings of a 19.2 m distance from the turnout front end to a girder joint, a mid-span pier with a fixed bearing, and a symmetrical bridge and crossover layout along their centerline, using the longitudinal linear rigidity of the fixed pier under the continuous girder of 400 kN/cm, 600 kN/cm, 800 kN/cm, 1,000 kN/cm, 2,000 kN/cm, 4,000 kN/cm, 8,000 kN/cm, 12,000 kN/cm, and 20,000 kN/cm respectively. Figure 4.79 shows the relationship of the additional expansion and contraction forces, deflection forces, and braking forces on the CWR turnout with the linear rigidity of the pier and abutment. Figure 4.80 shows the relationship between the displacement of critical components within the CWR turnout section and the linear rigidity of the pier/abutment.

As shown in Fig. 4.79, the additional expansion and contraction force on the rails increases with an escalation in pier rigidity, suggesting that increasing the pier rigidity leads to an unfavorable rise in the additional expansion and contraction forces on the rails. Conversely, the maximum additional deflection force on the rails decreases as the longitudinal horizontal rigidity of the fixed pier beneath the continuous girder increases. Moreover, the additional braking force on the rails diminishes with an increase in the longitudinal horizontal linear rigidity of the fixed pier under the continuous girder. For longitudinal horizontal linear rigidity values of the fixed bearing greater than 4,000 kN/cm, the maximum braking force on the rails tends to stabilize. For this rigidity of less than 1,000 kN/cm·(double-track), the additional braking force on the rails exhibits a linear relationship with pier rigidity, with a notable decline in the additional forces on the rails corresponding to escalating rigidity.

As shown in Fig. 4.80, under the influence of braking force, as the longitudinal horizontal linear rigidity of the fixed pier beneath the continuous girder increases, the maximum girder-rail quick displacement decreases. For rigidity values greater

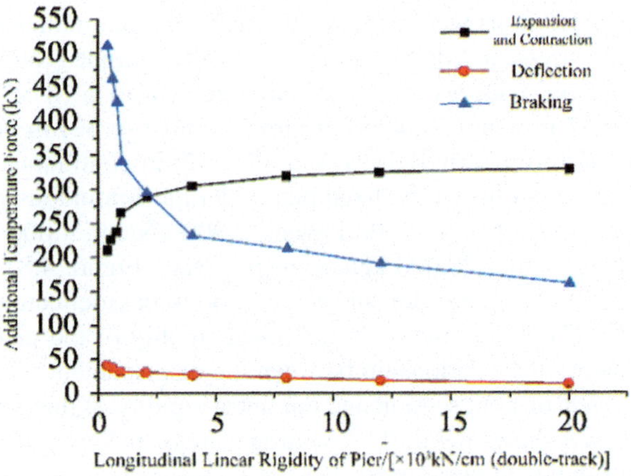

Fig. 4.79 Relationship between additional force and pier rigidity

Fig. 4.80 Relationship between girder-rail relative displacement and pier rigidity

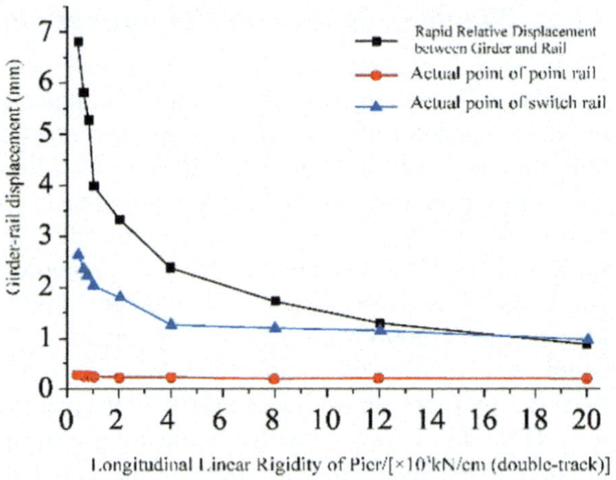

than 4,000 kN/cm, this displacement tends to stabilize. Furthermore, an increase in the longitudinal horizontal linear rigidity of the fixed pier underneath the continuous girder results in the maximum displacement of the switch diminishing. Similarly, the maximum displacement of the frog decreases as this rigidity increases.

To sum up, when the longitudinal horizontal linear rigidity of the pier is greater than 800 kN/cm, the overall stress of the rails remains below the permissible level, the girder-rail rapid relative displacement is kept under 4 mm, and both the switch and frog displacements meet the design requirements. Therefore, it is recommended to adopt a minimum horizontal linear rigidity of 800 kN/cm·(double-track) for the pier supporting the continuous girder that carries a single crossover.

4.6.3 Typical CWR Turnout Group on Bridge at Station

Calculations were conducted to examine the influence of varying linear rigidity on the stress and deformation of a CWR turnout group on a bridge at an overtaking station, with a bridge span arrangement of a (6 × 32) m continuous girder + a 2–32 m simply-supported girder + a (6 × 32) m continuous girder + a 2–32 m simply-supported girder + a (4 × 32) m continuous girder, using the longitudinal linear rigidity of the fixed pier under the continuous girders of 400 kN/cm, 600 kN/cm, 800 kN/cm, 1,000 kN/cm, 2,000 kN/cm, 4,000 kN/cm, 8,000 kN/cm, 12,000 kN/cm, and 20,000 kN/cm respectively. Figure 4.81 shows the relationship of the additional expansion and contraction forces, deflection forces, and braking forces on the CWR turnouts with the linear rigidity of the piers and abutments. Figure 4.82 shows the relationship between the displacement of critical components within the CWR turnout section and the linear rigidity of the piers and abutments.

As shown in Fig. 4.81, with an increase in the longitudinal horizontal linear rigidity of the fixed piers under the continuous girders, the expansion and contraction force,

Fig. 4.81 Relationship between additional force and pier rigidity

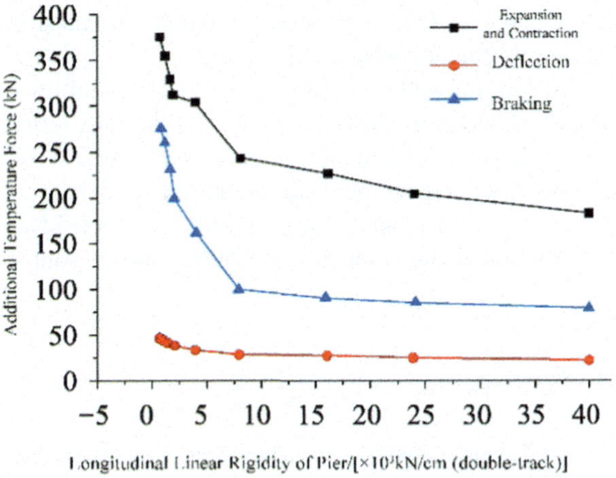

Fig. 4.82 Relationship between girder-rail relative displacement and pier rigidity

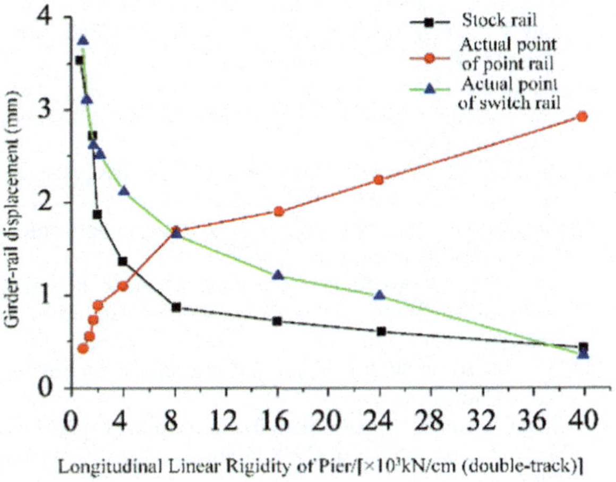

the maximum deflection force, and the braking force on the rails decrease. For rigidity values surpassing 8,000 kN/cm, the maximum braking force on the rails tends to stabilize. For rigidity values below 8,000 kN/cm, the maximum braking force on the rails exhibits an approximately linear relationship with pier rigidity.

As shown in Fig. 4.82, under the action of braking force, with an increase in the longitudinal horizontal linear rigidity of the fixed piers under the continuous girders, the maximum girder-rail rapid displacement decreases. For rigidity values greater than 8,000 kN/cm, the maximum girder-rail rapid displacement tends to stabilize. Furthermore, an increase in the longitudinal horizontal linear rigidity of the fixed pier underneath the continuous girder results in the maximum displacement of the switch diminishing. Conversely, the maximum displacement of the frogs increases with this rigidity enhancement. Careful consideration should be given to controlling the change in frog displacement within permissible limits when determining the minimum linear rigidity of the piers.

To sum up, when the longitudinal horizontal linear rigidity of the piers is greater than 1,600 kN/cm, the overall stress of the rails remains below the permissible level, the girder-rail rapid displacement is kept under 4 mm, and both the switch and frog displacements meet the design requirements. Therefore, it is recommended to adopt a minimum horizontal linear rigidity of 1,600 kN/cm for the piers supporting the continuous girders that carry a CWR turnout group at overtaking stations.

References

1. Rongshan Y (2008) Computational theory and experimental study on longitudinal force on CWR turnout on bridge [D]. Southwest Jiaotong University, Chengdu
2. Ping W (2003) Mechanism of longitudinal force transmission of CWR turnouts with different structures [J]. J Southwest Jiaotong Univ 38(4):371–374
3. Ping W (2002) Research on the influence of CWR turnout groups on rail displacement and longitudinal force [J]. J China Railw Soc 24(2):5
4. Juanjuan R (2009) Research on stress characteristics and structural optimization of longitudinally coupled ballastless track within CWR turnout section on bridge [D]. Southwest Jiaotong University, Chengdu
5. Li Q, Sun L, Yang Y (2008) Research on design approach of CWR turnout on bridge of passenger dedicated line [J]. J Railw Eng Soc
6. Ping W, Xueyi L (2003) Analysis of influencing factors on stress and deformation of CWR turnout [J]. China Railw Sci 24(2):9
7. Zhiping Z (2006) Study on expansion and contraction force of CWR turnout on high-speed railway bridge and spatial vibration of train-turnout-bridge system [D]. Central South University, Changsha
8. Ping W (2011) Design theory of CWR turnout on bridge [M]. Southwest Jiaotong University Press, Chengdu
9. Zhang L, Liu Y, Yang R (2010) Experiments on expansion and contraction force and simulation analysis using models of CWR turnouts on bridges of passenger dedicated lines [J]. Railw Eng 2010(10):135–138
10. Zhiping Z, Xiufang C, Guofan Z (2006) Computation of expansion and contraction force and displacement of CWR turnout on continuous girder bridge [J]. J Transp Eng 1:34–38

Chapter 5
Research on Structural Design of Girder Turnout

On conventional railways, it is standard practice to install CWR turnouts on bridges, leading to the establishment of well-defined standard design drawings for turnout girders in such systems. Typically, the turnout girders in these railways are constructed as ordinary reinforced concrete simply-supported T-girders, with a top width of either 1.6 or 1.7 m and a span ranging from 12 to 32 m. However, the design of CWR turnout girders on high-speed railways is currently confronted with new challenges. Specifically, there is a need for tailor-made bridge arrangements that integrate seamlessly with the turnouts. Accommodating the considerable width variations in the station throat area poses an additional design challenge for the bridge structure. Moreover, it is crucial to specifically study and rationally design the arrangements of longitudinal and transverse bearings for the turnout girders. The design of the bridge substructures presents its own set of challenges, particularly in addressing the significant longitudinal forces exerted by the CWR turnouts on the bridges. This chapter focuses on the study of turnout girders on high-speed railways, such as the fundamental principles governing the layout of turnouts and bridges, span arrangements, shear lag effects, and live load amplification coefficient. These analyses aim to effectively tackle the aforementioned challenges within the context of the Beijing–Shanghai High-Speed Railway.

5.1 Structural Design Overview of Turnout Girder

Due to the overall features, station arrangement, geological conditions, and topographical conditions of the Beijing–Shanghai High-Speed Railway, a large number of ballastless CWR turnouts are located on bridges. The Danyang–Kunshan Super Major Bridge crossing a total length of 164 km in the Xuzhou–Shanghai Section accommodates four elevated stations: Changzhoubei, Wuxidong, Suzhoubei, and Kunshannan. All turnouts for them are located on this bridge.

Q. Li et al., *Continuously Welded Turnouts on High-Speed Railway Bridge*, Advances in High-speed Rail Technology, https://doi.org/10.1007/978-981-97-6524-9_5

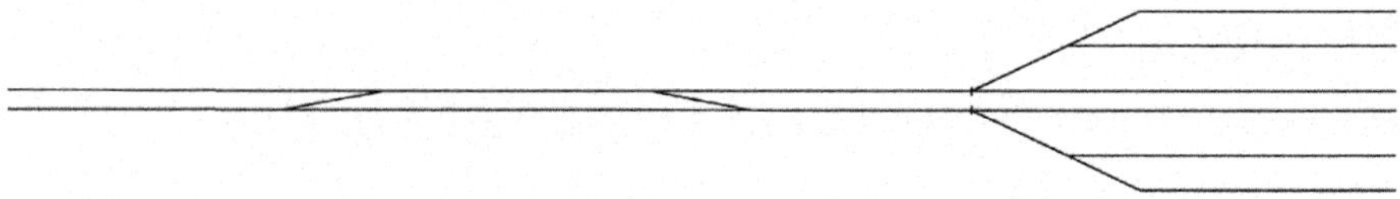

Fig. 5.1 Turnout layout plan at Beijing side throat area of Suzhoubei station

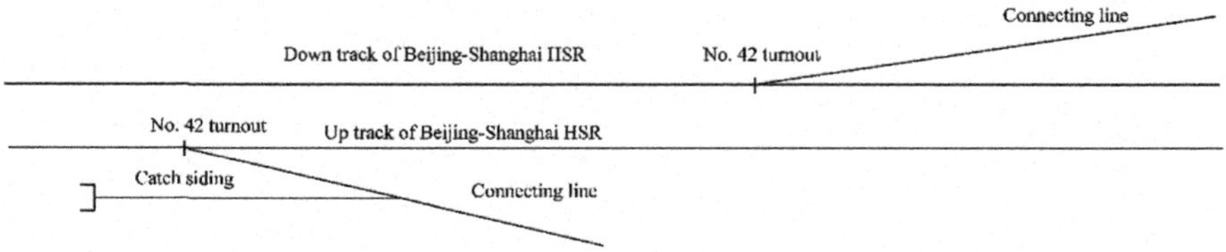

Fig. 5.2 Layout plan of turnouts on Huangdu connecting line

The turnout arrangements at the throat area on the Beijing side of Suzhoubei Station and Huangdu Connecting Line are taken as representative examples of the Beijing–Shanghai High-Speed Railway. The throat area of Suzhoubei Station features No. 18 turnouts, while the Huangdu Connecting Line is furnished with No. 42 turnouts. Their specific arrangements are shown in Figs. 5.1 and 5.2, respectively.

The computational analysis of turnout girders must incorporate the turnout-bridge interactions, as well as the influence of bridge arrangements on the tracks and turnouts. Bridges in the turnout area shall meet the following criteria: (1) complying with the bridge requirements in terms of structural strength, stress, stability, and cracking resistance safety factor to ensure the structural safety of the bridges; (2) complying with the track and turnout requirements in terms of strength, stability, displacement, and regularity to ensure the structural safety of the tracks and turnouts and minimize their maintenance workload; and (3) complying with the specification requirements of the "train-turnout-bridge" dynamic performance for the passage of high-speed trains at an hourly speed of 300–350 km through the turnout area to ensure the operational safety and riding quality of trains.

The key elements essential to the design of turnout bridges are outlined as follows: (1) implementing a rational bridge arrangement solution to minimize interaction forces between the bridges and tracks/turnouts and to decrease relative displacement between the turnouts and the bridges; (2) utilizing a sectional form that ensures strong integrity for the turnout girders to enhance both vertical and lateral rigidity within the section; and (3) appropriately enhancing the substructure rigidity of bridges within the turnout area.

5.2 Main Arrangement Principles of Turnout Girder

5.2.1 Arrangement Principle of Turnout

(1) The design speed of 350 km/h for the Beijing–Shanghai High-Speed Railway reflects high requirements for the regularity and geometry of turnouts. In scenarios of simply-supported girders, the turnouts are inevitably laid across girder joints. Any expansion/contraction and lateral dislocation of these joints, as well as rotation at girder ends, can potentially alter the geometry of the turnouts. Any irregularity of the high-speed turnouts exceeding permissible limits can trigger substantial dynamic reactions from trains, potentially leading to unforeseen consequences in terms of operational safety.

Foreign countries, such as Germany and France, have put forward consulting recommendations on installing high-speed turnouts on girders extending over entire spans. According to German regulations, positioning turnouts on movable expansion joints on bridges is unallowable. In line with French regulations, it is advisable to refrain from positioning turnouts across or near structural expansion joints to minimize stress on rails.

According to relevant Chinese research, turnouts on the main line of high-speed railways shall be entirely laid on a continuous girder rather than simply-supported girders. The distance from the beginning and end of the turnouts to any girder joint shall not be less than 18 m. The turnout-bridge relative positional relationship is shown in Fig. 5.3.

(2) The feasible bridge structures for the CWR turnout area on high-speed railways include concrete continuous girder bridges and small frame bridges with spans ranging from 6 to 12 m. Small-span frames with high rigidity and strong integrity are advantageous in reducing both displacements of the bridge and relative displacement between the turnout and bridge, thereby meeting the requirements for placing turnouts on bridges. Small-span frames offer another structural form choice for ballastless track bridges in the turnout area of railway sections

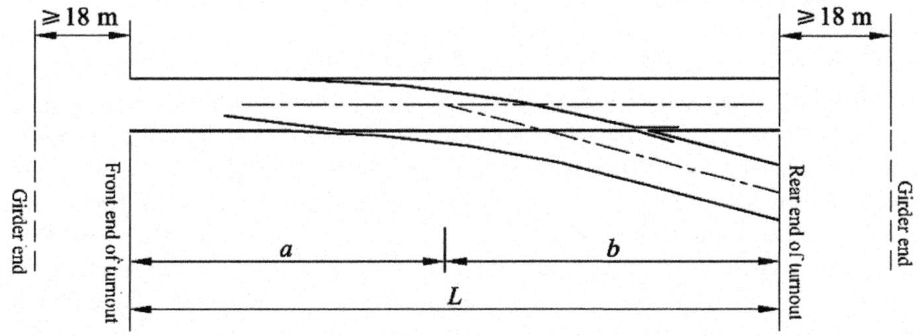

Fig. 5.3 Positional relationship between turnout and bridge

featuring dense turnouts, low bridge heights, and no interchange or navigation requirements.

(3) Simply-supported girders may be used to carry turnouts on station tracks, provided that the minimum distance from the actual ends and heel ends of the switch rails and point rails to any girder joint adheres to the necessary specifications to ensure the structural safety of the turnouts and bridges, as well as the smooth operation of the switches.

(4) To reduce relative displacements and interaction forces between turnouts and bridges, it is recommended to incorporate a simply-supported girder between two continuous girders under the turnouts.

5.2.2 Arrangement Principle of Bearing for Turnout Girder

1. Number of longitudinal fixed bearing

For scenarios of continuous girders to support turnouts, the incorporation of multiple fixed bearings helps reduce rail forces on each fixed pier. However, this setup can restrict the longitudinal displacement of girders due to factors like shrinkage, creep, and temperature variations, leading to significant longitudinal horizontal forces. When the piers are short and highly rigid, these horizontal forces outweigh the reduction in rail forces. Hence, when constructing continuous girders for turnouts, careful analysis and comparison are crucial before integrating multiple longitudinal fixed bearings into the design. Unless there are specific demands, it is advisable to typically utilize a single longitudinal fixed bearing.

2. Position of longitudinal fixed bearing

Elevated stations on the Beijing–Shanghai high-speed railway are typically equipped with multiple crossover turnouts and single turnouts at both ends, using multiple turnout girders. To mitigate temperature span, girder-rail relative displacement, and track force, it is advisable to position the longitudinal fixed bearing in the middle of the turnout girders, taking into account the longitudinal fixed bearing position of adjacent girders.

3. Position of transverse fixed bearing for turnout girder

To maintain the lateral track regularity, it is specified that in a ballastless track setting, the lateral displacement difference of the rails across the joint of two adjacent girders shall not exceed 1 mm. When the two adjacent girders exhibit uniformity in cross-sectional form and width, their aligned transverse fixed bearings guarantee adherence to the specified limit. Conversely, if a turnout girder differs in cross-sectional form and width from the neighboring girders, the transverse fixed bearings are misaligned. The transverse deformation of concrete structures due to temperature fluctuations leads to the displacement of transverse fixed bearings on adjacent girders, resulting in lateral dislocation of the rail crossing the girder joints. The transverse bearing misalignment of adjacent girders is shown in Fig. 5.4. Given the significant rigidity

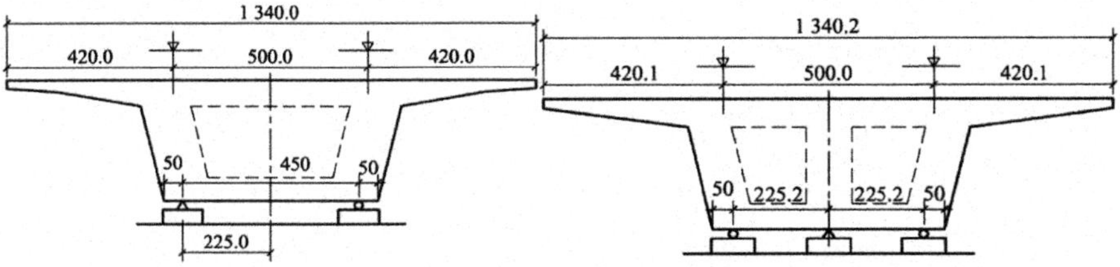

Fig. 5.4 Misalignment of transverse bearings of adjacent turnout girders (Unit: cm). *Note* △ denotes transverse fixed bearing; ○ denotes transverse movable bearing

in attaching ballastless track structures to the top slab of bridges and the common use of box girder sections for turnout girders, the calculations of lateral rail displacement can be simplified by calculating the spacing between the transverse fixed bearings of adjacent girders. The lateral displacement of the rail can then be determined by multiplying the spacing by the temperature variation and the linear expansion coefficient of the materials. Bearings should be placed with decreased spacing between transverse fixed bearings to maintain lateral displacement differences below 1 mm.

5.3 Study on Span Arrangements for Turnout Girder

5.3.1 Main Span for Continuous Girder Beneath Turnout

Simply-supported box girders featuring a 32 m span represent the prevalent girder type on the Beijing–Shanghai High-Speed Railway, showcasing the favorable economic performance of this span length. Continuous girders hosting turnouts across this railway line similarly span 32 m at locations without specific clearance constraints for interchanges and river navigation, ensuring uniformity in the bridge span length throughout. The number of spans for continuous girders beneath turnouts varies based on factors like turnout count, turnout length, and distance between turnouts and girders.

In places where a 32 m span does not satisfy the clearance needs of continuous girders supporting turnouts due to interchanges or river navigation, the implementation of a long-span continuous girder design becomes essential, incorporating a main span selected according to the specific conditions.

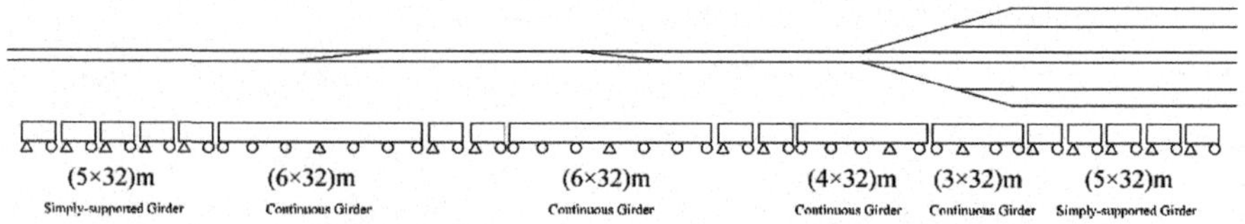

Fig. 5.5 Span arrangement at Beijing side throat area of Suzhoubei station

5.3.2 Span Arrangement of Turnout Girder at Throat Area of Elevated Station

The five elevated stations on the Beijing–Shanghai High-Speed Railway, i.e. Zhenjiangxi, Changzhoubei, Wuxidong, Suzhoubei, and Kunshannan, are all furnished with two platforms, six tracks, and a track arrangement at their throat areas featuring transitions from double tracks to four tracks and then to six tracks. Four No.18 single turnouts are placed at these throat areas, including two on the main line sections and two on the arrival and departure tracks.

Ensuring a distance from the beginning and end of the turnouts on the main line sections to a girder joint of not <18 m, the No. 18 turnout of 69 m in its full length on the main line is entirely placed on one continuous girder, covering a minimum length of (18 + 69 + 18) m = 105 m. Therefore, the (4 × 32) m continuous girders with a length of 130.7 m are erected to carry the turnouts on the main line sections, while the (3 × 32) m continuous girders with a length of 98 m are erected to carry the turnouts on the arrival and departure tracks. The bridge span arrangement at the throat area on the Beijing side of Suzhoubei Station is shown in Fig. 5.5.

5.3.3 Span Arrangement of Turnout Girder for Single Crossover

The No. 18 turnout in the crossover area is entirely placed on one continuous girder, covering a minimum length of (18 + 31.729 + 90 + 31.729 + 18) m = 189.458 m. Therefore, the 6 × 32 m continuous girders with a length of 196.0 m are erected to carry the single-crossover turnouts. The span arrangement for the girders is shown in Fig. 5.6.

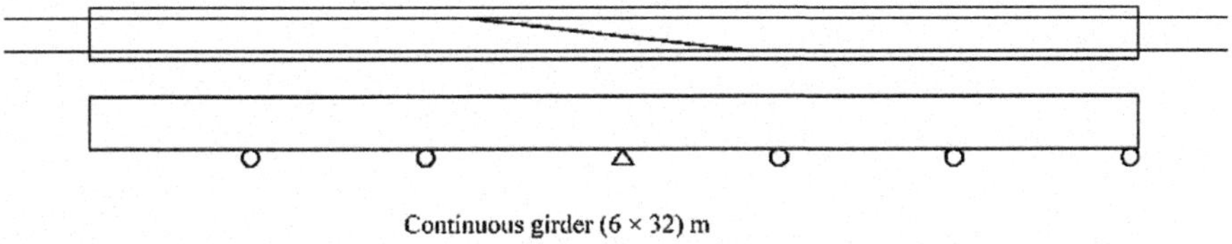

Fig. 5.6 Span arrangement of turnout girder for single crossover

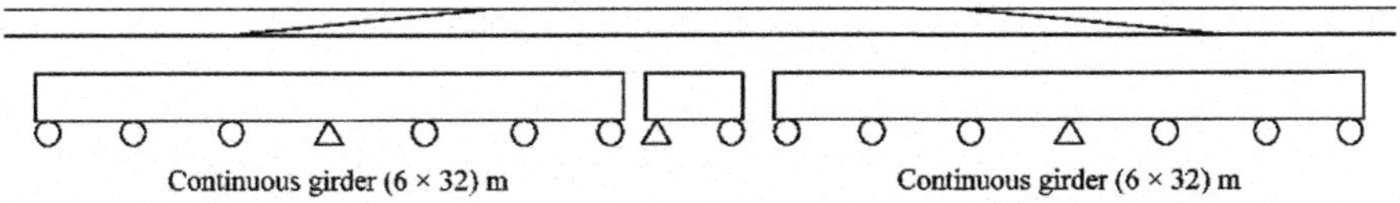

Fig. 5.7 Option I for span arrangement of turnout girder for double crossovers

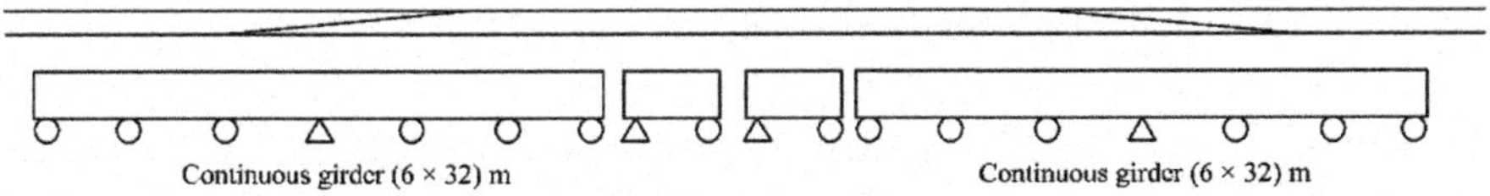

Fig. 5.8 Option II for span arrangement of turnout girder for double crossovers

5.3.4 Span Arrangement of Turnout Girder for Double Crossovers

The double crossovers consist of two single crossovers, with each crossover supported by a 6 × 32 m continuous girder mentioned above. Following the principle of inserting simple-supported girders between two continuous girders carrying turnouts, one- or two-span simply-supported girders are erected between two 6 × 32 m continuous girders. The specific span arrangements are shown in Figs. 5.7 and 5.8.

5.3.5 Turnout Girder on Huangdu Connecting Line

Huangdu Connecting Line diverges from the main line through via No.42 turnouts. Following the principle of placing an entire turnout on one continuous girder, maintaining a distance from the beginning and end of turnouts on the main line to a girder joint of not less than 18 m, and inserting simple-supported girders between two continuous girders carrying turnouts, a (24 + 4 × 32 + 24) m continuous girder is erected for the turnout on the up track of Huangdu Connecting Line, while a 6 × 32 m

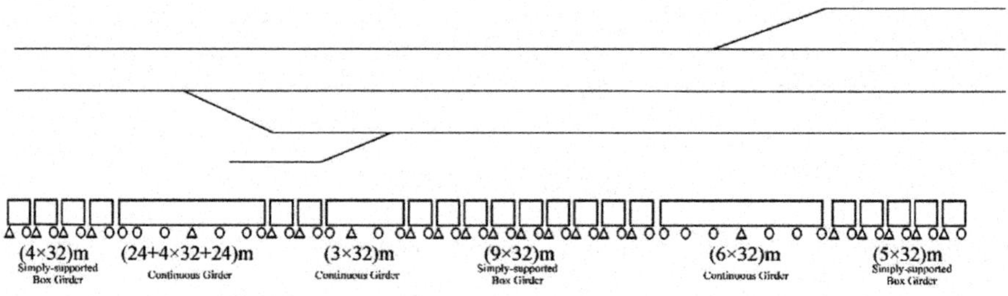

Fig. 5.9 Span arrangement of turnout girder on Huangdu connecting line

continuous girder is erected for the turnout on the down track. The span arrangements for the turnout girders on Huangdu Connecting Line is shown in Fig. 5.9.

5.3.6 Open Arrangements for Other Turnout Girders

Generally, 6 × 32 m continuous girders are used to carry crossover turnouts. However, the clearance requirements for interchanges and river navigation necessitate an increase in the bridge span, so long-span continuous girder design becomes essential. For example, Zhenjiangxi Station incorporates a (32 + 40 + 56 + 40 + 32) m continuous girder to cross the Changshanyin River. Wuxidong Station incorporates a (32 + 40 + 64 + 40 + 32) m continuous girder to cross Xihu Road and the Jiuli River. Suzhoubei Station spans Chengyang Road with a (32 + 40 + 64 + 40 + 32) m continuous girder. Kunshannan Station spans Bailu Road with a (32 + 40 + 72 + 40 + 32) m continuous girder.

5.4 Types and Typical Design of Turnout Girder

5.4.1 Main Types and Parameters of Turnout Girder

Turnouts are very sensitive to bridge displacement. To align turnout stress and displacement with specification requirements and enhance the dynamic performance of vehicle-bridge coupling in turnout bridges, it is preferred to use box girders with significant lateral and torsional rigidity, as well as robust structural integrity for supporting turnouts. Furthermore, enhancing the vertical rigidity within girder sections is recommended.

To accommodate turnouts and multiple tracks, turnout girders generally need a deck width greater than that of sectional girders carrying double tracks and may exhibit varying widths in certain instances. Turnout girders are optional in cross sections of single-box single-cell, single-box twin-cell, and single-box multi-cell configurations, primarily determined by the girder's width. Specifically, when the width of the turnout girder matches the standard girder width of 12 m, the preference generally leans toward the single-box and single-cell option. However, as the girder width increases, the box girders are subjected to heightened transverse bending moments. Consequently, the design of the single-box and single-cell configuration necessitates larger structural dimensions, particularly when reinforcing the top slab and enhancing the transverse reinforcement to counteract these bending moments. In such instances, opting for the more cost-effective and rational single-box and twin-cell configuration is advisable. As the girder width continues to increase beyond the practical limit for the single-box twin-cell design, the more suitable choice then becomes the single-box multi-cell configuration.

The height of continuous girders supporting turnouts is mainly determined by parameters such as span, secondary dead load, and the number of tracks simultaneously subjected to the ZK live load. Consequently, girder height stands as a pivotal factor in the structural design of turnout girders. It plays a crucial role in determining whether the stress, strength, deformation, and dynamic performance of turnout girders align with the specified requirements.

On the Beijing–Shanghai High-Speed Railway, continuous girders with medium and small spans that support turnouts are typically designed for construction span by span using full framing scaffolding. For the sake of simplicity in structure, ease of construction, and favorable economic benefits, equal-height girders are preferred. In terms of stress characteristics, continuous girders and simply-supported girders of the same height exhibit differences. Specifically, the absolute value of the bending moments sustained by continuous girders is lower than that of simply-supported girders over the same span. Additionally, the deflection experienced under train loads is lower in the continuous girder context. This results in a higher vertical rigidity for the bridge, enhancing its dynamic performance and ensuring better alignment with turnout requirements concerning bridge deformation and displacement. Hence, continuous girders supporting turnouts with a 32 m span are effectively designed with an equal height of 3.05 m. This choice maintains visual harmony with simply-supported girders spanning 32 m while adequately fulfilling the various requirements of both bridges and turnouts.

Long-span continuous girders supporting turnouts, under dead and live loads, experience a substantial negative bending moment at the middle fulcrum, often surpassing the midspan positive bending moment in absolute value. Therefore, using variable-height girders can offer better adherence to the internal force distribution law of girders. Furthermore, long-span continuous girders are typically built using a form traveler for cantilever pouring, making variable-height girders an ideal choice to correspond with the internal force conditions during construction. These variable-height continuous girders supporting turnouts can be designed with height variations following curve profiles such as arcs, quadratic parabolas, and the like, to guarantee an aesthetically pleasing and harmonious structure.

See Table 5.1 for the configurations and main parameters of continuous girders beneath turnouts on the Beijing–Shanghai High-Speed Railway.

Table 5.1 Configurations and parameters of continuous girders beneath turnouts

S/No.	Span (m)	Girder width (m)	Cross-sectional configuration	Equal-height girder or not	Girder height at fulcrum (m)	Midspan girder height (m)
1	6 × 32	12	Single-box single-cell	Equal-height	3.05	3.05
2	4 × 32	12–21.69	Single-box twin-cell/ triple-cell	Equal-height	3.05	3.05
3	3 × 32	5.84–11.32	Single-box single-cell	Equal-height	3.05	3.05
4	6 × 34	17.69–14.16	Single-box twin-cell	Equal-height	3.05	3.05
5	6 × 34	12.69–14.10	Single-box twin-cell	Equal-height	3.05	3.05
6	6 × 34	23.55–19.40	Single-box triple-cell	Equal-height	3.05	3.05
7	24 + 4 × 32 + 24	12.70–15.20	Single-box twin-cell	Equal-height	3.05	3.05
8	6 × 32	12.7–15.5	Single-box twin-cell	Equal-height	3.05	3.05
9	35.81 + 56 + 35.81	12–21.69	Single-box triple-cell	Variable-height	4.35	3.05
10	32 + 40 + 56 + 40 + 32	12	Single-box single-cell	Variable-height	4.35	3.05
11	32 + 40 + 64+40 + 32	12	Single-box single-cell	Variable-height	5.60	3.05
12	32 + 40 + 72 + 40 + 32	12	Single-box single-cell	Variable-height	6.20	3.05

To sum up, box girders stand out as the predominant girder type employed for continuous girders supporting turnouts on the Beijing–Shanghai High-Speed Railway. Depending on varying bridge deck widths, these box girders are categorized into single-box single-cell, single-box twin-cell, and single-box triple-cell cross sections. Equal-height girders are used to support turnouts over medium and short spans, while variable-height girders over long spans.

5.4.2 Research on Structural Design of Typical Continuous Girders Beneath Turnouts

1. (4 × 32) m continuous girder

The main girder features an equal-height inclined web section with a height of 3.05 m. The first and second spans at the narrower end are in a single-box twin-cell section, whereas the third and fourth spans are in a single-box triple-cell section. The top slab width varies between 12 and 21.69 m, while the bottom slab width ranges from 5.5 to 15.189 m.

Top slab: The top slab of the midspan section is 34 cm thick and is thickened near the fulcrum at the pier top to 60 cm.

Bottom slab: The bottom slab of the midspan section is 30 cm thick and is thickened near the fulcrum at the pier top to 60 cm.

Web: The midspan middle webs of the first and second spans are 65 cm thick and are thickened near the fulcrum at the pier top to 100 cm; those of the third and fourth spans are 45 cm thick and are thickened near the fulcrum at the pier top to 100 cm. The side web in the midspan section is 45 cm thick and is thickened near the fulcrum at the pier top to 100 cm.

Diaphragm: One diaphragm is set on the top of each main pier, amounting to a total of 5 diaphragms. These diaphragms are 2 m thick.

The computational model for this girder design is shown in Fig. 5.10, and the structure is shown in Fig. 5.11.

Under the effect of main forces: The section experiences maximum compressive stress of 14.18 MPa, minimum compressive stress of 1.10 MPa, maximum shear stress of 2.45 MPa, maximum principal tensile stress of −0.9 MPa, strength safety factor of 2.36, and cracking resistance safety factor of 1.53.

Under the effect of main forces + additional forces: The section experiences maximum compressive stress of 16.03 MPa, minimum compressive stress of 0.06 MPa, maximum shear stress of 2.50 MPa, maximum principal tensile stress of −0.95 MPa, strength safety factor of 2.14, and cracking resistance safety factor of 1.32.

Girder deformation: The maximum deflection under ZK static live load is 3.63 mm, and the deflection-span ratio is 1/8567. Under ZK static live load, the vertical

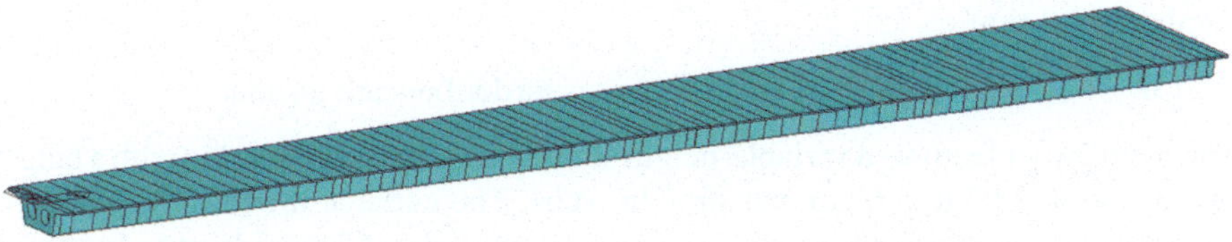

Fig. 5.10 Finite element model for (4 × 32) m continuous girder beneath turnout

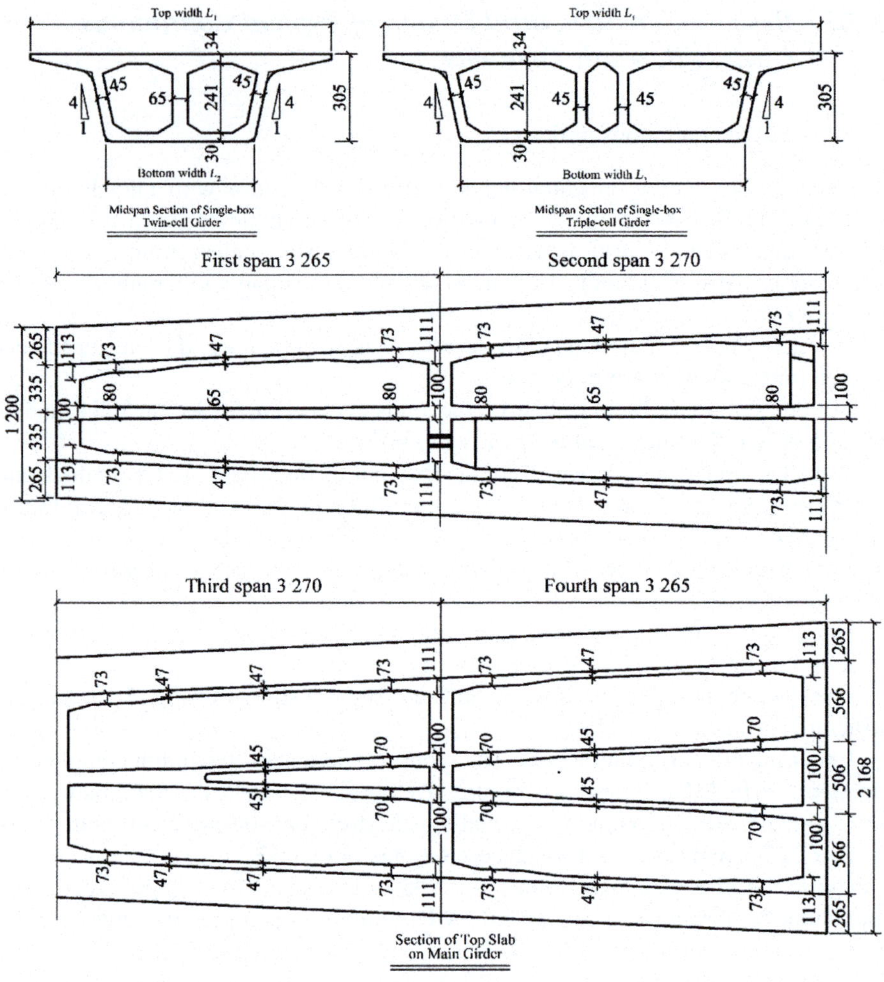

Fig. 5.11 Main girder structure of (4 × 32) m continuous girder beneath turnout (Unit: cm)

rotation angle at the girder end is 0.36‰rad. After track laying, the maximum creep camber is 1.4 mm.

2. (32 + 40 + 64 + 40 + 32) m continuous girder beneath turnout

The main girder features a variable-height single-box single-cell section with a height ranging from 3.05 to 5.60 m and straight webs. The curve at the lower edge of the curved section is expressed as $y = \frac{255}{2500^2}x^2$. The top slab is 12 m wide and is widened to 0.7 m on one side and 1.4 m on both sides due to the bridge deck widening to accommodate a switch. The bottom slab is 6.7 m wide.

Top slab: The top slab of the midspan section is 40 cm thick and is thickened near the fulcrum at the pier top to 55 and 70 cm.

Bottom slab: The bottom slab of the midspan section in the straight section over the side span is 40 cm thick and is thickened near the fulcrum at the pier top to 60 and 80 cm. The bottom slab of the midspan section in the middle span is 46 cm thick and is thickened near the fulcrum at the pier top to 160 cm.

Web: The midspan web is 45 cm thick and is thickened near the fulcrum at the pier top to 110 cm. The outer side of the web and the bottom edge of the cantilever slab are in an R20 cm arc transition, while the outer side of the web and the lower edge of the bottom slab are in an R10 cm arc transition. Such an arc transition is omitted between the outer side of the web and the lower edge of the bottom slab near the fulcrum.

Diaphragm: One diaphragm is set on the top of each pier, amounting to a total of 6 diaphragms. The diaphragm at the top of the middle pier is 2.5 m thick, that at the top of the secondary side pier is 2 m thick, and that of the side diaphragm is 1.5 m thick. A midspan diaphragm is set in the middle span, with a thickness of 0.6 m.

The computational model for this girder is shown in Fig. 5.12, and the structure is shown in Fig. 5.13.

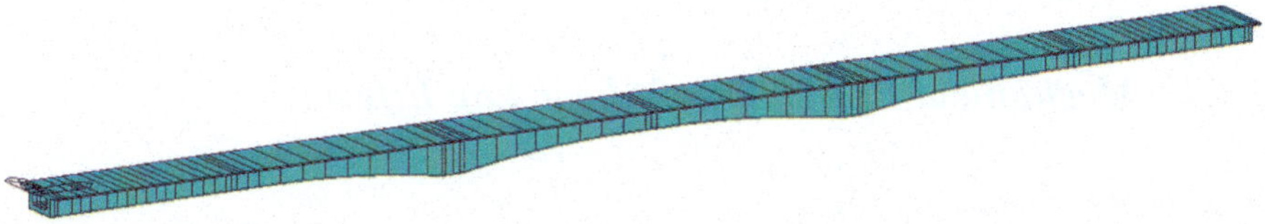

Fig. 5.12 Finite element model for (32 + 40 + 64 + 40 + 32) m continuous girder beneath turnout

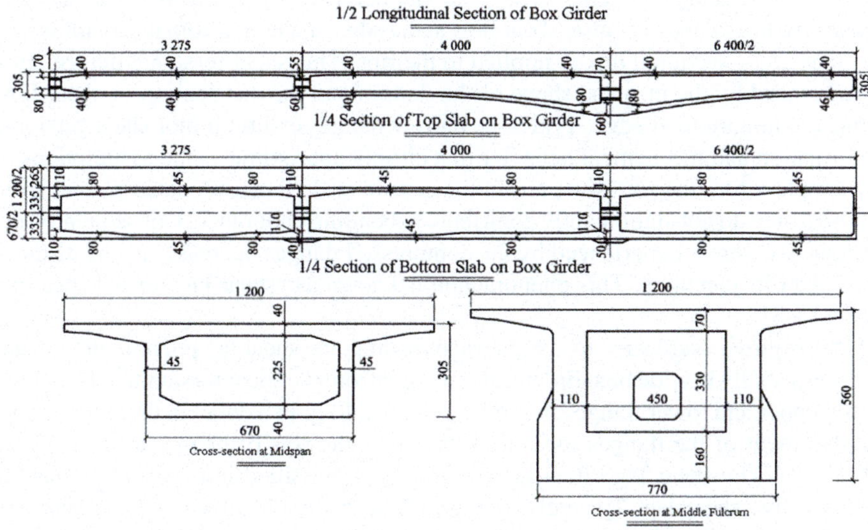

Fig. 5.13 Main girder structure of (32 + 40 + 64 + 40 + 32) m continuous girder beneath turnout (Unit: cm)

Under the effect of main forces: The section experiences maximum compressive stress of 13.18 MPa, minimum compressive stress of 1.02 MPa, maximum shear stress of 3.50 MPa, maximum principal tensile stress of −1.29 MPa, strength safety factor of 2.23, and cracking resistance safety factor of 1.54.

Under the effect of main forces + additional forces: The section experiences maximum compressive stress of 14.89 MPa, minimum compressive stress of 0.32 MPa, maximum shear stress of 3.52 MPa, maximum principal tensile stress of −1.30 MPa, strength safety factor of 2.02, and cracking resistance safety factor of 1.40.

Girder deformation: The maximum deflection under ZK static live load is 16.40 mm, and the deflection-span ratio is 1/3902. Under ZK static live load, the vertical rotation angle at the girder end is 0.42‰rad. After track laying, the maximum creep camber is 1.6 mm.

All calculation results related to the pertinent indexes comply with the specification requirements.

5.5 Study on Shear Lag Effect of Turnout Girder

5.5.1 Mechanical Principles of Shear Lag Effect

1. Concept of shear lag effect

When a T-girder is subjected to bending, transverse forces and shear flow are induced across the slab plane at the longitudinal edge of the flanges (specifically at the cut for ribs). The flanges undergo shear torsional deformation due to the effects of the transverse forces and eccentric shear flow at the edge, deviating from the assumptions of plane cross-sectional theory applied to the ribs. This shear torsional deformation is influenced by the in-plane shape of the flanges and the distribution of shear flow along the longitudinal edge. Typically, narrow flanges exhibit minor shear torsional deformation and mechanical behavior that closely approximates the assumptions of the simple girder theory. However, in the case of wide flanges located further from the ribs, they do not significantly contribute to bending endurance. Consequently, the compressive stress experienced by the compressed flanges decreases as the distance from the ribs increases. This phenomenon is known as "shear lag" or the shear lag effect [1–3].

In box girders subjected to symmetrical loading, the shear lag phenomenon manifests, especially in wide box girders, also arising from the shear torsional deformation of the upper and lower flanges in box girders, resulting in a delay in the longitudinal displacement of the flanges located farther from the ribs compared to those at the rib's edge. The shear lag effect in box girders is classified into positive shear lag and negative shear lag. Typically, the term "shear lag" denotes positive shear lag. Positive shear lag refers to a situation where the longitudinal stress in the flanges close to the web surpasses that at positions near the flanges' midpoint or the edge of the cantilever slab. Conversely, negative shear lag occurs when the longitudinal

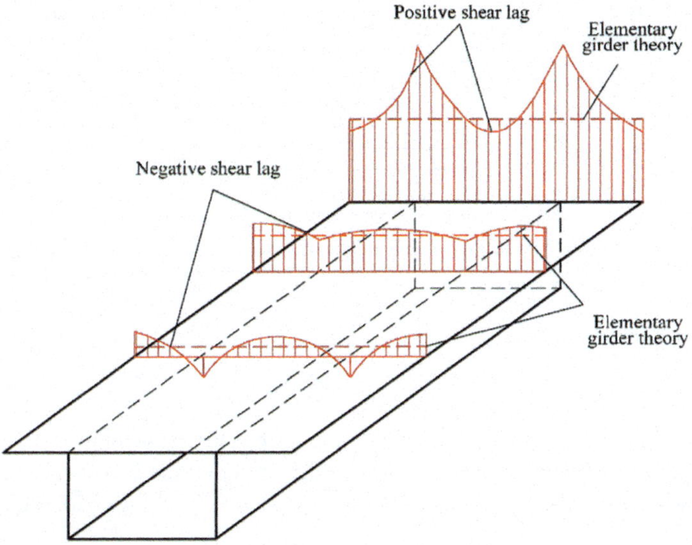

Fig. 5.14 Shear lag effect of box girder

stress at the flanges' midpoint or the cantilever slab's edge exceeds that in the flange adjacent to the web [4, 5]. These two scenarios are shown in Fig. 5.14.

2. Effective width of box girder

It can be quite intricate to accurately analyze the non-uniform distribution law of flange stress in box girders in a theoretical manner, making this analysis process particularly inconvenient for engineering design. To address this challenge, a concept known as the "effective width of box girder" is proposed. This concept aims to leverage the simple elementary formula of girder theory to capture stress values close to those encountered in practical scenarios to the greatest extent. The effective width of box girders is straightforwardly defined as the reduced width of box girders calculated using the elementary formula of girder theory, which closely approximates or matches the actual stress peak. This concept can be geometrically understood as follows: in Fig. 5.15, the shaded rectangle area, where the height corresponds to the peak value of actual stress, σmax, is equivalent to the area enclosed by the actual stress curve, so the side length of the shaded rectangle area denotes the effective distribution width of the flange.

3. Necessity of shear lag effect analysis

To accommodate turnouts and multiple tracks, turnout girders generally need a deck width greater than that of sectional girders carrying double tracks and may exhibit varying widths in certain instances. As the girder width expands, the width-span ratio escalates, leading to an intensified shear lag effect. Turnout girders with large widths may be designed in single-box twin-cell or single-box multi-cell cross sections.

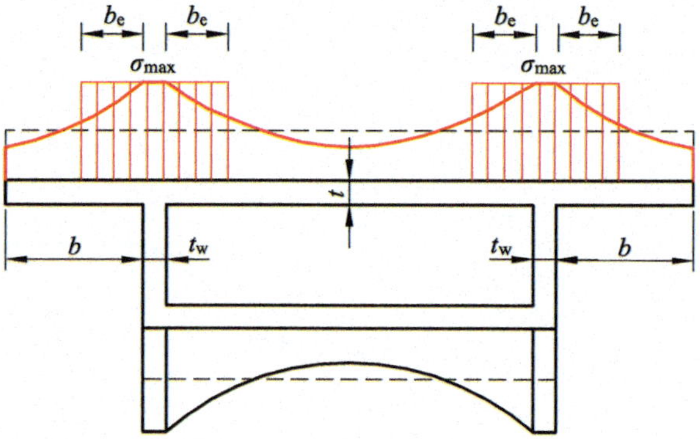

Fig. 5.15 Effective distribution width of box girder

Analyzing the stress distribution of the top slab and flange within twin-cell and triple-cell configurations is notably more intricate compared to single-cell girders. Hence, studying the shear lag effect of turnout girders is imperative for their structural design.

5.5.2 Analysis on Shear Lag Effect for Typical Turnout Girder

1. Computational model

This section delves into the shear lag effect observed in wide girders, using the example of (6 × 34) m continuous girders supporting turnouts on the Beijing–Shanghai High-Speed Railway. These (6 × 34) m box continuous girders, characterized by single-box twin-cell sections, feature a top width of 13.20 m, a bottom width of 7.80 m, and a height of 3.05 m. Three rows of bearings are positioned transversely, and a row of bearings is set at the center of each web. The girder structure is shown in Fig. 5.16.

A spatial solid finite element model was established to simulate these (6 × 34) m continuous girders, using ANSYS, a comprehensive finite element program [6, 7]. The model incorporates spatial 8-node hexahedron non-coordinating isoparametric solid elements for mesh division, with a total of 340,000 elements and 430,000 nodes. The finite element model is shown in Fig. 5.17.

2. Analysis of shear lag effect under dead weight

Stress calculations for these (6 × 34) m continuous girders under the action of dead weight were conducted by the finite element method, and the calculation results were compared with those derived from the elementary girder theory and those obtained after applying a reduction based on the effective width of box girders as outlined in the *Code for Design of High-Speed Railway* (TB 10621-2014).

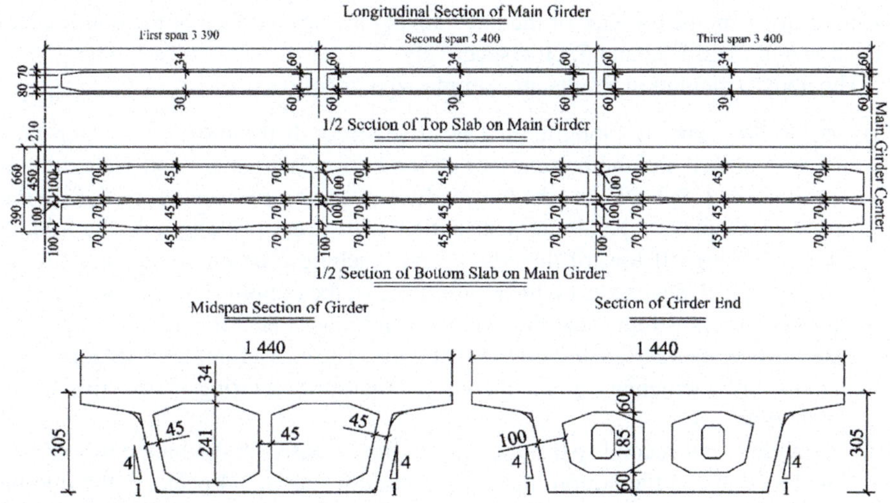

Fig. 5.16 Structure of (6 × 34) m continuous girder beneath turnout (Unit: cm)

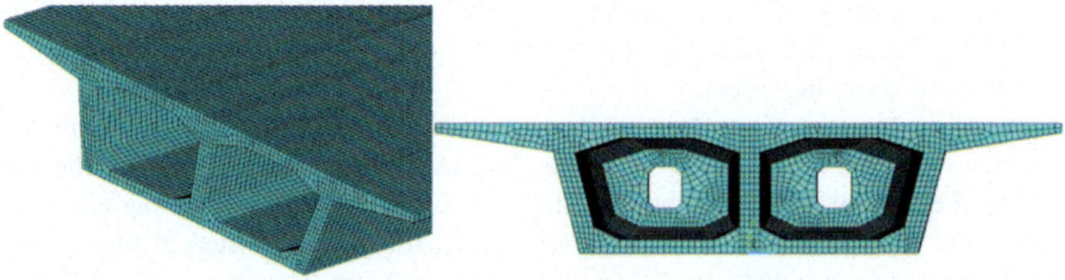

Fig. 5.17 Finite element computational model for (6 × 34) m turnout girder

Figures 5.18, 5.19, 5.20, 5.21, 5.22, 5.23 and 5.24 show the longitudinal positive stress distribution of these turnout girders for the first span. Figures 5.25, 5.26, 5.27, 5.28, 5.29, 5.30, and 5.31 show the longitudinal positive stress distribution of the turnout girders over the second span. The longitudinal positive stress distribution of the turnout girders over the third span is omitted due to its similarity to that over the second span. Spans 4–6 are also omitted owing to their symmetry with the first three spans. In the figures, curves labeled as the "spatial solid model" indicate the calculation results using the spatial finite element model. Curves designated as the "elementary girder (without reduction) model" indicate the calculation results derived from the elementary girder theory, neglecting the effective width reduction. Curves identified as the "elementary girder (with reduction) model" indicate the calculation results following the elementary girder theory and subsequent to the implementation of effective width reduction as outlined in the *Code for Design of High-Speed Railway* (TB 10621-2014). The longitudinal distribution of maximum positive stress on the top face of the top slab and the bottom face of the bottom slab are shown in Figs. 5.32 and 5.33, respectively.

The above calculation results are summarized as follows:

(1) In the first span, the longitudinal positive stress distributions on the top face of the top slab and the bottom face of the bottom slab at 1/4 and 1/2 exhibit a similar trend. Specifically, the longitudinal positive stress on the top face of the top slab displays a parabolic distribution with stress peaking near the middle web, tapering off toward the edges, and reaching a minimum at the extreme flange edge. Conversely, on the bottom face of the bottom slab, the longitudinal positive stress is highest at the side web and lowest near the middle web, with slight variations across the stress distribution. The maximum stress value closely aligns with elementary girder theory calculations, factoring in effective width reduction.

(2) In the first and second spans, the longitudinal positive stress on the top face of the top slab and the bottom face of the bottom slab at 3/4 peaks in the middle of the two webs, while becomes the lowest at the position close to the webs, reflecting a negative shear lag effect.

(3) At the middle fulcrum, the peak values of longitudinal positive stresses are relatively large at the top and bottom slabs positioned at the webs, and the stress decreases significantly toward the flange edges at both ends, reflecting an obvious shear lag effect.

(4) In the second span, the stress on the top face of the top slab at 1/4 is evident near the outer side of the side web, while obvious stress occurs near the middle web on the bottom face of the bottom slab, reflecting a negative shear lag effect.

(5) In the midspan section of the second span, the positive stress at the top and bottom slabs are distributed in a parabola-like shape. The stress peak at the top slab occurs at the middle web, while that at the bottom slab occurs at the side web.

(6) The maximum values of compressive stress and tensile stress resulting from the spatial finite element model are greater than those derived from the elementary

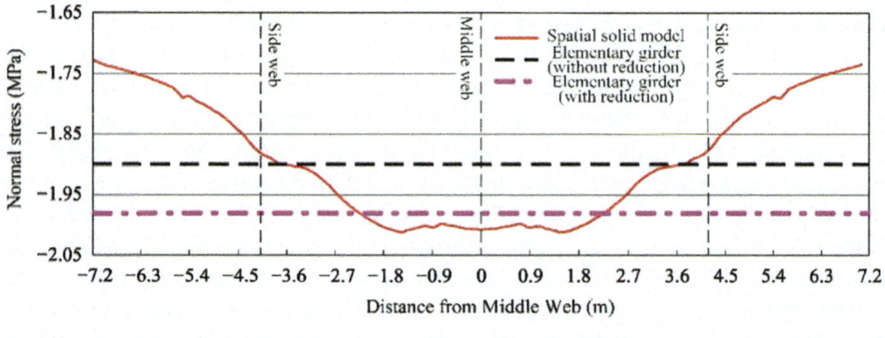

Fig. 5.18 Longitudinal normal stress distribution on top surface of top slab at 1/4 of the first span

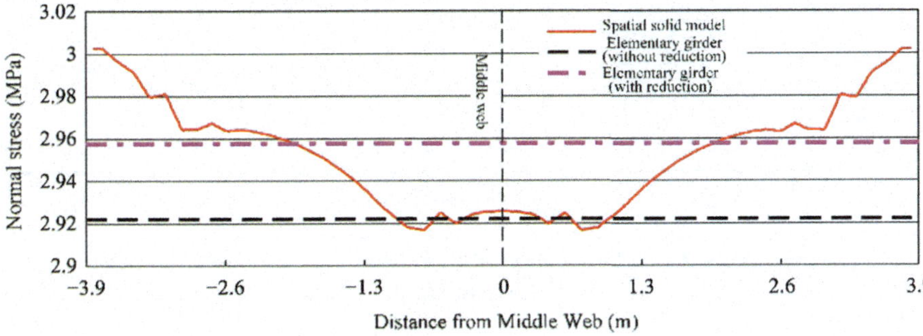

Fig. 5.19 Longitudinal normal stress distribution on bottom surface of bottom slab at 1/4 of the first span

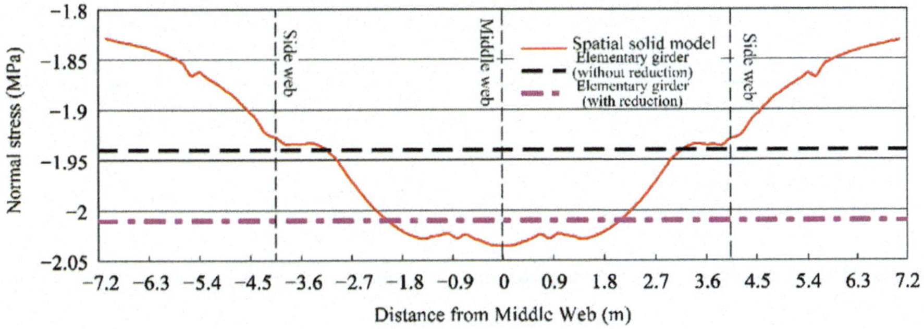

Fig. 5.20 Normal stress distribution on top surface of top slab at midspan of the first span

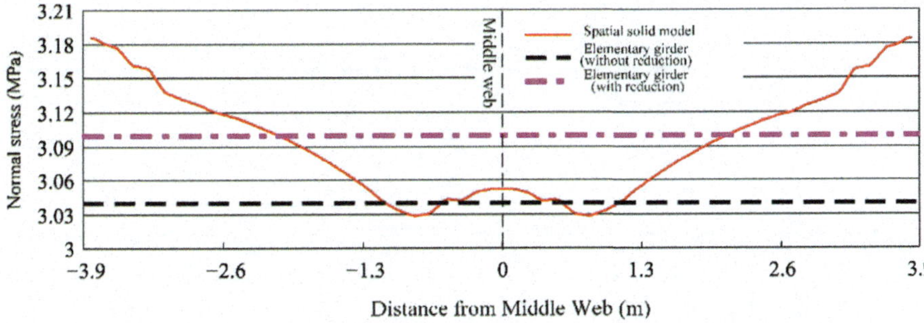

Fig. 5.21 Longitudinal normal stress distribution on bottom surface of base slab at midspan of the first span

girder model without effective width reduction, with an especially large increment near the middle fulcrum. The maximum values of compressive stress and tensile stress resulting from the spatial finite element model are close to the

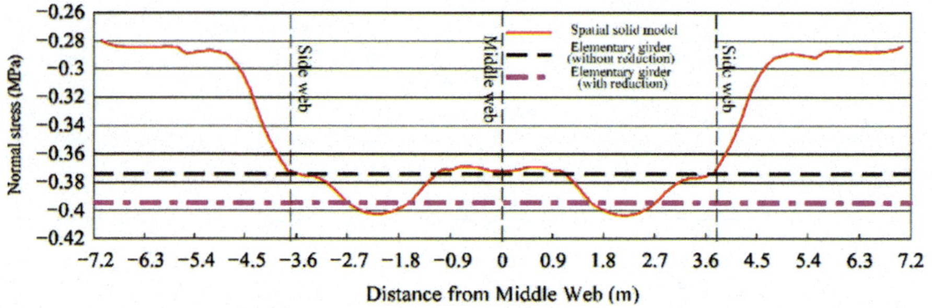

Fig. 5.22 Longitudinal normal stress distribution on top surface of top slab at 3/4 of the first span

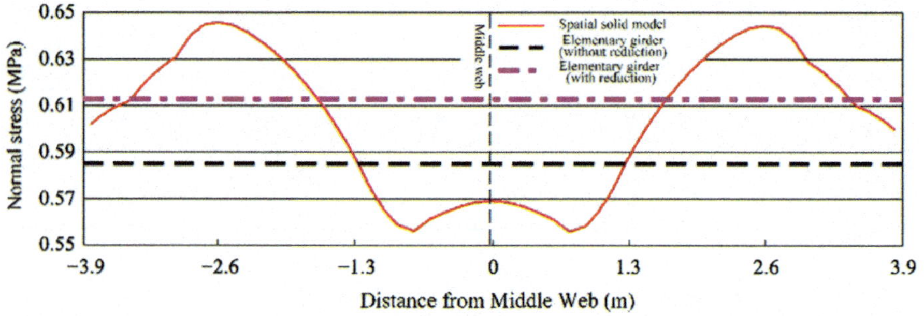

Fig. 5.23 Longitudinal normal stress distribution of bottom surface of bottom slab at 3/4 of the first span

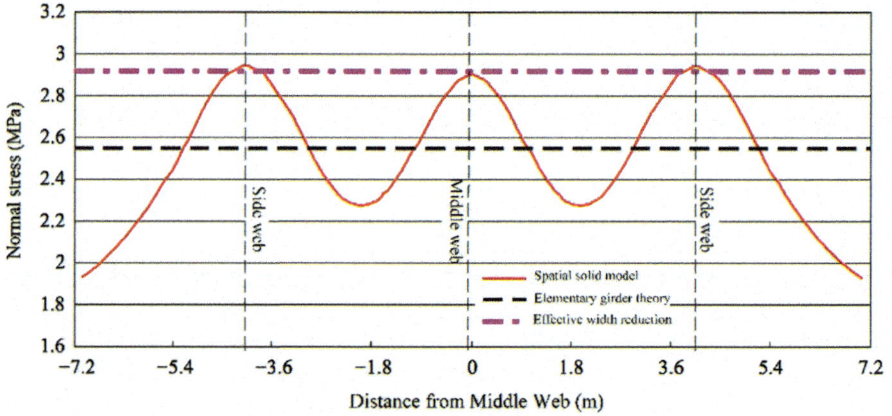

Fig. 5.24 Longitudinal normal stress distribution on top surface of top slab at middle fulcrum of the first span

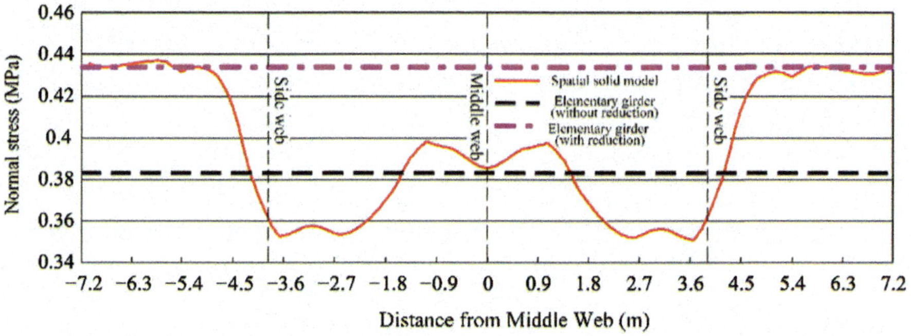

Fig. 5.25 Longitudinal normal stress distribution on top surface of top slab at 1/4 of the second span

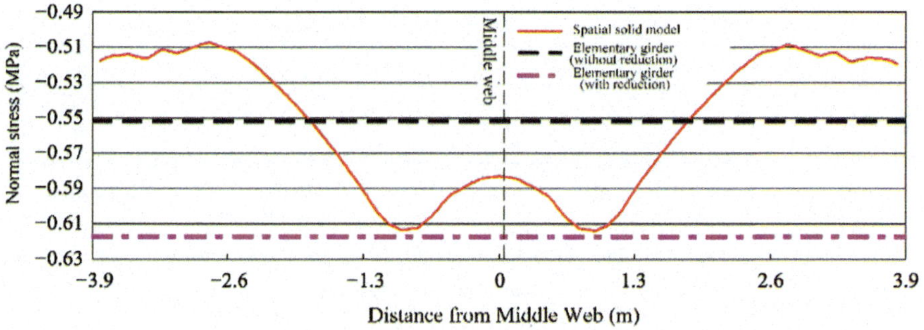

Fig. 5.26 Longitudinal normal stress distribution on bottom surface of bottom slab at 1/4 of the second span

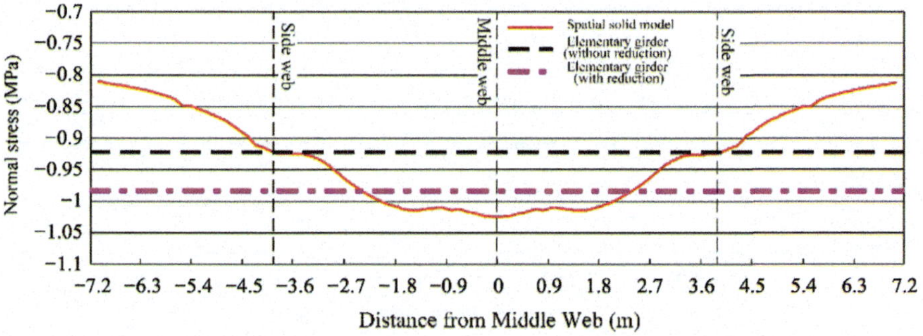

Fig. 5.27 Longitudinal normal stress distribution on top surface of top slab at midspan of the second span

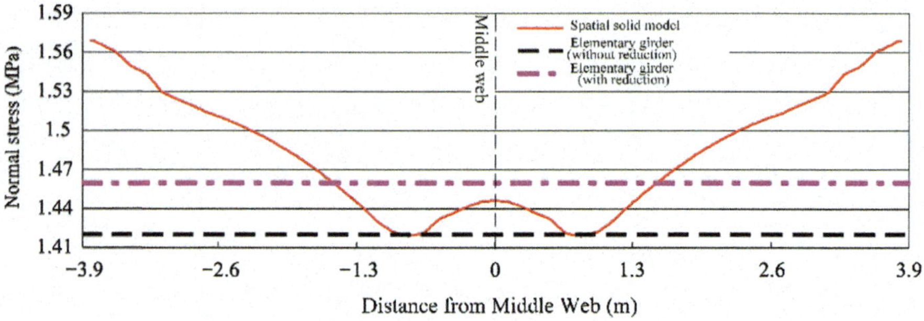

Fig. 5.28 Longitudinal normal stress distribution on bottom surface of base slab at midspan of the second span

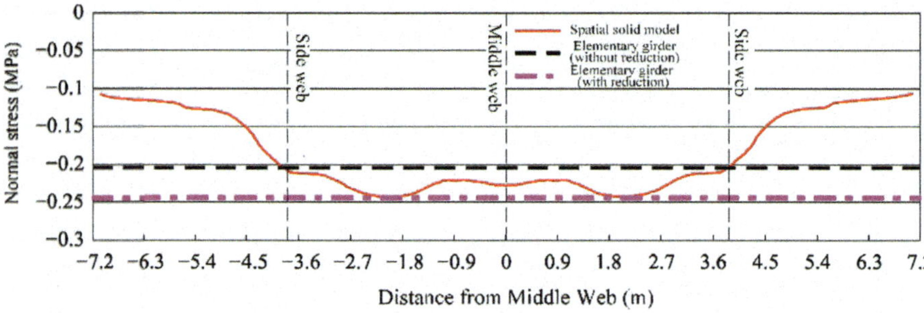

Fig. 5.29 Longitudinal normal stress distribution on top surface of top slab at 3/4 of the second span

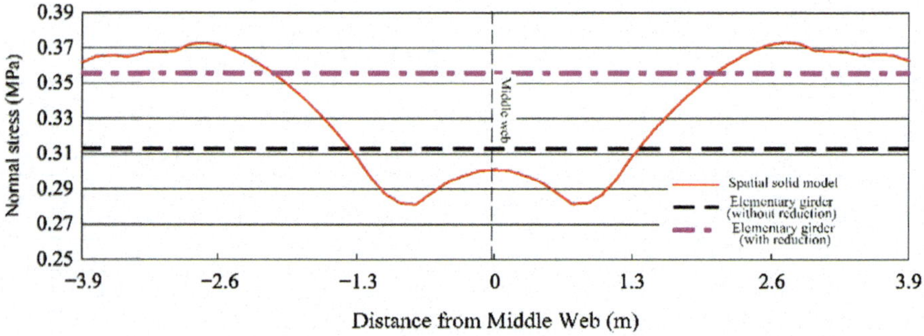

Fig. 5.30 Longitudinal normal stress distribution on bottom surface of bottom slab at 3/4 of the second span

calculation results based on the elementary girder model with effective width reduction as per the *Code for Design of High-Speed Railway* (TB10621-2014).

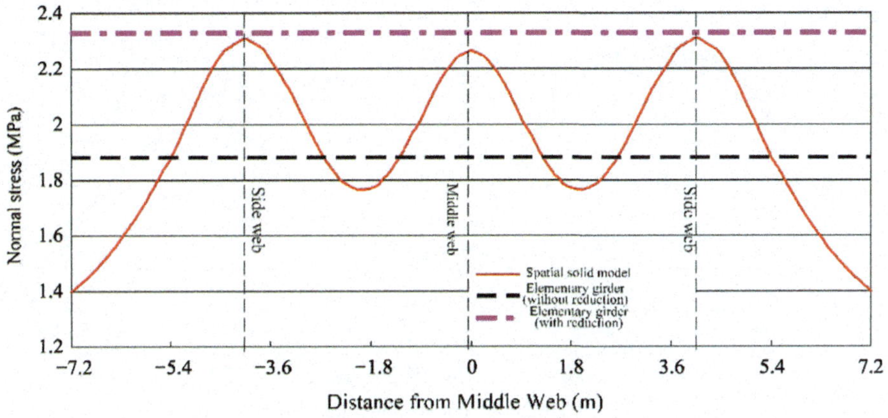

Fig. 5.31 Longitudinal normal stress distribution on top surface of top slab at fulcrum within the second span

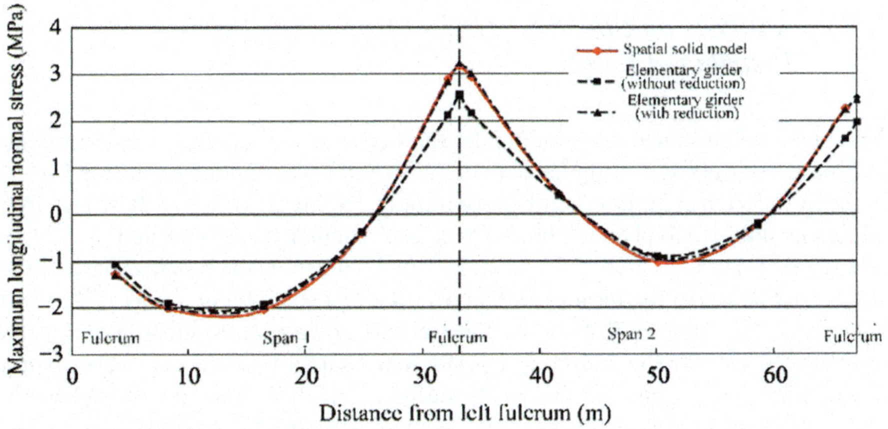

Fig. 5.32 Distribution curve of maximum longitudinal normal stress in longitudinal direction on the top surface of top slab on bridge

(7) The shear lag effect is obvious near the fulcrum, leading to a large shear lag coefficient. The shear lag coefficient in the midspan area is small. The maximum shear lag coefficient of 1.4 occurs at the middle fulcrum.

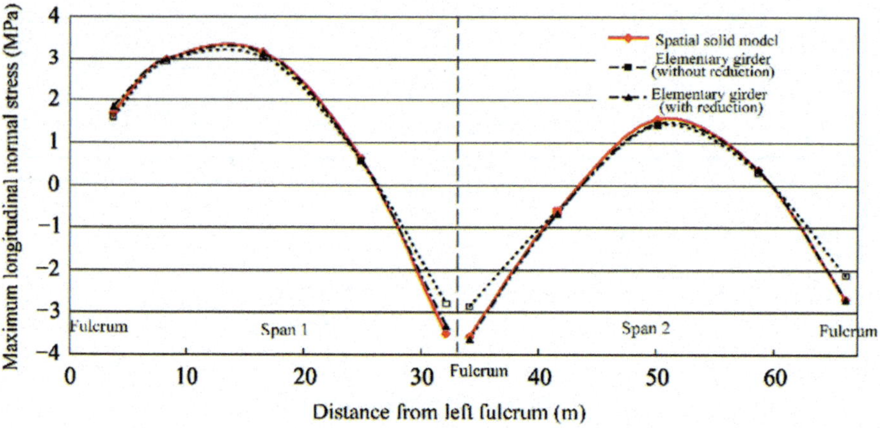

Fig. 5.33 Distribution curve of maximum longitudinal normal stress in longitudinal direction on the bottom surface of bottom slab on bridge

5.5.3 Analysis on Shear Lag Effect Under Uniformly Distributed Loads

To understand the shear lag effect under the track-related secondary dead load and live load, a uniformly distributed load of 7.78 kN/m^2 was applied on the top surface of the (6 × 34) m continuous girders for turnouts. Subsequently, the shear lag effect under this uniform load on the bridge deck was calculated and analyzed.

Figures 5.34, 5.35, 5.36, 5.37, 5.38, 5.39 and 5.40 show the longitudinal positive stress distribution of the turnout girders over the first span. Figures 5.41, 5.42, 5.43, 5.44, 5.45, 5.46 and 5.47 show the longitudinal positive stress distribution of the turnout girders over the second span. The longitudinal positive stress distribution in the third span is omitted due to its similarity to that of the second span. The longitudinal positive stress distributions in spans 4–6 are also omitted owing to their symmetry with the first three spans. Figures 5.48 and 5.49 show the longitudinal distribution of maximum positive stress on the top face of the top slab and the bottom face of the bottom slab, respectively.

The above calculation results are summarized as follows:

(1) When subjected to a uniformly distributed load on the bridge deck, the box girders demonstrate a noticeable shear lag effect, with the transverse flexural positive stress distribution primarily assuming a parabolic shape. In most instances, the flexural positive stress near the webs is higher compared to places further away, emphasizing a positive shear lag effect. However, certain cross sections, like those near 1/4 span or 3/4 span, depict a negative shear lag effect.
(2) The shear lag effect is obvious near the fulcrum, leading to a large shear lag coefficient. The shear lag coefficient in the midspan area is small. The maximum shear lag coefficient of 1.4 or so occurs at the middle fulcrum.

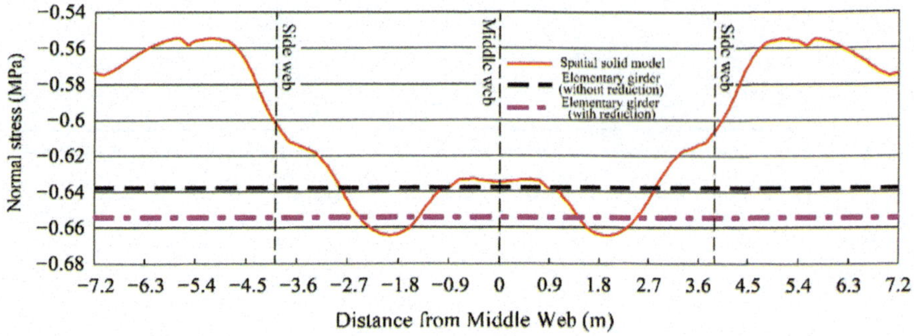

Fig. 5.34 Longitudinal normal stress distribution on top surface of top slab at 1/4 of the first span

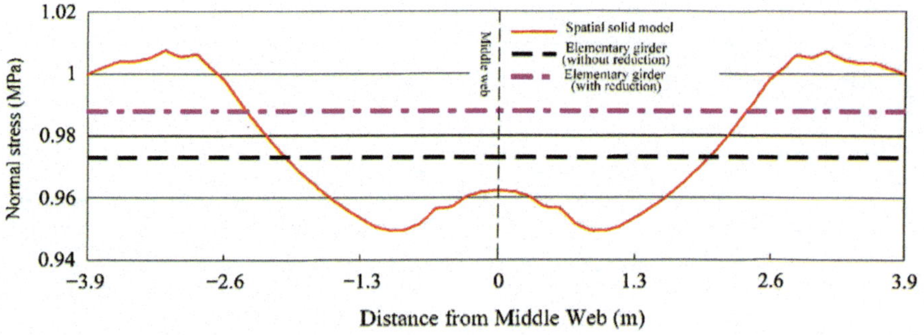

Fig. 5.35 Distribution of longitudinal normal stress in transverse direction on bottom surface of bottom slab at 1/4 span of the first span

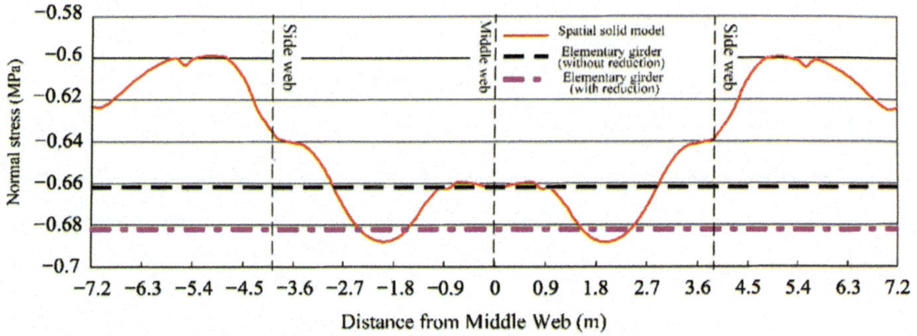

Fig. 5.36 Distribution of longitudinal normal stress in transverse direction on top surface of top slab at midspan of the first span

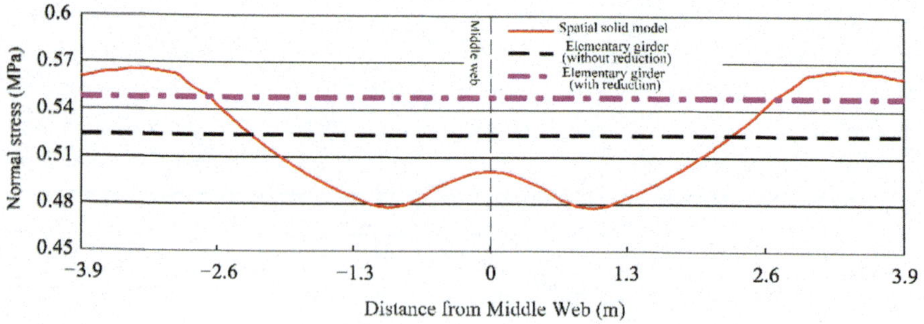

Fig. 5.37 Distribution of longitudinal normal stress in transverse direction on bottom surface of bottom slab at midspan of the first span

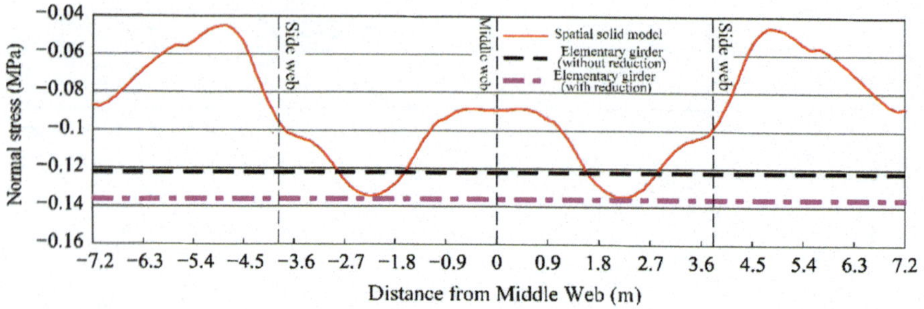

Fig. 5.38 Distribution of longitudinal normal stress in transverse direction on top surface of top slab at 3/4 of the first span

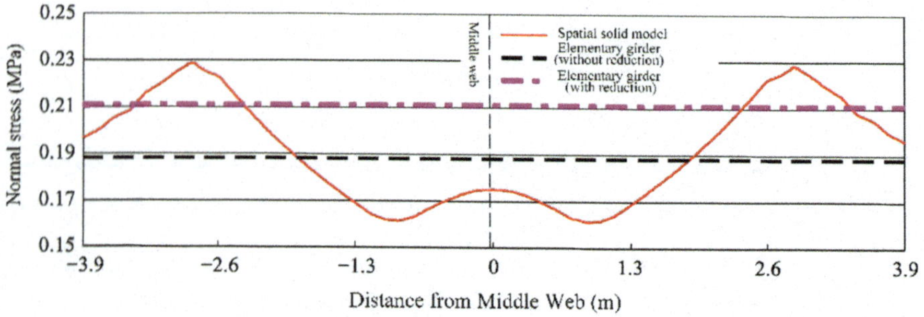

Fig. 5.39 Distribution of longitudinal normal stress in transverse direction on bottom surface of bottom slab at 3/4 span of the first span

(3) The maximum values of flexural compressive stress and flexural tensile stress resulting from the spatial finite element model are greater than those derived from the elementary girder model without effective width reduction, with an especially large increment near the middle fulcrum. The maximum values of

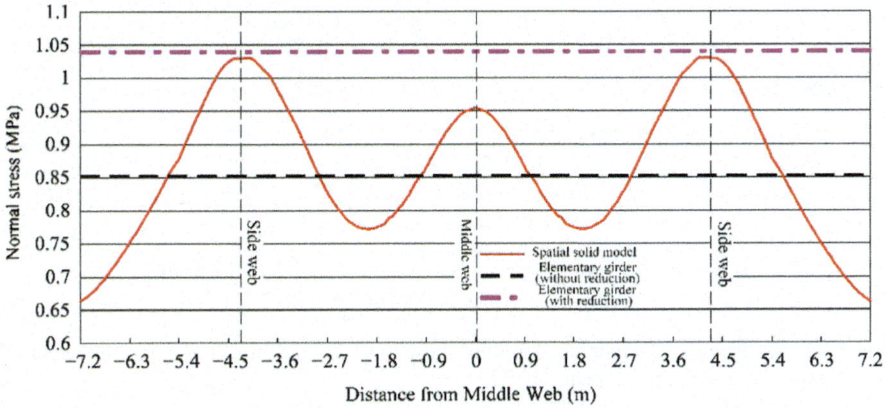

Fig. 5.40 Distribution of longitudinal normal stress in transverse direction on top surface of top slab at fulcrum within the first span

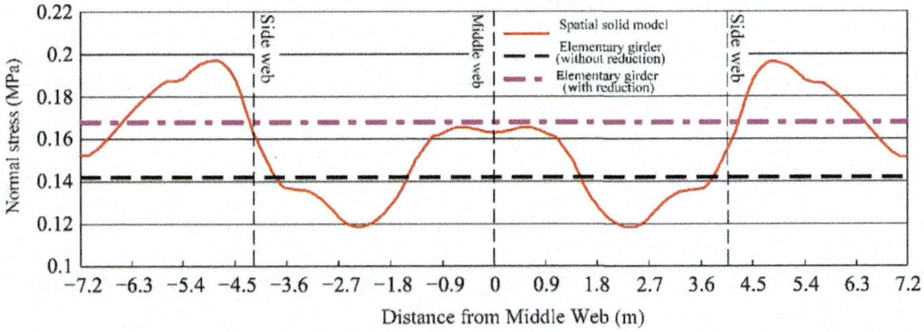

Fig. 5.41 Distribution of longitudinal normal stress in transverse direction on top surface of top slab at 1/4 of the second span

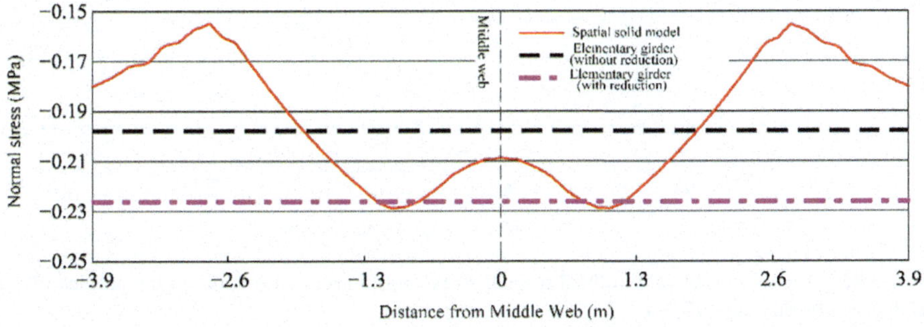

Fig. 5.42 Distribution of longitudinal normal stress in transverse direction on bottom surface of bottom slab at 1/4 of the second span

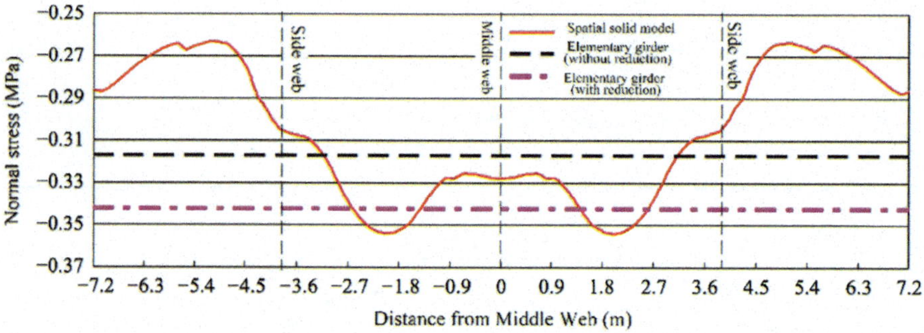

Fig. 5.43 Distribution of longitudinal normal stress in transverse direction on top surface of top slab at midspan of the second span

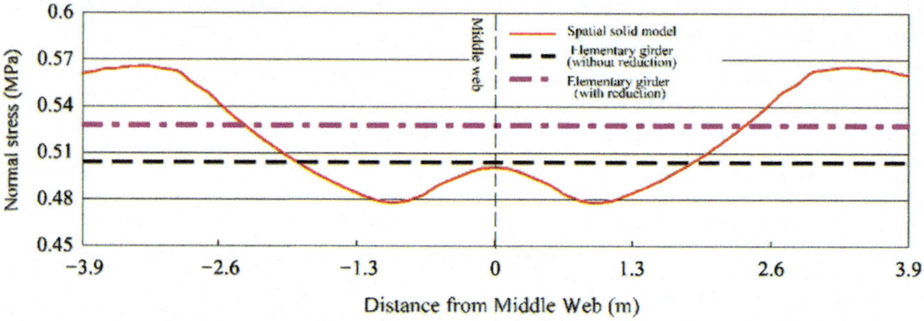

Fig. 5.44 Distribution of longitudinal normal stress in transverse direction on bottom surface of bottom slab at midspan of the second span

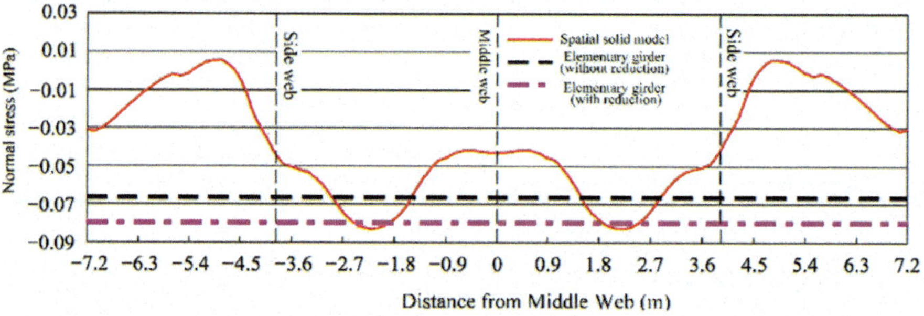

Fig. 5.45 Distribution of longitudinal normal stress in transverse direction on top surface of top slab at 3/4 of the second span

flexural positive stress resulting from the spatial finite element model are close to the calculation results based on the elementary girder model with effective width reduction as per the *Code for Design of High-Speed Railway* (TB 10621-2014).

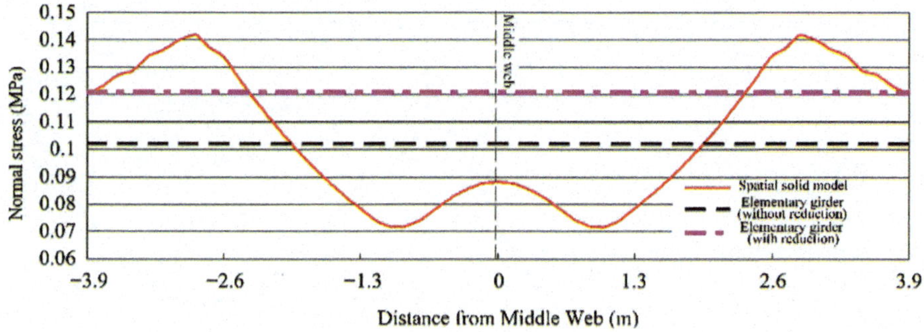

Fig. 5.46 Distribution of longitudinal normal stress in transverse direction on bottom surface of bottom slab at 3/4 of the second span

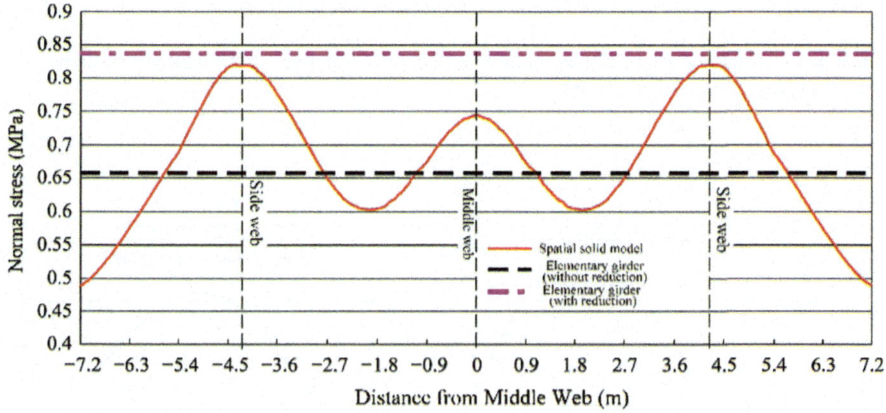

Fig. 5.47 Distribution of longitudinal normal stress in transverse direction on top surface of top slab at fulcrum within the second span

5.5.4 Conclusions

This section focuses on analyzing the shear lag effect of turnout girders using the spatial finite element method. It also includes a comparison with the calculation results obtained from the elementary girder model—both with and without effective width reduction. Conclusions are as follows:

(1) When under dead weight and subjected to a uniformly distributed load on the bridge deck, the box girders supporting turnouts demonstrate a noticeable shear lag effect, with the transverse flexural positive stress distribution primarily assuming a parabolic shape. In most instances, the flexural positive stress near the webs is higher compared to places further away, emphasizing a positive shear lag effect. However, certain cross sections, like those near 1/4 span or 3/

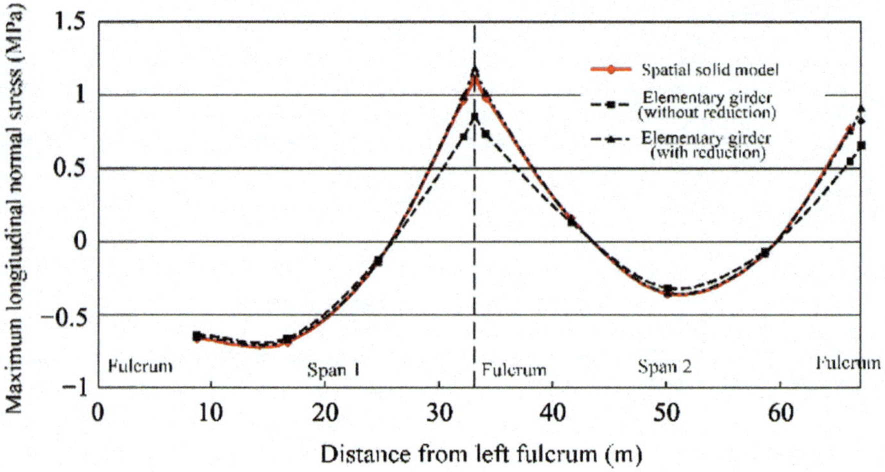

Fig. 5.48 Distribution curve of maximum longitudinal normal stress in longitudinal direction on the top surface of top slab on bridge

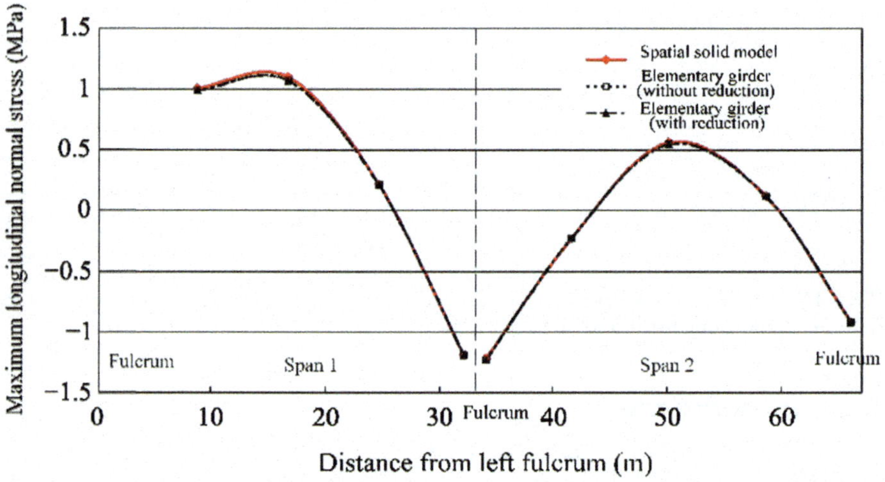

Fig. 5.49 Distribution curve of maximum longitudinal normal stress in longitudinal direction on the bottom surface of bottom slab on bridge

4 span, depict a negative shear lag effect, with smaller flexural positive stress near the webs compared to that at places further away.

(2) The shear lag effect is obvious near the fulcrum, leading to a large shear lag coefficient. The shear lag coefficient in the midspan area is small.

(3) The maximum values of flexural positive stress resulting from the spatial finite element model are greater than those derived from the elementary girder model without effective width reduction, with an especially large increment near the

middle fulcrum. The maximum values of flexural compressive stress and flexural tensile stress resulting from the spatial finite element model are close to the calculation results based on the elementary girder model with effective width reduction as per the *Code for Design of High-Speed Railway* (TB 10621-2014).

(4) The stress calculation results derived from the elementary girder model with effective width reduction based on the *Code for Design of High-Speed Railway* (TB 10621-2014) align closely with those from the spatial finite element model. Therefore, the effective width reduction method is feasible for simulating the shear lag effect of turnout girders.

5.6 Study on the Live Load Amplification Coefficient of Turnout Girder

5.6.1 Calculation Principle

Live loads act vertically on box girders in either symmetrical or asymmetrical modes, the latter known as eccentric loads. When subjected to an eccentric load, box girders undergo both flexural and torsional deformations. Any eccentric load applied to box girders can be broken down into symmetrical load and torque passing through the shear center of the section [8]. In the presence of symmetrical loading, the box girder exhibits symmetrical bending, leading to longitudinal positive stress σ_{m} and shear stress τ_{m} on the cross section. When torque is applied, the box girder experiences rigid torsion and distortion. Rigid torsion encompasses free torsion, generating free torsional shear stress τ_{k}, and constraint torsion, where longitudinal fiber deformation faces constraints, resulting in constrained torsional positive stress σ_{w} and constrained torsional shear stress τ_{w} on the cross section. Distortion describes peripheral deformations in torsion, causing distortional warping positive stress σ_{dw} and distortional shear stress τ_{dw}. The stress patterns of box girders under eccentric loads are shown in Fig. 5.50.

The stress of box girders generated under eccentric live load is expressed as follows:

Longitudinal positive stress $\sigma = \sigma_{\mathrm{m}} + \sigma_{\mathrm{w}} + \sigma_{\mathrm{dw}}$.
Shear stress $\tau = \tau_{\mathrm{k}} + \tau_{\mathrm{m}} + \tau_{\mathrm{w}} + \tau_{\mathrm{dw}}$.

Based on the above formulas, the longitudinal stress σ of box girders under eccentric loads surpasses the longitudinal stress σ_{m} under symmetrical loads. Therefore, to reflect the influence of eccentric loads on section stress, a load amplification coefficient (ξ) is introduced and defined as follows:

$$\xi = \frac{\sigma_m + \sigma_W + \sigma_{dW}}{\sigma_m}.$$

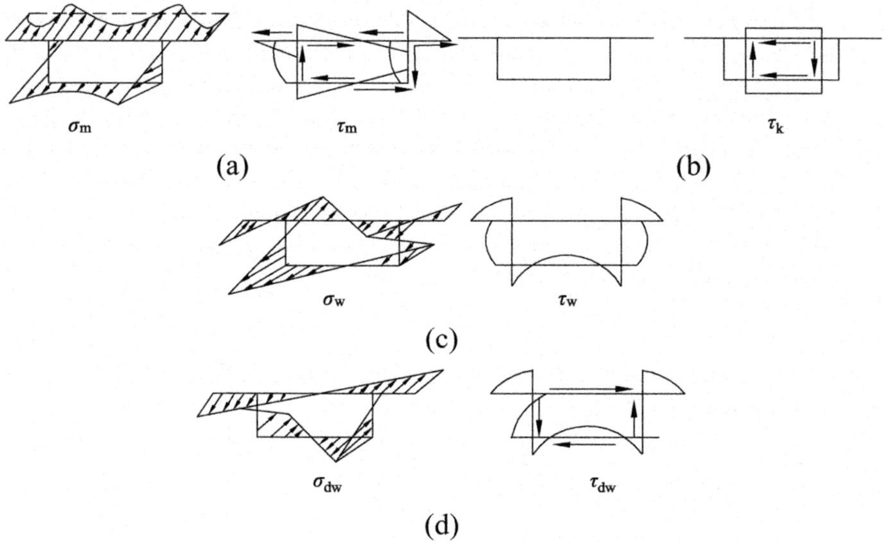

Fig. 5.50 Stress of box girder section under eccentric load

5.6.2 *Necessity of Study on Live Load Amplification Coefficient*

For straight girders with standard double-track configuration in sections, the double tracks are arranged symmetrically along the girder axis. When subjected to a transversely symmetrical double-track ZK live load along the girder axis, eccentric load is absent. The girder experiences even stress distribution on both sides of its axis, leading to a live load amplification coefficient of $\xi = 1$. However, when subjected to a single-track ZK live load, an eccentric load emerges with an eccentricity equal to half the distance between the track centers. In such cases, it becomes necessary to take into account a live load amplification coefficient induced by the eccentric load. Notably, although single-track loading triggers a live load amplification effect, the total effect is comparatively lower than that resulting from double-track loading. Consequently, for straight girders with standard double-track configuration in sections, the critical stress condition is the symmetric double-track loading. Therefore, it is unnecessary to include a live load amplification coefficient (ξ) in the design process [9–11].

For curved girders with standard double-track configuration in sections, simply-supported girders are generally designed with a straight configuration, in a bisecting midsagittal layout in the majority of cases and occasionally in a tangential layout. Continuous girders are usually designed with a curved configuration. In straight scenarios, eccentricity occurs between the track centerline and the girder axis. Consequently, even under double-track loading, there is a presence of eccentric loading, albeit of a minimal magnitude. In addition, girders on curves simultaneously endure transverse centrifugal force and vertical live load. The centrifugal force initiates eccentric loading, which in turn causes uneven girder stress on both sides of the girder axis. Therefore, a live load eccentric load coefficient should be considered in the design [12–14]. Concerning the live load eccentric load coefficient for curved girders, the *Technical Manual for Railway Engineering Design (Concrete Bridge)* provides in-depth calculations for live load amplification coefficients for simply-supported T-girders with ballasted decks on curves. These coefficients typically range from 1.0 to 1.15. As the curve radius increases, the amplification coefficient decreases. For curve radii equal to or exceeding 4000 m, the live load amplification coefficient is usually <1.02. High-speed railway sections commonly utilize box girders, which offer significant torsional rigidity, leading to a further reduction in the live load amplification coefficient resulting from eccentric loading. Additionally, high-speed railways are usually designed with curve radii exceeding 6000 m, resulting in very marginal live load amplification coefficients due to curve factors for standard curved girders in sections. These coefficients can either be omitted in the longitudinal design of these girders or considered appropriately based on past experience.

On high-speed railways, turnout girders often feature three or more tracks. Due to the specific positional relationship between tracks and turnouts, some turnout girders may not have symmetrical track positioning along the centerline of the girder axis, as shown in Fig. 5.51a. In certain instances, tracks on these turnout girders are not intended for simultaneous train operation, as illustrated in Fig. 5.51b. Conversely, some turnout girders are equipped with three or more tracks for simultaneous train operations. According to Article 7.2.6 of the *Code for Design of High-Speed Railway* (TB 10621-2014), design calculations must consider the most unfavorable conditions. Under Condition 1, two tracks are subjected to 100% ZK live load at the most unfavorable positions, while the remaining tracks remain unloaded. In Condition 2, all tracks endure 75% ZK live load at the most disadvantageous locations. In Condition 1, the two tracks are subjected to 100% ZK live load, potentially resulting in eccentric loading, which can become a critical stress condition when a live load amplification coefficient is taken into consideration.

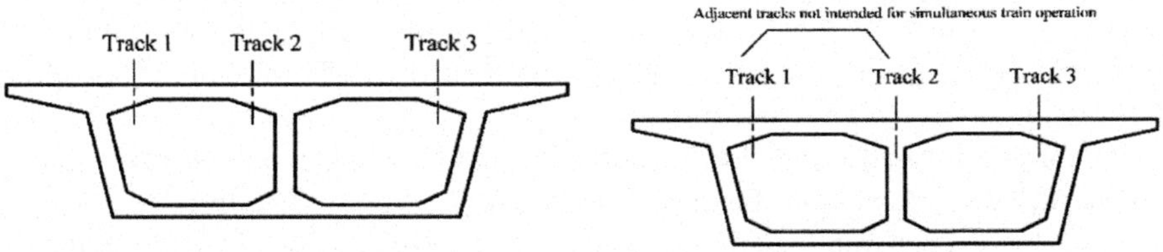

(a) Asymmetric track layout

(b) Adjacent tracks not intended for simultaneous train operation

Fig. 5.51 Schematic diagram for track arrangement on turnout girder

Therefore, for turnout girders on high-speed railways, it is crucial to study live load amplification coefficients under eccentric loading and analyze their calculation methodology for guidance and implementation in the design of turnout girders.

5.6.3 Analysis and Calculation of Live Load Amplification Coefficient

1. Computational model and conditions

This section focuses on (4 × 32) m continuous girder for turnouts for analyzing live load amplification coefficients. The theoretical span of these (4 × 32) m girders is 31.9 m + 2 × 32.7 m + 31.9 m, the distance between the center of their end bearings and the girder ends is 0.75 m, and the girder length is 130.70 m. They are equal-height box girders featuring a single-box twin-cell cross section, with inclined webs, and a girder height of 3.05 m. Their top and bottom slabs are 0.35 and 0.3 m thick at the midspan and thickened to 0.6 m and 0.6 m at the fulcrum, respectively. Their webs vary in thickness from midspan to fulcrum, transitioning from 0.45 m to 0.7 m and eventually 1.0 m. A 2 m wide diaphragm is positioned at the fulcrum of these girders. Their structure is shown in Fig. 5.52. These box girders of 19.75 m in top width are equipped with three tracks on the bridge deck. The leftmost track is positioned 4.125 m from the box girder edge, and the distance between the track centers from left to right is 5 and 6.5 m, respectively. The track arrangement on the bridge deck is shown in Fig. 5.53.

A three-dimensional solid finite element computational model was established to simulate a (4 × 32) m continuous girder for turnout accommodation, using ANSYS, a spatial finite element program. The computational model incorporates spatial 8-node solid elements (solid45) to ensure sufficient accuracy of the finite element meshing. Each track is subjected to transverse live loading for uniform distribution covering a 3.2 m width over the track slab, employing an influence surface loading approach. This girder carries three tracks, with Track 1 deviating from the structural center of the box girder by −5.75 m, Track 2 deviating from the structural center of the box girder by −0.75 m, and Track 3 deviating from the structural center of the box girder by 5.75 m.

According to single-track loading, double-track loading, and three-track loading among the three tracks on this girder, there are 7 loading conditions to consider: Track 1 loaded separately; Track 2 loaded separately; Track 3 loaded separately; Tracks 1 and 2 loaded concurrently; Tracks 1 and 3 loaded concurrently; Track 2 and 3 loaded concurrently; and Three tracks loaded concurrently. From the perspective of the maximum resultant force eccentricity of live load, the eccentricities are noticeable in three conditions, i.e. the resultant force eccentricity of −5.75 m in Condition 1 of separate loading on Track 1, resultant force eccentricity of −3.25 m in Condition 4 of concurrent loading on Tracks 1 and 2, and resultant force eccentricity of −0.25 m in Condition 7 of concurrent loading on three tracks, which are considered

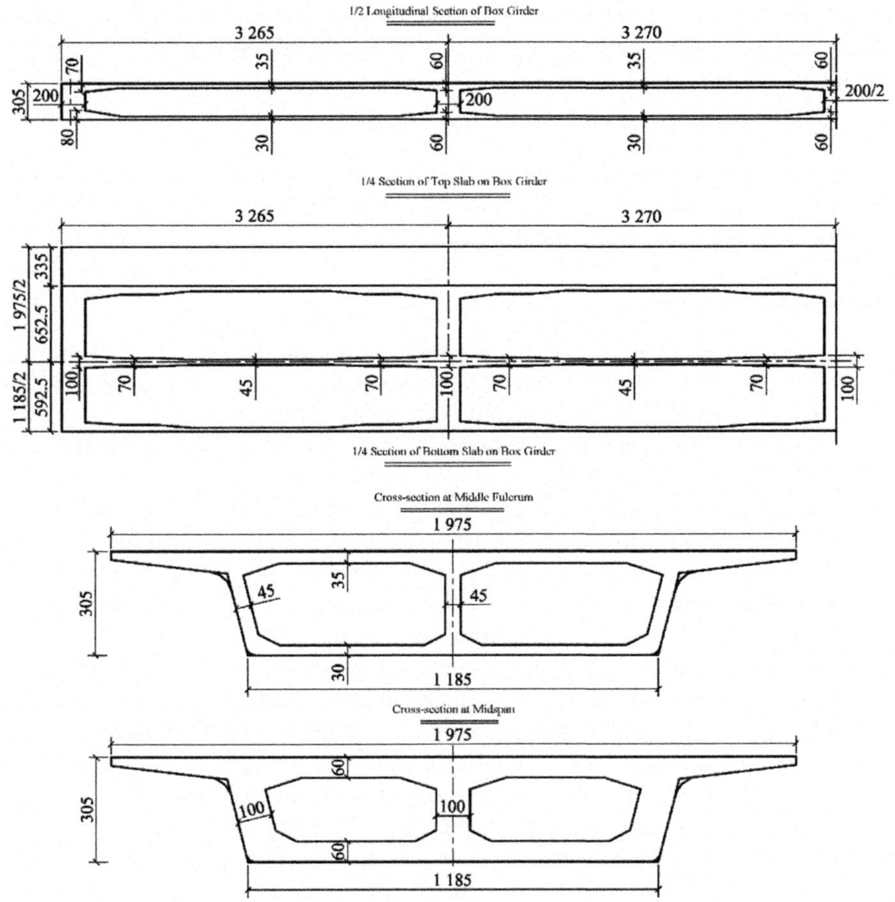

Fig. 5.52 Main girder structure of (4 × 32) m continuous girder beneath turnout (Unit: cm)

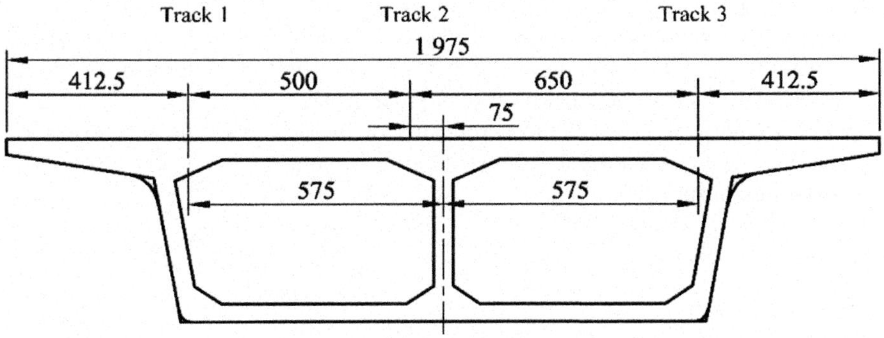

Fig. 5.53 Track arrangement on bridge deck (Unit: cm)

as critical conditions. Therefore, the computational analysis in this section concentrates on determining the live load amplification coefficients (ξ) under these specific conditions.

2. Computational analysis

Calculations were conducted, using the three-dimensional solid finite element model for the above-mentioned three conditions. See Table 5.2 for the calculation results of live load amplification coefficients for the main section under the above three critical conditions.

Conclusions are drawn from Table 5.2 as follows:

(1) For separate loading on Track 1, the live load eccentricity value is 5.75 m, with a corresponding live load amplification coefficient range of 1.367–1.543. For concurrent loading on Tracks 1 and 2, these values are 3.25 m and 1.069–1.222, respectively. For concurrent loading on three tracks, these values are 0.25 m and 1.001 to 1.030, respectively.
(2) The calculation results obtained from the modified eccentric pressure method closely align with those derived from the spatial finite element method. The discrepancy between them ranges from −8.2 to +11.0% for individual loading on Track 1, from −0.9 to +13.3% for concurrent loading on Tracks 1 and 2, and from −1.2 to +1.5% for simultaneous loading on all three tracks.
(3) The spatial finite element analysis method is effective in analyzing the influence of eccentric live load on turnout girders, boasting accurate calculation results, but the computational process is very complex. In contrast, the modified eccentric pressure method yields calculation results comparable to the spatial finite element method while simplifying the computational procedure. Therefore, the latter method is recommended for the design of turnout girders.
(4) While the empirical coefficient method lacks precision in reflecting live load amplification coefficients for turnout girders in different conditions, its results align closely with those of the spatial finite element method and modified eccentric pressure method under the most unfavorable conditions. Moreover, this method provides a simple process involving the direct multiplication of live loads by the empirical coefficient, thereby avoiding intricate calculations. As a result, the empirical coefficient method can serve as a simplistic approach suitable for conceptual and preliminary design stages.
(5) The eccentric pressure method produces notably inflated results as it overlooks the torsional characteristics of turnout girders; hence, it is not advisable for use in design purposes.

Table 5.2 Calculation results of live load amplification coefficient (ξ) for (4 × 32) m continuous girder beneath turnout

Position		Separate loading on Track 1				Concurrent loading on Tracks 1 and 2				Concurrent Loading on Three Tracks			
		Spatial Finite Element	Modified Eccentric Pressure Method	Eccentric Pressure Method	Empirical Coefficient Method	Spatial Finite Element	Modified Eccentric Pressure Method	Eccentric Pressure Method	Empirical Coefficient Method	Spatial Finite Element	Modified Eccentric Pressure Method	Eccentric Pressure Method	Empirical Coefficient Method
Side span	¼	1.367	1.373	2.597	1.15	1.107	1.211	1.903	1.15	1.001	1.016	1.069	1.15
	Midspan	1.495	1.373	2.597	1.15	1.222	1.211	1.903	1.15	1.029	1.016	1.069	1.15
	¾	1.543	1.373	2.597	1.15	1.182	1.211	1.903	1.15	1.010	1.016	1.069	1.15
	Secondary middle fulcrum	1.455	1.373	2.597	1.15	1.069	1.211	1.903	1.15	1.020	1.016	1.069	1.15
Middle span	1/4	1.531	1.4303	2.597	1.15	1.130	1.2432	1.903	1.15	1.008	1.0187	1.069	1.15
	Midspan	1.501	1.4303	2.597	1.15	1.211	1.2432	1.903	1.15	1.027	1.0187	1.069	1.15
	3/4	1.391	1.4303	2.597	1.15	1.097	1.2432	1.903	1.15	1.006	1.0187	1.069	1.15
	Middle fulcrum	1.462	1.4303	2.597	1.15	1.171	1.2432	1.903	1.15	1.030	1.0187	1.069	1.15

References

1. Li Q (2014) Research on shear lag effect of continuous variable cross-section box girder for railway turnout. J Shijiazhuang Tiedao Univ 27(S1):133–135
2. Chen L (2020) Research on design of single-box twin-cell continuous girder for railway turnout. Shijiazhuang Tiedao University, Shijiazhuang
3. Chen D (2012) Analysis of shear lag effect of single-cell box-shaped cross-section girder. Changsha University of Science and Technology, Changsha
4. Deng X, Xiang G (2011) Analysis on shear lag effect of single-box multi-cell skew box girder. Western China Commun Sci Technol (5):55–58, 64
5. Hu X (2010) A review of research on shear lag effect of box girder. Sci Technol Inform (23):853, 868
6. Zhou H, Zhu X (2008) Study on computational model of reinforced concrete continuous girder bridge with large width-span ratio. Shanghai Construct Sci Technol (4):20–23
7. Xue H (2008) Research on structural optimization design of 32 m simply-supported box girder for high-speed railway. Central South University, Changsha
8. Li Z (2007) Spatial effect analysis of bridge with wide box girder over long span. Harbin Institute of Technology, Harbin
9. Wei H, Hu S (2006) Bridge construction method and effective width of main girder. Struct Eng (2):21–24
10. Peng W (2017) Research on key technologies for design of long and wide continuous girder bridge over long spans for passenger dedicated railway. Railway Stand Des 61(6):88–92
11. Guo X, He C (2016) Discussion on eccentric load amplification coefficient of bridge with wide box girder. Construct Sci Technol (14):152–153
12. Chen G (2013) Discussion on eccentric load amplification coefficient of continuous wide box girder. J Highway Transp Res Dev 30(7):66–71
13. Wang Z, Xin L, Wu H, et al (2011) Calculation of internal force amplification coefficient under live load for variable cross-section continuous box girder. J Guangxi Univ 36(1):113–120
14. Cheng X, Shang C (2000) Review of formulas for internal force amplification coefficient under live load for box-shaped cross-section continuous girder. Highway (1):17–19

Chapter 6
Dynamic Simulation Analysis of Turnout on Bridge

High-speed railways (HSRs) are designed and constructed, incorporating numerous bridges that typically comprise a substantial portion of the infrastructure. These bridges often serve to accommodate turnouts in the throat area of the stations. When subjected to temperature changes, the bridge girders experience expansion and contraction along with the continuously welded rail (CWR) turnouts placed on them, leading to a coupling effect involving vertical, lateral, and longitudinal dynamics between them. When trains traverse these bridges, a coupling dynamic interaction system is further formed, encompassing the trains, turnouts, and bridges, and among them interaction and mutual influence occur. Incorporating the technical characteristics of high-speed trains, high-speed turnouts, and viaducts [1], this system gives rise to a myriad of technical challenges pertaining to the track dynamics of HSRs. Consequently, it stands as one of the crucial pressing issues to be addressed within the industry.

6.1 Models and Approaches for Train-Turnout-Bridge Dynamic Simulation Analysis

6.1.1 Overview of Train-Turnout-Bridge Coupling Dynamic Simulation

Among the core equipment of HSRs, high-speed turnouts necessitate high levels of safety, smoothness, comfort, and reliability. Since 2005, the former Ministry of Railways has mobilized nationwide efforts within the railway sector to develop high-speed turnouts capable of facilitating speeds of up to 250 and 350 km/h on the main route, as well as speeds of up to 160 km/h on the diverging route. These developed products have been successfully installed on various HSRs, including

Q. Li et al., *Continuously Welded Turnouts on High-Speed Railway Bridge*, Advances in High-speed Rail Technology, https://doi.org/10.1007/978-981-97-6524-9_6

Shijiazhuang–Taiyuan HSR, Ningbo–Taizhou–Wenzhou HSR, Wuhan–Guangzhou HSR, and Chengdu–Suining HSR. Through independent research and development of high-speed turnouts, China has established an initial theoretical system for the design of these turnouts. This system comprises computation theories in aspects such as turnout dynamics, CWR turnouts, and turnout transitions.

When high-speed turnouts are installed on bridges, significant deflection and amplitude of the bridges can impact their regularity, leading to a noticeable decrease in operational safety and riding quality of high-speed trains traversing them. The train-turnout-bridge coupling dynamic interaction will inevitably amplify wheel-rail dynamic responses within the turnout area. In severe cases, this effect can lead to insufficient strength allowance for switch rails and point rails and impair the riding quality and operational safety of trains. As a consequence, speed limits may be imposed when trains pass through turnouts on both main and diverging routes. Hence, it is crucial to incorporate the effects of train-turnout-bridge coupling vibration when designing bridge structures and determining turnout-bridge relative positions.

In general, stations are equipped with groups of turnouts located in the throat area, often resulting in arrangements with multiple tracks and groups of turnouts on bridges. As a result, the bridge deck tends to be relatively wide. To prevent twist and deformation not only on the diverging route of the turnouts but also on the piers and abutments of the bridges, the multi-track bridge deck should be designed as an integrated structure. Furthermore, any transverse deformation of this wide bridge deck can induce changes in the alignment of both the main and diverging routes of the turnouts, thereby affecting the train-turnout-bridge coupling vibration. Hence, it is essential to consider the transverse deformation of turnouts when analyzing the train-turnout-bridge coupling vibration.

In conclusion, to investigate the interaction behaviors between bridges and the CWR turnouts placed on them, it is necessary to consider not only the longitudinal static coupling effect between them but also their vertical and lateral interaction rules.

1. Progress in Research on Train-Track-Bridge Vibration

The interaction and mutual influence among trains, tracks, and bridges constitute a coupled dynamic system, where the interaction between wheels and rails, as well as that between bridges and tracks, acts as connecting links among trains, tracks, and bridges.

Extensive research over a span of more than 20 years has contributed to the development of an increasingly refined theory system for train-track coupling dynamics [2–6]. This system provides a fundamental approach for studying the dynamic interaction between rolling stock and track. From the initial applications to analyze dynamic impact actions such as rail joints or wheel flats, its application scope has been expanded to include the analysis of motion stability and riding quality of rolling stock [7–9], analysis of vibration characteristics and dynamic strength of track components, and dynamic analysis of trains passing through special sections such as turnouts and subgrade–bridge transitions.

The study of train-bridge interaction [10–15] has spanned over a hundred years, and such interaction has long been revealed as a connecting link between train vibration and bridge vibration. In previous research on train-bridge vibration, the track structures on bridges were often simplified as rails fixed to the girders, implying that the rails were assumed to possess the same deformation characteristics as the girders. Past studies primarily employed iterative methods to solve the wheel-rail interaction, considering the equilibrium conditions of forces and coordination conditions of displacement. However, few studies have focused on the coupling vibration between trains and bridges from the perspective of the real-world wheel-rail contact relationship. Even in rare instances where the wheel-rail contact was considered, it was typically treated in a planar pattern, or an assumption was made that the rail itself does not deform during vibration.

In the interactive and mutually influential dynamic system of trains, tracks, and bridges, the structural parameters of these components, as well as the track regularity, directly impact the dynamic performance of the system. In recent years, significant progress has been made in the study of train-track coupling dynamics [16–21], resulting in an increasingly refined theory system. This system has found extensive applications in all aspects of the dynamic interaction between rolling stock and track.

2. Progress in Research on Turnout Dynamics

The research on wheel-rail dynamics within the turnout section started relatively late in China, and the main findings can be summarized as follows. The wheel-rail interaction within the turnout section was analyzed using an elastic point-supported finite element articulated beam model. This model involves dividing the frog into multiple units using the finite element method and simulating the connection between the frog and rails with an elastic hinge. Additional efforts have been dedicated to studying the vertical vibration characteristics of turnouts with a stiff frog using a modal method, and the vertical vibration characteristics of movable-point frogs using the point-supported variable-section element beam method. However, these studies have predominantly focused on the vertical dynamic performance of the wheel-rail system within the frog area, resulting in a lack of research on the vibration characteristics of switches and connecting parts, as well as the lateral vibration characteristics of the wheel-rail system when trains pass through turnouts on diverging routes.

The Association of American Railroads has developed a wheel-rail computational model for the turnout section using NUCARS software. This model was utilized to investigate the lateral dynamic characteristics of the wheel-rail system when freight trains pass through ordinary AREA No. 20 turnouts and more advanced No. 20 tangential turnouts. Focusing on the lateral impact effect of trains on the switch rails, it still did not address the spatially coupled vibration of the wheel-rail system within the entire turnout section. Germany has developed SIMPACK dynamic analysis software primarily for investigating the vibration characteristics when trains pass through stiff frogs, rather than establishing a system model specifically for stress analysis of turnout structures.

As the railway industry in China progresses toward heavy-haul and high-speed railways, it has become an urgent task to investigate the wheel-rail interaction mechanism within the turnout section, understand the stress and deformation characteristics of track structures, and enhance the dynamic design theory and methodology for turnout structures. Since 1996, Southwest Jiaotong University has been conducting studies on wheel-rail interaction within the turnout section in response to China's demand for nationwide railway speed increases and the requirements of designing high-speed turnouts in recent years. These studies have led to the establishment of a preliminary dynamic theory for turnouts and the development of simulation analysis software, expanding the scope of research in wheel-rail system dynamics. The established theory has provided a theoretical foundation for the design and maintenance management of turnout structures, thereby facilitating China's transition from static strength-based design to dynamic strength-based design for turnouts [22–25].

The establishment and development of the turnout dynamic theory have facilitated China's full transition from the "geometry + experience" stage of static strength design to the "dynamic design" stage in turnout design.

6.1.2 Models and Computation Theories for Train-Turnout-Bridge Dynamic Analysis

1. Train-turnout-bridge Dynamic Analysis Models

To address the dynamic interaction among HSR trains, turnouts, and bridges, Chinese scholars have raised a comprehensive coupling system that integrates rolling stock, turnouts, and bridges, and further proposed a research idea rooted in finite element methods involving train dynamics, track dynamics, turnout dynamics, and bridge dynamics, with wheel-rail interaction and track-bridge interaction within the turnout area serving as connecting links. This idea aims to study the dynamic characteristics of turnout tracks and bridge structures under high-speed train operation conditions, as well as the associated operational safety and comfort by using the numerical simulation method.

In the established turnout dynamic model, where the interaction between ballasted tracks and bridges occurs at the sleepers and track beds, whereas, for ballastless tracks, the interaction takes place at the turnout slabs and track bed slabs, the track-bridge interaction is discretized into a series of point-to-point interactions, with linear springs and dampers connecting the corresponding action points. By defining the vibration displacement and velocity of each track component and the bridge, the track-bridge interaction force can be obtained. Figure 6.1 depicts an end view of the train-turnout-bridge dynamic model.

The wheel-rail contact geometry revealed within the turnout section and the motion analysis of individual wheels and wheelsets passing through turnouts show significant differences between the wheel-rail contact geometry in the turnout section and that in the ordinary track section, due to the unique structures of switch rails and

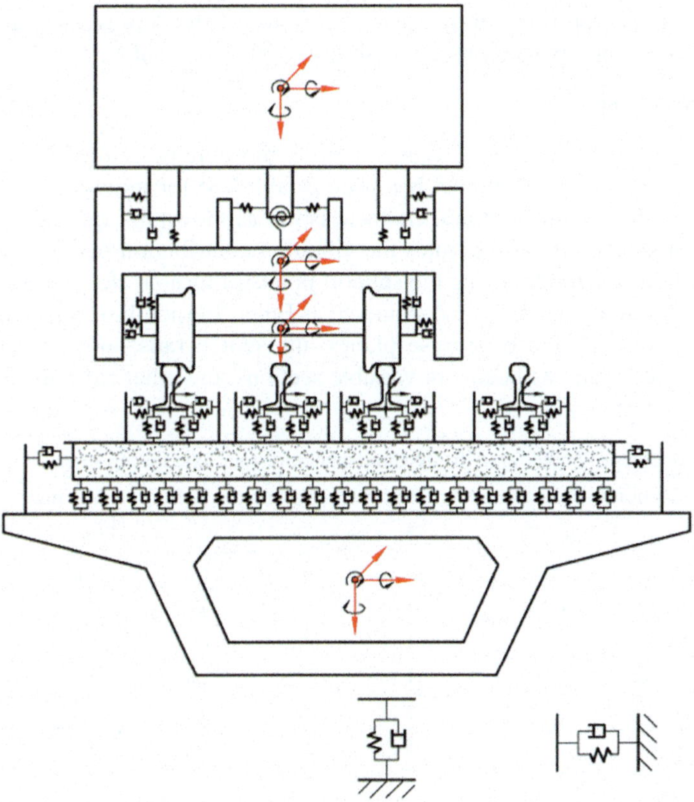

Fig. 6.1 Model of train-turnout-bridge dynamic theory

point rails, which guide the wheels to steer and cross the turnouts. Such variations result in vertical and lateral structural irregularities due to changes in wheel-rail contact points. These irregularities, in turn, cause significant vertical and lateral vibrations of the wheels, thereby impacting the riding quality of the trains passing through turnouts. In severe cases, running safety can even be compromised. To accurately reflect the vibration behavior of the train-turnout system, it is essential to delve deeper into the dynamic wheel-rail contact geometry within the turnout section on the basis of the static wheel-rail contact geometry and establish a detailed dynamic model and corresponding computation theory for the train-turnout system.

(1) Train model

A dynamic model has been developed specifically for four-axle rolling stock, which is extensively utilized in HSRs and is equipped with bogies that lack a bolster, swing bolster, and side bearing. This four-axle rolling stock model comprises seven rigid bodies, which include the car body, frames, and wheelsets. Each of these rigid bodies is considered to have five degrees of freedom: lateral, vertical, rolling, yawing, and pitching. Therefore, each train has 35 degrees of freedom in total, as shown in Table 6.1. The train dynamic model is shown in Figs. 6.2 and 6.3.

(2) Turnout model

Taking 60 kg/m No. 18 single turnouts with a movable point for 350 km/h HSRs as an example, an overall model has been established for movable-point turnouts according to their main structural characteristics, as shown in Figs. 6.4 and 6.5.

In the model which incorporates the vibration contribution of each rail, the rail structure is discretized by using the support points of turnout sleepers as the nodes. Within this system, the rails are considered as Euler–Bernoulli beams that can bend in both the vertical and transverse planes; the switch rails, movable points, and wing rails are treated as beams in variable sections; the other rails are regarded as

Table 6.1 Degrees of Freedom in the dynamic model for four-axle rolling stock

Degree of Freedom	Longitudinal	Lateral	Vertical	Rolling	Yawing	Pitching
Car body	–	Y_c	Z_c	φ_c	ϕ_c	β_c
Front frame	–	Y_{t1}	Z_{t1}	φ_{t1}	ϕ_{t1}	β_{t1}
Rear frame	–	Y_{t2}	Z_{t2}	φ_{t2}	ϕ_{t2}	β_{t2}
First wheelset	–	Y_{w1}	Z_{w1}	φ_{w1}	ϕ_{w1}	β_{w1}
Second wheelset	–	Y_{w2}	Z_{w2}	φ_{w2}	ϕ_{w2}	β_{w2}
Third wheelset	–	Y_{w3}	Z_{w3}	φ_{w3}	ϕ_{w3}	β_{w3}
Fourth wheelset	–	Y_{w4}	Z_{w4}	φ_{w4}	ϕ_{w4}	β_{w4}

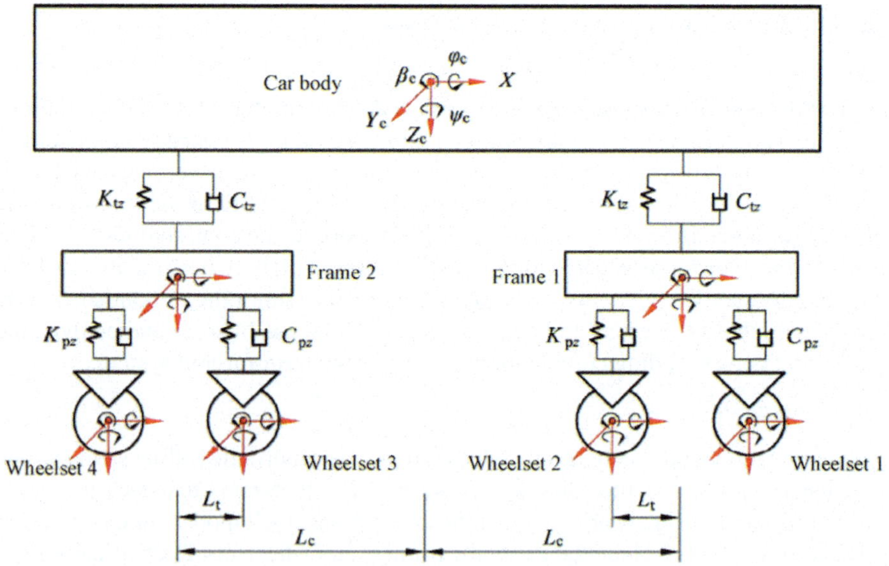

Fig. 6.2 Side view of dynamic model for four-axis rolling stock

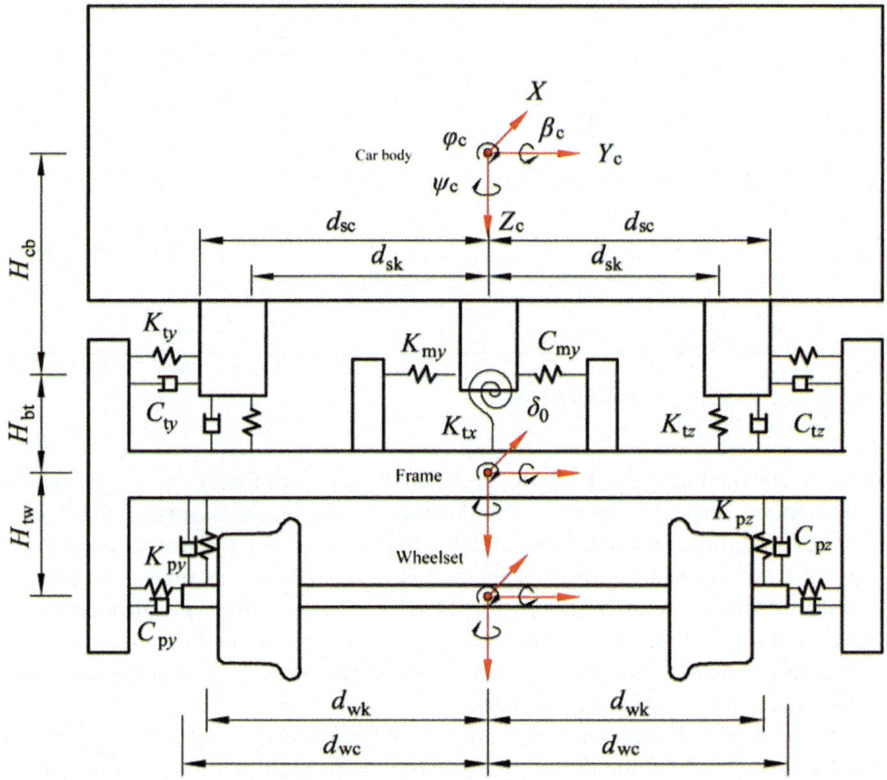

Fig. 6.3 End view of dynamic model for four-axis rolling stock

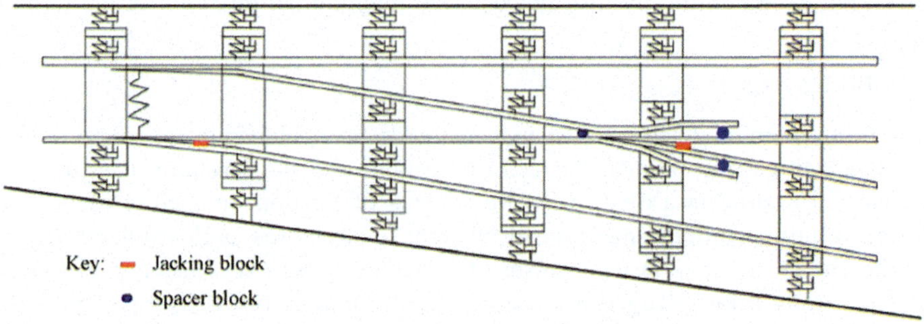

Fig. 6.4 Plan of movable-point turnout

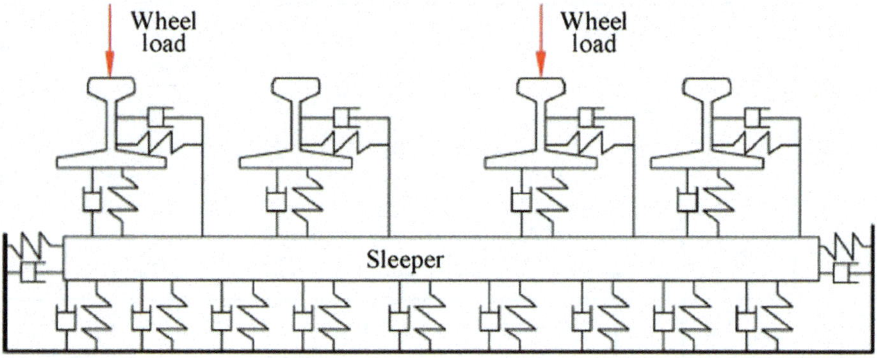

Fig. 6.5 Elevation of movable-point turnout

beams in constant sections. Each rail node has 4 degrees of freedom, namely vertical displacement, vertical deflection, lateral displacement, and transverse deflection.

To account for eccentric loading and flexural deformation, the structure of turnout sleepers is discretized by utilizing the action points of rails as nodes. Within this system, the turnout sleepers are considered as Euler–Bernoulli beams that can flex in the vertical plane, while they are treated as rigid mass blocks in the transverse plane. Each turnout sleeper node has 3 degrees of freedom, namely vertical displacement, vertical deflection, and lateral displacement.

The connection between the rails and turnout sleepers is considered as a spring–damper system, where the elasticity and damping properties vary according to the supporting conditions. The connection between the turnout sleepers and track beds is also treated as a spring–damper system, with the supporting elasticity and damping of the track bed uniformly distributed in the longitudinal direction of the turnout sleepers.

(3) Bridge model

The bridges are modeled using spatial two-node beam elements, where each node possesses three translational degrees of freedom and three rotational degrees of freedom, resulting in a total of twelve degrees of freedom in each element. Due to the complex damping mechanism inherent to structures, a clear understanding of the energy dissipation mechanism of structural vibration is lacking. However, it is generally acknowledged that the causes of damping primarily encompass the viscous damping of structural materials (proportional to velocity), hysteretic damping of materials (proportional to displacement), friction between solids, and the incorporation of artificial energy dissipation devices on structures. Numerous mathematical models of material damping exist, each applicable within specific scopes and with certain limitations. From a practical standpoint, the Rayleigh damping model is extensively utilized in engineering applications and is mathematically expressed as follows:

$$\boldsymbol{C} = \alpha_0 \boldsymbol{M} + \alpha_1 \boldsymbol{K} \tag{6.1}$$

where $\boldsymbol{M}$ and $\boldsymbol{K}$ denote the total mass matrix and total rigidity matrix of the bridge structure;

α_0 and α_1 denote the Rayleigh damping constants.

When the natural frequency and damping ratio of a second-order vibration mode are known, the damping constants can be computed using the following equation:

$$\begin{aligned} \alpha_0 &= \frac{2(\xi_j \omega_i - \xi_i \omega_j)\omega_i \omega_j}{\omega_i^2 - \omega_j^2} \\ \alpha_1 &= \frac{2(\xi_i \omega_i - \xi_j \omega_j)}{\omega_i^2 - \omega_j^2} \end{aligned} \tag{6.2}$$

where ω_i and ω_j denote the natural vibration frequencies of the ith- and jth-order vibration modes of the bridge;

ξ_i and ξ_j denote the damping ratio of the ith- and jth-order vibration modes of the bridge.

In bridge dynamic analysis, the damping coefficient is often computed based on the first-order and second-order natural vibration frequencies. Typically, the damping ratio for concrete bridges falls within the range of 0.01–0.05.

2. Wheel-rail Interaction Model

The trajectory method is commonly employed for computing the geometry of wheel-rail spatial contact that is typically asymmetric. The fundamental principle is that the collection of wheel-rail spatial contact points creates a spatial curve on the wheel tread, referred to as "the track of wheel-rail contact points on the wheel tread", or simply the trace. By utilizing this trace, the entire tread area of the wheel can be represented, allowing for scanning within a one-dimensional space curve instead of a two-dimensional curved surface. As a result, the computational workload is significantly reduced. Figure 6.6 illustrates the spatial wheel-rail contact.

The normal force between wheels and rails is determined using Hertz's nonlinear elastic contact theory:

$$P_{\mathrm{N}}(t) = \left[\frac{1}{G}\delta N(t)\right]^{\frac{2}{3}} \tag{6.3}$$

where $\delta N(t)$ denotes the normal elastic compression of the wheel-rail contact point;

G denotes the wheel-rail contact constant, taken as $G = 4.57R^{-0.149} \times 10^{-8}\mathrm{m/N^{2/3}}$ for conical treads and $G = 3.86R^{-0.115} \times 10^{-8}\mathrm{m/N^{2/3}}$ for worn treads, in which R refers to the wheel's rolling circle radius (m).

According to Kalker's linear creep theory, the linear creep force between wheels and rails can be obtained. This creep force can be further corrected using Shen–Hedrick–Elkin's theory to account for nonlinearity. As a result, the computation of

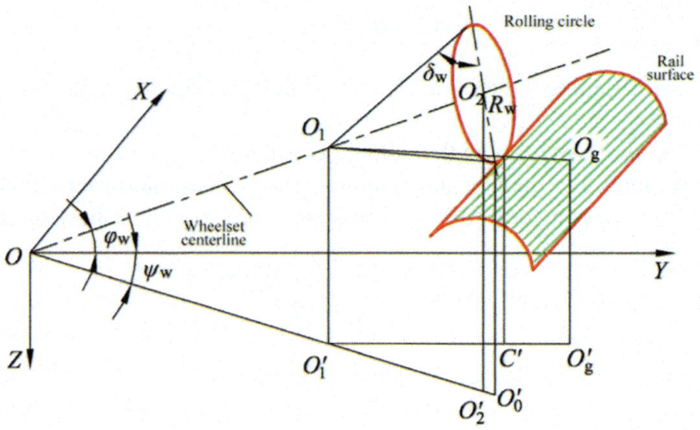

Fig. 6.6 Wheel-rail spatial contact

the creep force becomes applicable to a wide range of creep rate values, allowing for simulation computation of wheel-rail interaction in real-world conditions.

The resulting wheel-rail interaction force can be put into the train-track dynamic equation as the reaction force on the wheels and the external load on the track, respectively.

3. Turnout-bridge Interaction

Assuming the transverse displacement, vertical displacement, and rotation angle of the bridge centroid at the ith supporting section at a certain moment are y_{bh}, z_{bv}, and ϕ_b respectively, the transverse displacement of the turnout sleeper is y_{si}, and the vertical displacement of the jth supporting point of the turnout sleeper is z_{sij}, the lateral interaction force F_{shi} of the bridge turnout at the ith supporting section and the vertical interaction force F_{svij} of the bridge turnout at the jth supporting point of the turnout sleeper can be expressed as follows:

$$\begin{aligned} F_{shi} &= K_{sh}(y_{si} - y_{bh} - \phi_b H_b) + C_{sh}(\dot{y}_{si} - \dot{y}_{bh} - \dot{\phi}_b H_b) \\ F_{svij} &= K_{sv}(z_{sij} - z_{bv} - \phi_b x_{sij}) + C_{sv}(\dot{z}_{sij} - \dot{z}_{bv} - \dot{\phi}_b x_{sij}) \end{aligned} \tag{6.4}$$

where K_{sh} and C_{sh} denote the transverse restraint rigidity and damping of the turnout sleeper;

K_{sv} and C_{sv} denote the rigidity and damping of the support beneath the turnout sleeper;

x_{sij} denotes the transverse distance between the jth supporting point beneath the turnout sleeper and the bridge centroid.

As bridges are modeled using finite elements, it is not necessary for the discrete supporting points beneath the turnout sleepers to align with the bridge nodes. Consequently, the bridge displacement and velocity at the supporting point of the turnout sleepers can be interpolated using a cubic spline function based on the displacement and velocity of the corresponding bridge node. If the bridge deck is modeled using plate elements, interpolation should be conducted in both the longitudinal and transverse directions of the bridges.

The resulting turnout-bridge interaction force can be put into the turnout-bridge dynamic equation as the supporting reaction force on the turnout and the external load on the bridge, respectively.

4. Formulation of Vibration Equation

Once the train-turnout-bridge coupling vibration model is established, it becomes essential to formulate a coupled vibration equation for this system.

In this regard, this book introduces the vibration differential equation for the wheel-rail system in the turnout section, which is derived based on the principle of minimum potential energy in a variational form. Following deriving the variational expressions and displacement coordination conditions of various energies within the system, the mass matrix, rigidity matrix, and damping matrix of the system can be generated through computer matching, and by utilizing the geodetic coordinate reference system and selecting the zero point where both rails and trains are not subjected to external forces. Thus, the vibration differential equation set can be formulated for the train-turnout-bridge system as follows:

$$\boldsymbol{M}\boldsymbol{u}(t) + \boldsymbol{C}\dot{\boldsymbol{u}}(t) + \boldsymbol{F}(t, \boldsymbol{u}) = \boldsymbol{P}(t, \boldsymbol{u}, \dot{\boldsymbol{u}}) \tag{6.5}$$

The trains are positioned differently within the turnout structure at various moments. Moreover, the contact state between the wheel flanges and jacking blocks, as well as the support state of the turnout sleepers beneath the switch rails, may vary across integration time steps. Consequently, the rigidity matrix and load array are subject to time-varying changes.

5. Experimental Verification of Dynamic Simulation Analysis Theory

In order to verify the structural safety of high-speed turnouts independently developed by China, particularly assess their dynamic performance and verify the correctness of their design theory, the former Ministry of Railways has organized more than 10 rounds of dynamic tests on high-speed turnouts, using real trains on HSRs, such as Qingdao–Jinan HSR, Suining–Chongqing HSR, Wuhan–Guangzhou HSR, Ningbo–Taizhou–Wenzhou HSR, Wenzhou–Fuzhou HSR, Shanghai–Nanjing HSR, and Shanghai–Hangzhou HSR. The test results demonstrate the correctness of the design theory for these high-speed turnouts and validate their structural rationality and adherence to operational safety and riding quality requirements within permissible limits. The test results verify the correctness of the dynamic design theory of these high-speed turnouts in the following two aspects.

(1) Comparison of dynamic response

See Table 6.2 for the comparison between the results of each dynamic test using a real train and numerical simulation results.

Table 6.2 Comparison of dynamic response

Test location		Jiaozhoubei Station of Qingdao–Jinan HSR		Caijia Station of Suining–Chongqing HSR		Wulongquan Station of Wuhan–Guangzhou HSR	
Turnout type		250 km/h No. 18 ballasted turnout		250 km/h No. 18 ballastless turnout		350 km/h No. 18 ballastless turnout	
Type of turnout crossing		Main route	Diverging	Main route	Diverging	Main route	Diverging
Train type		EMU	Freight train	EMU	EMU	EMU	EMU
Maximum speed (km/h)		250	90	200	90	350	90
Derailment coefficient	Computed value	0.38	0.47	0.53	0.49	0.58	0.36
	Measured value	0.21	0.43	0.46	0.35	0.54	0.43
Wheel unloading rate	Computed value	0.48	0.49	0.36	0.47	0.28	0.47
	Measured value	0.26	0.40	**0.55**	0.30	0.50	0.26
Lateral wheelset force on switch rail (kN)	Computed value	39.4	63.9	28.4	55.3	26.4	26.9
	Measured value	17.9	52.8	7.0	38.3	23.0	28.3
Stress of point rail (MPa)	Computed value	79.3	121.1	87.5	80.7	81.1	90.4
	Measured value	46.0	123.0	–	53.3	76.1	–
Opening of switch rail (mm)	Computed value	0.76	0.47	0.77	0.46	0.78	0.47
	Measured value	0.64	0.49	0.28	0.60	0.75	0.33
Opening of point rail (mm)	Computed value	1.04	0.93	1.28	0.84	1.17	0.79
	Measured value	0.65	0.89	1.34	1.34	0.35	0.29
Acceleration of turnout sleeper (×g)	Computed value	15.4	22.0	2.4	0.8	3.5	0.8
	Measured value	14.5	19.4	1.3	–	8.5	0.6

It can be seen from Table 6.2 that the computed values for each dynamic response parameter are roughly equivalent to the corresponding measured values. This pattern persists consistently across different options such as running speed, turnout crossing type (main or diverging route), turnout type, and substructure type. The minor disparities between the computed and measured values can be attributed to variances in manufacturing and assembly errors, as well as the actual laying conditions of turnouts, verifying the reliability of applying the turnout dynamic theory to evaluate the safety and dynamic performance of turnouts.

(2) Comparison of wheel load transition

In the tests using the rail web compression method, the transition ranges of wheel load on the points arranged on switch rails were examined by measuring the vertical loads supported by the switch rails with varying sectional widths. The distribution pattern of wheel load on switch rails was examined through tests conducted with the EMUs traversing on the main route across No. 18 ballasted turnouts (250 km/h) on the Qingdao–Jinan HSR and No. 18 ballastless turnouts (350 km/h) on the Wuhan–Guangzhou HSR. The measured values were then compared to the computed values, illustrated in Figs. 6.7 and 6.8, respectively.

It can be seen from Figs. 6.7 and 6.8 that the computed and measured wheel load transition ranges well coincide and follow a consistent variation pattern, substantiating the reliability of the turnout dynamic simulation results and validating the correctness of the computation theory.

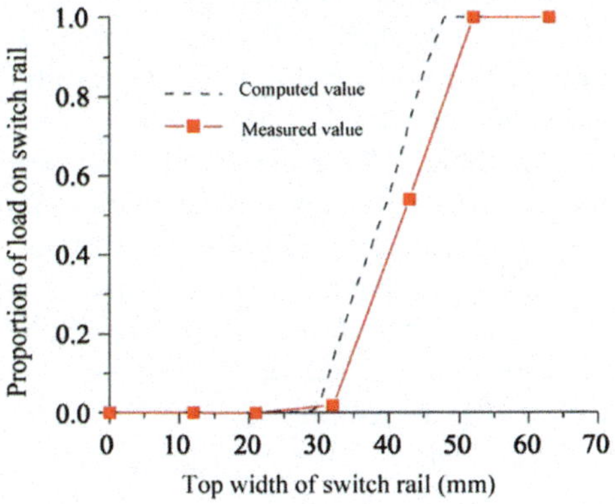

Fig. 6.7 Wheel load transition range on turnouts of Jiaozuo–Jinan HSR

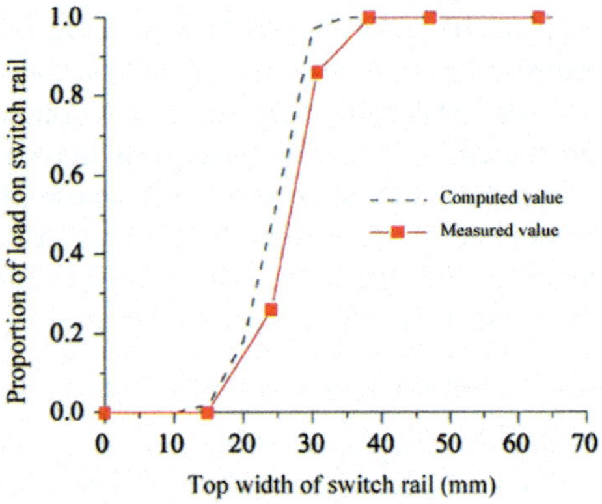

Fig. 6.8 Wheel load transition range on turnouts of Wuhan–Guangzhou HSR

6.2 Evaluation Criteria on Dynamic Safety and Comfort for Train-Turnout-Bridge System

6.2.1 Evaluation Criteria on Safety

In the field of locomotive dynamics, a comprehensive set of indexes is commonly employed to assess the operational safety of rolling stock in China and abroad, including derailment coefficient, wheel unloading rate, and overturning coefficient.

In accordance with the *Specification for Dynamic Performance Assessment and Testing Verification of Rolling Stock* (GB/T 5599-2019), the derailment coefficient of the wheels on the climbing side of rolling stock shall conform to the requirements given in Table 6.3.

See Table 6.4 for the derailment coefficient limits of locomotives specified in the *Appraisal Method and Evaluation Standard for Dynamic Performance Test of Railway Locomotive* (TB/T 2360-1993).

Additional requirements specified in the *Specification of Strength and Dynamic Performance for Motor Car of High-Speed Test Train* (95 J01-L) and *Specification of Strength and Dynamic Performance for Passenger Car of High-Speed Test Train* (95 J01-M) are as follows:

Motor car:

Table 6.3 Derailment coefficient of wheels at climbing side of rolling stock (as per GB/T 5599-2019)

Type of car	Derailment coefficient, Q/P	
	Curve (250 m $\le$ R $\le$ 400 m)	Other sections (R > 400 m)
Passenger car, EMU	$\le$1.0	$\le$0.8
Locomotive	$\le$0.9	$\le$0.8
Freight car	$\le$1.2	$\le$1.0

Table 6.4 Derailment coefficient limits for locomotive (as per TB/T 2360-1993)

Item	Excellent	Good	Acceptable
Maximum $(Q/P)_{max}$	0.6	0.8	0.9
Common maximum $(Q/P)_{m\cdot M}$	0.4	0.6	0.7

$$\begin{cases} (Q/P)_{\max} \leq 0.8 \\ Q/P)_{\max} \leq 0.056\frac{1}{t} \text{(time with the derailment coefficient greater than } 0.8 \leq 0.07 \text{ s)} \end{cases} \tag{6.6}$$

Passenger car:

$$\begin{cases} Q/P \leq 0.8 \text{ (action time of lateral force } t > 0.05 \text{ s)} \\ Q/P \leq 0.04\frac{1}{t} \text{ (action time of lateral force } t < 0.05 \text{ s)} \end{cases} \tag{6.7}$$

The wheel unloading rate serves as a safety index for assessing the risk of derailment resulting from significant unloading of the wheels on one side, particularly in situations where the wheelset lateral force acting on the wheelsets is zero or approaching zero. In accordance with the *Railway Vehicles—Specification for Evaluation the Dynamic Performance and Accreditation Test* (GB 5599-1985), the wheel unloading rate of rolling stock shall conform to the following requirements:

$$\begin{cases} \Delta P/\overline{P} \leq 0.65 \text{ (First limit, criteria for acceptance)} \\ \Delta P/\overline{P} \leq 0.60 \text{ (Second limit, criteria with increased safety margin)} \end{cases} \tag{6.8}$$

The following requirement is stipulated in the *Specification of Strength and Dynamic Performance for Motor Car of High-Speed Test Train* (95 J01-L) and *Specification of Strength and Dynamic Performance for Passenger Car of High-Speed Test Train* (95 J01-M):

$$\Delta P/P \leq 0.60 \tag{6.9}$$

The derailment coefficient and wheel unloading rate serve as measures for assessing the risk of derailment under two different conditions of wheelsets: wheelset lateral force $H > 0$ and $H = 0$, respectively, based on the stress equilibrium condition of the wheels. Typically, the derailment coefficient is utilized to evaluate operational safety, while the wheel unloading rate may be employed as an evaluation criterion under static or quasi-static conditions. In high-speed train operations, with speeds exceeding 30 km/h, a wheel unloading rate not exceeding 0.8 can be considered a safety threshold.

6.2.2 Evaluation Criteria on Riding Quality

The maximum vibration acceleration A_{max} and the effective value of cab vibration weighted acceleration A_w are used for riding quality evaluation, as stipulated in the *Appraisal Method and Evaluation Standard for Dynamic Performance Test of Railway Locomotive* (TB/T 2360-93) and *Specification of Strength and Dynamic Performance for Motor Car of High-Speed Test Train* (95 J01-L). See Table 6.5 for the specific requirements.

The Sperling index is adopted for assessing the riding quality of rolling stock in China, and the corresponding computing equation is as follows:

$$W = 7.08\left[\frac{A^3}{f}F(f)\right]^{0.1} \tag{6.10}$$

where W denotes the Sperling index;

A denotes the vibration acceleration of the car body (g);

f denotes the vibration frequency (Hz);

$F(f)$ denotes the frequency correction factor (Table 6.6).

See Table 6.7 for the riding quality rating of rolling stock stipulated in the *Appraisal Method and Evaluation Standard for Dynamic Performance Test of Railway Locomotive* (TB/T 2360-93) and *Railway Vehicles—Specification for Evaluation the Dynamic Performance and Accreditation Test* (GB 5599-1985), and the same rating system is utilized for evaluating both the vertical and lateral aspects of performance.

Table 6.5 Evaluation rating of riding quality for locomotives based on vibration acceleration in China

Rating	A_{max}/(m/s^2)		A_w/(m/s^2)	
	Vertical	Lateral	Vertical	Lateral
Excellent	2.45	1.47	0.393	0.273
Good	2.95	1.96	0.586	0.407
Acceptable	3.63	2.45	0.840	0.580

Table 6.6 Frequency correction factors for computing equation of Sperling index in China

Vertical vibration		Lateral vibration	
0.5–5.9 Hz	$F(f) = 0.325f^2$	0.5–5.4Hz	$F(f) = 0.8f^2$
5.9–20 Hz	$F(f) = 400f^2$	5.4–26Hz	$F(f) = 650f^2$
>20 Hz	$F(f) = 1$	>26Hz	$F(f) = 1$

Table 6.7 Riding quality rating of rolling stock in China

Riding quality rating	Assessment	Locomotive	Passenger car	Freight train
Grade 1	Excellent	<2.75	<2.5	<3.5
Grade 2	Good	2.75–3.10	2.5–2.75	3.5–4.0
Grade 3	Acceptable	3.10–3.45	2.75–3.0	4.0–4.25

6.2.3 Evaluation Criteria on Dynamic Interaction in Turnout Structure

On railways with speed increases and high-speed lines, the dynamic interaction between wheels and rails intensifies as train speeds rise. Excessive wheel-rail vertical and lateral forces not only cause damages and failures to various components such as rails, fasteners, and sleepers but also have the potential to sharply amplify track irregularities, increasing maintenance workload and associated costs. In severe instances, they can even jeopardize the safety of train operations. In light of these concerns, it is crucial to control the dynamic interactions between wheels and rails.

1. Wheel-rail Vertical Force

It is explicitly specified in the Japanese specification for the speed-up test of existing railways that the maximum wheel load shall be less than the design load of track components. The design load should incorporate the wheel-rail dynamic impact caused by wheel flats. Revealed through investigation and analysis, the wheel load dynamic–static ratio due to wheel flats is approximately 1.7 and is taken as 2.0 to consider a certain safety factor. The maximum wheel load under the effect of wheel flats may be taken as triple the static wheel load, which is adopted as the design load. Adhering to the P standard live load corresponding to an axle load of 170 kN, Shinkansen lines are designated with a design load of 270 kN. On the other hand, existing lines are applied with the K live load corresponding to an axle load of 160 kN, and their design load is set at 255 kN. The design dynamic wheel load is specified in the *Code for Design of Railway Track* (TB 10082-2017) as the following:

$$P_{\mathrm{d}} = \alpha P_{\mathrm{j}} \tag{6.11}$$

where P_{d} denotes the vertical design load;
P_{j} denotes the static wheel load;
α denotes the dynamic load coefficient, taken according to TB 10082-2017.

2. Wheel-rail Lateral Force

Due to the specific function and structure of turnouts, trains passing through switches, lead curves, and frog areas introduce a heightened level of lateral dynamic interaction between wheels and rails compared to regular sectional tracks. In situations where trains traverse turnouts positioned on bridges, additional considerations arise related to the vibration and deformation in the transverse direction of the bridges, particularly the piers and abutments. Excessive lateral forces exerted by the wheels on the rails can not only jeopardize the connection between the track and girders, leading to lateral shift and instability of the track panels on the bridges but also widen the gauge, thereby compromising the effectiveness of maintaining proper track geometry in the turnout section. Therefore, the wheel-rail lateral force must be brought under control. The evaluation criteria for lateral force Q specified in the *Railway Vehicles—Specification for Evaluation the Dynamic Performance and Accreditation Test* (GB 5599-1985) are as follows:

$$\begin{cases} Q \leq 29 + 0.3P & \text{Hazard limit} \\ Q \leq 19 + 0.3P & \text{Permissible limit} \end{cases} \tag{6.12}$$

where P denotes the axle load (kN).

This equation is applicable to the safety assessment on the track gauge widening.

It is stipulated in the *Provisional Specification on Design of New Passenger and Freight Mixed Railway (200 km/h)* that the wheel-rail lateral force Q shall be less than or equal to 80 kN.

3. Wheelset Lateral Force

Based on previous studies on track stability, it has been established that the primary cause of lateral shift of track panels, dynamic instability of continuously welded rail (CWR) tracks, and track buckling is the presence of excessive lateral forces exerted by the wheelsets.

The limit of this lateral force on tracks was originally obtained by the National Society of the French Railways (SNCF) from the results of freight train derailment tests. The research on the limit of wheelset lateral force conducted by Prud'homme has established the following equations for calculating this limit, which are adopted by European and American countries, as well as the International Union of Railways (UIC):

$$\begin{cases} Q \leq 10 + \frac{P}{3} & \text{(Limit, value)} \\ Q \leq 0.85\left(10 + \frac{P}{3}\right) & \text{(Recommended value)} \end{cases} \tag{6.13}$$

where Q denotes the wheelset lateral force (kN);

P denotes the axle load (kN).

The requirements stipulated in the *Specification of Strength and Dynamic Performance for Motor Car of High-Speed Test Train* (95 J01-L) and *Specification of Strength and Dynamic Performance for Passenger Car of High-Speed Test Train* (95 J01-M) are consistent with the UIC standards.

4. Opening of Switch Rail and Point Rail

The opening of switch rails serves to measure their separation from the stock rails. It is stipulated in Chinese specifications that the dynamic opening of switch rails and point rails during turnout switching should be <4 mm, to ensure the track, communication, and signaling control and prevent wheels from colliding with the tip of switch rails and point rails when trains pass through turnouts in a facing-point movement.

5. Dynamic Stress of Switch Rail and Point Rail

Switch rails and point rails, being susceptible components of turnouts, are directly supported on slide chair plates without any connection to turnout sleepers through fasteners. Any breakage in these components can lead to safety accidents during train operation. Therefore, it is crucial to control their dynamic stress level, which is typically required not to exceed 374 MPa—the fatigue strength limit.

6. Track Deformation

The vertical displacement of tracks should be less than or equal to ±3 mm (with trains passing through turnouts on main and diverging routes). The lateral elastic displacement of rail components should be less than or equal to ±1.5 mm (with trains passing through turnouts on main route).

6.2.4 Evaluation Criteria on Dynamic Performance of Bridge Structure

1. Deflection (Rigidity) of Bridge

According to the *Code for Design on Railway Bridge and Culvert* (TB 10002-2017), the vertical deflection of girders under the static live load of trains shall not exceed the permissible values given in Table 6.8.

When trains pass through bridges, the train-bridge system produces vibration in both vertical and lateral directions. There are relatively few studies on the lateral rigidity limit of bridges in China and abroad, and only relatively general requirements are currently available in this regard. Typical requirements for the lateral rigidity of bridges are given in Table 6.9.

2. Natural Vibration Frequency of Bridge

The requirements of natural vibration frequency of bridges in China and abroad usually take into account the running speed of trains. According to the *Provisional Specification on Design of Beijing–Shanghai High-Speed Railway*, the vertical

Table 6.8 Limits of vertical deflection for bridges (As per TB 10002-2017)

Railway design criteria	Design load (km/h)	Span range		
		$L \leq 40$ m	40 m $< L \leq 80$ m	$L > 80$ m
HSR	350	L/1 600	L/1 900	L/1 500
	300	L/1 500	L/1 600	L/1 100
	250	L/1 400	L/1 400	L/1 000

Table 6.9 Representative limits for lateral rigidity of bridges

Index	Standard title	Limit	Remarks
Deflection	UIC776-2R	$L/4000$	$v = 120–200$ km/h
	Japanese standards for Shinkansen lines	$L/3\,600–L/5000$ (half in vertical direction)	
	SNCF standards	$L/4000$	
	Code for design of high speed railway (TB 10621-2014) of China	$L/4000$	$v \leq 350$ km/h
Width-span ratio	AREA-1983 of the United States	Through type: $B/L \geq 1/20$ Deck type: $B/L \geq 1/15$	$v \leq 130$ km/h
	JNR standards	$B/L \geq 1/20$	
	Code for design on railway bridge and culvert (TB 10002-2017)	Through type simply-supported steel girder: $B/L \geq 1/20$ Girder side span of continuous steel truss: $B/L \geq 1/20$ Non-girder side span of continuous steel truss: $B/L \geq 1/25$ Simply-supported slab-girder: $B/L \geq 1/15$, $B \geq 2$m	$v \leq 140$ km/h

natural vibration frequency (Hz) of simply-supported girders with a span not greater than 80 m shall not be lower than the following limits:

$$f = \begin{cases} 120/L & (L \leq 40\text{ m}) \\ 23.58L^{-0.592} & (40\text{ m} < L \leq 80\text{ m}) \end{cases} \tag{6.14}$$

See Table 6.10 for the vertical natural vibration frequency limits of simply-supported girders in common spans.

3. Transverse Amplitude of Bridge

With reference to foreign experience, it is proposed that the maximum lateral amplitude at bridge midspans shall meet the requirements given by using the following equation:

Table 6.10 Natural vibration frequency limits of simply-supported girders in common spans

Span (L/m)	16	20	24	32	40	48	56
Natural vibration Frequency limit (Hz)	7.5	6.0	5.0	3.75	3.0	2.38	2.18

$$A_{\max} \leq \frac{L}{16.5} \tag{6.15}$$

where A_{max} denotes the half-peak maximum of transverse amplitude at bridge midspan (mm);

L denotes the bridge span (m).

4. Vibration Acceleration of Bridge

The vertical acceleration limit of bridges should be 0.35 *g*, and the vertical vibration acceleration limit of ballastless decks should be 0.5 *g*. The lateral vibration acceleration limit of bridge span structures on the plane of loading should be 0.14 *g*.

5. Displacement of Structures

According to the *Code for Design of High-Speed Railway* (TB 10621-2014), for girder structures with a span >80 m, the vertical rotation angle at the girder end shall not be >2‰ and the horizontal rotation angle shall not be >1‰ under the static action of ZK live load. It is also specified that the maximum permissible torsion angle of girder bodies under ZK live load shall be 1‰.

6.3 Computing Parameters for Dynamic Simulation of CWR Turnouts on Bridges

6.3.1 Computing Parameters of EMUs

See Tables 6.11 and 6.12 for the computing parameters of CRH_2 and CRH_3 EMUs designed for operations on HSRs. The computation takes into account multiple-car EMUs passing through turnouts.

6.3.2 Structural Parameters of Turnouts

High-speed turnouts can be categorized in various ways as follows. Firstly, they can be classified into two types based on the permissible speeds for negotiation on the main route: 250 and 350 km/h. Secondly, there are four types based on the permissible speeds for negotiation on the diverging route: 80, 120, 160, and 220 km/h. Thirdly, turnouts can be categorized into three types according to their functions: station track turnouts, crossover turnouts, and connecting track turnouts. Finally, they can be further divided into two types based on the substructures: ballasted track and ballastless track.

The turnouts for a speed of up to 350 km/h designed by China independently, as well as by Germany and France for HSRs in China, are the same in coding and horizontal alignment as those designed by them with a design speed of up to 250 km/

Table 6.11 Computing parameters of CRH_2 EMU

Parameter	Value	Unit	Parameter	Value	Unit
M_c	51.96	t	M_b	2.62	t
M_w	1.0	t	$J_{b\varphi}$	2.247	t·m^2
$J_{b\psi}$	2.806	t·m^2	$J_{b\xi}$	5.045	t·m^2
L_0	1.35	m	r_0	0.46	m
Δ	0.007	m	λ	0.05	
b_1	0.813	m	b_2	0.978	m
H_1	1.1	m	H_3	0.27	m
H_4	0.1	m	K_{1x}	2.8×10^4	kN/m
K_{1y}	4.0×10^3	kN/m	K_{1z}	1.216×10^3	kN/m
C_{1x}	0.0	kN·s/m	C_{1y}	0.0	kN·s/m
C_{1z}	17.7	kN·s/m	K_{2x}	0.145×10^3	kN/m
K_{2y}	0.205×10^3	kN/m	K_{2z}	0.148×10^3	kN/m
C_{2x}	343.0	kN·s/m	C_{2y}	24.5	kN·s/m
C_{2z}	31.6	kN·s/m	μ_1	0.3	
μ_2	0.2		v	80 (diverging)/250 (main)	km/h

Table 6.12 Computing parameters of CRH_3 EMU

Parameter	Value	Unit	Parameter	Value	Unit
M_c	48	t	M_b	3.2	t
M_w	2.4	t	$J_{b\varphi}$	7.3	t·m^2
$J_{b\psi}$	3.2	t·m^2	$J_{b\xi}$	6.8	t·m^2
L_0	1.25	m	r_0	0.46	m
δ	0.007	m	λ	0.05	
b_1	0.748	m	b_2	0.978	m
H_1	1.7	m	H_3	0.28	m
H_4	0.14	m	K_{1x}	0.9×10^4	kN/m
K_{1y}	0.3×10^4	kN/m	K_{1z}	0.104×10^4	kN/m
C_{1x}	0.0	kN·s/m	C_{1y}	0.0	kN·s/m
C_{1z}	5.0	kN·s/m	K_{2x}	0.24×10^3	kN/m
K_{2y}	0.24×10^3	kN/m	K_{2z}	0.4×10^3	kN/m
C_{2x}	10.0	kN·s/m	C_{2y}	30.0	kN·s/m
C_{2z}	6.0	kN·s/m	μ_1	0.3	
μ_2	0.2		v	80 (diverging)/350 (main)	km/h

h, including No. 18, 30, 42, and 62 turnouts for passenger dedicated lines, No. 18, 41, and 58 turnouts for CZ technology series, and No. 18, 39.113, and 50 turnouts for CN technology series.

See Table 6.13 for the comparison among turnouts for passage on diverging routes at a speed of up to 80 km/h in terms of horizontal alignment and related parameters. It can be seen from the table that the turnouts with a speed of up to 80 km/h for passage on diverging routes, designed for passenger dedicated lines, and the CZ and CN series are significantly similar in horizontal alignment. The separated style horizontal alignment design distinctive to the passenger dedicated line series is also incorporated into the CZ series, to enhance the robustness and resistance to wear of switch rails, enabling them to effectively adapt to the operational conditions on mixed passenger and freight railways.

See Table 6.14 for the comparison among turnouts for passage on diverging routes at a speed of up to 160 km/h in terms of horizontal alignment and related parameters. It can be seen from the table that Germany employs a design approach that combines a transition-circular-transition alignment with FAKOP gauge widening to enhance the robustness of switch rails. China adopts the same circular-transition alignment strategy as France. However, China's No. 42 turnout offers better riding comfort for passing on diverging routes due to its larger lead curve radius when compared to the radii adopted by Germany and France.

Table 6.13 Comparison of horizontal alignment among turnouts (80 km/h on diverging route)

Type	Passenger dedicated line series	CZ series	CN series
Design parameter	① Unbalanced centrifugal acceleration ≤ 0.5 m/s^2 ② Variation rate of unbalanced centrifugal acceleration ≤ 0.4 m/s^3 ③ Tip of switch rail to be checked against 1.3 m/s^3	① Unbalanced centrifugal acceleration ≤ 0.5 m/s^2 ② Variation rate of unbalanced centrifugal acceleration ≤ 0.4 m/s^3 ③ Tip of switch rail to be checked against 1.3 m/s^3	① Unbalanced centrifugal acceleration ≤ 0.45 m/s^2 ② Unbalanced centrifugal acceleration time-varying rate of 0.59m/s^3 at the tip of switch rail
Code	18	18	18
Angle	3°10′47.4″	3°10′47.4″	3°10′47.4″
Alignment	Single circular curve (12 mm apart) *R*1 100 m	Single circular curve (12 mm apart) *R*1 100 m	Single circular curve *R*1 100 m (outer track)
Total length (m)	69	69	69
Foreside length (m)	31.729	31.729	31.729
Tail length (m)	37.271	37.271	37.271

Table 6.14 Comparison of horizontal alignment among turnouts (160 km/h on diverging route)

Type	Passenger dedicated line series	CZ series	CN series
Code	42	41	39.113
Angle	1°21′50.13″	1°23′50″	1°27′52.39″
Alignment	Circular curve *R*5 000 m + parabola	Circular curve *R*4 500 m + parabola	Transition + circular + transition curve 10,000 / 4000 m/∞
Total length (m)	157.2	140.599	136.945
Foreside length (m)	60.573	51.319	62.862
Tail length (m)	96.627	84.280	74.083

6.4 Dynamic Simulation Analysis of No. 18 Single Crossover on (6 x 32) m Continuous Girder Bridge at Xuzhoudong Station

6.4.1 Computing Parameters

1. Introduction to Site

In the arrangement of the Suihe Super Major Bridge in the direction of Shanghai at the throat area of Xuzhoudong Station shown in Fig. 6.9, this single-crossover bridge consists of a (6 × 32) m continuous girder + a 24 m simply-supported girder. A No. 18 crossover is positioned on the continuous girder bridge with a distance between centers of tracks of 5 m.

2. Main Technical Standards

Design speed for passenger trains in the section: 350 km/h;

Form of track structure: CRTS II slab ballastless track, and turnout-dedicated slab ballastless track in the turnout section;

Railway classification: high-speed railway;

Number of main lines: double tracks;

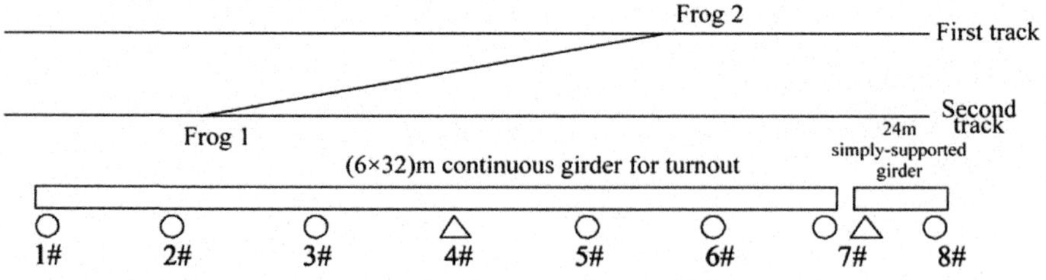

Fig. 6.9 Arrangement of CWR turnouts on Suihe super major bridge

Distance between centers of tracks on the main line: 5.0 m;
Traction type: electric power; Locomotive type: EMU;
Train operation mode: automatic control;
Train operation command mode: centralized traffic control.

3. Computing Parameters (Train)

Relevant parameters of CRH-380A EMUs were used, and the train length was taken based on an 8-car consist.

4. Computing Parameters (Turnout)

The computing parameters of No. 18 ballasted turnouts produced by Chinese New Turnout Technologies Co., Ltd. (CNTT) were used. The turnout structure is shown in the corresponding structural design drawing. The foreside length is 31.729 m and the tail length is 37.271 m. The vertical rigidity of the fastener system is 25 kN/mm.

5. Computing Parameters (Bridge)

The prestressed concrete continuous girder bridge is a double-track integral structure, featuring a single-box single-cell section for the girder body. The concrete used in its construction possesses a strength rating of C50 and an elastic modulus of 35.5 GPa. The continuous girder at the midspan is in a cross-sectional area of 12.706 m^2, with a vertical moment of inertia of 15.908 m^4 and a transverse moment of inertia of 300.26 m^4. At the support, it is in a cross-sectional area of 22.73 m^2, with a vertical moment of inertia of 21.196 m^4 and a transverse moment of inertia of 400.07 m^4.

The turnouts are placed on a substructure comprising turnout slabs, which are 240 mm thick and connected to the base as a unified unit through pins and a mortar adjustment course. The mortar adjustment course, made of asphalt and cement, and approximately 3 cm thick, exhibits an elastic modulus ranging from 7000 to 10,000 MPa, along with a surface bearing rigidity of 21 GPa/m for the turnout slabs. The base slabs form a longitudinally coupled structure within the bridge, featuring a strength rating of C30 and a thickness of 180 mm in general sections. Each girder span is equipped with a shear notch consolidation mechanism positioned over the fixed bearing. Additionally, ordinary clamping-type lateral stop blocks are installed in the longitudinal direction of the bridge, while rigid foamed plastic plates are situated at the girder joints. Furthermore, friction plates, end thorns, and transition plates are placed behind the abutments at both ends of the bridge for the subgrade.

To better reflect the dynamic influence of the substructure on the train-turnout-bridge system, the dynamic model of the train-turnout-slab-bridge system was used for analysis. For comparative purposes, a CWR turnout was also examined on a longitudinally coupled slab ballastless track in a subgrade section. The surface bearing rigidity of the turnout slabs was assumed to be 85 MPa/m, correlating with a subgrade bed thickness of 0.7 m.

6.4.2 *Dynamic Response Analysis of CWR Turnouts on Continuous Girder Bridge*

The model used for the dynamic simulation analysis of the No. 18 crossover turnouts on the Suihe Super Major Bridge at Xuzhoudong Station of Beijing–Shanghai HSR is shown in Fig. 6.10.

Two sets of No. 18 crossover turnouts capable of handling speeds of up to 350 km/h are symmetrically positioned on the continuous girder bridge with longitudinally coupled base slabs. For the main route scenario, consideration is given to train operation through the turnouts concurrently on both the up and down tracks.

1. Main Route Scenario
(1) Dynamic wheel load

Figure 6.11 illustrates the dynamic wheel load distribution of the inner wheel on the front axle of the EMU's first bogie. As depicted in the figure, under the combined influence of turnout irregularity and bridge vibration, the dynamic wheel load fluctuations were more significant compared to those observed at the CWR turnout in the subgrade section, and the maximum dynamic wheel load was approximately 178.4 kN. Additionally, the dynamic wheel load of the bogie's rear axle exhibited similar fluctuations to those of the front axle.

(2) Wheel unloading rate

The switch section exhibits a maximum unloading rate of approximately 0.15 for the inner wheel on the front axle, while the frog section records a maximum unloading rate of around 0.78 for the same wheel. Similarly, the switch section shows a maximum unloading rate of approximately 0.15 for the inner wheel on the rear axle, whereas the frog section exhibits a maximum unloading rate of about 0.74 for the same wheel. Comparatively, the fluctuations of unloading rates for the outer

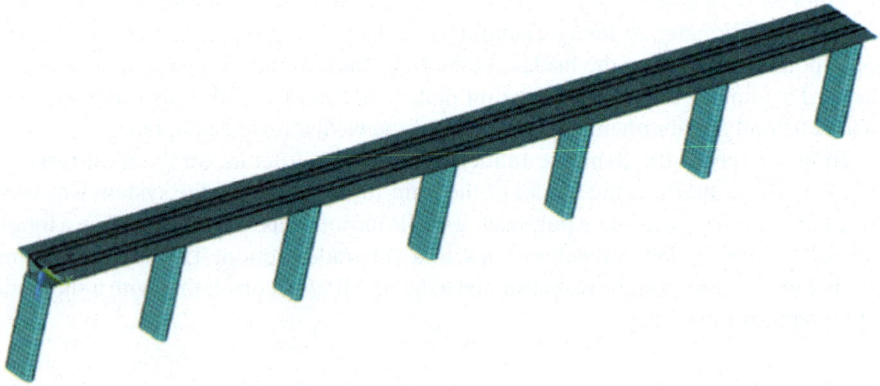

Fig. 6.10 Model of (6 × 32) m continuous girder

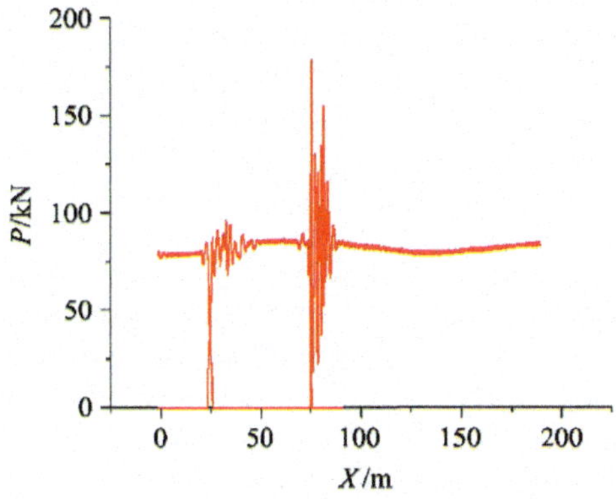

Fig. 6.11 Distribution of dynamic wheel load

wheels, both on the front and rear axles, are small, with a maximum of approximately 0.32. All these results fall below the permissible limit of 0.8. The unloading rates of the inner wheels at both the front and rear axles in the switch section are equivalent to those observed at the CWR turnout in the subgrade section. However, in the frog section, the unloading rate is significantly higher compared to that observed at the CWR turnout in the subgrade section. This discrepancy is primarily due to the vertical vibration of the bridge, which intensifies the fluctuations of the wheel load in the frog section. Figure 6.12 illustrates the unloading rate distribution caused by dynamic wheel load fluctuations of the inner wheel on the front axle.

(3) Derailment Coefficient

In the switch section, the maximum derailment coefficient for the front axle is approximately 0.38, while for the frog section, it is approximately 0.67. Regarding the rear axle, the maximum derailment coefficient in the switch section is around 0.21, and in the frog section, it is approximately 0.61. The derailment coefficient in the switch section is equivalent to that observed at the CWR turnout in the subgrade section, while that in the frog section is larger. All these results fall within the permissible

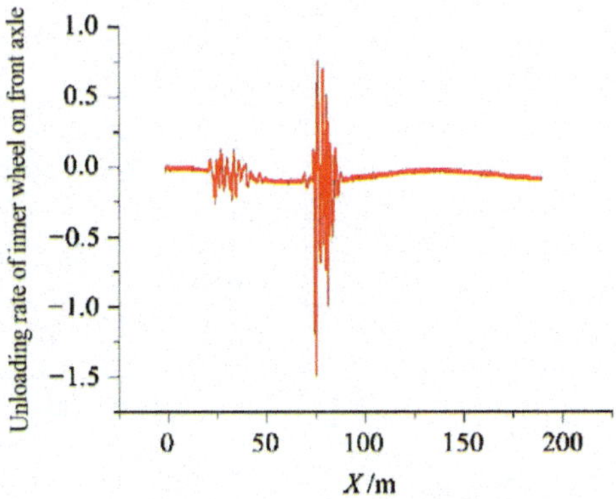

Fig. 6.12 Unloading rate of inner wheel on front axle

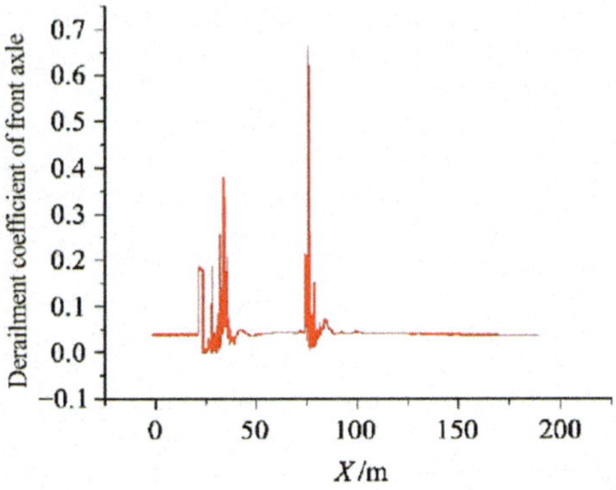

Fig. 6.13 Derailment coefficient of front axle

limit of 0.8. Figure 6.13 illustrates the derailment coefficient distribution of the front axle.

(4) Lateral acceleration of car body

Figure 6.14 illustrates the lateral vibration acceleration distribution of the car body. The maximum lateral acceleration of the car body in the switch section was about 0.037 *g*, and that in the frog section was about 0.012 *g*. The lateral vibration accelerations of the car body in both the switch and frog sections are equivalent to those observed at the CWR turnout in the subgrade section.

(5) Opening of switch rail and point rail

Figure 6.15 illustrates the dynamic opening at the tips of the switch rail and point rail when the EMU's first bogie passes through. The maximum opening at the tip of the switch rail is around 0.41 mm and that of the point rail is approximately 0.45 mm, which both are equivalent to those observed at the CWR turnout in the subgrade section and fall within the permissible range.

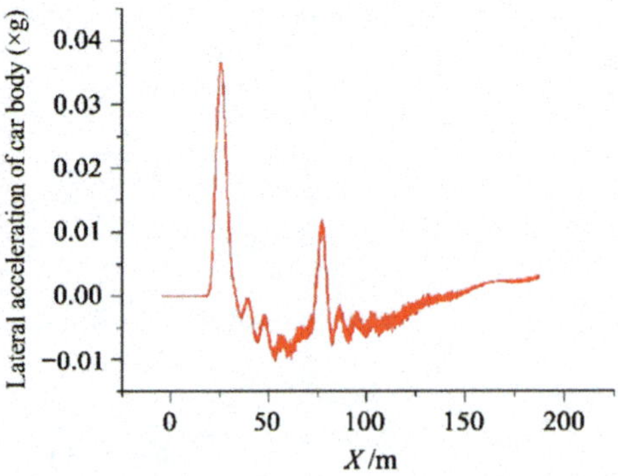

Fig. 6.14 Lateral acceleration of car body

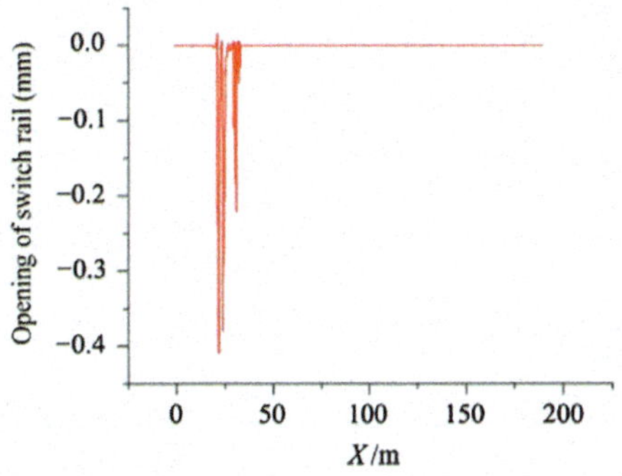

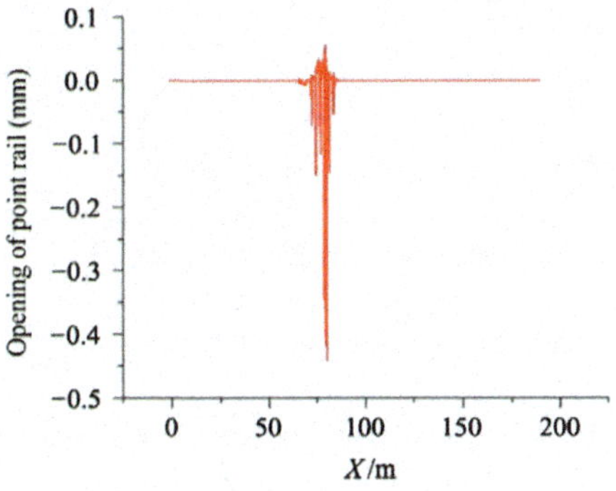

Fig. 6.15 Opening of switch rail and point rail

(6) Dynamic stress of rail

The dynamic stress distribution of rails under the inner wheel is shown in Fig. 6.16. As depicted in the figure, the dynamic stress of the rails in the switch and frog sections experienced a slight increase due to the combined influence of dynamic wheel load, flange force, as well as vertical and lateral deformation of the bridge. The maximum dynamic stress for the switch rail was approximately 177.2 MPa, while for the point rail, it was roughly 130.7 MPa. Conversely, the dynamic stress of the rails in the lead curve section was relatively low.

(7) Vibration acceleration of bridge

Figure 6.17 illustrates the vertical and lateral vibration acceleration distribution of the bridge beneath the front axle. The vibration acceleration experienced by the bridge under a straight and level track was very small. However, when subjected to turnout irregularity, the maximum vertical vibration acceleration reached approximately 0.375 *g*, while the maximum lateral vibration acceleration measured around 0.078 *g*.

Fig. 6.16 Distribution of dynamic stress of rail

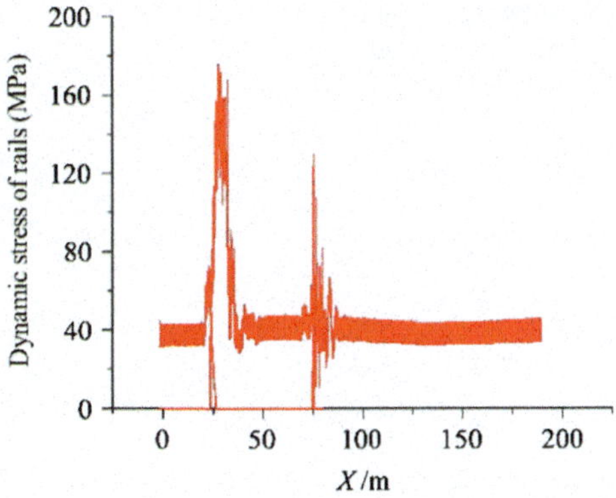

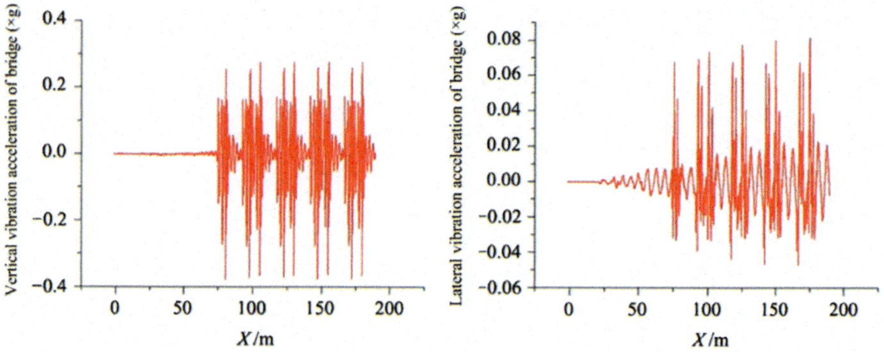

Fig. 6.17 Vertical and lateral vibration acceleration of bridge

2. Diverging Route Scenario
(1) Dynamic wheel load

Figure 6.18 illustrates the dynamic wheel load distribution on the outer wheel of the front axle when the EMU traverses the turnout on diverging routes. Under the joint excitation of turnout irregularity and bridge vibration, the dynamic wheel load still fluctuated to a certain extent in the switch and frog sections. However, due to the low speed, the fluctuations of wheel load only slightly increased compared to those observed at the CWR turnout in the subgrade section, and the maximum dynamic wheel load in the frog section was approximately 118.1 kN.

(2) Wheel unloading rate

The switch section exhibits a maximum unloading rate of approximately 0.29 for the inner wheel on the front axle, while the frog section records a maximum unloading rate of around 0.35 for the same wheel. Similarly, the switch section shows a maximum unloading rate of approximately 0.28 for the inner wheel on the rear axle, whereas the frog section exhibits a maximum unloading rate of about 0.42 for

Fig. 6.18 Distribution of dynamic wheel load

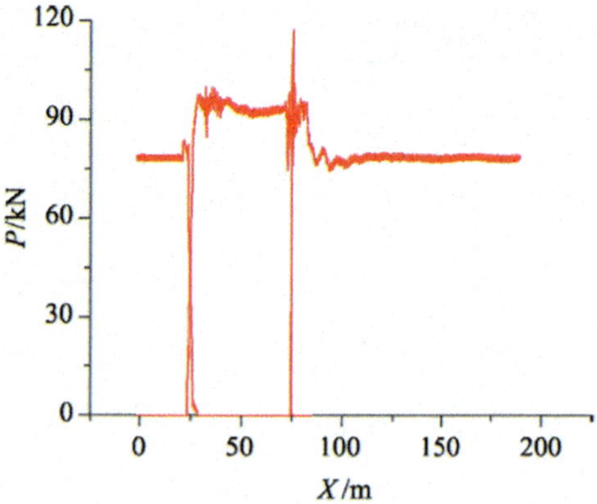

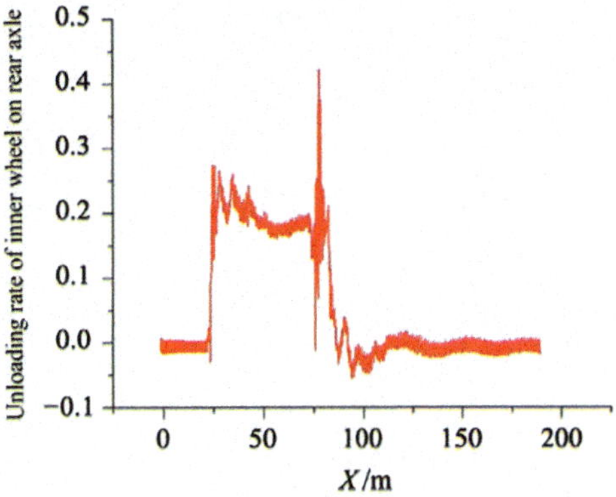

Fig. 6.19 Unloading rate of inner wheel on rear axle

the same wheel. Comparatively, the fluctuations of the unloading rates for the outer wheels, both on the front and rear axles, are small, with a maximum of approximately 0.10. All these results fall below the permissible limit of 0.8 and are equivalent to those observed at the CWR turnout in the subgrade section. Figure 6.19 illustrates the unloading rate distribution on the inner wheel on the rear axle.

(3) Derailment coefficient

In the switch section, the maximum derailment coefficient for the front axle is approximately 0.57, while for the frog section, it is approximately 0.32. Regarding the rear axle, the maximum derailment coefficient in the switch section is around 0.28, and in the frog section, it is approximately 0.31. All these results fall within the permissible limit of 0.8 and are equivalent to those observed at the CWR turnout in the subgrade section. Figure 6.20 illustrates the derailment coefficient distribution on the front axle of the bogie.

(4) Lateral vibration acceleration of car body

Figure 6.21 illustrates the lateral vibration acceleration distribution of the car body. The maximum lateral acceleration of the car body in the switch section is around

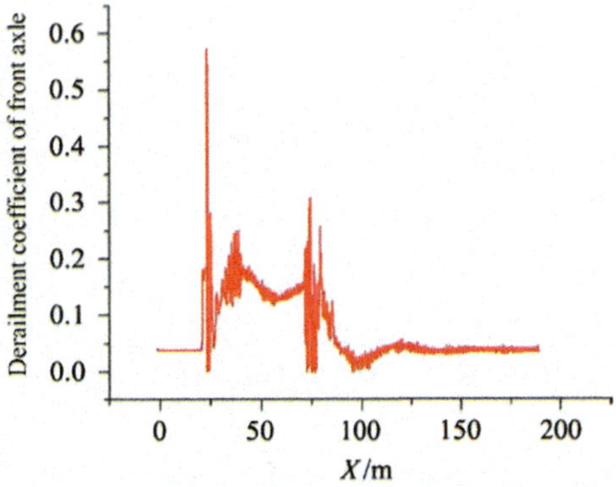

Fig. 6.20 Derailment coefficient of front axle

Fig. 6.21 Lateral vibration acceleration of car body

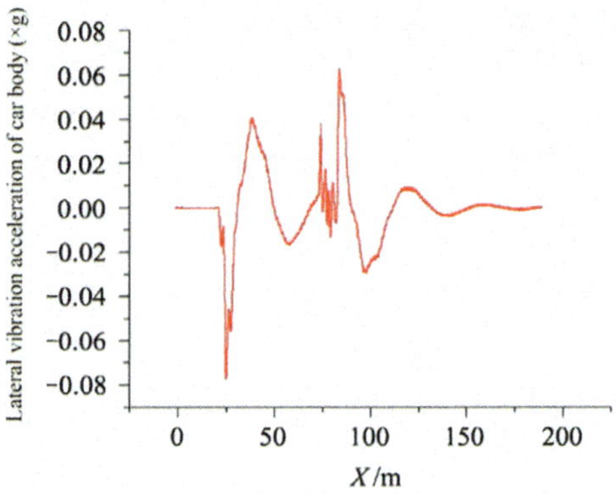

0.078 *g*, and that in the frog section is approximately 0.064 *g*, which both fall within the permissible limit and are equivalent to those observed at the CWR turnout in the subgrade section.

(5) Opening of switch rail and point rail

Figure 6.22 illustrates the opening at the tips of the switch rail and point rail. The maximum opening at the tip of the switch rail is around 0.63 mm, and that of the point rail is approximately 0.18 mm, which both fall within the permissible limit and are equivalent to those observed at the CWR turnout in the subgrade section.

(6) Dynamic stress of rail

Figure 6.23 illustrates the dynamic stress distribution of rails under the outer wheel of the front axle. Under the combined influence of dynamic wheel load, flange force, and bridge deformation, the dynamic stress of the rails in the switch section experienced an increase compared with that observed at the CWR turnout in the subgrade section. The maximum dynamic stress for the switch rail was approximately 178.4 MPa, while for the point rail, it was roughly 100.6 MPa.

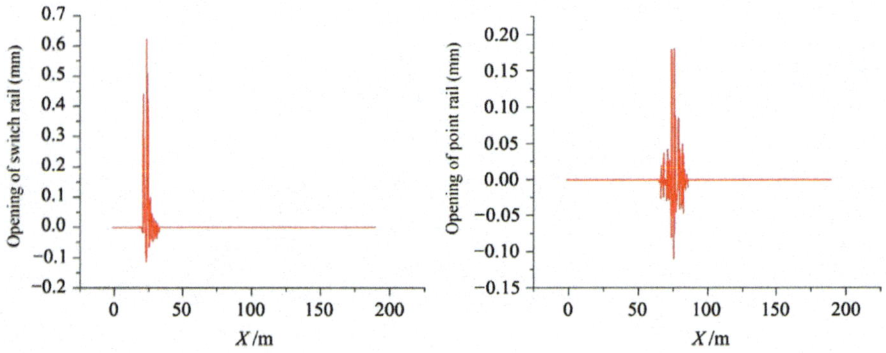

Fig. 6.22 Opening of switch rail and point rail

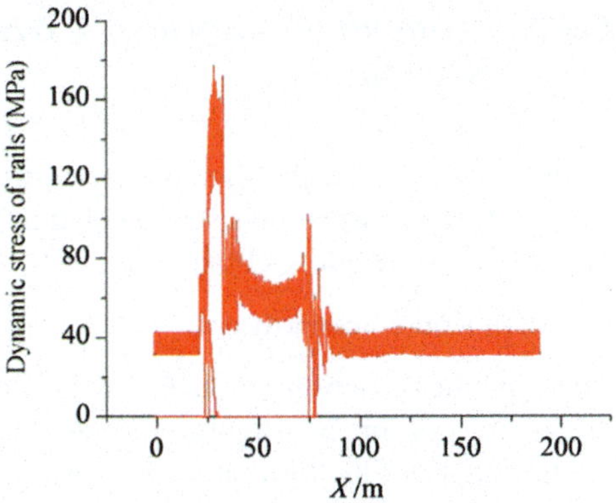

Fig. 6.23 Dynamic stress of rail

(7) Vibration acceleration of bridge

Figure 6.24 illustrates the vertical and lateral vibration acceleration distribution of the bridge beneath the front axle. The vibration acceleration experienced by the bridge under a straight and level track was very small. Under the stimulation of turnout irregularity, the maximum vertical vibration acceleration of the bridge was around 0.046 *g*, significantly smaller compared to that observed when passing through the turnout on the main route due to the lower speed. The maximum lateral vibration acceleration of the bridge was approximately 0.12 *g*, which surpassed that experienced when passing through the turnout on the main route, mainly due to the substantial wheel-rail lateral force.

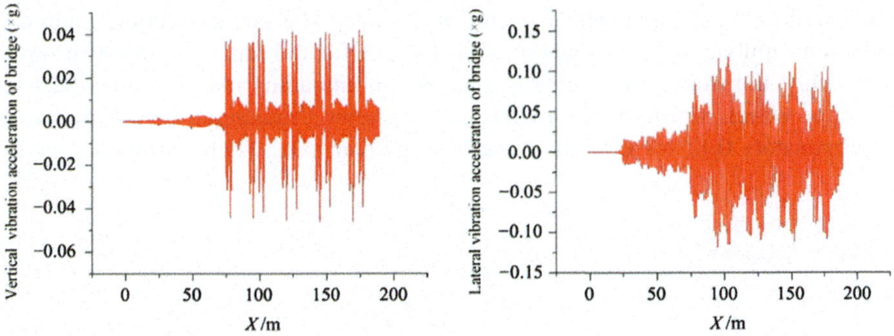

Fig. 6.24 Vertical and lateral vibration acceleration of bridge

6.4.3 Dynamic Response of CWR Turnout in Subgrade Section

In order to compare the dynamic performance between CWR turnouts on bridges and subgrade sections, this section elaborates on the dynamic simulation of a CWR turnout in a subgrade section.

1. Main Route Scenario

The dynamic responses when the EMU passes through the turnout on the main route are as follows. In each figure presented below, the origin of the abscissa indicates the beginning of the turnouts.

(1) Dynamic wheel load

Figure 6.25 illustrates the dynamic wheel load distribution of the inner wheel (the side passing through the straight switch rail) on the front axle of the EMU's first bogie. As depicted in the figure, the dynamic wheel load fluctuated within the turnout range, due to the structural irregularity of the turnout. The fluctuations were particularly noticeable in the switch and frog sections. The frog section, characterized by a short wavelength of structural irregularity, showed more significant fluctuations in the dynamic wheel load. The maximum dynamic wheel load reached approximately 165.1 kN. Additionally, the dynamic wheel load of the bogie's rear axle exhibited similar fluctuations to those of the front axle.

(2) Wheel unloading rate

Figure 6.26 illustrates the unloading rate distribution caused by dynamic wheel load fluctuations of the inner wheel on the front axle. The switch section exhibits a maximum unloading rate of approximately 0.13 for the inner wheel on the front axle, while the frog section records a maximum unloading rate of around 0.51 for the same wheel. Similarly, the switch section shows a maximum unloading rate of approximately 0.17 for the inner wheel on the rear axle, whereas the frog section

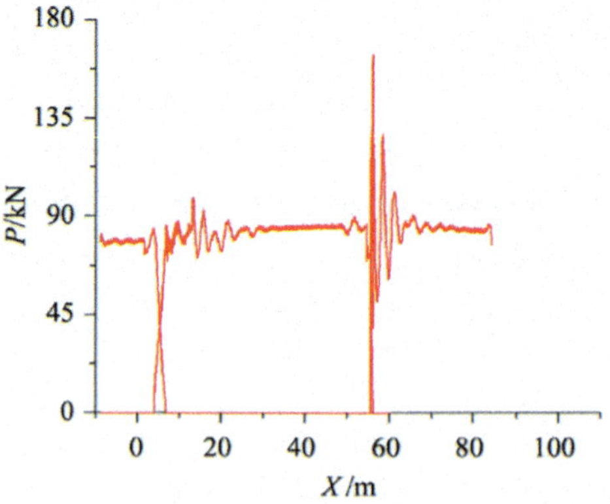

Fig. 6.25 Distribution of dynamic wheel load

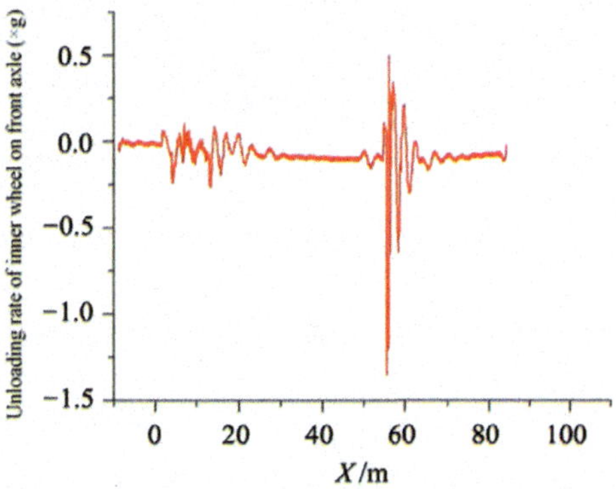

Fig. 6.26 Unloading rate of inner wheel on front axle

exhibits a maximum unloading rate of about 0.37 for the same wheel. Comparatively, the fluctuations of unloading rates for the outer wheels, both on the front and rear axles, are small, with a maximum of approximately 0.15. All these results fall below the permissible limit of 0.8.

(3) Derailment coefficient

Figure 6.27 illustrates the derailment coefficient of the rear axle. In the switch section, the maximum derailment coefficient for the front axle is approximately 0.40, while for the frog section, it is approximately 0.45. Regarding the rear axle, the maximum derailment coefficient in the switch section is around 0.25, and in the frog section, it is approximately 0.46. All these results fall within the permissible limit of 0.8.

(4) Lateral vibration acceleration of car body

The lateral vibration acceleration of the car body represents the comfort of the EMU passing through turnouts, as shown in Fig. 6.28. The maximum lateral acceleration of the car body in the switch section was about 0.037 *g*, and that in the frog section was about 0.012 *g*. Due to the structural irregularity of turnouts, the EMUs passing

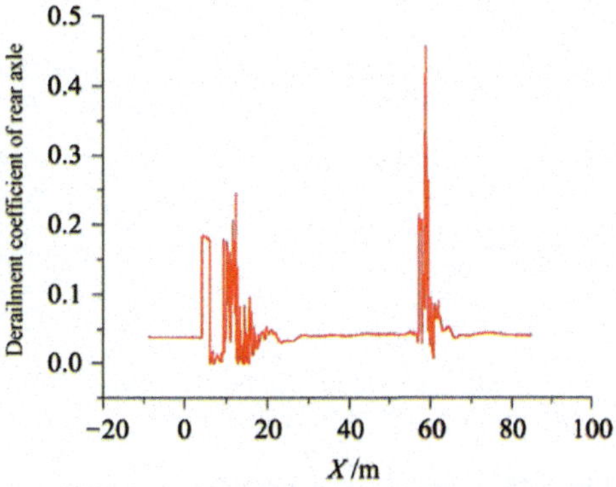

Fig. 6.27 Derailment coefficient of rear axle

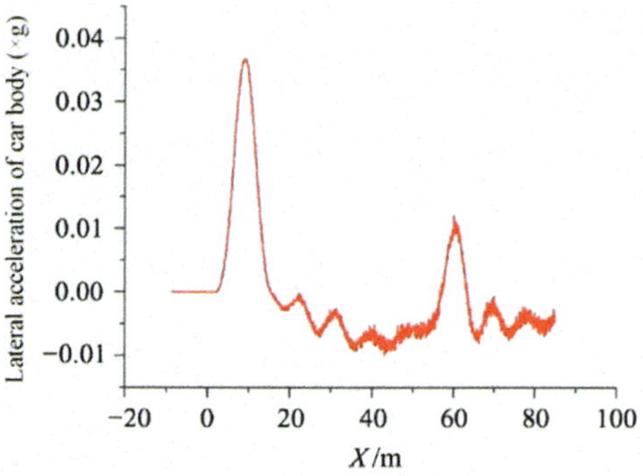

Fig. 6.28 Lateral acceleration of car body

through the switch section of turnouts inevitably experience noticeable rolling, which can only be reduced as much as possible by optimizing the wheel-rail relationship.

(5) Opening of switch rail and point rail

Figure 6.29 illustrates the dynamic opening at the tips of the switch rail and point rail when the EMU's first bogie passes through. The maximum opening at the tip of the switch rail is around 0.41 mm and that of the point rail is approximately 0.43 mm, which both fall within the permissible range.

(6) Dynamic stress of rail

The dynamic stress distribution of rails under the inner wheel is shown in Fig. 6.30. As depicted in the figure, the dynamic stress of the rails in the switch and frog sections was relatively high due to the influence of dynamic wheel load and flange force. The maximum dynamic stress for the switch rail was approximately 164.0 MPa, while for the point rail, it was roughly 128.9 MPa. Conversely, the dynamic stress of the rails in the lead curve section was relatively low.

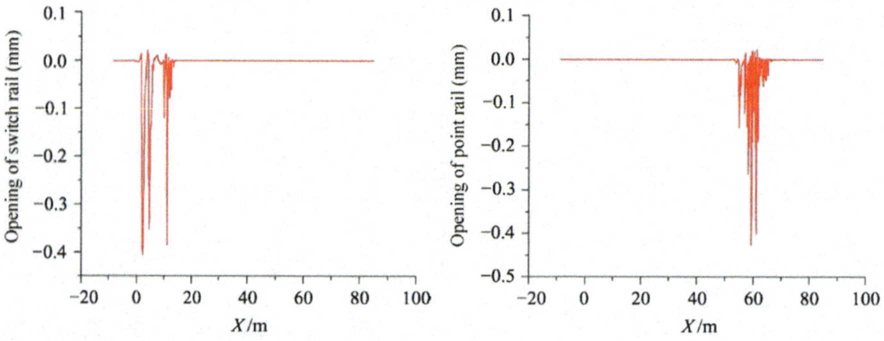

Fig. 6.29 Opening of switch rail and point rail

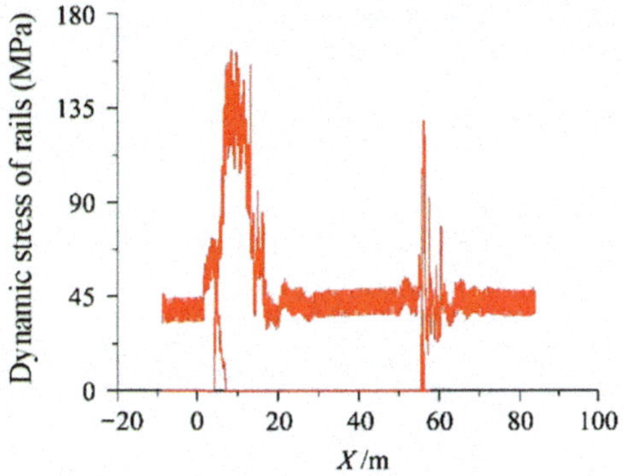

Fig. 6.30 Distribution of dynamic stress of rail

2. Diverging Route Scenario
(1) Dynamic wheel load

Figure 6.31 illustrates the dynamic wheel load distribution of the outer wheel (the side passing through the outer rail of the lead curve section) on the front axle when the EMU traverses the turnout on diverging routes. Due to the unbalanced centrifugal force, the dynamic wheel load of the outer wheel on the lead curve surpassed its corresponding static wheel load. Due to the excitation of turnout irregularity, the dynamic wheel load also fluctuated to a certain extent in the switch and frog sections. However, owing to the lower speed in these sections, the fluctuations of the wheel load were significantly smaller compared to those when passing through turnouts on the main route. The maximum dynamic wheel load in the frog section was approximately 113.8 kN.

(2) Wheel unloading rate

The bogie's inner wheels are in a constant unloading state. Figure 6.32 illustrates the unloading rate distribution on the inner wheel of the rear axle. The switch section exhibits a maximum unloading rate of approximately 0.27 for the inner wheel on the front axle, while the frog section records a maximum unloading rate of around 0.33

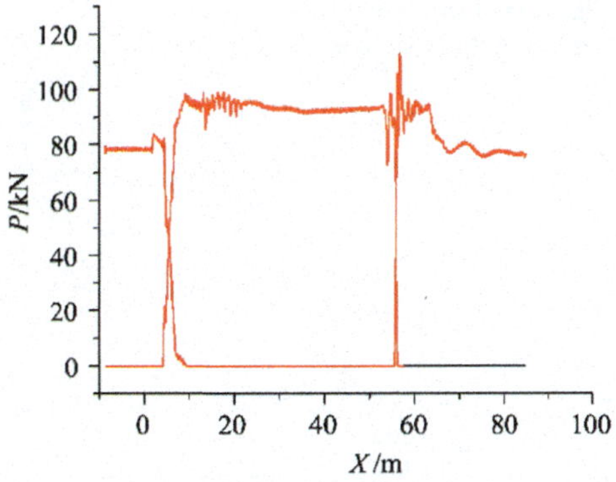

Fig. 6.31 Distribution of dynamic wheel load

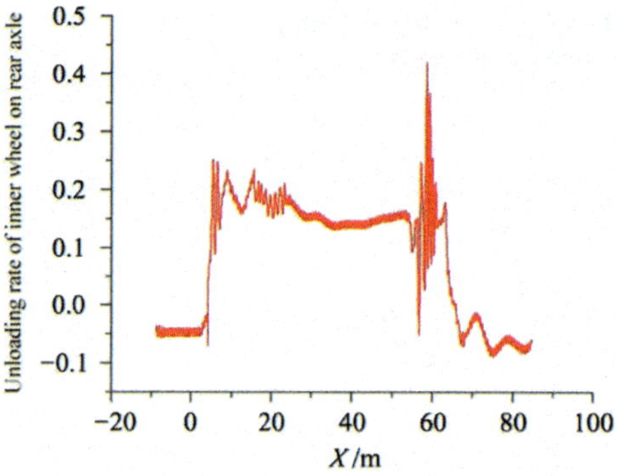

Fig. 6.32 Unloading rate of inner wheel on rear axle

for the same wheel. Similarly, the switch section shows a maximum unloading rate of approximately 0.25 for the inner wheel on the rear axle, whereas the frog section exhibits a maximum unloading rate of about 0.43 for the same wheel. Comparatively, the fluctuations of unloading rates for the outer wheels, both on the front and rear axles, are small, with a maximum of approximately 0.11. All these results fall below the permissible limit of 0.8.

(3) Derailment coefficient

In the switch section, the maximum derailment coefficient for the front axle is approximately 0.58, while for the frog section, it is approximately 0.31. Regarding the rear axle, the maximum derailment coefficient in the switch section is around 0.29, and in the frog section, it is approximately 0.31. All these results fall within the permissible limit of 0.8. Figure 6.33 illustrates the derailment coefficient distribution on the front axle of the bogie.

(4) Lateral vibration acceleration of car body

Figure 6.34 illustrates the lateral vibration acceleration distribution of the car body. The maximum value is approximately 0.079 *g*, falling within the permissible limit.

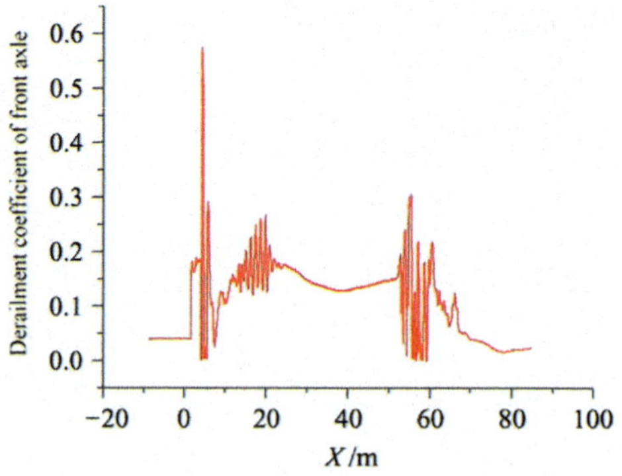

Fig. 6.33 Derailment coefficient of front axle

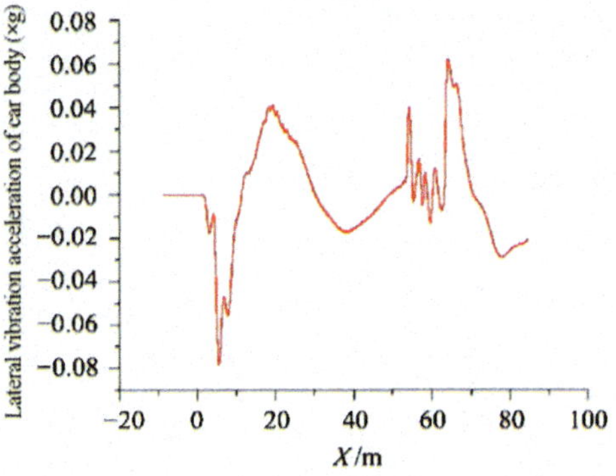

Fig. 6.34 Lateral vibration acceleration of car body

In the switch section, the car body encounters lateral vibrations, while in the check rail section, there is vibration acceleration toward the check rail. This observation indicates that while the lateral wear-resistant check rail may help reduce wear on the point rail, it negatively affects the lateral riding quality.

(5) Opening of switch rail and point rail

Figure 6.35 illustrates the opening at the tips of the switch rail and point rail when the EMU's first bogie passes through. The maximum opening at the tip of the switch rail is around 0.65 mm, while the presence of the lateral wear-resistant check rail decreases the opening at the tip of the point rail to 0.23 mm. All these results fall within the permissible range.

(6) Dynamic stress of rail

The dynamic stress distribution of rails under the outer wheel of the front axle is shown in Fig. 6.36. The dynamic stress of the rails in the switch and frog sections was relatively high due to the influence of dynamic wheel load and flange force. The maximum dynamic stress for the switch rail was approximately 165.1 MPa, while

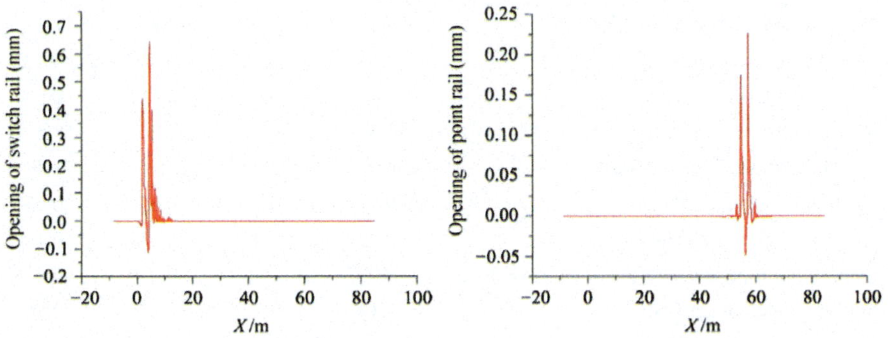

Fig. 6.35 Opening of switch rail and point rail

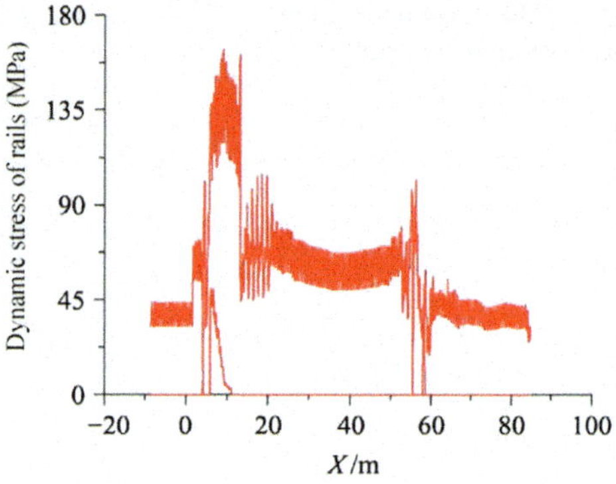

Fig. 6.36 Dynamic stress of rail

for the point rail, it was roughly 103.0 MPa. Conversely, the dynamic stress of the rails in the lead curve section was relatively low.

6.4.4 Comparative Analysis

See Table 6.15 for a comprehensive comparison of the dynamic responses observed when the EMUs traverse the CWR turnouts on continuous girder bridges on both main and diverging routes, as well as in subgrade sections.

Comparison results:

(1) In the case of turnouts located on bridges, the presence of inherent structural irregularities further stimulates the vibration of the bridge. As a result, a coupling

Table 6.15 Comparison of simulation results

Simulation item	Main route scenario of EMU		Diverging route scenario of EMU	
	Subgrade	Bridge	Subgrade	Bridge
Max. dynamic wheel load (kN)	165.1	178.4	113.8	118.1
Wheel unloading rate	0.51	0.78	0.43	0.42
Derailment Coefficient	0.46	0.67	0.58	0.57
Wheel-rail lateral force (kN)	56.5	58.8	59.2	58.4
Dynamic stress of rails (MPa)	164.0	177.2	165.1	178.4
Opening of switch rail (mm)	0.41	0.41	0.65	0.63
Opening of point rail (mm)	0.43	0.45	0.23	0.18
Lateral vibration acceleration of car body (×g)	0.037	0.037	0.079	0.078
Sperling index of car body	2.14	2.35	2.27	2.56
Vertical vibration acceleration of bridge (×g)	–	0.375	–	0.46
Lateral vibration acceleration of bridge (×g)	–	0.078	–	0.12

vibration system emerges among the EMUs, turnouts, turnout slabs, bases, and bridges, leading to an amplification of each dynamic response.

(2) When the EMUs pass through the turnouts on the main route, due to the high speed, the dynamic wheel load, unloading rate, and derailment coefficient largely increase in the frog section, and the dynamic stress of the switch rail also increases in the switch section. These changes result in slight negative effects on the riding quality of the EMUs, noticeable increases in the vertical and lateral vibration accelerations of the bridge compared to those experienced on straight and level track sections, and relatively unchanged wheel-rail lateral force. It can be concluded that the vertical deformation and vibration of bridges have noticeable impacts on the EMUs passing through turnouts on the main route, and these two parameters need to be focused on in the design of bridge structures.

(3) When the EMUs pass through the turnouts on diverging routes, due to the low speed and substantial transverse flexural rigidity offered by the integral box girder structure of the bridge, bridge vibration exerts little impact on the dynamic responses such as dynamic wheel load, unloading rate, derailment coefficient, and lateral vibration acceleration of the car body. However, a noticeable effect is observed on the dynamic stress of the switch rail.

6.5 Dynamic Simulation Analysis of No. 42 Simple Turnout on (6 × 32) m Continuous Girder Bridge of Yangzhou Connecting Track

6.5.1 Computing Parameters

1. Introduction to Site

The No. 42 turnout on the up track of the Yangzhou Connecting Track, part of the Beijing–Shanghai HSR, is placed on a (6 × 32) m continuous girder bridge with intermediate fixed bearings, and CRTS II slab ballastless track is laid on the bridge, as shown in Fig. 6.37.

2. Main Technical Standards

Design speed for passenger trains in the section: 350 km/h;

Form of track structure: CRTS II slab ballastless track;

Railway classification: high-speed railway;

Number of main lines: double tracks;

Distance between centers of tracks on the main line: 5.0 m;

Traction type: electric power;

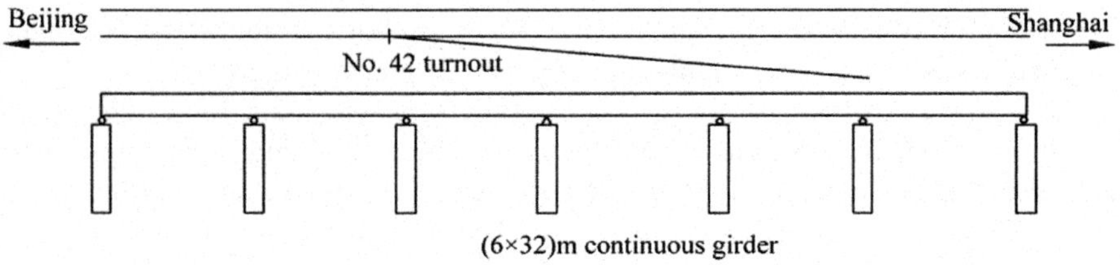

Fig. 6.37 Turnout-bridge arrangement

Locomotive type: EMU;
Train operation mode: automatic control;
Train operation command mode: centralized traffic control.

3. Computing Parameters (Train)

Relevant parameters of CRH_3 EMUs were used, and the train length was taken based on an 8-car consist.

4. Computing Parameters (Turnout)

The No. 42 turnout, designed to accommodate speeds of up to 350 km/h, spans a total length of 157.2 m. Its lead curve consists of a circular curve with a radius of R = 5000 m plus a transition curve. This turnout is equipped with a horizontal tip-hidden frog featuring a double-arm flexible bendable point rail and a high elastic vulcanized rubber pad fastener system. Thanks to the utilization of elastic fasteners and rigidity homogenization technology throughout the entire length of the turnout, a nearly uniform rigidity is achieved across all track components.

6.5.2 Dynamic Simulation Analysis

The permissible negotiation speed plus 10%, i.e. 385 km/h, is taken as input for analysis in the case of trains passing through the turnout on the main route, while concurrent operations on both the up and down tracks are considered.

1. Distribution of Dynamic Wheel Load

With the left abutment as the origin of the coordinates, the dynamic wheel load distribution of the inner wheel on the front axle of the first motor car's first bogie is shown in Fig. 6.38. As a result of the vibration contribution from the bridge, the dynamic wheel load undergoes more noticeable fluctuations when compared to those observed in the subgrade section. This is primarily due to the intensified vertical displacement change of the wheelsets resulting from the combined influence of bridge vibration and inherent structural irregularities present in the turnout, as shown in Fig. 6.39. The maximum dynamic wheel load in the frog section was around 127.1 kN, which was slightly greater than that in the subgrade section, but still falling within the permissible limit.

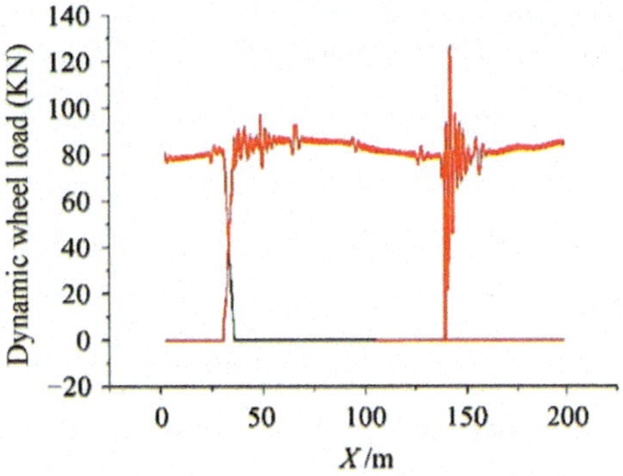

Fig. 6.38 Dynamic wheel load distribution under wheel

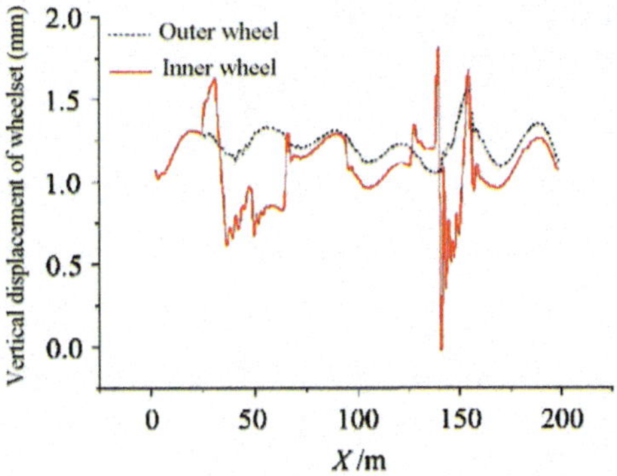

Fig. 6.39 Vertical displacement of wheelset

2. Flange Force Distribution

The contact between the wheelset's inner wheel and the switch rail, as well as the point rail, occurs within a limited range. The flange force distribution is shown in Fig. 6.40. The maximum flange force in the switch section was 16.9 kN, and the maximum instantaneous lateral impact force in the frog section was approximately 13.4 kN, exhibiting a slight decrease compared to those observed at the CWR turnout in the subgrade section and both falling within the permissible range. The main reason lies in the lateral displacement of the bridge caused by the flange force, resulting in an effective enhancement to the lateral elasticity of the track.

3. Lateral Shift of Wheelset and Bogie

Figure 6.41 illustrates the lateral shift distribution of the front axle, rear axle, and frame of the leading bogie. As depicted in the figure, the maximum lateral shift of the front axle in the switch section is around 7.1 mm, and that in the frog section is approximately 6.7 mm, exhibiting a slight increase compared to those observed at CWR turnouts in the subgrade section, which indicate that the riding quality of the EMU at this turnout is lower than that at the CWR turnout in the subgrade section.

Fig. 6.40 Flange force distribution of inner wheel

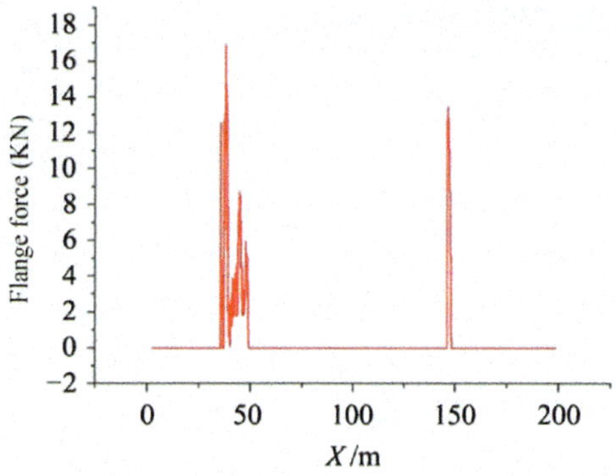

Fig. 6.41 Lateral shift of wheelset

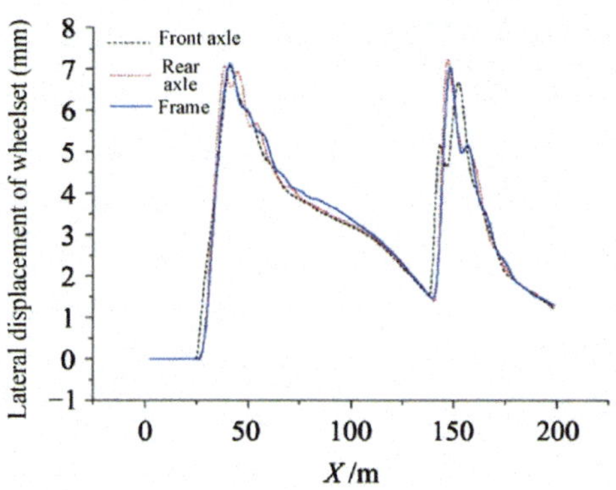

4. Derailment Coefficient

The derailment coefficient of the wheels is represented by the ratio of the sum of wheel-rail lateral forces to the dynamic wheel load, and its distribution is shown in Fig. 6.42. As depicted in the figure, the maximum derailment coefficients of the front axle in the switch and frog sections were 0.18 and 0.20 respectively, while those of the rear axle in the switch and frog sections were 0.18 and 0.26 respectively, all falling within the safety limit.

5. Wheel Unloading Rate

The unloading rate on wheels is represented by the ratio of the unloading amount of dynamic wheel load to static wheel load, and its distribution is shown in Fig. 6.43. As depicted in the figure, the unloading rates of the inner wheel on the front axle in the switch and frog sections were 0.01 and 0.71 respectively, while those of the inner wheel on the rear axle in the switch and frog sections were 0.01 and 0.70 respectively, which still fell within the safety limit.

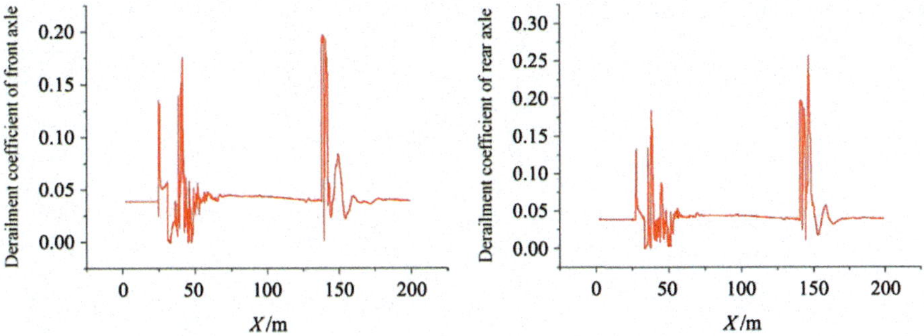

Fig. 6.42 Wheel derailment coefficient distribution

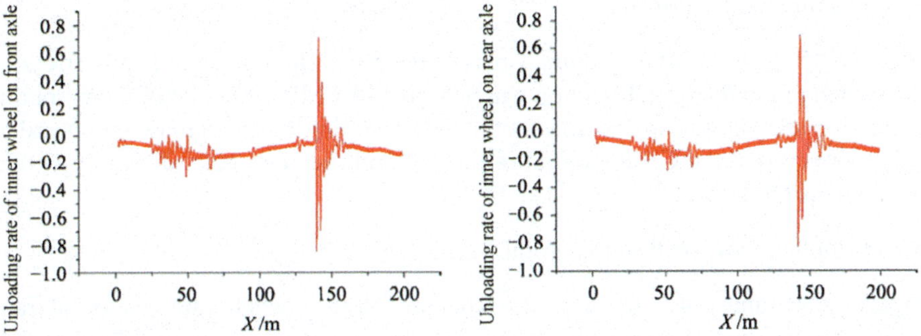

Fig. 6.43 Unloading rates of wheels

6. Riding Quality of Car Body

Figure 6.44 illustrates the vertical and lateral vibration accelerations of the car body. The maximum lateral vibration acceleration in the switch section was around 0.025 g, and that in the frog section was approximately 0.037 g, both falling within the permissible limit.

Fig. 6.44 Lateral vibration acceleration of car body

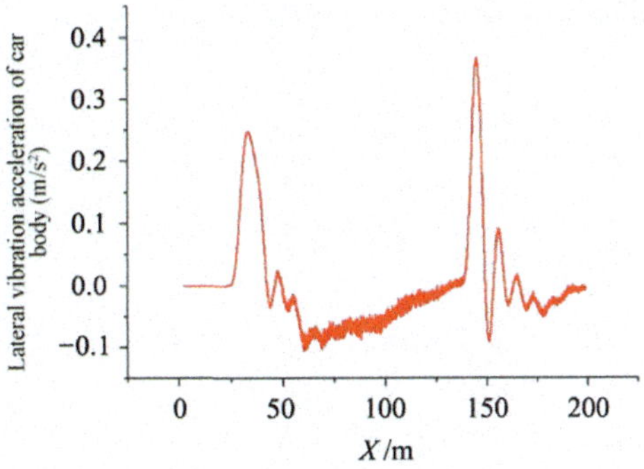

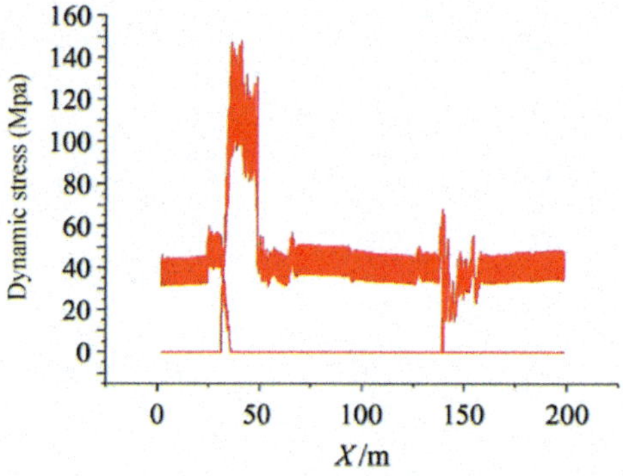

Fig. 6.45 Dynamic stress of rail

7. Dynamic Stress of Rail

Figure 6.45 illustrates the dynamic stress of the rails. The impact dynamic stress of the switch rail in the switch section was around 148.1 MPa, and that of the point rail in the frog section was approximately 68.2 MPa, exhibiting an increase compared to those observed at the CWR turnout in the subgrade section, but still falling within the permissible limit.

8. Opening of Switch Rail and Point Rail

Figures 6.46 and 6.47 illustrate the opening of the switch rail and point rail, which were recorded around 0.33 and 0.01 mm respectively, both falling within the permissible limit.

9. Bridge Displacement

Figure 6.48 illustrates the corresponding relationship between the vertical dynamic displacement of the bridge under the front axle/deformation of the bridge and the unloading rate. As depicted in the figure, under the load effect from the EMU, the maximum dynamic displacement of the 32 m girder span is around 0.52 mm, and the frog is positioned at a location on the bridge where the rate of change of dynamic

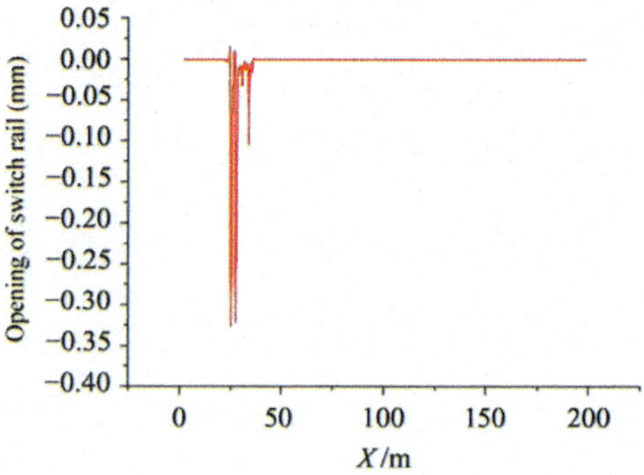

Fig. 6.46 Opening of switch rail

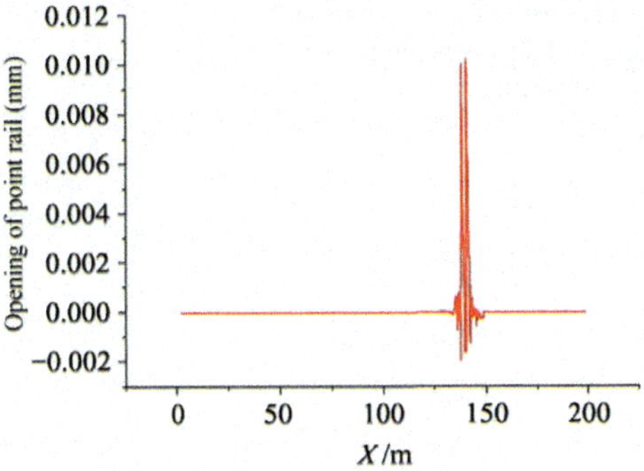

Fig. 6.47 Opening of point rail

deflection is relatively small. Thus, the relative positions of the turnout and bridge can be considered reasonably rational. Figure 6.49 illustrates the lateral dynamic displacement of the bridge under the front axle. Under the sole action of turnout structural irregularities, the lateral dynamic displacement of the bridge did not exceed 0.02 mm, which was considered unnoticeable.

Figure 6.50 illustrates the vertical rotation angle distribution at the bearings in middle spans, and the maximum dynamic rotation angle was approximately 0.03%.

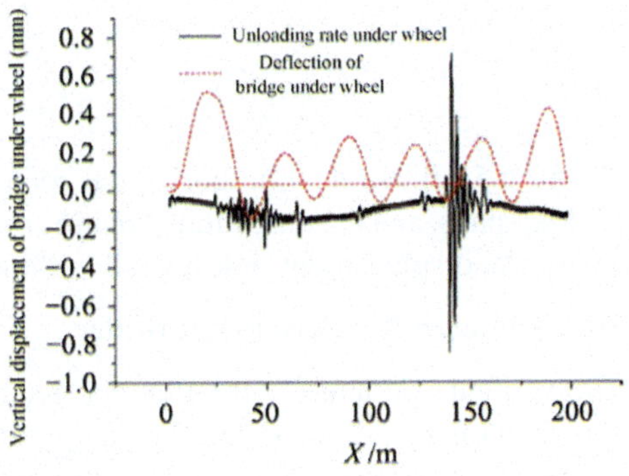

Fig. 6.48 Vertical displacement of bridge under wheel

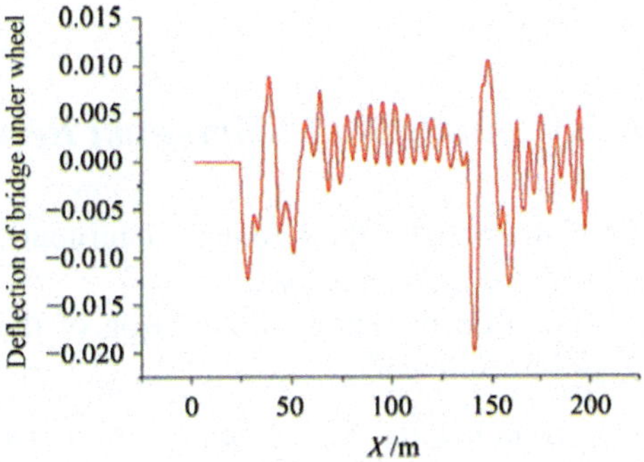

Fig. 6.49 Lateral displacement of bridge under wheel

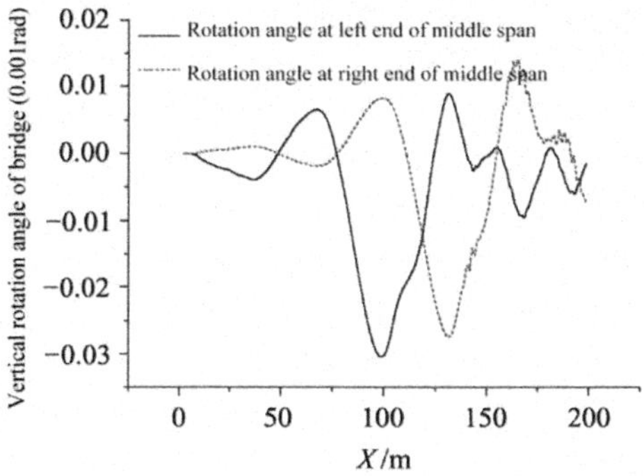

Fig. 6.50 Vertical rotation angle distribution at bearings in middle spans

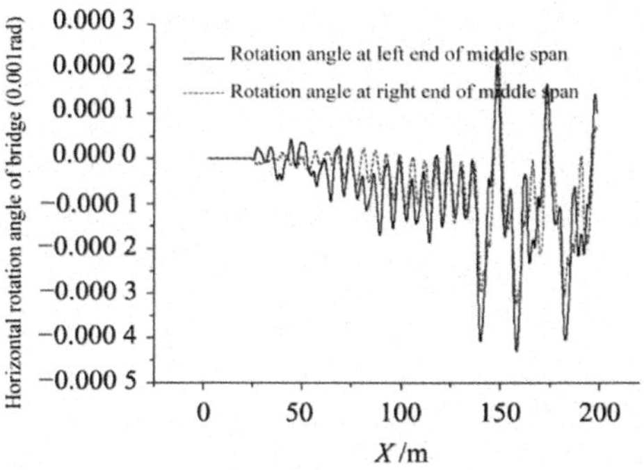

Fig. 6.51 Horizontal rotation angle distribution at bearings in middle spans

Figure 6.51 illustrates the horizontal rotation angle distribution at the bearings in middle spans, and the maximum dynamic rotation angle was approximately 0.0004%.

All the displacement indexes fall within the permissible limits.

10. Vibration Acceleration of Bridge

The vertical and lateral vibration accelerations of the bridge under the wheels are shown in Figs. 6.52 and 6.53. The maximum vertical vibration acceleration is around 0.057 *g* and the maximum lateral vibration acceleration is approximately 0.016 *g*, both falling within the permissible limit.

6.5.3 Analysis of Simulation Results and Conclusions

See Table 6.16 for the summary of turnout simulation results under various scenarios considered in the analysis.

Based on the above comparison of the computing results, conclusions can be drawn as follows:

(1) Although the Nos. 42 and 18 turnouts are equivalent in the amplitude of structural irregularity of their switches and frogs, the change rate of the former

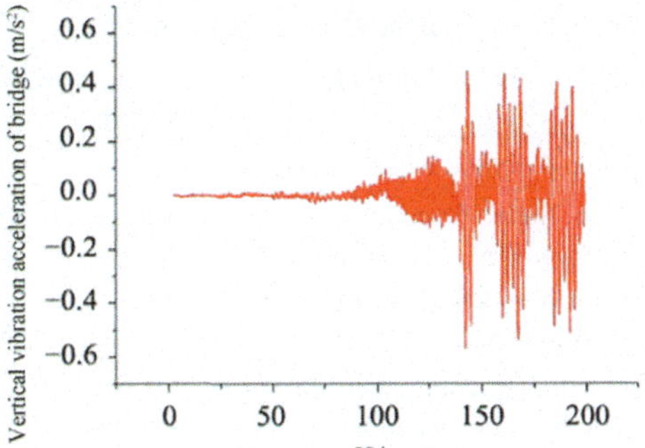

Fig. 6.52 Vertical vibration acceleration of bridge

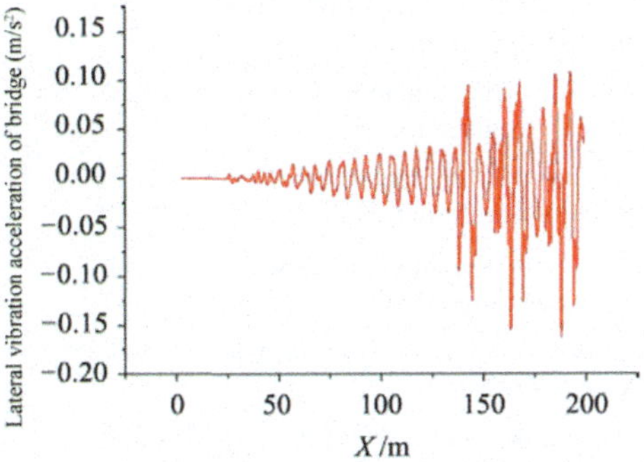

Fig. 6.53 Lateral vibration acceleration of bridge

is considerably smaller, as shown in Figs. 6.54 and 6.55. Therefore, various dynamic responses experienced at the No. 42 turnout, whether in subgrade sections or on bridges, are lower compared to those observed at the No. 18 turnout. Figures 6.56 and 6.57 respectively illustrate the dynamic wheel load distributions when the 350 km/h EMUs pass through the Nos. 42 and 18 turnouts in the subgrade section under identical conditions.

(2) When the EMUs pass on the main and diverging routes through the No. 42 turnout placed on a widened continuous girder, due to its minor structural irregularity, various dynamic responses increase slightly compared to those observed at the CWR turnout in the subgrade sections. However, these increases are significantly smaller in magnitude compared to the No. 18 turnout placed on a continuous girder. Additionally, the vertical and lateral vibration accelerations experienced by the bridge are also considerably lower than those encountered at the No. 18 turnout. These results indicate that the coupling vibration among the No. 42 turnout, EMUs, and the bridge is small.

(3) Thanks to the implemented design measures, such as significantly increasing the vertical rigidity of the bridge and widening the girder at the turnout laying position, the relative positions of the turnout and bridge are considered rational. This means that the turnout switch and frog are strategically placed in areas with relatively small deflection deformations and change rates on the bridge.

Table 6.16 Summary of computing results

Arrangement of turnout	No. 42 turnout in subgrade section	No. 42 turnout on continuous girder
Max. dynamic wheel load (kN)	110.6	127.1
Max. flange force in switch section (kN)	16.5	16.9
Max. flange force in frog section (kN)	13.8	13.4
Derailment coefficient of front axle in switch section	0.16	0.18
Derailment coefficient of front axle in frog section	0.21	0.20
Derailment coefficient of rear axle in switch section	0.18	0.18
Derailment coefficient of rear axle in frog section	0.21	0.26
Max. unloading rate of front axle in switch section	0.01	0.01
Max. unloading rate of front axle in frog section	0.44	0.71
Max. unloading rate of rear axle in switch section	0.01	0.01
Max. unloading rate of rear axle in frog section	0.44	0.70
Lateral vibration acceleration of car body (×g)	0.028	0.037
Max. dynamic stress of switch rail (MPa)	117.5	148.1
Max. dynamic stress of point rail (MPa)	61.7	68.2
Max. opening of switch rail (mm)	0.32	0.33
Max. opening of point rail (mm)	0.01	0.01
Max. vertical displacement of bridge (mm)	–	0.52
Max. lateral displacement of bridge (mm)	–	0.02
Vertical vibration acceleration of bridge (×g)	–	0.057
Lateral vibration acceleration of bridge (×g)	–	0.016

Note The computing parameters used for the No. 42 turnout in the subgrade section are the same as those used for the turnout on the bridge

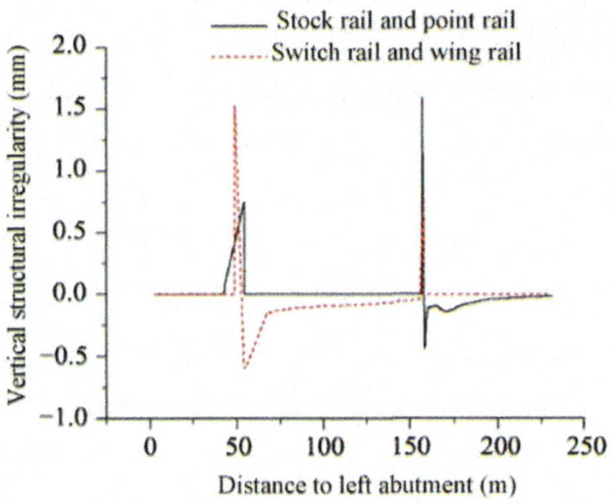

Fig. 6.54 Vertical structural irregularity of No. 42 turnout

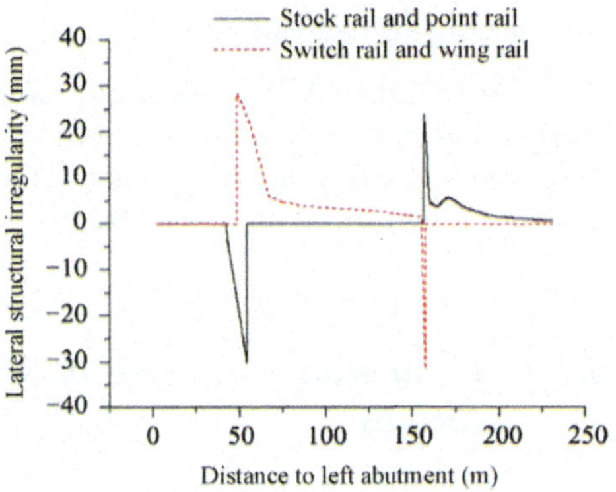

Fig. 6.55 Lateral structural irregularity of No. 42 turnout

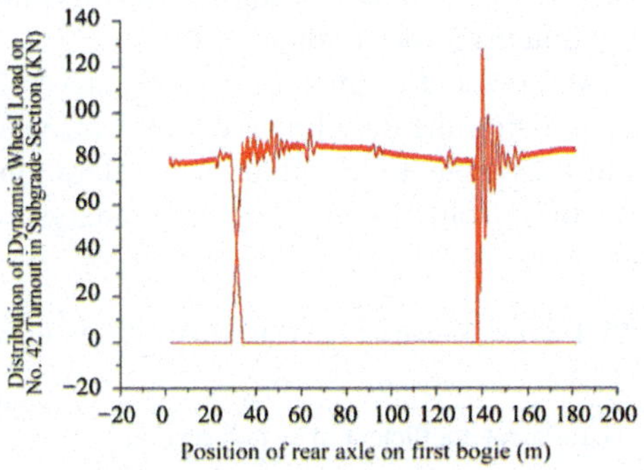

Fig. 6.56 Distribution of dynamic wheel load of No. 42 turnout in subgrade section

Moreover, the structural irregularity of the No. 42 turnout is relatively minor, and the transition range of wheel load on the wing rail and point rail is notably elongated. As a result, various dynamic responses of the EMUs passing through the CWR turnout on the bridge experience only slightly increase compared to

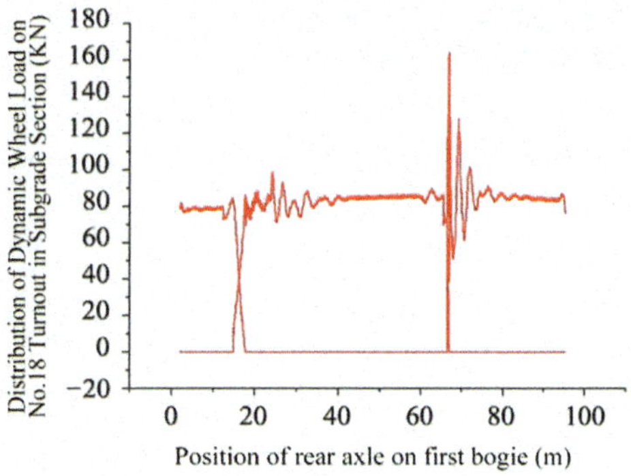

Fig. 6.57 Distribution of dynamic wheel load of No. 18 turnout in subgrade section

those observed when traversing the turnout in the subgrade section. It is worth mentioning that various indexes fall within the permissible ranges.

In general, the CWR turnout laid on the (6 × 32) m continuous girder of the Chunhe Super Major Bridge of the Beijing–Shanghai HSR can meet the requirements of operational safety and riding quality for the EMUs passing at 385 km/h through the No. 42 high-speed turnout on the main route.

6.5.4 Comparative Analysis of Simulation and Measured Results

See Table 6.17 for the summary of the dynamic simulation and measured results for the passage of the 350 km/h EMUs through the No. 42 turnout on the (6 × 32) m continuous girder bridge.

It is evident from Table 6.15 that several dynamic indexes align remarkably well, comprising the derailment coefficient, wheel unloading rate, switch rail and point rail stress, and bridge vibration acceleration. This coincidence reflects that the results obtained from dynamic simulation analysis of turnouts on bridges can aptly mirror the

Table 6.17 Summary of simulation and measured results

Item of comparison	Simulation item	Field measurement
Derailment coefficient in switch section	0.26	0.21
Wheel unloading rate in switch section	0.093	0.086
Dynamic stress at switch rail with a top width of 30 mm (MPa)	116.7	108.9
Dynamic stress at point rail with a top width of 35 mm (MPa)	24.9	17.0
Vertical vibration acceleration in third bridge span (×g)	0.045	0.029
Lateral vibration acceleration in third bridge span (×g)	0.010	0.007

dynamic characteristics of the train-turnout-bridge system under the dynamic load of high-speed trains. It should be noted that the operational conditions at the site and constraints in positioning measuring points contribute, to some extent, to biased and random measurements. Consequently, it becomes challenging to fully unveil the distribution pattern of responses throughout the entire process of trains crossing turnouts on bridges using measurement techniques. In addition, accurately characterizing the physical model of the huge and complex train-turnout-bridge coupling dynamic system, and precisely predicting the dynamic performance of this system, pose considerable challenges. Further enhancements in the computational model and analytical methodologies, leading to the gradual refinement and practical implementation of the simulation technology, can present valuable prospects for its application in the evaluation of dynamic performance in turnout bridge structures.

6.6 Dynamic Simulation Analysis of No. 18 Simple Turnout on (4 x 32) m Continuous Girder Bridge at Suzhou Station

Under the chosen characteristic scenario with a single No. 18 turnout with 60 kg/m rails (station number: 9 and 13; frog chainage: DK1227 + 408.49) placed on a (4 × 32) m continuous girder at New Suzhou Station, the train-turnout-bridge coupling dynamic simulation analysis was performed at speeds of 350 km/h and 420 km/h respectively. This bridge features a single-box double-cell section, monolithic deck, and (4 × 32) m prestressed concrete continuous girder double-track integral structure. The bridge girder is constructed using C55 concrete, with an elastic modulus of 3.60 × 104 N/mm^2 and a unit weight of 26.5 kN/m^3. The bridge deck is surfaced with CRTS II ballastless track. The train parameters are considered based on CRH$_3$ EMU. The specific turnout bridge arrangement is shown in Fig. 6.58.

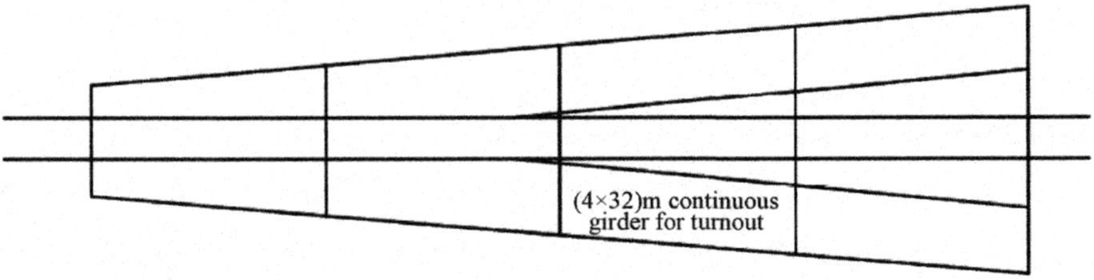

Fig. 6.58 Characteristic turnout bridge arrangement at New Suzhou station

6.6.1 Dynamic Analysis Results of EMUs Passing Through Turnout on Main Route at 350 km/h

1. Distribution of Dynamic Wheel Load

With the beginning of the turnout as the origin of the coordinates, the dynamic wheel load distribution of the inner wheel on the front axle of the first motor car's first bogie is shown in Fig. 6.59. As depicted in the figure, the presence of structural irregularities in the switch and frog sections results in significant fluctuations in wheel load when the wheelset's inner wheel traverses these particular areas of the turnout. These fluctuations are shown in Figs. 6.60 and 6.61. Within the switch section, the wheel load exhibits subdued fluctuations, characterized by an elongated transition range. Conversely, in the frog section, the wheel load fluctuation range is comparatively brief. Additionally, the vibration of the inner wheel induces wheel load fluctuations in its outer counterpart at corresponding locations. The front axle's outer wheel of the bogie displays substantial wheel load fluctuations within the switch section, while experiencing subdued fluctuations within the frog section. This behavior is primarily attributed to the elongated wavelength of structural irregularity in the switch section. On the other hand, the outer wheel on the rear axle of the bogie exhibits significant fluctuations in both the switch and frog sections, primarily resulting from the superposition of vibration from the front and rear axles during passage through the frog. The wheelset's inner wheel exhibits similar wheel load fluctuations on both the switch rail and point rail, with a maximum dynamic wheel load of approximately 167.5 kN, which falls within the permissible limit.

2. Flange Force Distribution

The contact between the wheelset's inner wheel and the switch rail, as well as the point rail, occurs within a limited range. The flange force distribution is shown in Fig. 6.62. The maximum flange force in the switch section was 35.4 kN, and the maximum instantaneous lateral impact force in the frog section was approximately 60.6 kN, both falling within the permissible range.

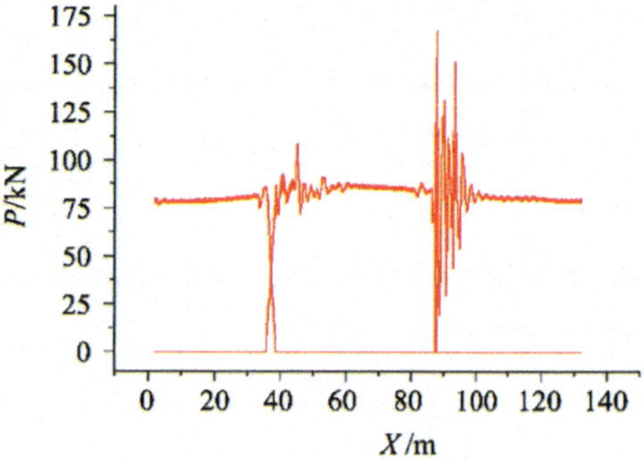

Fig. 6.59 Dynamic wheel load distribution under wheel

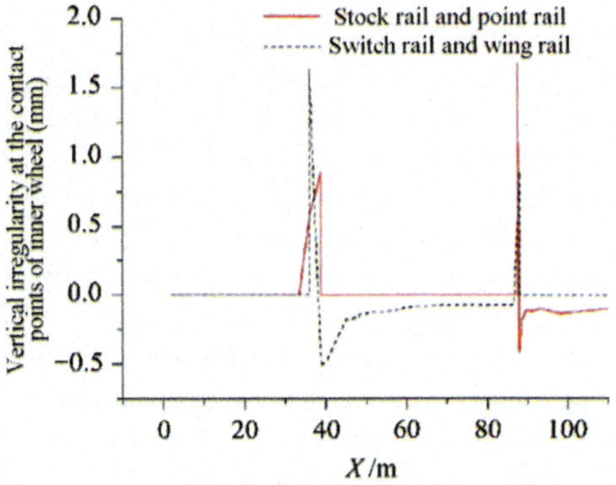

Fig. 6.60 Vertical irregularity at the contact point between inner wheels and rails

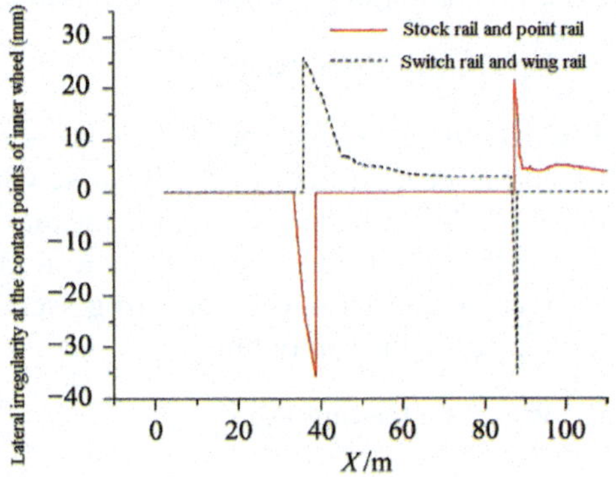

Fig. 6.61 Lateral irregularity at the contact point between inner wheels and rails

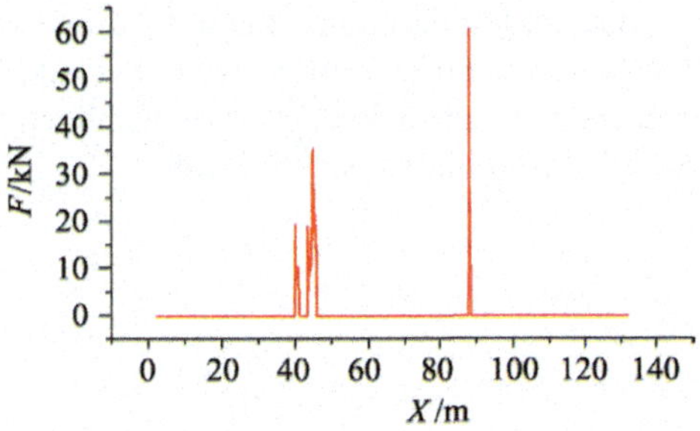

Fig. 6.62 Flange force distribution of inner wheel

3. Lateral Shift of Wheelset and Bogie

Figure 6.63 illustrates the lateral shift distribution of the front axle, rear axle, and frame of the leading bogie. As depicted in the figure, when the EMU passes through the turnout on the main route, both the front and rear wheelsets of the bogie shift laterally toward the switch rail and point rail, which is also mainly due to the lateral structural irregularity of the switch and frog. The maximum lateral shift of the front axle

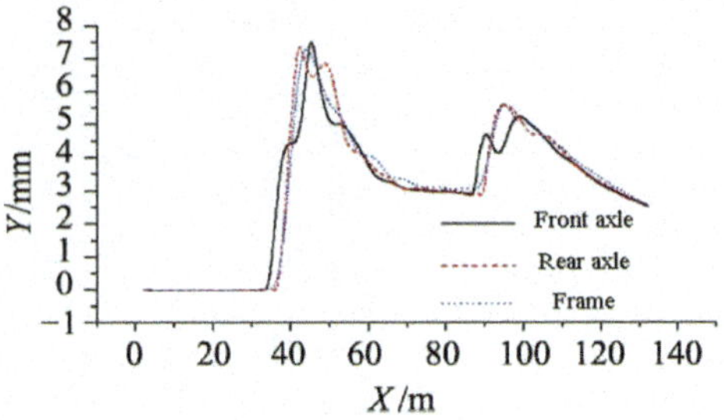

Fig. 6.63 Lateral shift of bogie and wheelset

in the switch section is around 7.5 mm, and that in the frog section is approximately 5.3 mm, both falling within the permissible range.

4. Derailment Coefficient

The derailment coefficient of the wheels is represented by the ratio of the sum of wheel-rail lateral forces to the dynamic wheel load, and its distribution is shown in Fig. 6.64. As depicted in the figure, the maximum derailment coefficients of the front axle in the switch and frog sections were 0.31 and 0.64 respectively, while those of the rear axle in the switch and frog sections were 0.40 and 0.62 respectively, all falling within the safety limit.

5. Wheel Unloading Rate

The unloading rate on wheels is represented by the ratio of the unloading amount of dynamic wheel load to static wheel load, and its distribution is shown in Fig. 6.65. As depicted in the figure, the unloading rates of the inner wheel on the front axle in the switch and frog sections were 0.16 and 0.77 respectively, while those of the inner wheel on the rear axle in the switch and frog sections were 0.16 and 0.76 respectively, all falling within the safety limit.

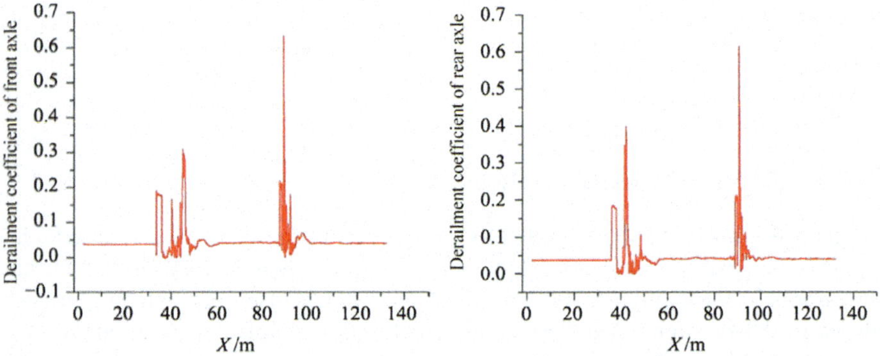

Fig. 6.64 Wheel derailment coefficient distribution

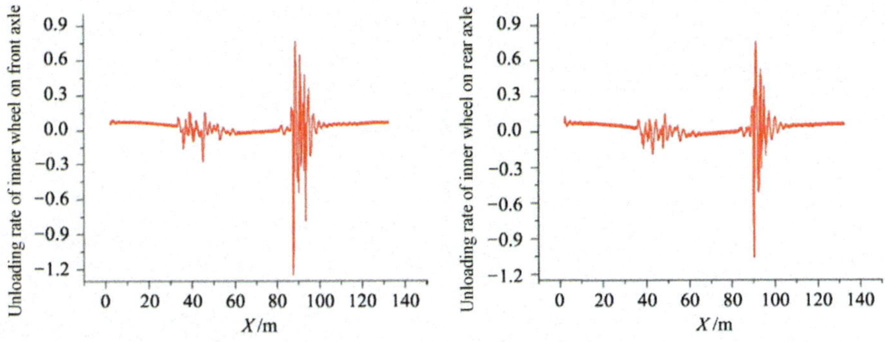

Fig. 6.65 Unloading rates of wheels

6. Vertical and Lateral Vibration Acceleration of Car Body

Figures 6.66 and 6.67 illustrate the vertical and lateral vibration accelerations of the car body. The maximum vertical vibration acceleration was around 0.149 *g*; the maximum lateral vibration acceleration in the switch section was around 0.038 *g*, and that in the frog section was approximately 0.012 *g*. All these results fell within the permissible limit.

7. Dynamic Stress of Rail

Figure 6.68 illustrates the simulation analysis results on the dynamic stress of the rails. The impact dynamic stress of the switch rails in the switch section was around 214.2 MPa, and that of the point rail in the frog section is approximately 124.8 MPa, both falling within the permissible limit.

8. Opening of switch rail and point rail

Figures 6.69 and 6.70 illustrate the opening of the switch rail and point rail, which were recorded around 0.42 and 0.52 mm respectively, both falling within the permissible limit.

Fig. 6.66 Vertical vibration acceleration of car body

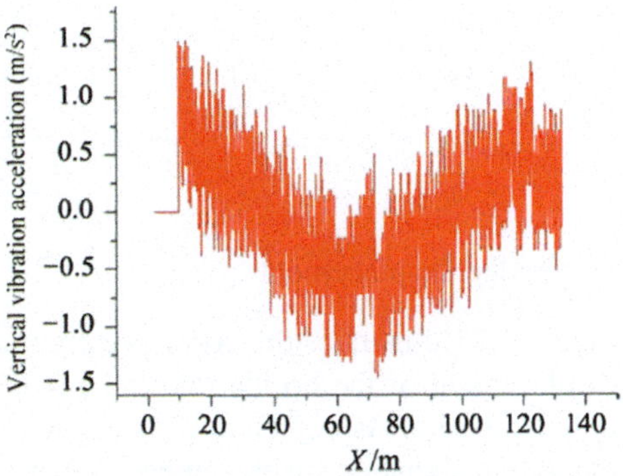

Fig. 6.67 Lateral vibration acceleration of car body

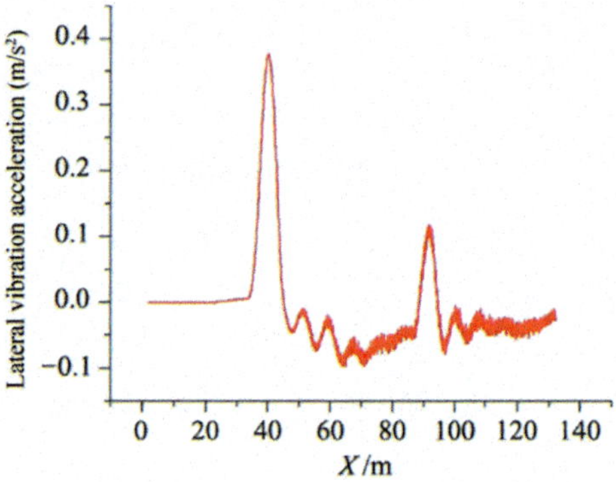

Fig. 6.68 Dynamic stress of rail

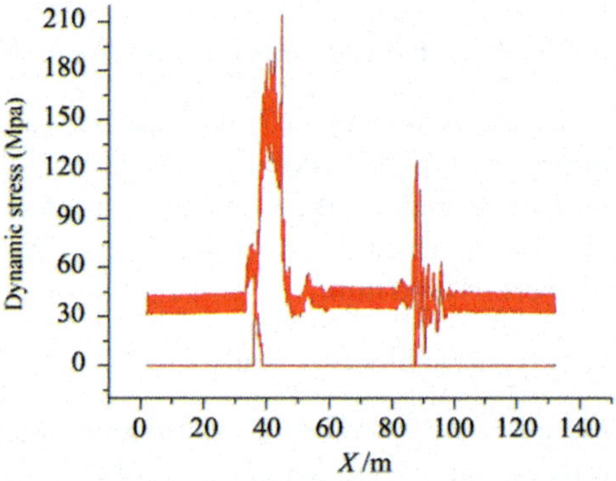

Fig. 6.69 Opening of switch rail

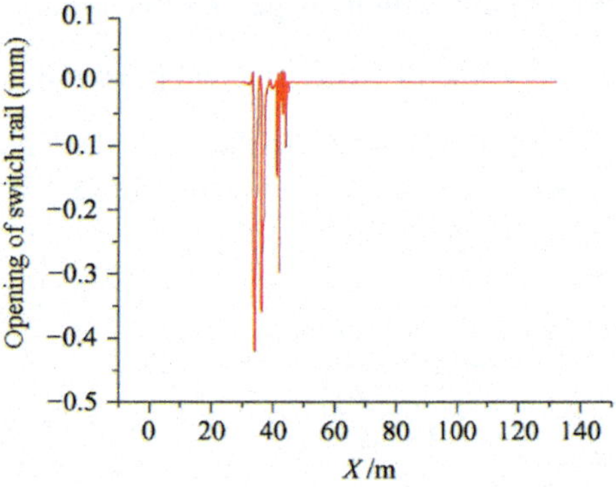

9. Bridge Displacement

Figure 6.71 illustrates the corresponding relationship between the vertical dynamic displacement of the bridge under the front axle/deformation of the bridge and the unloading rate. As depicted in the figure, under the load effect from the EMU, the maximum dynamic displacement of the girder span is around 0.48 mm, and the

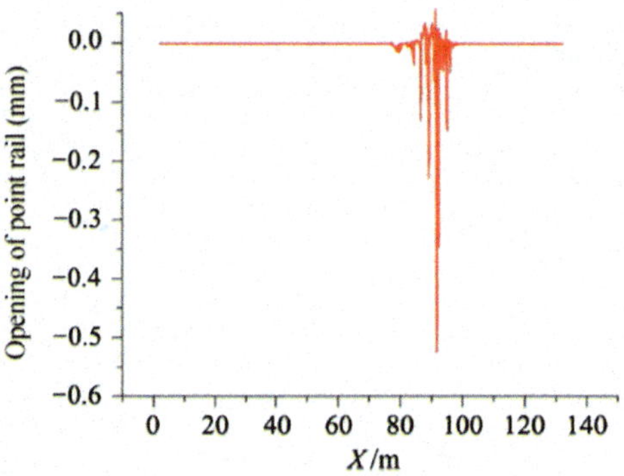

Fig. 6.70 Opening of point rail

frog is positioned at a location on the bridge where the rate of change of dynamic deflection is relatively small. Thus, the relative positions of the turnout and bridge can be considered reasonably rational. Figure 6.72 illustrates the lateral dynamic displacement of the bridge under the front axle, which did not exceed 0.02 mm and was considered unnoticeable.

Figure 6.73 illustrates the vertical displacement distribution throughout the whole bridge when the EMU passes through the end of the turnout, with a maximum dynamic displacement of around 0.56 mm. Figure 6.74 illustrates the vertical rotation angle distribution at the bearing of the second girder span, with a maximum dynamic rotation angle of around 0.045‰. Figure 6.75 illustrates the horizontal rotation angle distribution at the bearing of the second girder span, and the maximum dynamic

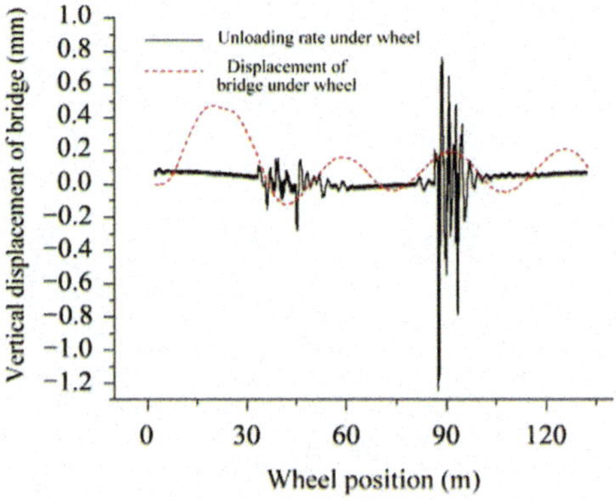

Fig. 6.71 Vertical displacement of bridge under wheel

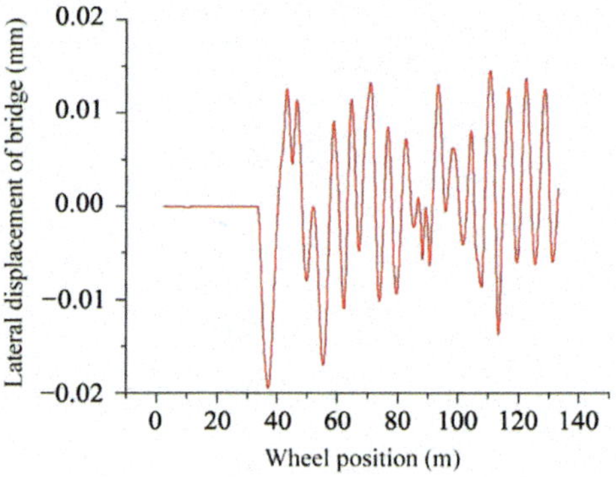

Fig. 6.72 Lateral displacement of bridge under wheel

rotation angle was approximately 0.00056%. All the displacement indexes fall within the permissible limits.

10. Vibration Acceleration of Bridge

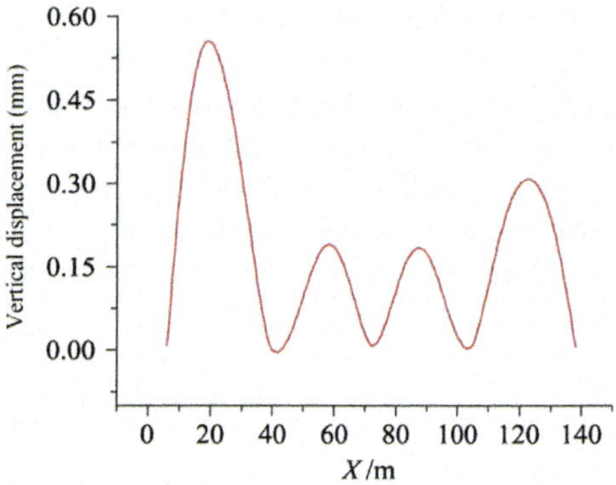

Fig. 6.73 Vertical displacement of bridge during passage of EMU through the end of the turnout

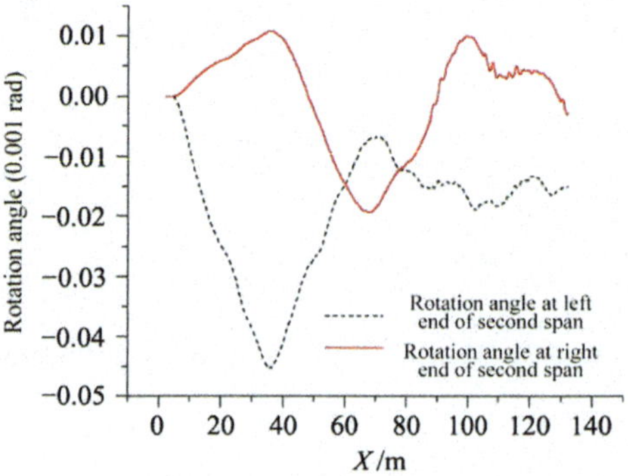

Fig. 6.74 Vertical rotation angle distribution at bearings in middle spans

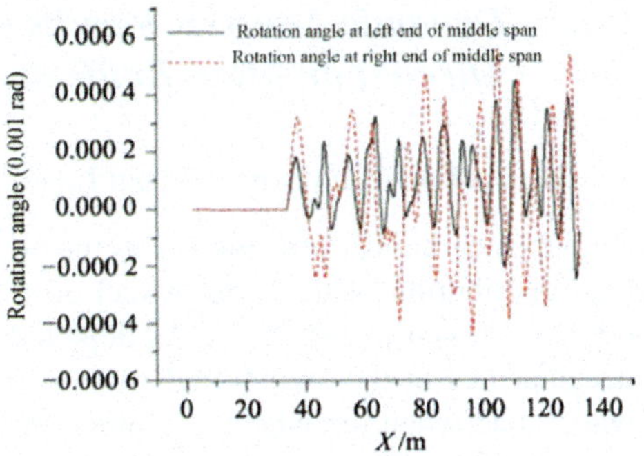

Fig. 6.75 Horizontal rotation angle distribution at bearings in middle spans

The vertical and lateral vibration accelerations of the bridge under the wheels are shown in Figs. 6.76 and 6.77. The maximum vertical vibration acceleration was about 0.07g and the maximum lateral vibration acceleration was about 0.029g, both falling within the permissible limit.

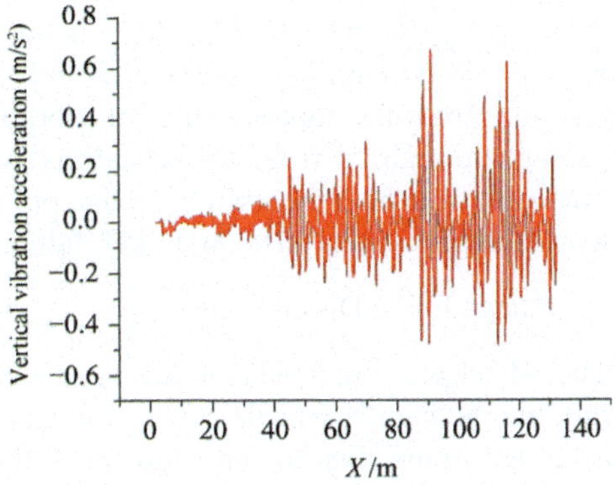

Fig. 6.76 Vertical vibration acceleration of bridge

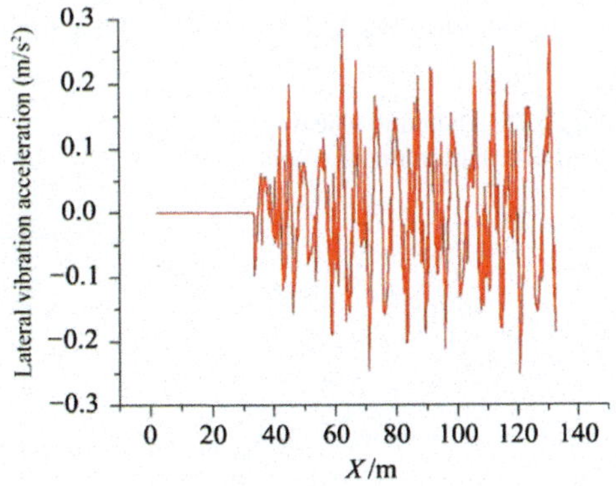

Fig. 6.77 Lateral vibration acceleration of bridge

6.6.2 Dynamic Analysis Results of EMUs Passing Through Turnout on Main Route at 420 km/h

1. Distribution of Dynamic Wheel Load

With the beginning of the turnout as the origin of the coordinates, the dynamic wheel load distribution of the inner wheel on the front axle of the first motor car's first bogie is shown in Fig. 6.78. As depicted in the figure, the presence of structural irregularities in the switch and frog sections results in significant fluctuations in wheel load when the wheelset's inner wheel traverses these particular areas of the turnout. These fluctuations are shown in Figs. 6.79 and 6.80. Within the switch section, the wheel load exhibits subdued fluctuations, characterized by an elongated transition range. Conversely, in the frog section, the wheel load fluctuation range is comparatively brief. Additionally, the vibration of the inner wheel induces wheel load fluctuations in its outer counterpart at corresponding locations. The outer wheel on the front axle of the bogie displays substantial wheel load fluctuations within the switch section, while experiencing subdued fluctuations within the frog section. This behavior is primarily attributed to the elongated wavelength of structural irregularity in the switch section. On the other hand, the outer wheel on the rear axle of the bogie exhibits significant fluctuations in both the switch and frog sections, primarily resulting from the superposition of vibration from the front and rear axles during passage through the frog. The wheelsets' inner wheels exhibited similar wheel load fluctuations on both the switch rail and point rail, with the maximum dynamic wheel load of approximately 188.8052 kN, falling within the permissible limit.

2. Flange Force Distribution

The wheelsets' outer wheels are kept away from contact with the straight stock rail, but their inner wheels get in contact with the switch rail and point rail within a limited range. Figure 6.81 illustrates the flange force distribution. The contact between the flange and the point rail in the frog section lasted for approximately 0.006 s, representing an instantaneous impact action. The maximum flange force in the switch section was 23.269 kN, and the maximum instantaneous lateral impact force

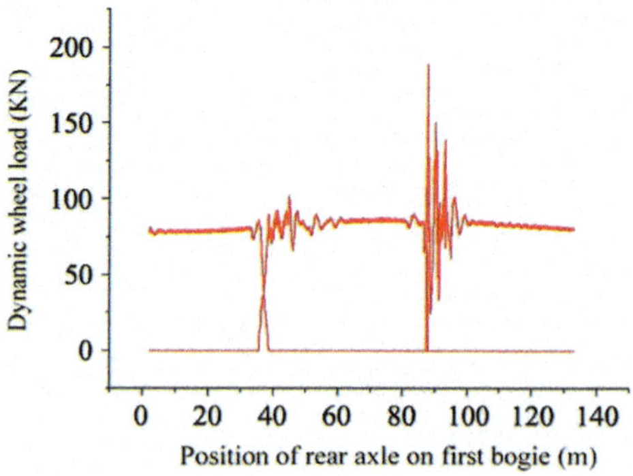

Fig. 6.78 Dynamic wheel load distribution under wheel

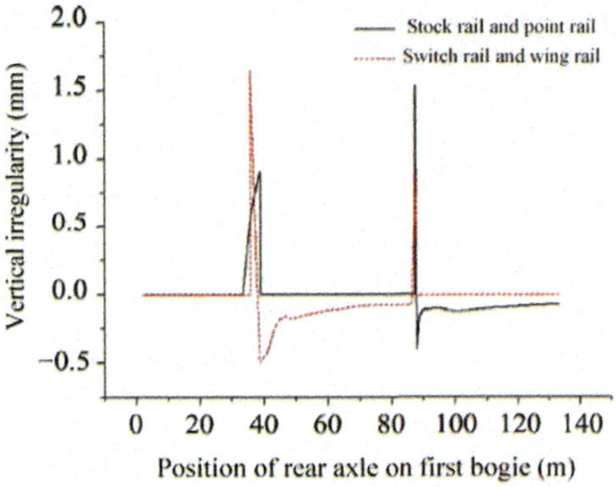

Fig. 6.79 Vertical irregularity at the contact point between inner wheels and rails

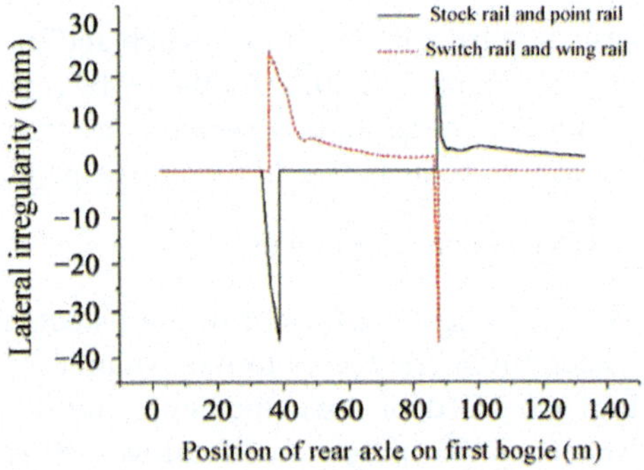

Fig. 6.80 Lateral irregularity at the contact point between inner wheels and rails

in the frog section was approximately 67.414 kN, both falling within the permissible range.

3. Lateral Shift of Wheelset and Bogie

Figure 6.82 illustrates the lateral shift distribution of the front axle, rear axle, and frame of the leading bogie. As depicted in the figure, when the EMU passes through the turnout on the main route, both the front and rear wheelsets of the bogie shift laterally toward the switch rail and point rail, which is also mainly due to the lateral

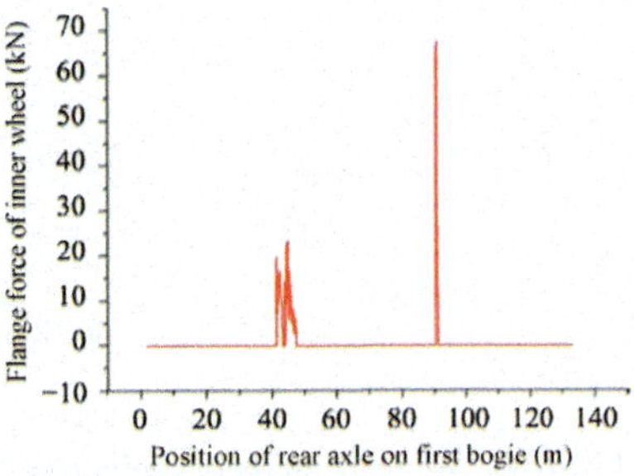

Fig. 6.81 Flange force distribution of inner wheels

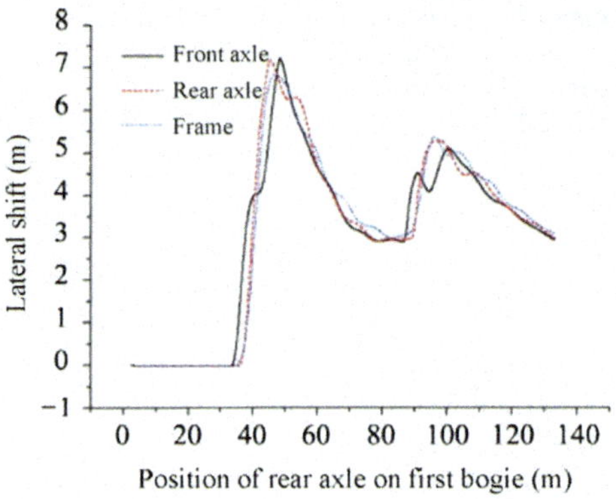

Fig. 6.82 Lateral shift of bogie and wheelset

structural irregularity of the switch and frog. The maximum lateral shift of the front axle is around 7.19 mm. As the EMUs traverse the turnout on the main route, they exhibit a hunting motion within a specific amplitude, which is attributed as one of the factors contributing to the impairment of driving comfort.

4. Derailment Coefficient

The derailment coefficient of the wheels is represented by the ratio of the sum of wheel-rail lateral forces to the dynamic wheel load, and its distribution is shown in Fig. 6.83. As depicted in the figure, the derailment coefficient of the wheelsets' outer wheels on the straight stock rail was <0.1; the maximum derailment coefficients of the front axle in the switch and frog sections were 0.497 and 0.697 respectively, while those of the rear axle in the switch and frog sections were 0.252 and 0.736 respectively, all falling within the safety limit.

5. Wheel Unloading Rate

The unloading rate on wheels is represented by the ratio of the unloading amount of dynamic wheel load to static wheel load, and its distribution is shown in Fig. 6.84.

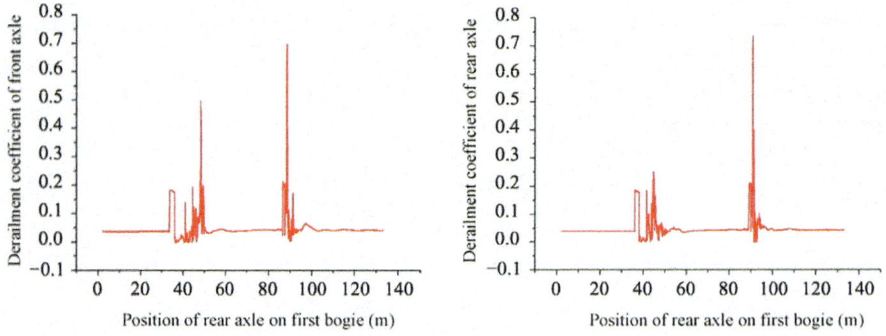

Fig. 6.83 Wheel derailment coefficient distribution

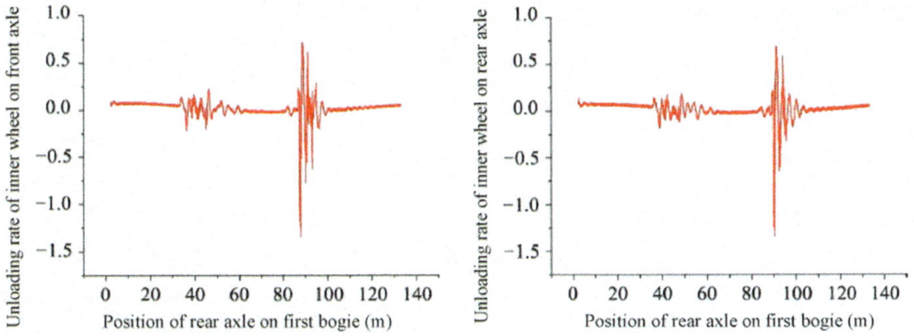

Fig. 6.84 Unloading rates of wheels

As depicted in the figure, the unloading rates of the inner wheel on the front axle in the switch and frog sections were 0.224 and 0.713 respectively, while those of the inner wheel on the rear axle in the switch and frog sections were 0.189 and 0.689 respectively, all falling within the permissible limit.

6. Riding Quality of Car Body

Figures 6.85 and 6.86 illustrate the vertical and lateral vibration accelerations of the car body. The maximum vertical vibration acceleration was around 0.149 g, while the maximum lateral vibration acceleration was approximately 0.036 g, representing noticeable hunting motions. However, these values still fell within the permissible limit.

7. Dynamic Stress of Rail

Figure 6.87 illustrates the dynamic stress of the rails. The impact dynamic stress of the switch rail in the switch section was around 197.608 MPa, and that of the point rail in the frog section was approximately 137.865 MPa, both falling within the permissible limit.

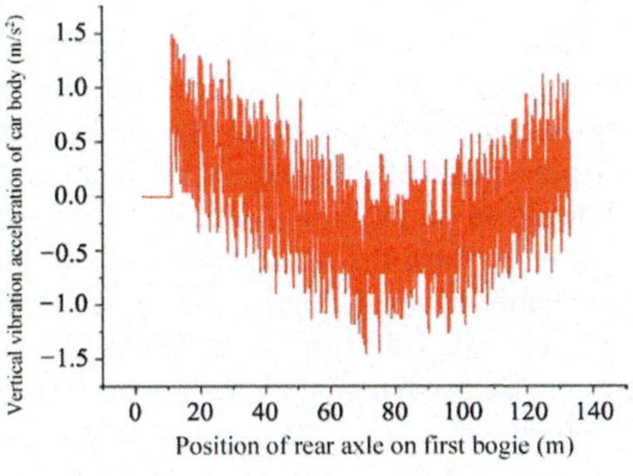

Fig. 6.85 Vertical vibration acceleration of car body

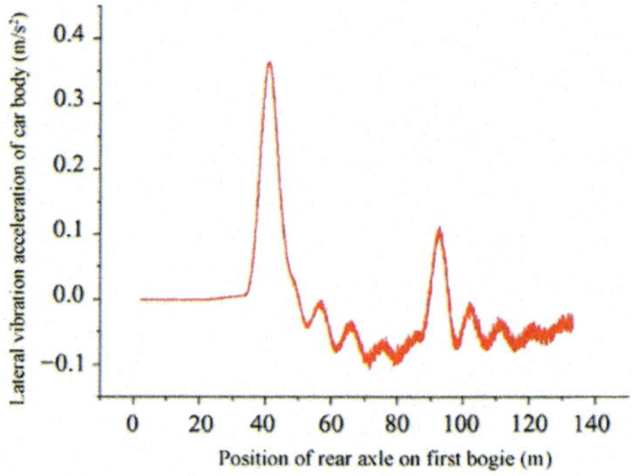

Fig. 6.86 Lateral vibration acceleration of car body

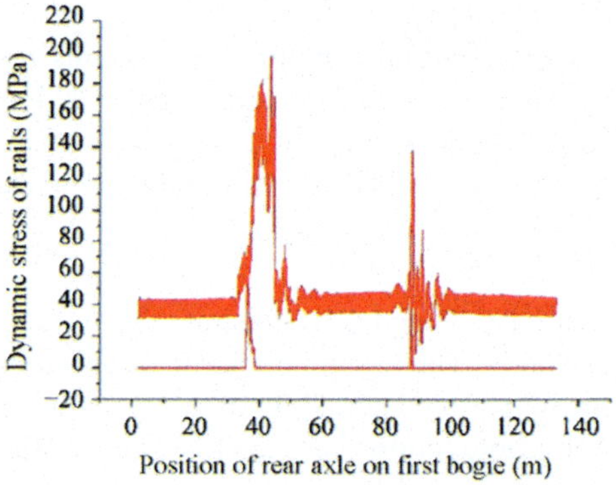

Fig. 6.87 Dynamic stress of rail

8. Opening of Switch Rail and Point Rail

Figures 6.88 and 6.89 illustrate the opening of the switch rail and point rail, which were recorded around 0.409 and 0.867 mm respectively, both falling within the permissible limit.

9. Bridge Displacement

Figure 6.90 illustrates the corresponding relationship between the vertical dynamic displacement of the bridge under the front axle/deformation of the bridge and the unloading rate. As depicted in the figure, under the load effect from the EMU, the maximum dynamic displacement of the girder span is around 0.472 mm, and the frog is positioned at a location on the bridge where the rate of change of dynamic deflection is relatively small. Thus, the relative positions of the turnout and bridge can be considered reasonably rational. Figure 6.91 illustrates the lateral dynamic displacement of the bridge under the front axle. Under the sole action of turnout structural irregularities, the lateral dynamic displacement of the bridge reached 0.021 mm at the maximum, which was considered unnoticeable.

Figure 6.92 illustrates the vertical rotation angle distribution at the bearing of the third span over the continuous girder, and the maximum dynamic rotation angle

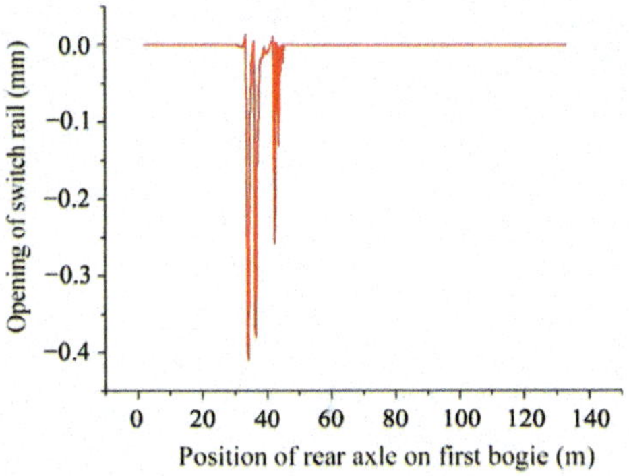

Fig. 6.88 Opening of switch rail

Fig. 6.89 Opening of point rail

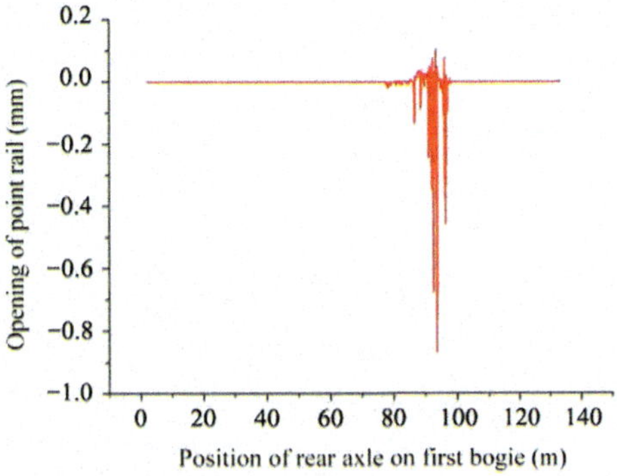

Fig. 6.90 Vertical displacement of bridge under wheel

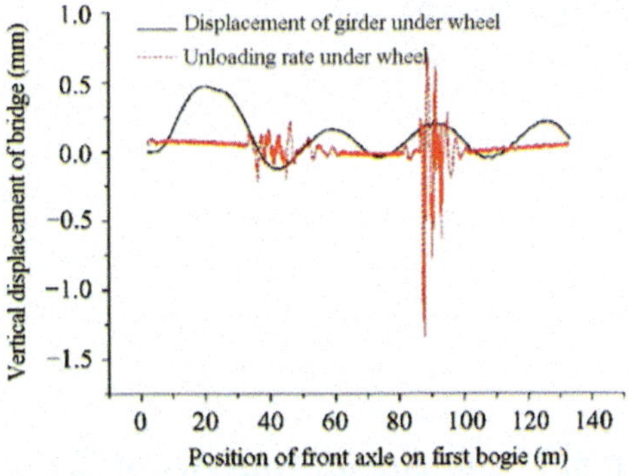

was approximately 0.02186%. Figure 6.93 illustrates the horizontal rotation angle distribution at the bearing of the second girder span, and the maximum dynamic rotation angle was approximately 0.00043%. All the displacement indexes fall within the permissible limits.

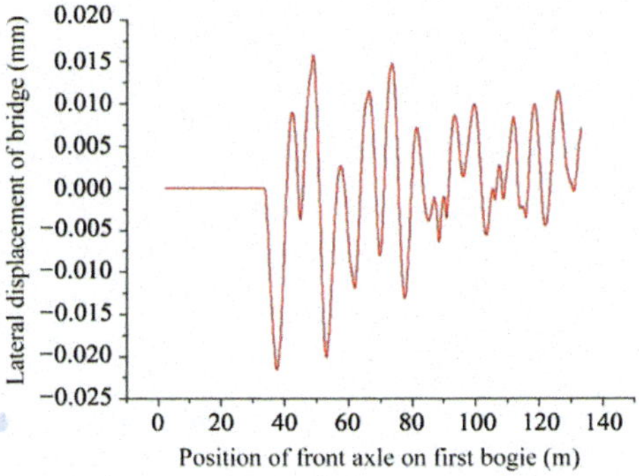

Fig. 6.91 Lateral displacement of bridge under wheel

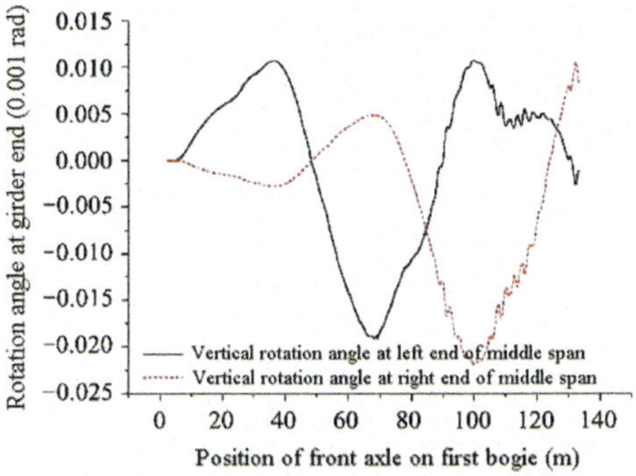

Fig. 6.92 Vertical rotation angle distribution at bearings in middle spans

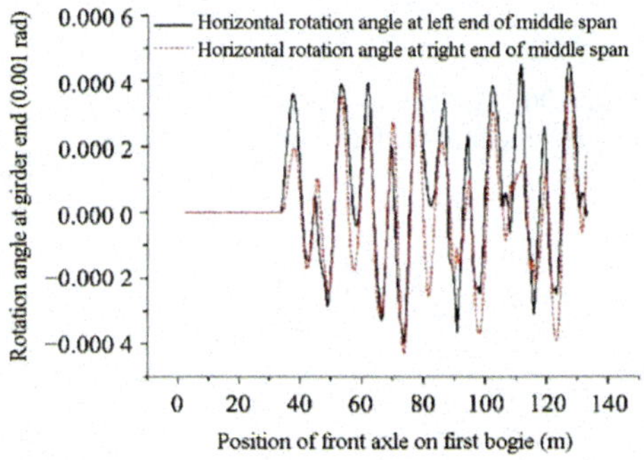

Fig. 6.93 Horizontal rotation angle distribution at bearings in middle spans

6.6.3 Statistical Analysis of Train-Turnout-Bridge Dynamic Simulation Results

See Table 6.18 for the results from the dynamic simulation analysis on the train-turnout-bridge system at New Suzhou Station.

Table 6.18 Comparison of simulation results

Item	Main route, 350 km/h	Main route, 420 km/h	Assessment criteria
Max. dynamic wheel load (kN)	167.5	188.805	170
Max. flange force in switch section (kN)	35.4	23.270	80
Max. flange force in frog section (kN)	60.6	67.414	80
Derailment coefficient of front axle in switch section	0.31	0.497	0.8
Derailment coefficient of front axle in frog section	0.64	0.697	0.8
Derailment coefficient of rear axle in switch section	0.40	0.252	0.8
Derailment coefficient of rear axle in frog section	0.62	0.736	0.8
Max. unloading rate of front axle in switch section	0.16	0.224	0.6
Max. unloading rate of front axle in frog section	0.77	0.713	0.6
Max. unloading rate of rear axle in switch section	0.16	0.189	0.6
Max. unloading rate of rear axle in frog section	0.76	0.689	0.6
Vertical vibration acceleration of car body (×g)	0.149	0.149	0.13
Lateral vibration acceleration of car body (×g)	0.038	0.036	0.10
Max. dynamic stress of switch rail (MPa)	214.2	197.608	350
Max. dynamic stress of point rail (MPa)	124.8	137.865	350
Max. opening of switch rail (mm)	0.42	0.409	4
Max. opening of point rail (mm)	0.52	0.867	4
Max. vertical displacement of bridge (mm)	0.48	0.472	32/1 500
Max. lateral displacement of bridge (mm)	0.020	0.021	2

(continued)

Table 6.18 (continued)

Item	Main route, 350 km/h	Main route, 420 km/h	Assessment criteria
Vertical rotation angle at girder end	0.045‰	0.021 86‰	1‰
Horizontal rotation angle at girder end	0.000 56‰	0.000 43‰	1‰

(1) Under the condition where the only consideration is the structural irregularity of the turnout, the dynamic responses during the passage of the EMUs at 350 km/h through the turnout on the bridge are generally not very significant. This can be attributed to the light axle load of the EMUs and the minor structural irregularity of the turnout optimized. Additionally, the vertical and lateral rigidity of the bridge satisfactorily meet the requirements for EMUs passing through the turnout. However, when the EMUs pass through the turnout on the main route at a higher speed of 420 km/h, the dynamic analysis results indicate that the wheel-rail lateral force and derailment coefficient approach the permissible limits.
(2) Attributed to the relatively high rigidity of the girder section (a reduced girder section was used for computational analysis compared with the actual girder section), all dynamic response values fall within the permissible limits, demonstrating the capacity to substantially satisfy the requirements of the train-turnout-bridge dynamic coupling under high-speed conditions.
(3) Based on the study results of train-turnout-bridge coupling dynamics, the relative positions of the turnout and bridge exert a significant impact on the vibration response of the system. Notably, when the switch or frog section aligns with or is in close proximity to a location with a relatively high deflection rate of the bridge, the interaction between the turnout and bridge becomes noticeably amplified.

References

1. Xin T (2014) Study on dynamic characteristics of ballastless turnout on high-speed railway bridge. Beijing Jiaotong University, Beijing
2. Chen L (2016) Analysis of vibration response of different ballastless track structures to train-track coupling system. Soochow University, Suzhou
3. Li G (2018) Research on optimization of vertical vehicle-track coupling interaction on high-speed railway. China Academy of Railway Sciences, Beijing
4. Zhai W (2015) Vehicle-track coupled dynamics. China Science Publishing & Media Ltd., Beijing
5. Wang F (1994) Vehicle system dynamics. China Railway Publishing House, Beijing
6. Lian S (2003) Track dynamics. Tongji University Press, Shanghai
7. Zhai W (1996) Locomotive-and-track system coupling dynamics and its application to the study of locomotive performance. China Railway Science 2:58–73

8. Lei X (2015) High-speed railway track dynamics: model, algorithm and application. China Science Publishing & Media Ltd., Beijing
9. Li T, Zhang J, Zhang W (2011) Coupling dynamic performance of vehicle-track under cross wind. Journal of Traffic and Transportation Engineering 11(5):55–60
10. Liu T (2020) Research on vehicle-bridge coupling vibration and structural noise on high-speed railway. Shijiazhuang Tiedao University, Shijiazhuang
11. Wu D, Tu J, Su W et al (2022) Study on the dynamic coefficient law of typical bridges of high-speed railway based on vehicle-bridge coupling vibration. J East China Jiaotong Univ 39(2):17–26
12. Yu Z, He H, Jiang L et al (2017) Dynamics and key technology research on high-speed railway track-bridge system under multiple dynamic sources. Chin Civil Eng J 50(11):1–9
13. Cui S, Liu P, Cao Y et al (2017) Simulation study on multiline vehicle-bridge coupled vibration. J Southwest Jiaotong Univ 52(5):835–843
14. Li K, Ban X, Su Y (2021) Study on the influence of deflection-span ratio of simply-supported girder bridge on vehicle-bridge dynamic response. Railw Eng 61(6):6–10
15. Zhang N, Xia H (2013) A vehicle-bridge interaction dynamic system analysis method based on inter-system iteration. China Railw Sci 34(5):32–38
16. Gao Y (2012) Research on spatial coupling vibration of the vehicle-track-bridge system on high-speed railway. South China University of Technology, Guangzhou
17. Pan J, Gao M (2008) Vehicle-railway-bridge system dynamic analysis. China Railway Publishing House, Beijing
18. Zhu Y (2006) Preliminary study on coupling vibration theory of vehicle-track-bridge system on high-speed railway. Southwest Jiaotong University, Chengdu
19. Chen G (2000) Random vibration analysis of vehicle-track coupling system. Southwest Jiaotong University, Chengdu
20. Xie G (2013) Dynamic simulation of railway vehicle-track vertical coupling system. Kunming University of Science and Technology, Kunming
21. Zhai W, Xia H (2011) Train-track-bridge dynamic interaction: theory and engineering application. China Science Publishing & Media Ltd., Beijing, pp 1–50
22. Shi H, Wu P, Luo R (2014) Coupled vibration characteristics of flexible car body and equipment of EMU. J Southwest Jiaotong Univ 49(4):693–699
23. Chen H, Lu D (2021) Kinetic analysis of high-speed railway turnout and research of its key technologies. High Speed Railw Technol 12(3):40–46
24. Chen X, Wang P, Chen R (2008) Spatial coupling vibration properties of high-speed vehicle-turnout. J Southwest Jiaotong Univ 43(4):453–458
25. Fu Y, Luo Y (2011) Analysis of the influence of uneven rigidity under turnout rail on dynamic characteristics of wheel-rail system. Railw Eng 51(4):139–141

Chapter 7
Status Monitoring of CWR Turnout on Bridge

High-speed train operation on high-speed railways imposes higher requirements for the track system in terms of stability, safety, and maintenance. On the one hand, due to high track utilization rates, dense traffic, and limited maintenance windows, the track structure and substructure should be maintained with reduced workloads or be designed to be maintenance-free. On the other hand, due to the combined influence of various factors such as train impacts, fatigue, temperature fluctuations, foundation settlement, and material aging, track structures inevitably experience accumulated damage, resulting in the degradation of load-bearing capacities. Resulting failures, especially in severe cases, have the potential to impact the quality of train rides and safety during operations. To assess the operational status of CWR turnouts on bridges, this chapter discusses studies conducted on No. 18 slab turnouts on the Lei Bridge of the Wuhan–Guangzhou HSR, the Haihang Super Major Bridge of the Shanghai–Hangzhou HSR, and the bridge at Xuzhoudong Station of the Beijing–Shanghai HSR. These studies involve long-term monitoring of various factors such as temperature, stress, and displacement. The monitoring results provide valuable insights for the design theory, operation, and maintenance practices of CWR turnouts on bridges in China.

7.1 Operational Status Monitoring of CWR Turnouts on Lei Bridge of Wuhan–Guangzhou HSR

7.1.1 Brief Introduction to Monitoring Site

The first experimental section in China and even worldwide for CWR turnouts with slab ballastless tracks featuring longitudinally coupled base slabs on bridges is located on the Lei Bridge at Chenzhouxi Station of the Wuhan–Guangzhou HSR

Q. Li et al., *Continuously Welded Turnouts on High-Speed Railway Bridge*, Advances in High-speed Rail Technology, https://doi.org/10.1007/978-981-97-6524-9_7

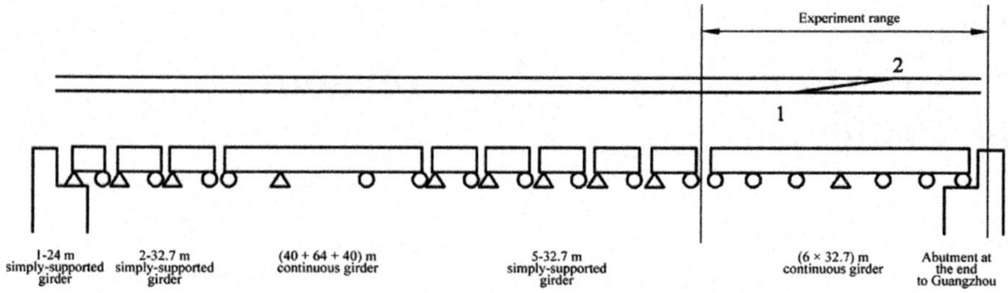

Fig. 7.1 Range of measuring points for experiment on (6 × 32) m continuous girder

[1]. This bridge is 595.62 m long totally and features a span arrangement of 1–24 m simply-supported girder + 2–32 m simply-supported girder + (40 + 64 + 40) m continuous girder + 5–32 m simply-supported girder + (6 × 32) m prestressed concrete continuous girder. The slab ballastless track for the No. 18 turnout is located on the (6 × 32) m continuous girder [1]. The range of measuring points covers the slab ballastless track for the No. 18 single-crossover turnout on the (6 × 32) m continuous girder, as shown in Fig. 7.1.

At this bridge location, the maximum rail temperature reaches 61.3 °C, while the minimum rail temperature goes down to −9 °C. The design stress-free rail temperature is set at 26 °C ± 5 °C, and the stress-free slab temperature of the base slabs is maintained at 20 °C.

7.1.2 Monitoring Items and Monitoring System

1. Monitoring Items

Two types of sensors were used for monitoring, as shown in Tables 7.1 and 7.2

2. Monitoring Principle

Table 7.1 Monitoring items of fiber grating sensor

No	Monitoring item	Monitoring principle	Sensor
1	Rail temperature	Principle of optics	Fiber grating sensor
2	Rail strain	Principle of optics	Fiber grating sensor
3	Relative displacement between rail and substructure	Principle of optics	Fiber grating sensor
4	Relative displacement between base slab and bridge	Principle of optics	Fiber grating sensor

Table 7.2 Monitoring items of vibrating wire sensor

No	Monitoring item	Monitoring principle	Remarks
1	Base slab temperature	Principle of vibrating wire	Vibrating wire sensor
2	Base slab concrete strain	Principle of vibrating wire	Vibrating wire sensor
3	Base slab reinforcement stress	Principle of vibrating wire	Vibrating wire sensor

(1) Working principle of fiber grating sensor

Fiber grating is a kind of optical fiber passive device. Fiber grating sensors, commonly embedded with Fiber Bragg Gratings (FBG), offer several advantages including immunity to electromagnetic interference, high-temperature resistance, compact size, tolerance to high temperatures, and corrosion resistance. These sensors modulate sensing signals based on wavelengths, ensuring stable measurement signals that are unaffected by factors like fiber bending losses, connection losses, light source fluctuations, and detector aging. With a robust multiplexing capability, multiple FBGs can be arranged in series on a single optical fiber to monitor multiple targets simultaneously, enabling quasi-distributed measurements [2, 3].

FBGs are formed by altering the lattices of silicon dioxide materials through irradiation on the polished side of single-mode germanium-doped fibers with ultraviolet excimer laser beams. The core refractive index of the fibers after grating shows periodic distribution fringes along the grating length direction, leading to the Bragg grating effect. FBG fibers act as narrow-band optical filters centered at the resonant wavelength when viewed from a fundamental optical properties perspective [4–7]. The Bragg wavelength of the gratings is expressed below [8]

$$\lambda_B = 2n\Lambda$$

where n denotes the effective refractive index of core mode; Λ denotes the grating period.

When a broadband light source emits a beam of light into an FBG through a fiber coupler, according to the grating theory, total reflection occurs when the Bragg condition is met, and the reflection spectrum peaks at the Bragg wavelength, λ_B. Factors in the external physical field like stress, strain, and temperature impact the grating periods, Λ, leading to changes in the wavelength of the backward retroreflected light, λ_B. By analyzing the deviation in wavelength, λ_B, any physical changes in the monitored component can be understood [9, 10]. Grating sensors are illustrated in Figs. 7.2 and 7.3.

(2) Working principle of vibrating wire strain and stress sensor

Vibrating wire strain gauges typically consist of a steel wire secured between end blocks or the components being measured. Physical quantities like tension or strain in the steel wire are determined by monitoring frequency variations in the tensioned steel wire [11]. Their setup is shown in Fig. 7.4.

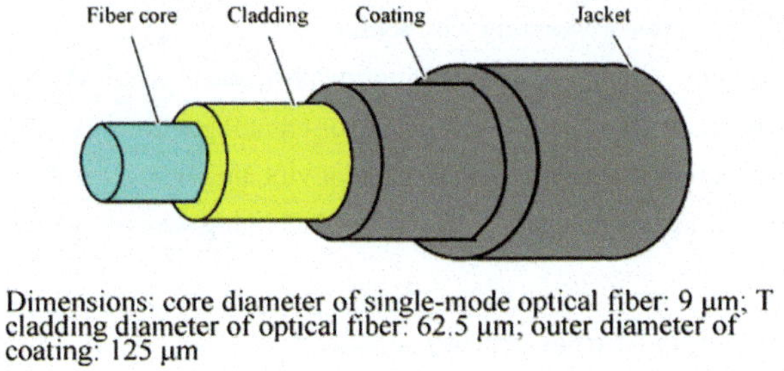

Fig. 7.2 Structure of grating

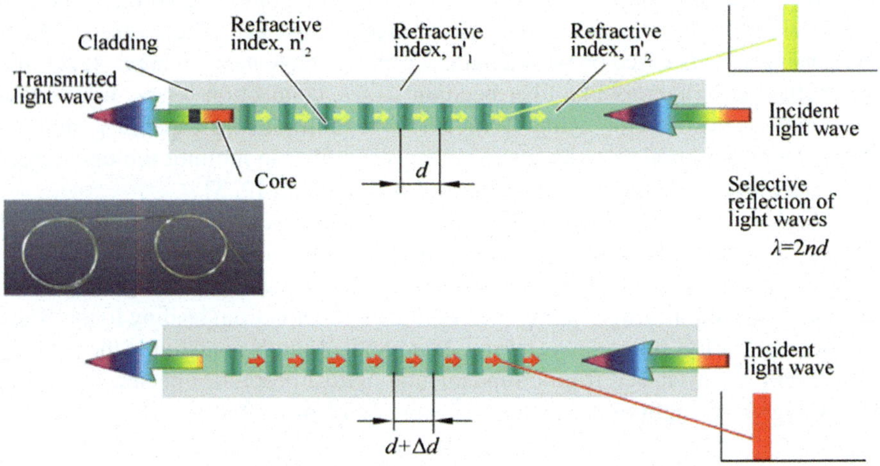

Fig. 7.3 Measurement principle of grating sensing

(3) Comparison of experiment techniques

Various methods are available for measuring the temperature, strain, stress, and displacement of structures. Commonly utilized approaches include the resistance type and vibrating wire type. The optical fiber technology developed in recent years has been increasingly employed in structural measurements.

(1) Resistance sensor

Resistance sensors use Wheatstone bridges to translate alterations in the resistance of the resistance wire caused by different strains and temperatures into voltage signals for strain measurements [12]. This method is suitable for both static and dynamic strain of structures. Resistance sensors are typically used for a short service life because they are susceptible to damage owing to their simple packaging technology. Moreover, temperature variations can induce changes in the resistance of transmission cables, leading to measurement errors and inaccurate results. Consequently, resistance strain gauges are primarily employed for short-term structural

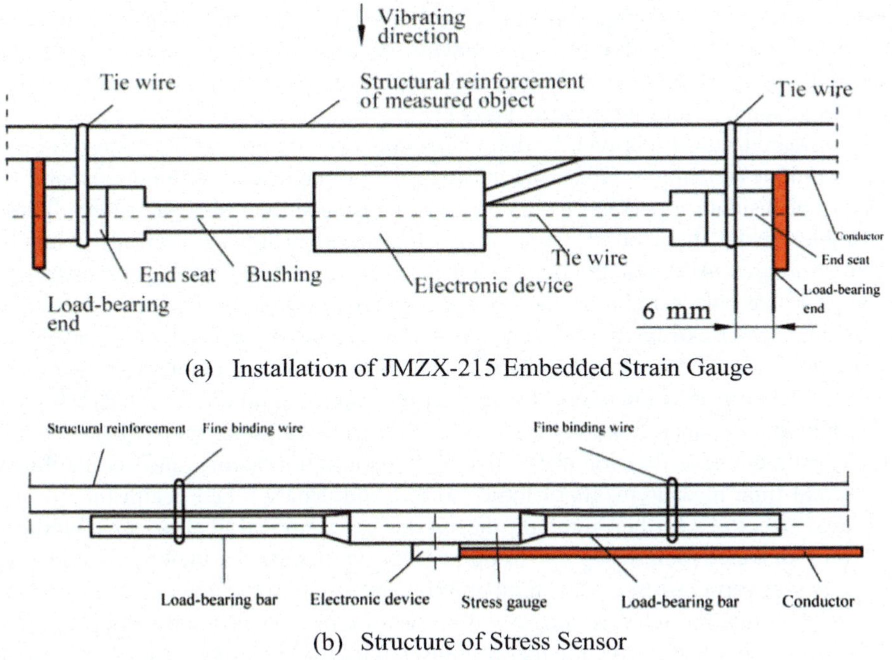

Fig. 7.4 Principle of vibrating wire strain and stress sensor

measurements rather than long-term monitoring, especially for extended static strain measurements.

(2) Vibrating wire sensor

Vibrating wire sensors typically consist of a steel wire secured between end blocks or the components being measured. Physical quantities like tension or strain in the steel wire are determined by monitoring frequency variations in the tensioned steel wire [11]. Their long-term stability is influenced by various factors. Key considerations include the choice of materials for steel wires and associated components, along with the securing method of steel wires, aimed at reducing wire creep caused by temperature and strain variations, as well as preventing corrosion. High-quality vibrating wire sensors are designed to ensure excellent operational characteristics and minimal long-term drifts. They are generally unsuitable for dynamic strain measurements in structures. However, they can also be used for such purposes at ultra-low frequencies below 1 Hz. Vibrating wire sensors concentrate on the wire frequency, thereby ensuring measurement accuracy is unaffected by external factors such as transmission cables. They are well-suited for on-site measurements of large structures, such as dams and bridges. In addition, without voltage signal drift, they can well function for long-term monitoring of structures, provided there is appropriate wire selection, secure fastening, and proper installation.

(3) Optical fiber strain-gauge sensor

Representing an emerging application technology, optical fiber sensors can keep normal functions in adverse environments of extreme temperatures, corrosion, vacuum, and dangers, providing a new technical means to solve many tough challenges in previous monitoring tasks [13].

Compared with traditional bridge inspection sensors, optical fiber sensors offer many unique advantages. These include anti-electromagnetic interference capability, electrical insulation performance, corrosion resistance, and intrinsic safety. These attributes make them highly suitable for long-term health monitoring of intelligent structures. Additionally, optical fiber sensors are known for their light weight, compact size, and variable shape, as well as their minimal influence on measured mediums. Their extremely high sensitivity and resolution, coupled with convenient multiplexing and networking capabilities, enable the formation of telemetry networks and optical fiber sensor networks using existing optical communication technologies. The low maintenance cost over the service life, wide transmission frequency bands for convenient time-division or frequency-division multiplexing, and compatibility with real-time measurements of large-capacity information and health monitoring of large structures further highlight the versatility of optical fiber sensors. Optimal for long-distance monitoring due to optical fiber transmission, their applications as sensors—covering stress, strain, temperature, and vibration—showcase their diverse utility. Stability through wavelength coding ensures minimal impact from light source intensity fluctuations, while the incorporation of multiple gratings in a single optical fiber allows for distributed automatic online monitoring.

(4) Comparison of measurement methods

These different types of sensors are compared in their measurement methods, as shown in Table 7.3.

4. Remote online real-time monitoring system

The remote online real-time monitoring system for the CWR turnouts with ballastless tracks on the Lei Bridge of the Wuhan–Guangzhou HSR consists of sensors, data acquisition modules, wireless transmission modules, and a remote monitoring center, as shown in Fig. 7.5. This system can realize continuous signal acquisition and wireless transmission, real-time monitoring, automatic alarm, and other functions in terms of temperature, strain, and displacement of the CWR turnouts, ballastless tracks, and CWR tracks, featuring a high degree of automation and unattended operation to ensure personnel safety.

7.1.3 Arrangement of Measuring Points

1. Arrangement of temperature, strain, and displacement measuring points for fiber grating sensor

Table 7.3 Comparative analysis of measurement methods

Measurement method	Technical features	Scope of Application	Price
Vibrating wire type	Structural strains are measured based on wire frequency. This technique is mature and widely used. The selection of frequency is crucial in this traditional method, as errors in frequency selection can frequently occur	Static strain measurements, long-term strain monitoring of structures	The unit price of these sensors is typically moderate, ranging from around RMB 300 to 1,000. Acquisition equipment, on the other hand, generally costs tens of thousands of yuan, but the price can be less than RMB 10,000 if portable options are chosen
Optical fiber type	Various measurement techniques are utilized due to the subdivision of these sensors, encompassing grating fibers and Brillouin fibers. These measurements rely on the structural strains generated by the sensors in varying environmental conditions, such as strain and temperature	These sensors can perform both static and dynamic strain measurements, along with long-term monitoring of structures. However, due to factors such as price, installation complexity, and technical stability, this type of sensor has not been widely popularized at present	The sensors are relatively costly, typically priced at around RMB 1,000. Similarly, the acquisition equipment is expensive, with prices generally ranging from hundreds of thousands to millions of yuan
Resistance type	The strain of structures is determined by measuring the voltage produced through a Wheatstone bridge setup, leveraging the resistance variations of these sensors under specific environmental conditions like strain and temperature. Baseline drift poses a common challenge in these measurements	Although these sensors are primarily employed for dynamic measurements, they are also suitable for static measurements. However, in static applications, it is advisable not to extend the measurements beyond one week	These sensors are typically cost-effective, with well-packaged options available at a moderate price. Acquisition equipment is also moderately priced, ranging from thousands to tens of thousands of yuan for static testing equipment and from tens of thousands to hundreds of thousands of yuan for dynamic testing equipment

See Table 7.4 for the temperature, strain, and displacement measurement items and numbers of fiber grating sensors, and see Figs. 7.6, 7.7, 7.8, 7.9, 7.10, 7.11 for the placement scenes.

2. Arrangement of measuring points for vibrating wire strain and stress sensor

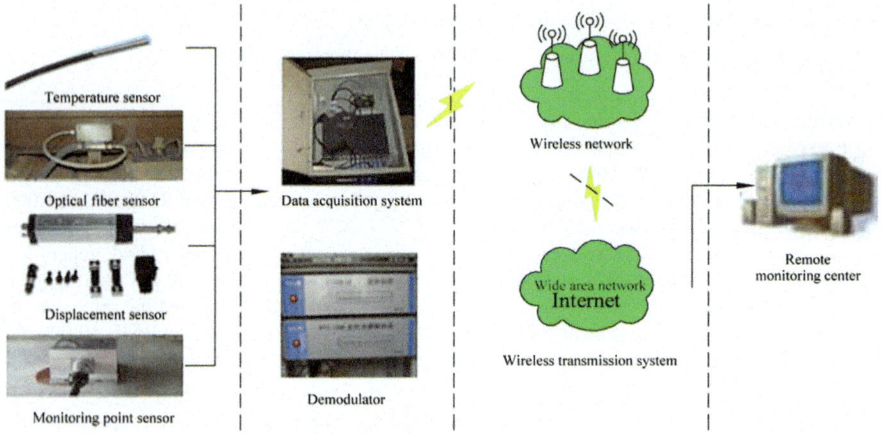

Fig. 7.5 Remote online real-time monitoring system

Table 7.4 Measurement items and number of FBG sensors

No	Monitoring Item	Number	Measuring point No
1	Rail temperature	6	87#~92#
2	Rail strain	32	1#~32#
3	Relative displacement between rail and substructure	32	33#~64#
4	Relative displacement between base slab and bridge	22	65#~86#
Total number of measuring points		92	1#~92#

Fig. 7.6 Temperature sensor

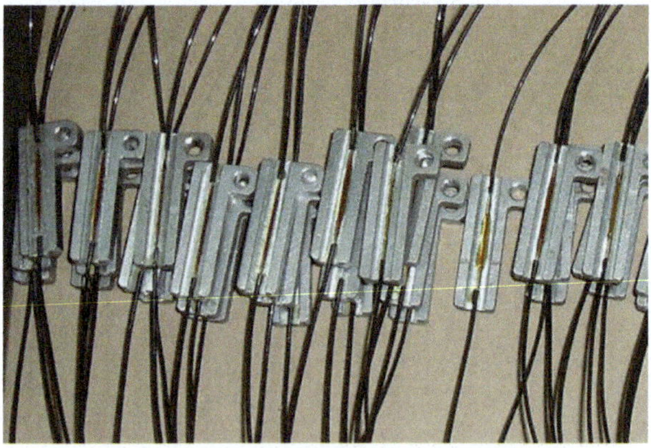

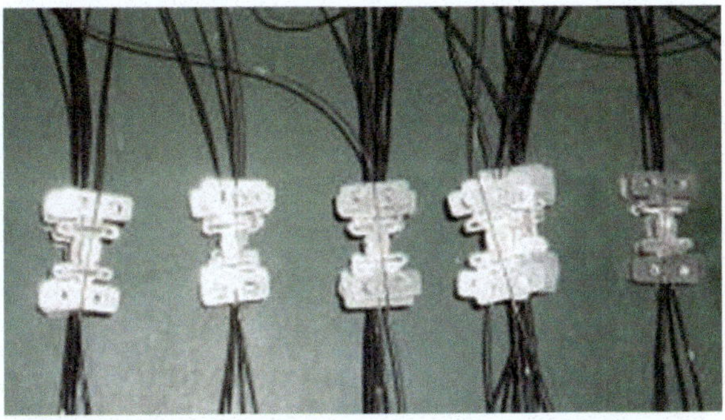

Fig. 7.7 Strain sensor

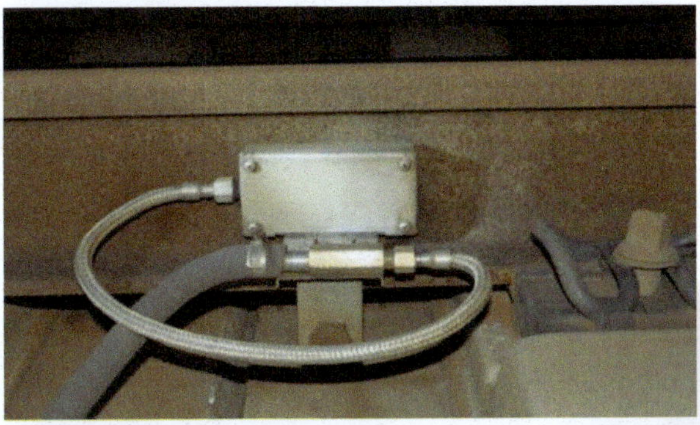

Fig. 7.8 Rail temperature and strain sensor

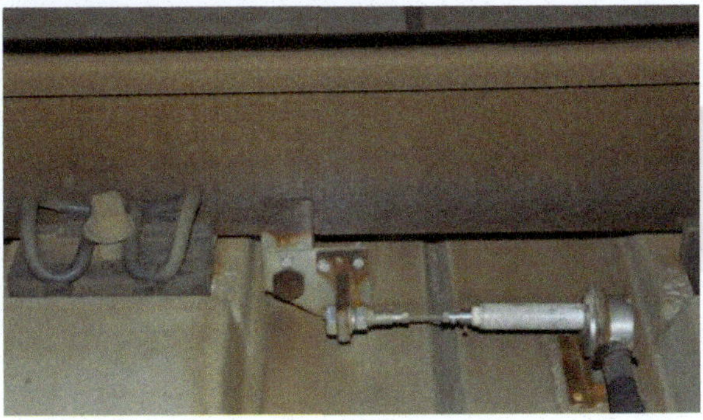

Fig. 7.9 Longitudinal displacement sensor

The arrangement of measuring points for vibrating wire strain and stress sensors is shown in Fig. 7.12.

Concrete strain measurement for base slab: Measuring points are placed every 16 m longitudinally on the base slabs across the whole bridge, and a total of 28 measuring points are placed.

Concrete strain measurement for post-cast strip base: four measuring points in total.

Fig. 7.10 Fiber grating sensor

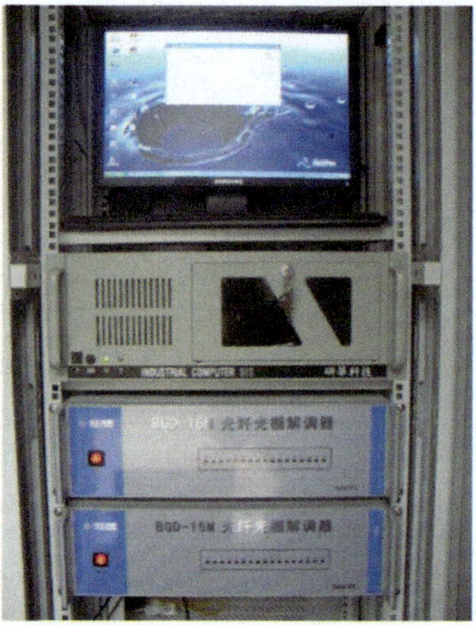

Fig. 7.11 Acquisition system of fiber grating sensor

Reinforcement stress measurement for base slab at girder end: eight measuring points located at the end of the ballastless turnout girder (four measuring points on the base slab at the girder end for each track).

Concrete strain measurement of end thorn base slab: Four measuring points located on the end thorn base slab (two measuring points on the base slab at the girder end for each track).

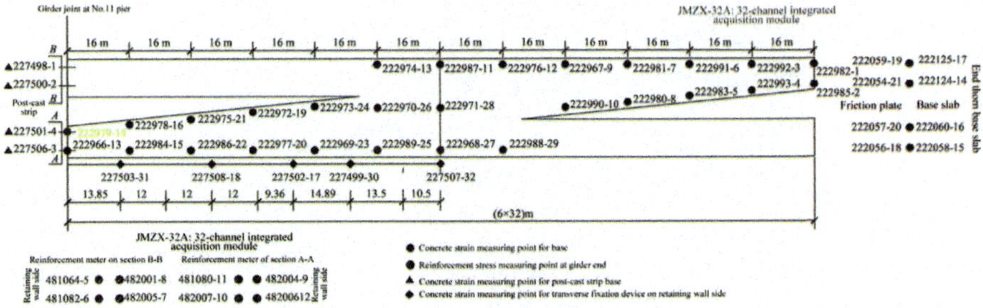

Fig. 7.12 Arrangement of measuring points for vibrating wire strain and stress sensors

Fig. 7.13 Concrete strain sensor embedded in base slab

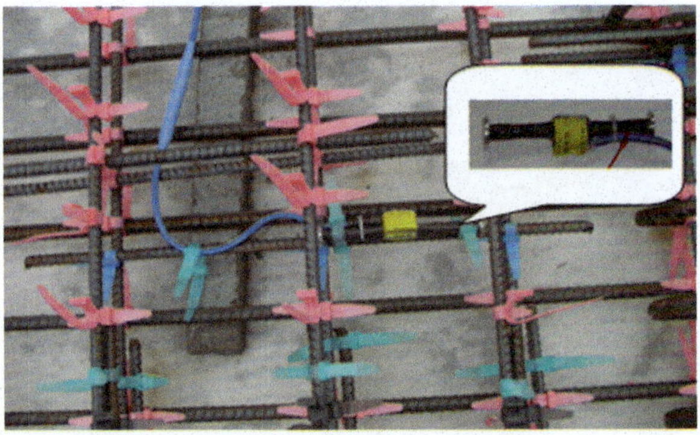

Concrete strain measurement of end thorn friction plate: Four measuring points located on the end thorn friction plate (two measuring points on the friction plate at the girder end for each track).

Concrete strain measurement for transverse fixation device on the retaining wall side: five measuring points in total.

The placement scenes of vibrating wire strain and stress sensors are shown in Figs. 7.13, 7.14, 7.15, respectively.

7.1.4 Analysis of Monitoring Data from Optical Fiber Sensing

Sensors were placed and optical cables were laid during the window time frame for integrated commissioning and testing of Wuhan–Guangzhou HSR from October 20 to November 18, 2009.

1. Rail temperature measurements

Fig. 7.14 Reinforcement stress sensor in base slab

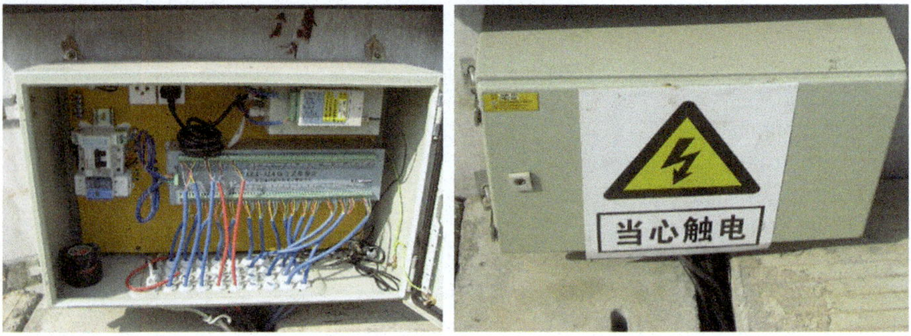

Fig. 7.15 Acquisition system of vibrating wire strain and stress sensors

The monitoring results pertaining to the maximum rail temperature from June to September 2010 and the minimum rail temperature from November 2010 to February 2011 are shown in Figs. 7.16 and 7.17.

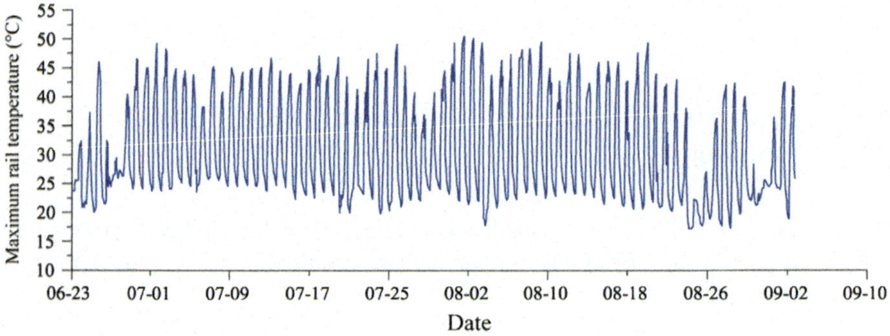

Fig. 7.16 Maximum rail temperature from June to September 2010

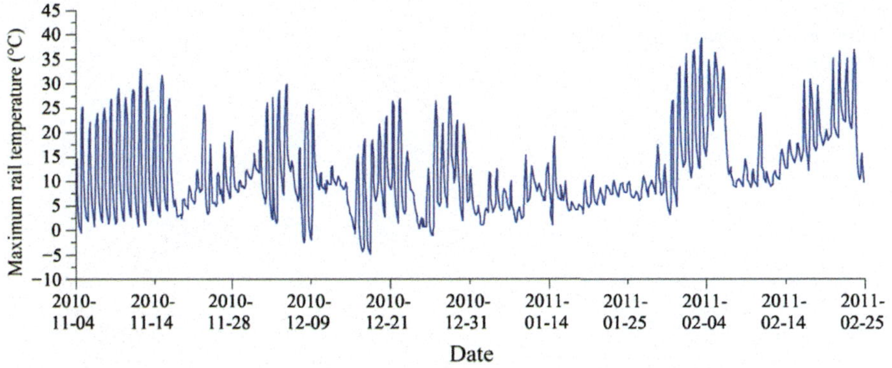

Fig. 7.17 Minimum rail temperature from November 2010 to February 2011

The monitoring results pertaining to the maximum air temperature from June to September 2010 and the minimum air temperature from November 2010 to February 2011 are shown in Figs. 7.18 and 7.19.

The air temperature and rail temperature data above show that during the summer months from June to September in 2010, the maximum air temperature recorded was 40.3 °C, while the highest rail temperature reached 50.4 °C, resulting in a temperature difference of 10.1 °C. In the winter season from June 2010 to February 2011, the minimum air temperature was −3.9 °C, and the lowest rail temperature observed was −3.5 °C. °These results verify the relationship between rail temperature and air temperature to ensure safety, as detailed in the *Code for Design of Railway Continuous Welded Rail* (TB 10,015–2012), that is, maximum rail temperature =

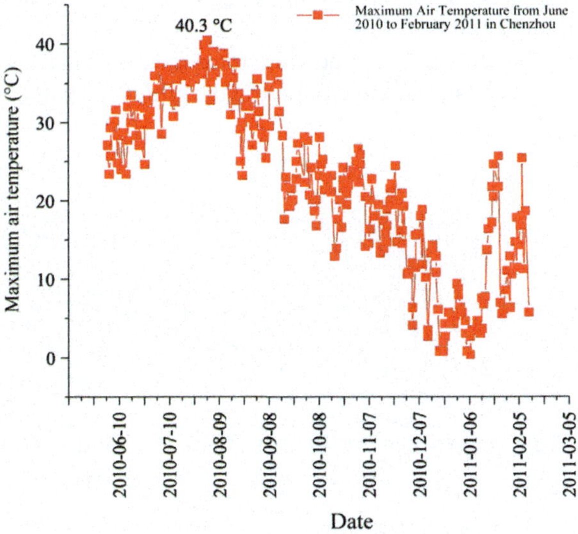

Fig. 7.18 Maximum temperature from June 2010 to February 2011

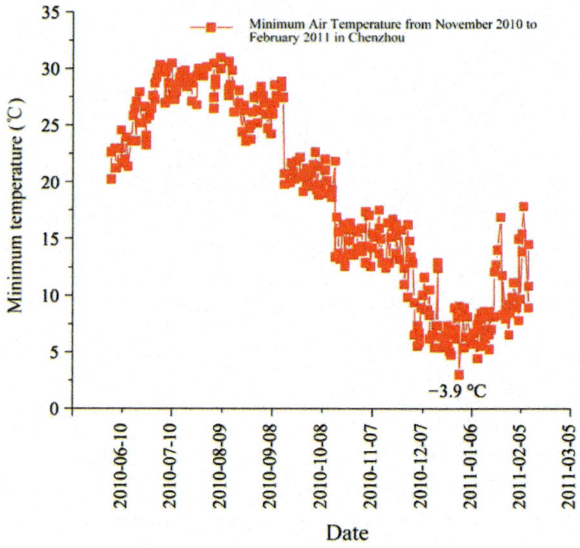

Fig. 7.19 Minimum Temperature from June 2010 to February 2011

maximum air temperature + 20 °C and minimum rail temperature = minimum air temperature.

2. Rail stress monitoring

(1) For the measurements from June to September 2010, taking Measuring Point D_1 as an example, the stress of the straight stock rail within No. 1 turnout is shown in Fig. 7.20.

The stress of the straight stock rail within No. 1 turnout measured on August 3, 2010 is shown in Fig. 7.21.

The additional temperature force measured at the girder end was 117 kN, whereas the calculated additional force at the girder end was 270 kN. The measured result was

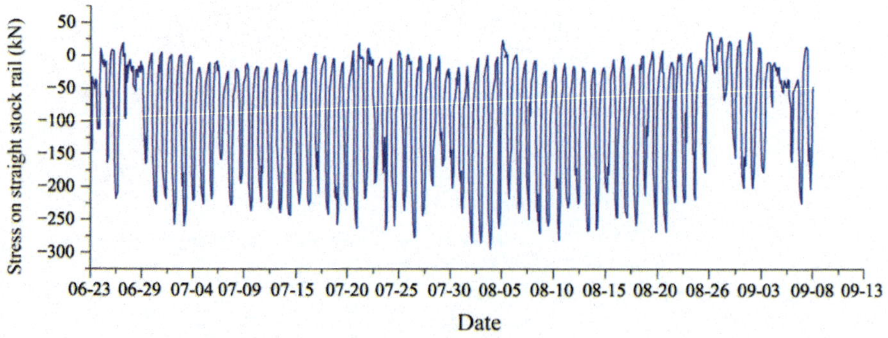

Fig. 7.20 Stress on straight stock rail within No. 1 turnout from June to September 2010

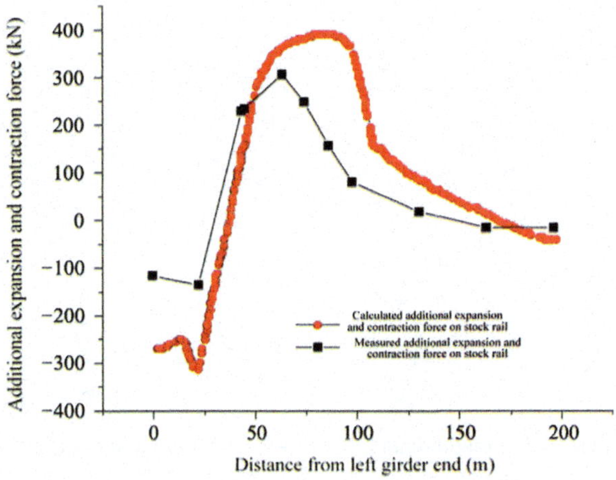

Fig. 7.21 Longitudinal force on straight stock rail within No. 1 turnout

lower than the calculated value. The peak additional force was observed preceding the turnout, and there was minimal variation in stress preceding and following the stopper.

(2) For the measurements from June to September 2010, taking Measuring Points D12 and D15 as examples, the stresses of the curved closure rail and long point rail within No. 1 turnout are shown in Figs. 7.22 and 7.23.

The stresses of the curved closure rail and long point rail of No. 1 turnout measured on August 3, 2010, are shown in Fig. 7.24.

The expansion/contraction experienced by the curved closure rail caused the stopper to fail to close properly, subsequently resulting in a progressive increase in stress on the curved closure rail due to the fastener resistance. The force transmitted to the curved closure rail through the spacing block located following the frog reached 52.14 kN, and the additional force gradually rose thereafter.

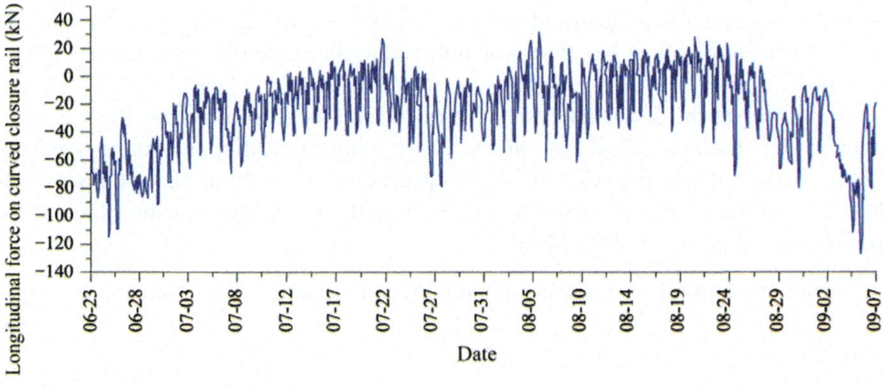

Fig. 7.22 Longitudinal force on curved closure rail at measuring point D_{12} within No. 1 turnout

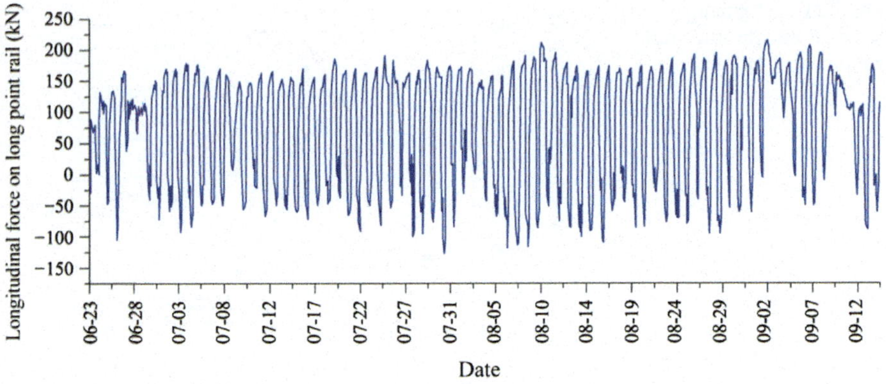

Fig. 7.23 Longitudinal force on long point rail at measuring point D_{15} within No. 1 turnout

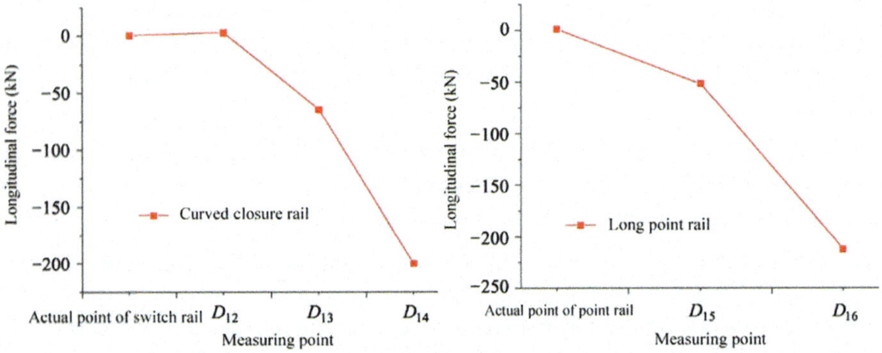

Fig. 7.24 Longitudinal force on curved closure rail and long point rail within No. 1 turnout

The measuring points are asymmetrically positioned for No. 1 and No. 2 turnouts. Due to the proximity of No. 1 turnout and No. 2 turnout to the simply-supported girder and subgrade, variations occurred in the number of measuring points and stress levels. The measuring points at No. 2 turnout adhered to the same pattern as those at No. 1 turnout.

According to the above analysis of stress on the measuring points, the measured data was less than the calculated values. The maximum additional compressive force on the stock rail was recorded at 137 kN preceding the turnout. The stopper at the heel end of the frog was stress-free longitudinally, while the spacing block at the frog experienced stress at 52.14 kN.

3. Monitoring of relative displacement between rail and substructure

For the measurements from June 23 to September 6, 2010, taking Measuring Point D_{33} as an example, the monitoring results of relative displacements between the rail and track slab or turnout slab are shown in Fig. 7.25.

Based on the measured data, the maximum relative displacements between the rails and substructure are shown in Table 7.5.

As shown in Table 7.5, the maximum relative displacement between the girder-end rail and track slab was approximately 1 mm, with a slightly larger displacement observed at the left girder end. The maximum relative displacement between the rail and turnout slab preceding the turnout was 0.78 mm, and the maximum relative displacement between the rail and track slab of the other track preceding the turnout varied between 1 mm and 1.5 mm. The relative displacement between the rail and turnout slab at the heel end of the frog was 1.26 mm. The maximum relative displacement between the curved closure rail and track slab ranged from 1.1 mm to 1.3 mm. The relative displacement between the rail and the turnout slab following the turnout was 1.58 mm.

4. Monitoring of relative displacement between base slab and bridge

Based on the measurements on the up track of the Lei Bridge on Wuhan–Guangzhou HSR from June 1 to August 6, 2010, taking Measuring Point D_{65} as an example, the maximum relative displacement between the base slab and the bridge is shown in Fig. 7.26.

Based on the measured data, the maximum relative displacements between the base slabs and the bridge deck are shown in Table 7.6.

As shown in Table 7.6, the relative displacements between the girder end base slabs and the bridge deck ranged from 8 to 10 mm, with a slightly larger displacement observed at the left girder end. The maximum relative displacements between the base slabs and the bridge deck preceding the turnouts ranged from 5 to 7 mm, while the relative displacements between the base slabs and the bridge deck following the turnouts ranged from 2 to 4 mm. The relative displacements between the track slabs and base slabs corresponding to No. 13 pier and No. 16 pier ranged from 5 to 6 mm.

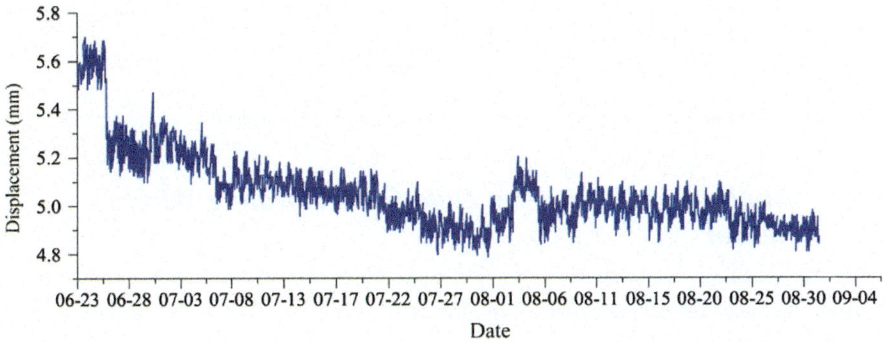

Fig. 7.25 Relative displacement between rail and track slab

Table 7.5 Maximum relative displacements between rails and substructure

Measuring point	Position	Measured Displacement (mm)	Zero-point Displacement (mm)	Relative Displacement (mm)
D_{33}	Left girder end	5.89	4.8	1.09
D_{34}	Front end of turnout	3.14	3.92	−0.78
D_{36}	Following heel end of frog	5.89	4.63	1.26
D_{45}	Curved closure rail	6.66	7.77	−1.11
D_{42}	Location of No. 16 pier	4.56	3.11	1.45
D_{43}	Right girder end	4.26	5.21	−0.95
D_{54}	Following turnout	10.01	8.43	1.58
D_{58}	Location of No. 12 pier	6.03	7.13	−1.1
D_{62}	Straight closure rail	7.53	8.8	−1.58

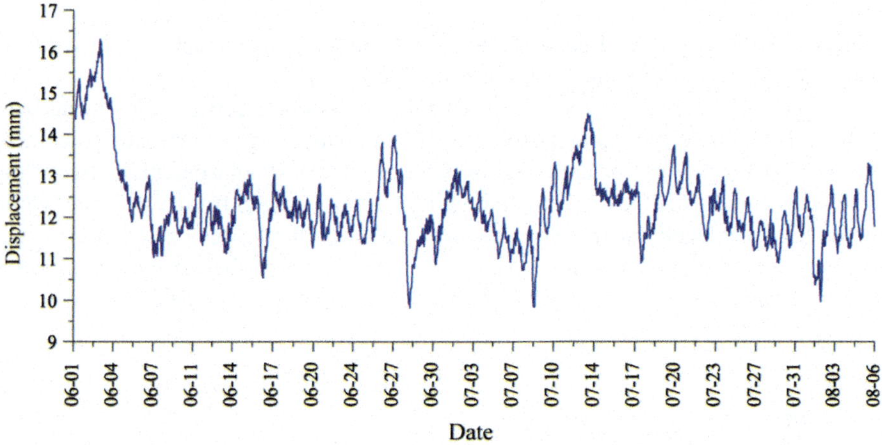

Fig. 7.26 Maximum Relative Displacement between Base Slab and Bridge

7.1.5 *Analysis of Monitoring Data from Vibrating Wire Sensor*

During the base slab construction of the Lei Bridge in May 2009, vibrating wire strain gauges and stress meters were placed at the base slabs within the turnout section, end thorn friction plates, and post-cast strips of the base slabs on the bridge, to monitor the

Table 7.6 Maximum relative displacements between base slabs and bridge deck

Measuring point	Position	Measured Displacement (mm)	Zero-point Displacement (mm)	Relative Displacement (mm)
D_{65}	Left girder end	14.4	5.3	9.1
D_{66}	Preceding No. 1 turnout	11.7	5.2	6.5
D_{74}	No. 16 pier	10.8	4.8	6
D_{75}	Right girder end	13.5	4.9	8.6
D_{76}	Right girder end	10.6	2.4	8.2
D_{77}	Following No. 2 turnout	6.16	0.36	5.8
D_{81}	Following turnout	6.4	3.8	2.6
D_{84}	No. 13 pier	11.8	6.6	5.2
D_{86}	Left girder end	12.2	3.3	8.9

concrete temperature, strain, and reinforcement internal force of concrete structures such as base slabs within the turnout section and friction plates. The monitoring results are summarized as follows.

1. Concrete temperature monitoring for base slab

According to the temperature monitoring data, concrete temperatures from July 2009 to January 2010 are shown in Fig. 7.27 and the temperature statistics are shown in Table 7.7.

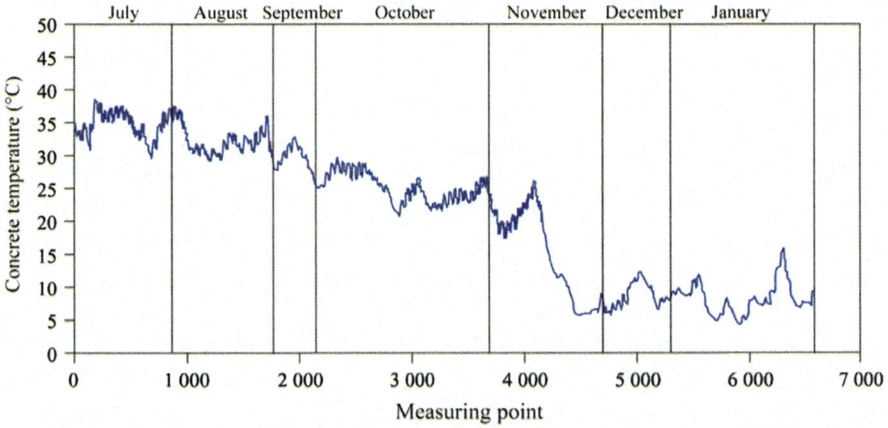

Fig. 7.27 Concrete Temperatures from July 2009 to January 2010. *Note* Measuring points indicated in the figure represent concrete temperature measurements collected every 30 min by the temperature acquisition equipment.

Table 7.7 Statistics of concrete temperature monitoring results

Month	°Average temperature of concrete (°C)	°Maximum temperature of concrete (°C)	°Minimum temperature of concrete (°C)
7	34.7	38.5	29.7
8	32.6	37.5	29
9	29.3	33.4	25
10	25.1	29.6	21
11	15.3	26.8	5.7
12	8.8	12.7	5.7
1	1	15.6	4

It is evident from Fig. 7.27 that the concrete temperature fluctuated every month, presenting a gradual descending trend. The maximum concrete temperature was recorded at 38.5 °C in July and the minimum temperature at 4 °C in January.

Table 7.7 shows monthly maximum and minimum concrete temperatures from July 2009 to January 2010.

The comparison between concrete temperature and air temperature from July to August 2009 and the comparison for December 2009 to January 2010 are shown in Figs. 7.28 and 7.29, respectively.

As shown in Figs. 7.28 and 7.29, the concrete temperatures in July and August were higher than the air temperatures overall, with a temperature difference of up to 6.2 °C. The concrete temperatures in December and January were also higher than the minimum air temperatures, with a temperature difference of up to approximately 6 °C.

2. Concrete strain monitoring for turnout slab-base

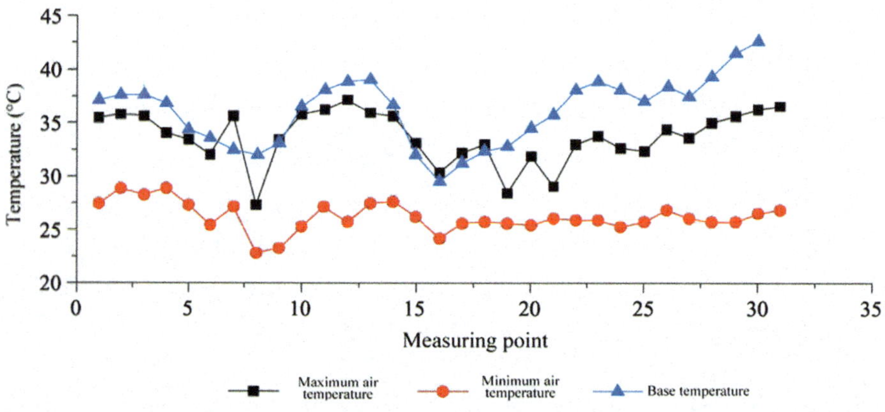

Fig. 7.28 Comparison between concrete temperature and air temperature in July and August

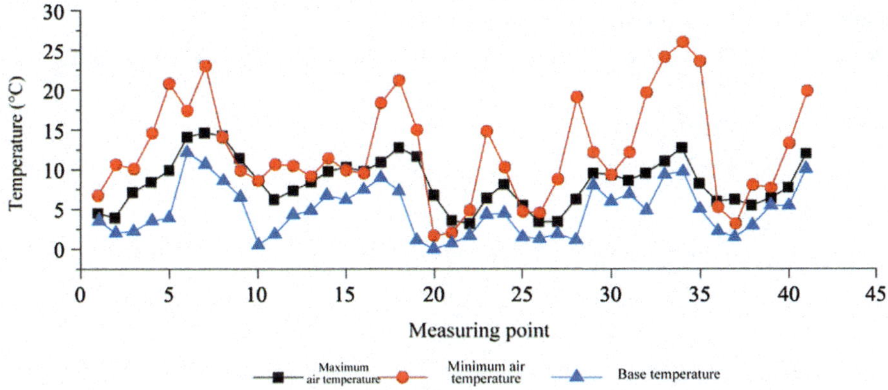

Fig. 7.29 Comparison between concrete temperature and air temperature in December and January

The concrete strain measurements for the turnout slab-base from July 2009 to January 2010 are shown in Fig. 7.30.

From the measured values, the concrete strain of the base increased gradually over time. Compressive strain was noted from July to October, transitioning to tensile strain from October to January. Through an analysis corresponding to the measured concrete temperatures, the strain of the base gradually increased, with decreasing concrete temperature. Since the stress-free temperature for the turnout slab-base was set at 20 °C, the concrete strain was observed to approach 0 $\mu\varepsilon$ in October.

The maximum compressive strain of the base was recorded at 175.3 $\mu\varepsilon$ in July, and the maximum tensile strain of the base was recorded at 126.9 $\mu\varepsilon$ in January. Refer to Table 7.8 for the statistics of concrete strain for the base slab from July 2009 to January 2010.

Figures 7.31 and 7.32 show the concrete strain variations of the base slab corresponding to different locations of measuring points from July to August 2009 and from December 2009 to January 2010, respectively.

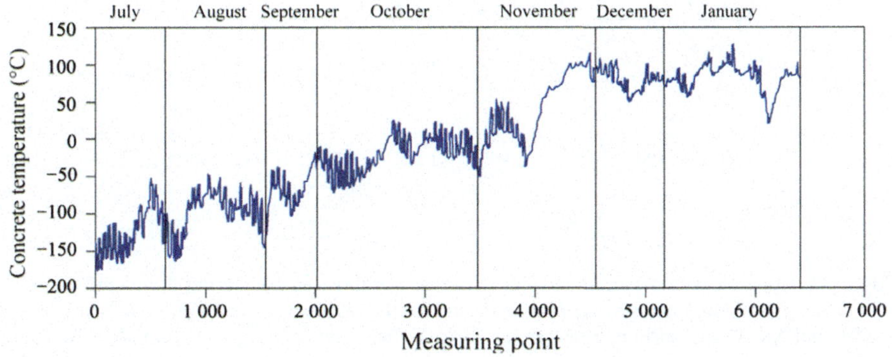

Fig. 7.30 Concrete strain of base from July 2009 to January 2010

Table 7.8 Statistics of concrete strain monitoring results

Month	Maximum compressive strain (με)	Maximum tensile strain (με)
7	175.3	–
8	162.2	–
9	133.3	–
10	70	25.2
11	49	116.1
12	–	108.7
1	–	126.9

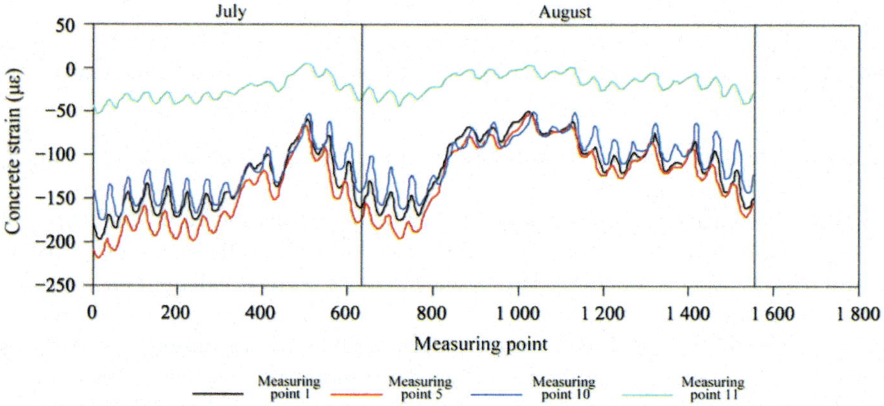

Fig. 7.31 Base concrete strain variations with different locations of measuring points in July and August

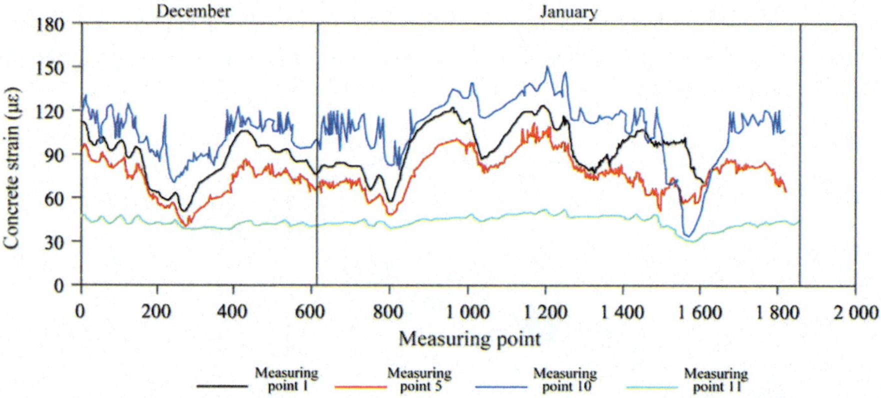

Fig. 7.32 Concrete Strain Variations with Different Locations of Measuring Points in December and January. *Note* No. 1 measuring point was located at the Guangzhou end of the (6 × 32) m turnout girder, and No. 11 measuring point was located over the fixed bearings of the middle continuous girder beneath the turnout, with a spacing of 32 m between measuring points

As shown in Figs. 7.31 and 7.32, the strain variations monitored at Measuring Points 1, 5, and 10 followed the rules that are basically consistent, resulting in minimal variance in strain amplitude variations. Measuring Point 11 displayed relatively stable strain values with lower measurements compared to other measuring points.

Calculations were performed using the stress formula for the base as follows:

$$F = E \times \varepsilon \times A$$

where F denotes the concrete stress of base slab;

E denotes the concrete elasticity modulus of base slab;

ε denotes the monitored concrete strain of base slab;

A denotes the sectional area of base slab concrete.

Through calculations, the internal compression of the concrete was 2.79 MN, corresponding to a compressive strain of the base concrete at 175.3 $\mu\varepsilon$, while the internal tension of the concrete was 2.02 MN, corresponding to a tensile strain of the concrete at 126.9 $\mu\varepsilon$.

3. Concrete strain monitoring for post-cast strip base slab

The concrete strain of the post-cast strip is shown in Fig. 7.33.

As shown in Fig. 7.33, the concrete strain of the post-cast strip fluctuated exhibiting a gradual increase trend. Based on a comparison with the measured concrete temperatures, the concrete strain of the post-cast strip concrete increased gradually through fluctuations alongside decreasing concrete temperatures, and the average concrete strain reached 150 $\mu\varepsilon$ during the period from November to January.

The maximum concrete tensile strain of the post-cast strip was recorded at 250 $\mu\varepsilon$ in November. Derived from the stress calculation formula, the internal concrete tension was 4.43 MN, corresponding to a concrete tensile strain of the post-cast strip at 250 $\mu\varepsilon$.

4. Reinforcement stress monitoring for base slab at girder end

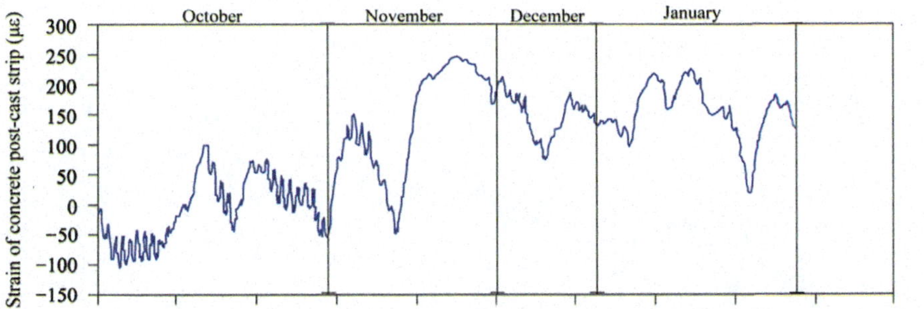

Fig. 7.33 Concrete strain of post-cast strip from October to January

The reinforcement stress of the base slab at the girder end is shown in Fig. 7.34.

From the measured values, the reinforcement stress of the base slab at the girder end gradually increased over time. Based on a comparison with the measured concrete temperatures, the reinforcement stress of the base slab at the girder end gradually increased alongside decreasing concrete temperatures, and the average reinforcement tensile stress reached 30 MPa during the period from November to January.

The maximum reinforcement tensile stress of the base slab at the girder end occurred in November, reaching 47 MPa. Derived from the reinforcement stress calculation formula, the reinforcement tensile stress of the base slab at the girder end was 0.856 MN, corresponding to a reinforcement tensile stress at 47 MPa.

5. Concrete strain monitoring for end thorn base slab

The concrete strain of the end thorn base slab is shown in Fig. 7.35.

From the measured values, the concrete strain of the end thorn base slab gradually increased over time. Compressive strain was noted from July to October, transitioning to tensile strain from October to January. Compared with the measured concrete

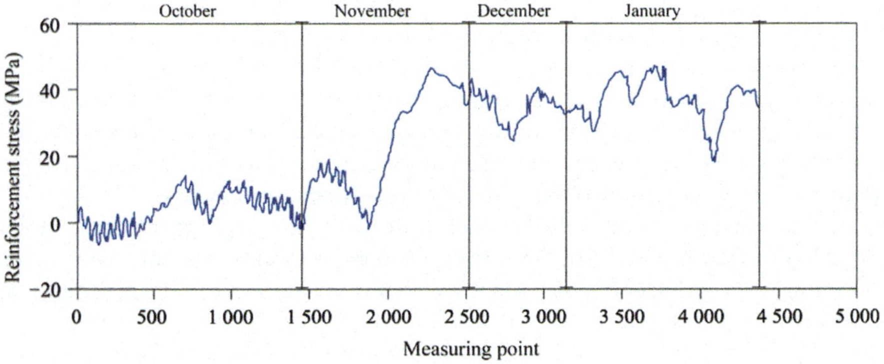

Fig. 7.34 Reinforcement stress of base slab at girder end from October to January

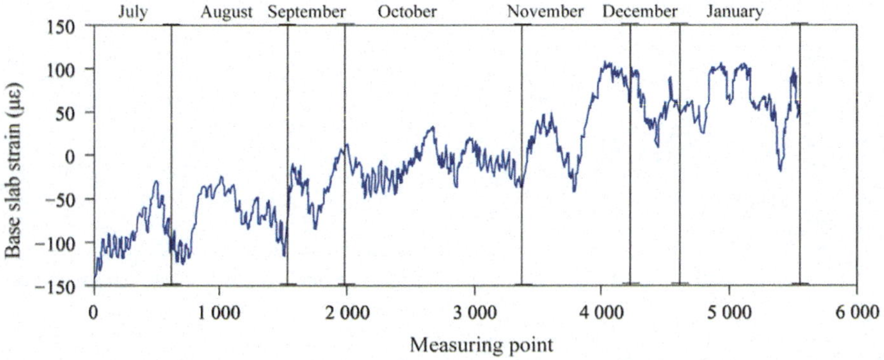

Fig. 7.35 Concrete strain of end thorn base slab from July to January

temperatures, the strain gradually increased, corresponding to decreasing concrete temperatures. The strain of the base slab fluctuated around 0 με in October.

The maximum concrete compressive strain of the end thorn base slab occurred in July, peaking at 140.2 με, and the maximum tensile strain occurred in January, reaching 109.1 με.

Derived from the stress calculation formula for base slabs, the internal compression of the concrete was 2.48 MN, corresponding to a compressive strain of the base slab concrete at 140.2 με, while the internal tension of the concrete was 1.93 MN, corresponding to a tensile strain of the base slab concrete at 109.1 με.

6. Concrete strain monitoring for end thorn friction plate

The concrete strain of the end thorn friction plate is shown in Fig. 7.36.

From the measured values, the concrete strain of the end thorn friction plate increased gradually over time. Compressive strain was noted from July to October, transitioning to tensile strain from October to January. Through a comparison with the measured concrete temperatures, the strain gradually increased, with decreasing concrete temperatures. The strain of the friction plate was observed to fluctuate around 0 με in October.

The maximum concrete strain of the end thorn friction plate occurred in July, peaking at 53.3 με, and the maximum tensile strain occurred in January, reaching 79 με.

Derived from the stress calculation formula for base slabs, the internal compression of the friction plate was 6.48 MN, corresponding to a compressive strain of the friction plate concrete at 53.3 με, while the internal tension of the friction plate was 0.96 MN, corresponding to a tensile strain of the friction plate concrete at 79 με. (The elastic modulus reduction coefficient of concrete was taken as 0.1.)

7. Concrete strain monitoring for transverse fixation device on retaining wall side

The concrete strain of the transverse fixation device on the retaining wall side is shown in Fig. 7.37.

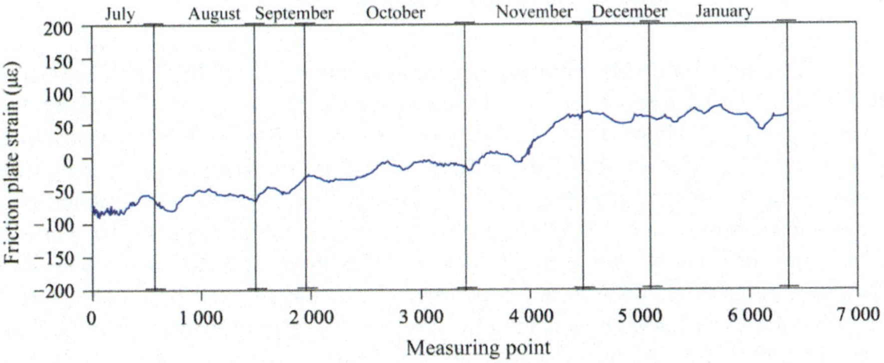

Fig. 7.36 Concrete strain of end thorn friction plate from July to January

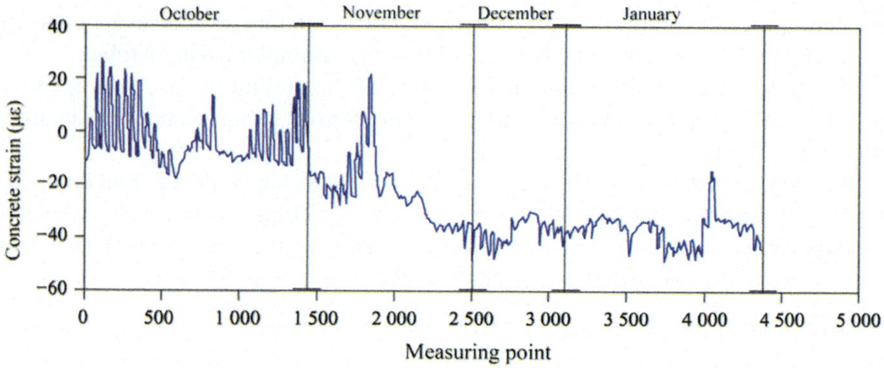

Fig. 7.37 Concrete strain of transverse fixation device on retaining wall side from October to January

From the measured values, the concrete strain of the transverse fixation device on the retaining wall side gradually decreased over time. Compared with the measured concrete temperatures, the strain gradually increased, corresponding to decreasing concrete temperatures. The maximum compressive strain occurred in January, peaking at 49.1 με.

Derived from the concrete stress calculation formula, the internal compression was 0.265 MN, corresponding to a concrete strain of the transverse fixation device on the retaining wall side at 49.1 με.

7.2 Operational Status Monitoring of CWR Turnouts on Haihang Super Major Bridge of Shanghai–Hangzhou HSR

7.2.1 Brief Introduction to Monitoring Site

Shanghai–Hangzhou HSR offers an application scenario of CRTS II slab ballastless tracks. The No. 42 turnout section at Jianqiao Block Post on the track to Hangzhou of Haihang Super Major Bridge was located on a $(32 + 3 \times 43 + 32)$ m continuous girder. Trains pass through the diverging route of the turnout and enter Hangzhou Station via Jianqiao Connecting Line. To anchor the longitudinally coupled base slabs within the turnout section, temporary end thorns are placed for a 780 m range from the fourth simply-supported girder span following the slab ballastless track for the turnout as a substitution for permanent end thorns. The base of the CRTS II slab ballastless track ends at the $(44 + 72 + 36)$ m continuous girder (DK149 + 341) across Shiji Avenue. The bridge arrangement is shown in Fig. 7.38. The close distance between the turnout section and the temporary end thorn zone on the

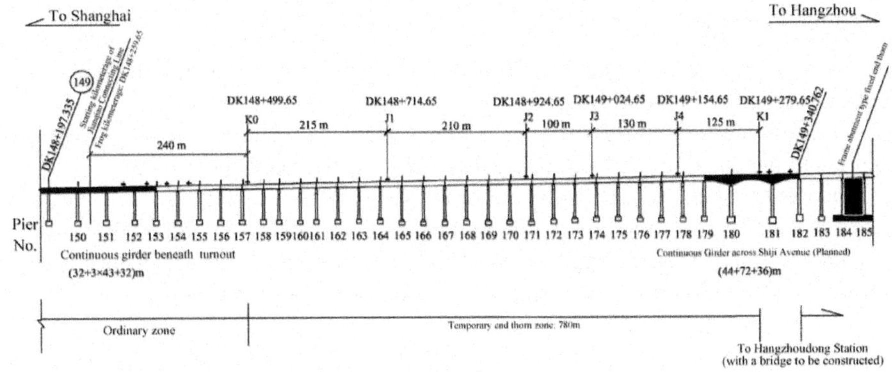

Fig. 7.38 Bridge arrangement

bridge results in more intricate local stress and deformation. To verify the reliability of the solution of temporary end thorns in replace of permanent ones and ensure the safety and riding comfort for trains passing through on the diverging route of the No. 42 turnout on the bridge during the integrated commissioning & test period of Shanghai–Hangzhou HSR, stress and deformation monitoring were conducted for the slab ballastless track within the No. 42 turnout section and the base slabs of the ballastless track within the temporary end thorn zone of Haihang Super Major Bridge on Shanghai–Hangzhou HSR. This move holds great significance in improving the design studies and application level of CRTS II slab ballastless tracks for high-speed railways in China.

7.2.2 Monitoring Items

The monitoring tasks consist of two parts: strain and displacement of the slab ballastless track within the No. 42 turnout section on the bridge at Jianqiao Block Post, and strain and displacement of the base slabs within the temporary end thorn zone.

1. No. 42 turnout section on the bridge

The strain and displacement monitoring items for the slab ballastless track within the No. 42 turnout section on the bridge at Jianqiao Block Post are mainly as follows:

Observe longitudinal relative displacement between the base slab and the bridge to prevent the shear notch from being damaged due to excessive stress;

Observe vertical relative displacement between the base and the bridge to prevent arching-induced instability of the base slab;

Observe strain of the base slab to prevent concrete cracking due to excessive strain on the base slab;

Observe longitudinal displacement at the pier top to prevent pier instability and failure caused by longitudinal displacement due to excessive stress.

2. Temporary end thorn zone

The strain and displacement monitoring items for the base slab within the temporary end thorn zone at Jianqiao Block Post are mainly as follows:

Observe longitudinal relative displacement between the base slab and the bridge to prevent the shear notch from being damaged due to excessive stress;

Observe vertical relative displacement between the base and the bridge to prevent the base slab from arching up, which will affect the regularity of the track;

Observe longitudinal displacement at the pier top to prevent pier instability and failure caused by longitudinal displacement due to excessive stress.

7.2.3 Arrangement of Measuring Points

1. No. 42 turnout section on the bridge

(1) Longitudinal relative displacement between base slab and bridge

(1) A total of 10 measuring points are placed for monitoring longitudinal relative displacements between the base slab and the bridge, respectively at the left girder end, shear notch, right girder end of the turnout girder, as well as the shear notch in the first simply-supported girder span adjacent to the turnout girder, and the base slab of the shear notch in the second simply-supported girder span adjacent to the turnout girder.

(2) Arrangement of measuring points: five measuring points, numbered from 11 to 15 for the No. 42 turnout on the down track; five measuring points, numbered from 16 to 20 for the No. 42 turnout on the up track.

(3) Measuring instruments: displacement meters (automatic acquisition and wireless transmission), as shown in Figs. 7.39 and 7.40.

(2) Vertical relative displacement between base and bridge

Fig. 7.39 Displacement meter

Fig. 7.40 Data acquisition unit

(1) A total of 8 measuring points were placed for measuring vertical relative displacements between the base and the bridge, respectively located over the movable bearing in the left 32.6 m span, at the center of the middle 43.6 m span, at the center of the right 43.6 m span, and at the center of the right 32.6 m span of the (32.6 + 43.6 × 3 + 32.6) m continuous turnout girder.
(2) Arrangement of measuring points: four measuring points, numbered from 21 to 24 for the No. 42 turnout on the down track; four measuring points, numbered from 31 to 34 for the No. 42 turnout on the up track.
(3) Measuring instruments: eddy current displacement sensor, as shown in Figs. 7.41 and 7.42.

(3) Base slab strain

(1) A total of 12 measuring points were placed for measuring strain of the base slab, respectively located at the left girder end of the turnout girder, as well as over the movable bearings 149, 151, 152, 153, and 154.

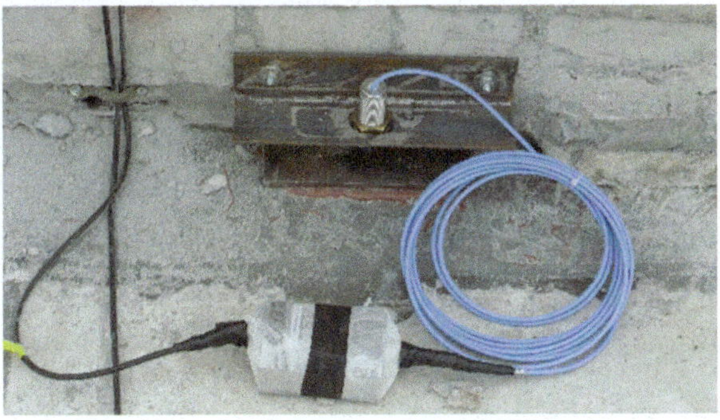

Fig. 7.41 Eddy current displacement sensor

Fig. 7.42 Data acquisition unit and remote wireless transmission controller

(2) Arrangement of measuring points: six measuring points, numbered from 41 to 46 for the No. 42 turnout on the down track; six measuring points, numbered from 51 to 56 for the No. 42 turnout on the up track.
(3) Measuring instrument: vibrating wire strain gauge (automatic acquisition and wireless transmission), as shown in Fig. 7.43.

(4) Longitudinal displacement at pier top

(1) Three measuring points were arranged for longitudinal displacements at the pier top, respectively located at the fixed bearing under the turnout girder, the fixed bearing under the first span of the simply-supported girder adjacent to the turnout girder, and the fixed bearing under the second span of the simply-supported girder adjacent to the turnout girder.
(2) Arrangement of measuring points: numbered from 61 to 63 to measure longitudinal absolute displacements along the line at the top of No. 150, 153, and 154 piers.

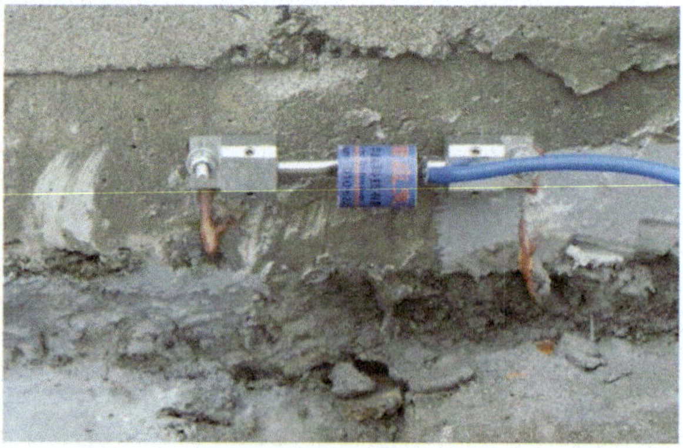

Fig. 7.43 Vibrating Wire Strain Gauge

Fig. 7.44 Fixed inclinometer and installation

(3) Measuring instrument: fixed clinometer (automatic acquisition and wireless transmission), as shown in Fig. 7.44.

2. Temporary end thorn zone

(1) Longitudinal relative displacement between base slab and bridge

  (1) Sixteen measuring points were arranged to measure longitudinal relative displacements between the base slab at the shear notch and the bridge, located at the shear notches of No. 171, 172, 174, 175, 176, 177, 178, and 180 piers, respectively.
  (2) Arrangement of measuring points: eight measuring points on the simply-supported girder of the down track, numbered from 1 to 8; eight measuring points on the simply-supported girder of the up track, numbered from 9 to 16, as shown in Fig. 7.45.

(3) Measuring instrument: displacement meter (automatic acquisition and wireless transmission).

(2) Vertical relative displacement between base and bridge

(1) Eight measuring points were placed for vertical relative displacements between the base and the bridge, located at the center of the No. 172 ~ 173, 174 ~ 175, 176 ~ 177, and 178 ~ 179 girder-pier structures, respectively.
(2) Arrangement of measuring points: four measuring points numbered from 23 to 26 at the center of the No. 172~173, 174~175, 176~177, and 178~179 girder-pier structures on the down track; four measuring points numbered from 27 to 30 at the center of the No. 172 ~ 173, 174 ~ 175, 176 ~ 177, and 178 ~ 179 girder-pier structures on the down track.
(3) Measuring instrument: eddy current displacement sensor (automatic acquisition, remote control, and wireless transmission of stored data).

(3) Longitudinal displacement at pier top

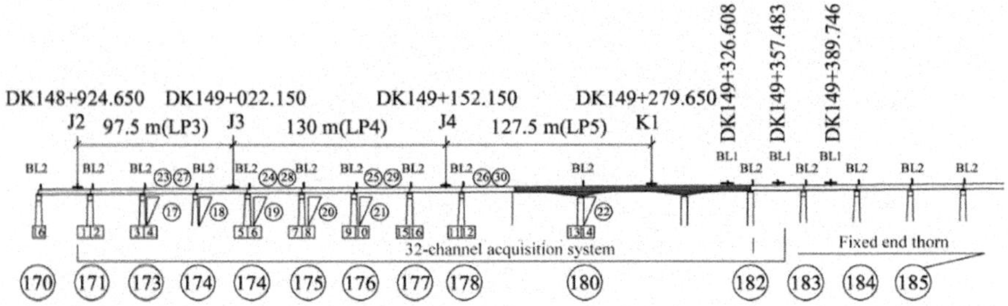

Fig. 7.45 Arrangement of measuring points within temporary end thorn zone. *Note* ① A total of 16 measuring points numbered from 1 to 8 and 9 to 16 (corresponding mileage positions of No. 172 to 182 piers) are placed for relative displacements between the base slab at the shear notch and the bridge. ② Six measuring points numbered from 17 to 22 with inclination sensors are placed on the top platform of the pier. ③ Four measuring points numbered from 23 to 26 are placed at the center of the No. 173 ~ 174, 175 ~ 176, 177 ~ 178, and 178 ~ 179 girder-pier structures on the down track, as shown in the above figure. Four measuring points numbered from 27 to 30 are placed at the center of the No. 173 ~ 174, 175 ~ 176, 177 ~ 178, and 178 ~ 179 girder-pier structures on the up track, as shown in the above figure

(1) Six measuring points were arranged to measure longitudinal displacements at the pier top, located at the fixed bearings of the simply-supported girders and continuous girders, respectively.
(2) Arrangement of measuring points: numbered from 17 to 22 to measure longitudinal absolute displacements along the line at the central pier top of the No. 172, 173, 174, 175, 176, and 180 girder-pier structures.
(3) Measuring instrument: fixed clinometer (automatic acquisition and wireless transmission).

7.2.4 Analysis of Monitoring Data

1. Monitoring results and data analysis of Jianqiao No. 42 turnout section
(1) Temperature monitoring results

According to the temperature monitoring data, the temperature statistics in September, October, and November are shown in Table 7.9.

Table 7.9 Statistics of temperature monitoring results

Month	Average air temperature (°C)	Maximum air temperature (°C)	Minimum air temperature (°C)
Sep	25.8	39.5	12.1
Oct	18.3	27.4	9.9
Nov	15.3	24.2	7.7

The monthly average air temperature in Hangzhou is 25.8°C and the maximum is 39.5°C in September; the minimum is 7.7°C in November.

(2) Longitudinal relative displacement between the base slab and bridge within No. 42 turnout section on the bridge

Among the measuring points for relative displacements between the base slabs and the bridge within the No. 42 turnout on the down track, No. 11, 13, 16, and 18 are arranged to monitor relative displacements between the movable bearing base slabs and the bridge, while No. 12, 14, 15, 17, 19, and 20 are arranged to monitor relative displacements between the fixed bearing base slabs and the bridge.

The maximum and minimum relative displacements between the base slab and the bridge at the movable bearing from September to November are shown in Fig. 7.46.

As shown in the figure, the variation amplitude of the relative displacement between the base slab and the bridge gradually narrows with decreasing temperature differences of the base slab.

See Table 7.10 for the maximum and minimum relative displacements between the base slab and the bridge in daily monitoring from September to November.

As shown in the table, the longitudinal relative displacements between the base slab and the bridge at the movable bearings are within 15 mm, while there is almost no relative displacement between the base slab and the bridge at the fixed bearings. These monitoring data also verify the stability of the anchorage with a shear notch for the slab ballastless track within the No. 42 turnout section.

The time-varying curves of the temperatures for 24 h and the relative displacements between the base slab and the bridge at the No. 13 measuring point are shown in Figs. 7.47 and 7.48.

As shown in the figure, the longitudinal relative displacements between the base slab and the bridge vary with temperature. Specifically, as the temperature rises from 9:00 to 15:00, the elongation displacements occur at the movable bearing ends of the bridge, and the relative displacements between the base slab and the bridge

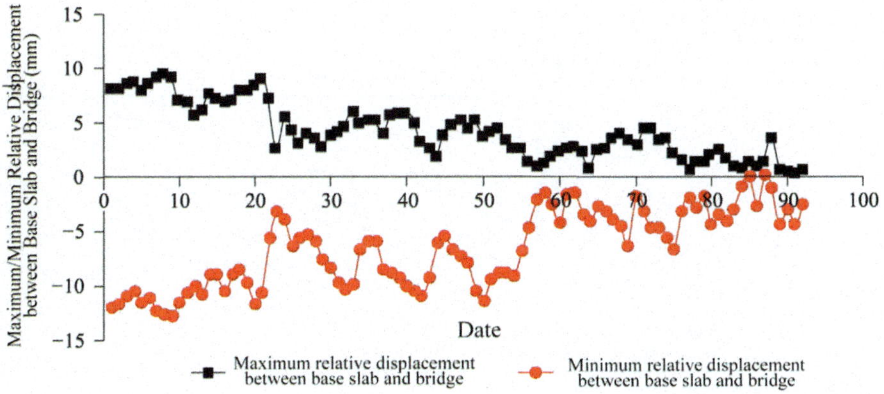

Fig. 7.46 Variations in maximum/minimum relative displacement between base slab and bridge

Table 7.10 Longitudinal relative displacement between base slab and bridge within turnout section

Monitoring Period	Extreme Value	No. 11 and 16	No. 13 and 18	No. 12 and 17	No. 14 and 19	No. 15 and 20
Sept	Minimum (mm)	−12.08	−14.76	−0.063	−0.062	−0.065
	Maximum (mm)	8.58	9.62	0.059	0.061	0.065
Oct	Minimum (mm)	−10.76	−13.58	−0.055	−0.057	−0.057
	Maximum (mm)	6.03	7.18	0.051	0.054	0.053
Nov	Minimum (mm)	−6.26	−6.63	−0.032	−0.032	−0.033
	Maximum (mm)	4.78	5.53	0.042	0.044	0.047

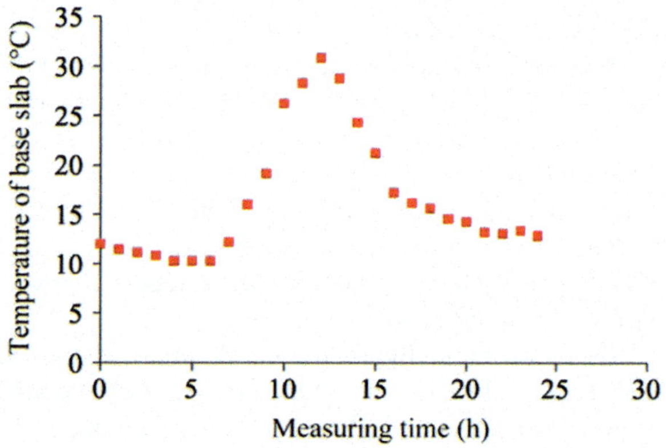

Fig. 7.47 Temperature variations of base slab with time

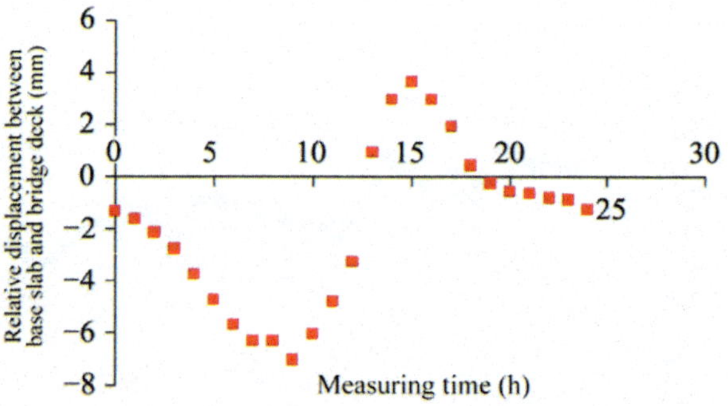

Fig. 7.48 Relative displacement variations between base slab and bridge with time

gradually increase. As the temperature falls from 15:00 to 20:00, the expansion and contraction displacements at the movable bearing ends of the bridge decrease, and the relative displacements between the base slabs and the bridge gradually decrease. As the temperature further falls from 20:00 to 9:00 the next day, the contraction displacements occur at the movable bearing ends of the bridge, and the relative displacements between the base slab and the bridge gradually increase.

(3) Vertical relative displacement between the base slab and the bridge within No. 42 turnout section on the bridge

Among the measuring points for vertical relative displacements between the base slabs and the bridge, No. 21 to 24 measuring points and No. 31 to 34 measuring points are arranged respectively to monitor vertical relative displacements between the base slabs and the bridge preceding (over the movable bearing of 32.6 m girder), within (at the center of 43.6 m girder), and following (at the centers of 43.6 m girder and 32.6 m girders) the turnout.

The maximum and minimum vertical displacements of the base slab in daily monitoring from September to November are shown in Fig. 7.49.

As shown in the figure, the variation amplitude of the vertical relative displacement between the base slab and the bridge at the girder center gradually narrows with decreasing temperature differences of the base slab.

See Table 7.11 for the maximum and minimum vertical displacements of the base slabs in daily monitoring from September to November.

The maximum and minimum vertical relative displacements between the base slabs and the bridge deck are 0.17 mm and -0.12 mm, respectively. According to the monitoring data, the vertical relative displacements between the base slabs and the bridge are insignificant, which verifies the stability of the base slabs. Therefore, the vertical displacement due to temperature change is negligible for the base slabs in the turnout section.

The time-varying curves of the temperatures at the base slab for 24 h and the vertical displacements of the base slab at the No. 23 measuring point are shown in Figs. 7.50 and 7.51.

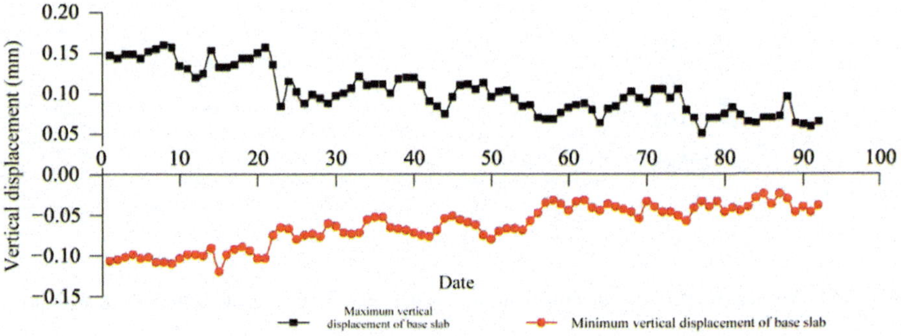

Fig. 7.49 Variations in maximum/minimum vertical displacement of base slab

Table 7.11 Vertical displacement statistics of base slabs

Monitoring period	Extreme Value	No. 21 and 31	No. 22 and 32	No. 23 and 33	No. 24 and 34
Sep	Minimum (mm)	−0.10	−0.12	−0.08	−0.07
	Maximum (mm)	0.15	0.14	0.17	0.16
Oct	Minimum (mm)	−0.03	−0.04	−0.05	−0.04
	Maximum (mm)	0.11	0.14	0.13	0.15
Nov	Minimum (mm)	−0.05	−0.06	−0.04	−0.04
	Maximum (mm)	0.12	0.10	0.12	0.11

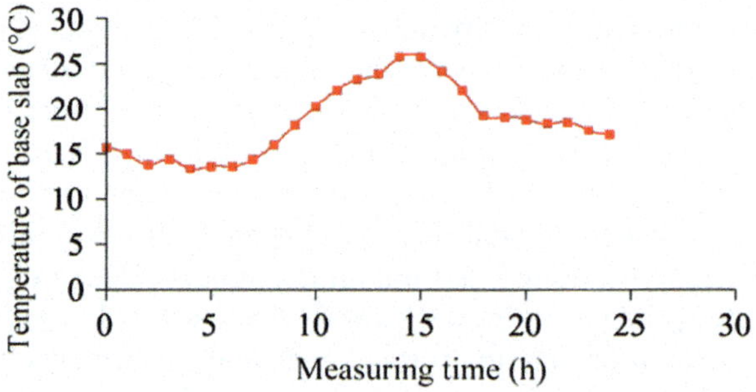

Fig. 7.50 Temperature variations of base slab with time

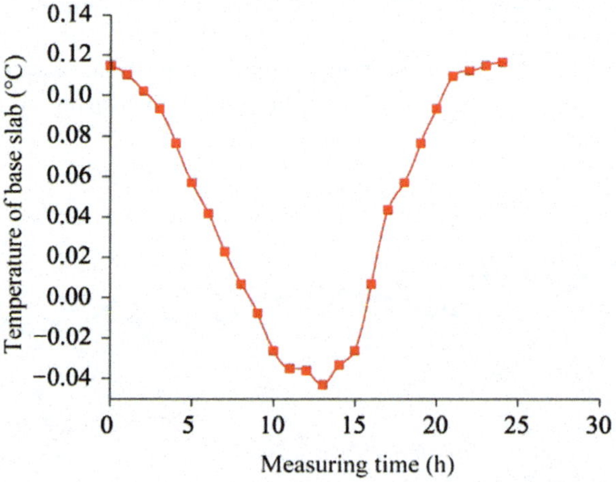

Fig. 7.51 Vertical relative displacement variations between base slab and bridge deck with time

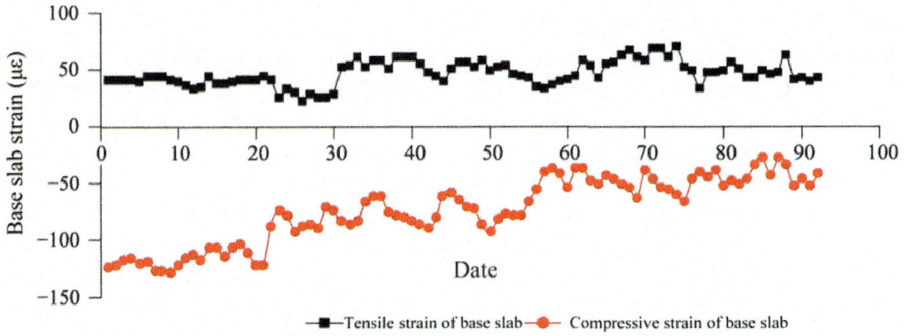

Fig. 7.52 Variations in maximum/minimum strain of base slab

As shown in the figure, the vertical displacement of the base slab varies with the changing temperature (24 h). Specifically, as the temperature rises, the vertical displacement of the base slab gradually decreases. As the temperature falls, the vertical displacement of the base slab gradually increases.

(4) Strain of base slab within No. 42 turnout section on the bridge

The strain measuring points of the base slab were respectively located at the left girder end of the turnout girder, and over the No. 149, 151, 152, 153, and 154 movable bearings.

The maximum compressive strain and minimum tensile strain of the base slab in daily monitoring from September to November are shown in Fig. 7.52.

As shown in the figure, the variation amplitude of the maximum tensile and compressive strains of the base slab gradually narrows with decreasing temperature differences of the base slab.

The maximum tensile strains and compressive strains of the monitored base slabs from September to November are shown in Table 7.12.

The maximum compressive strain on the base slab surface is 135 $\mu\varepsilon$ and the maximum tensile strain is 64 $\mu\varepsilon$. The monitoring results indicate the base slab is free of cracking at the monitored position, thus verifying the safety at the critical part of the base slab.

The time-varying curves of the temperatures for 24 h and the strains at the No. 43 measuring point are shown in Figs. 7.53 and 7.54.

As shown in Figs. 7.53 and 7.54, the strain of the base slab varies with temperature. Specifically, as the temperature of the base slab increases, the strain of the base slab gradually increases. As the temperature of the base slab decreases, the strain of the base slab gradually decreases.

(5) Longitudinal displacement at the pier top within No. 42 turnout section on the bridge

The pier inclinometers within the turnout section were located at No. 150, No. 153, and No. 154 piers, respectively. The specific monitoring data is as follows.

Table 7.12 Strain statistics of base slabs

Monitoring Period	Strain	No. 41 and 51	No. 42 and 52	No. 43 and 53	No. 44 and 54	No. 45 and 55	No. 46 and 46
Sep	Maximum compressive strain (με)	132.00	130.00	135.00	134.00	125.00	130.00
	Maximum tensile strain (με)	32.20	36.80	34.50	41.40	32.20	29.90
Oct	Maximum compressive strain (με)	70.00	72.00	68.00	69.00	73.00	71.00
	Maximum tensile strain (με)	55.25	63.75	63.75	49.30	58.65	59.50
Nov	Maximum compressive strain (με)	35.70	36.72	34.68	35.19	37.23	36.21
	Maximum tensile strain (με)	46.07	53.16	53.16	41.11	48.91	49.62

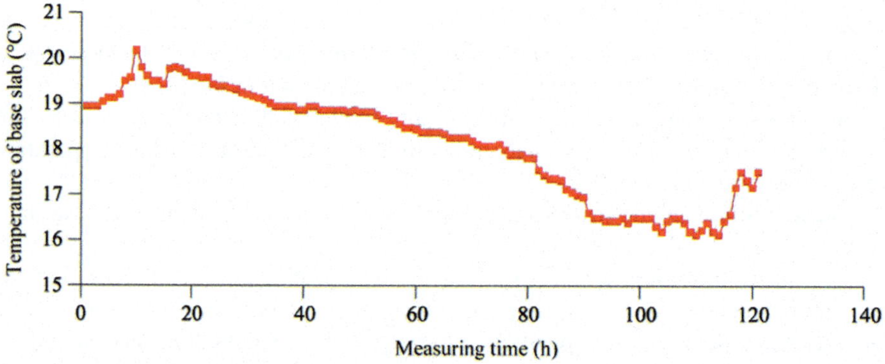

Fig. 7.53 Temperature variations of base slab with time

The longitudinal displacement of the pier (height: 17 m) measured with the inclinometer on the No. 150 pier from September to November is shown in Fig. 7.55.

As shown in the figure, the longitudinal displacement of the pier/abutment is associated with temperature, indicating the longitudinal displacement varies with changing temperature. Specifically, when the temperature changes at a rising rate, the longitudinal displacement increases. When the temperature changes at a lowering rate, the longitudinal displacement decreases. The longitudinal displacement of the

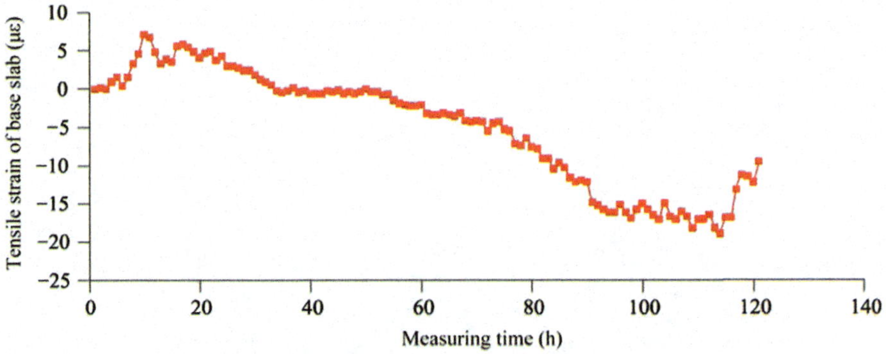

Fig. 7.54 Strain variations of base slab with time

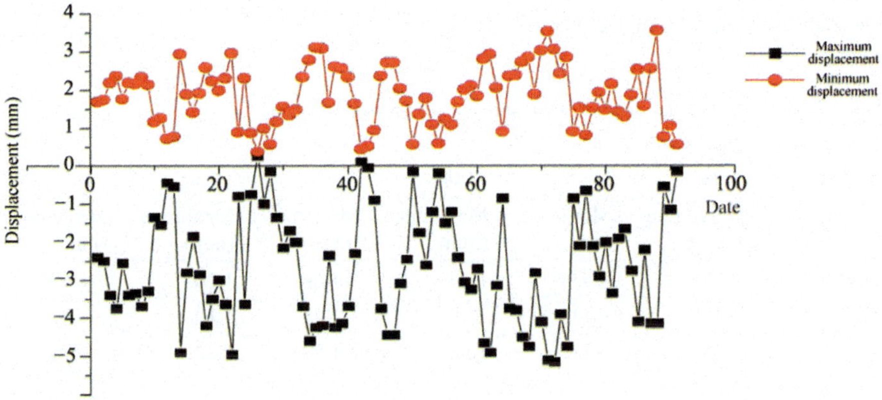

Fig. 7.55 Longitudinal displacement of No. 150 Pier

No. 150 pier varies from -5 mm to 4 mm, falling within the permissible range of design.

The longitudinal displacement of the pier (height: 18 m) measured with the inclinometer on the No. 154 pier from September to November is shown in Fig. 7.56.

As shown in the figure, the longitudinal displacement of the pier/abutment is associated with temperature, indicating the longitudinal displacement varies with changing temperature. Specifically, when the temperature changes at a rising rate, the longitudinal displacement increases. When the temperature changes at a lowering rate, the longitudinal displacement decreases. The longitudinal displacement of the No. 154 pier varies from -2 mm to 2 mm, falling within the permissible range of design.

See Table 7.13 for the maximum and minimum longitudinal displacements at the pier top monitored with inclinometers from September to November.

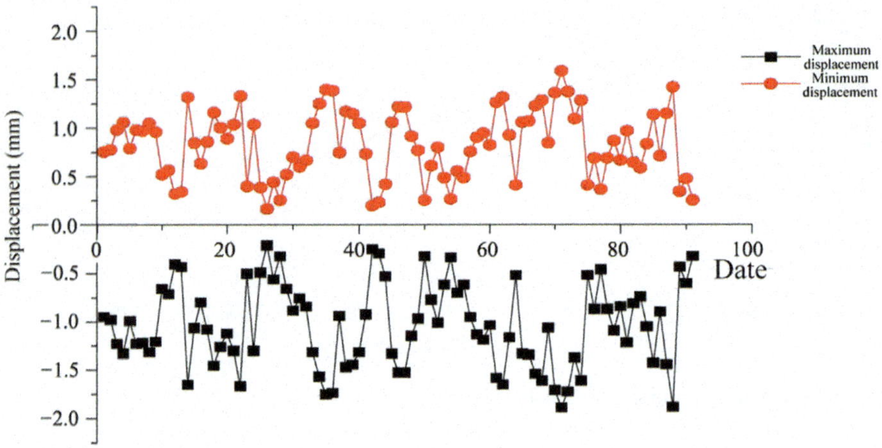

Fig. 7.56 Longitudinal displacement of No. 154 Pier

Table 7.13 Longitudinal displacement statistics of piers/abutments

Monitoring period	Pier No					
	150#		154#		153#	
	Maximum (mm)	Minimum (mm)	Maximum (mm)	Minimum (mm)	Maximum (mm)	Minimum (mm)
Sep	2.97	−4.95	1.33	−1.66	1.20	−1.45
Oct	3.125	−4.6	1.39	−1.78	1.10	−1.52
Nov	3.55	−5.0	1.58	1.88	1.32	−1.74

The longitudinal displacements at the top of the fixed piers under the continuous girder and simply-supported girder vary from -5 mm to 4 mm and from −2 mm to 2 mm, both falling within the permissible ranges of design. The above results verify the design rationality and safety of the bridge piers/abutments.

2. Monitoring results and data analysis of temporary end thorn zone
(2) Longitudinal relative displacement between the base slab and the bridge within temporary end thorn zone at Jianqiao Block Post

The measuring points for relative displacements between the base slabs and the bridge were located at the shear notch positions of the No. 171, 172, 174, 175, 176, 177, and 178 piers under the simply-supported girder and the shear notch position of the No. 180 pier under the continuous girder respectively. Concrete is not poured at the shear notch positions of the base slabs of the simply-supported girder and continuous girder.

The maximum and minimum relative displacements between the base slab and the bridge at the movable bearing from September to November are shown in Fig. 7.57.

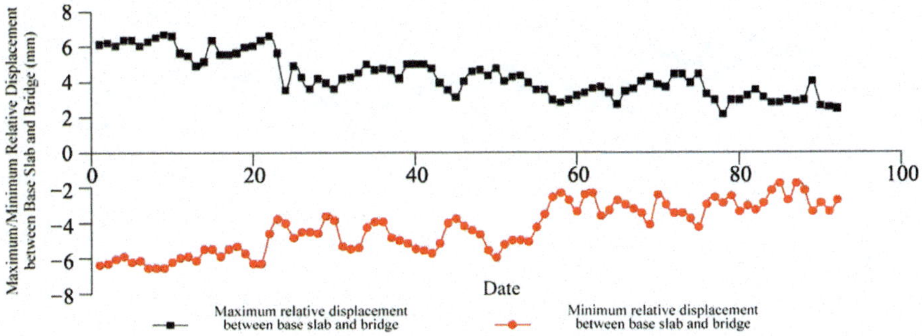

Fig. 7.57 Variations in maximum/minimum relative displacement between base slab and bridge

As shown in the figure, the variation amplitude of the relative displacement between the base slab and the bridge gradually narrows with decreasing temperature differences of the base slab.

See Table 7.14 for the maximum and minimum vertical displacements of the base slabs in daily monitoring from September to November.

As shown in the table, the longitudinal relative displacements between the base slabs and the bridge are in a range of -6.5 mm to 4.5 mm at the shear notch positions without concreting on the simply-supported girder and fall within a range of −7.2 mm to 6.8 mm at the shear notch positions without concreting on the continuous girder. Despite relative displacements of approximately 1 cm between the base slab and

Table 7.14 Monitoring values of relative displacement between base slabs and bridge within temporary end thorn zone

Monitoring period	Extreme value	At shear notch of continuous girder	At shear notch of Simply-supported girder						
		No. 13 and 14	No. 1 and 2	No. 3 and 4	No. 5 and 6	No. 7 and 8	No. 9 and 10	No. 11 and 12	No. 15 and 16
Sep	Minimum (mm)	−7.20	−3.90	−4.20	−4.50	−5.00	−6.50	−5.40	−6.40
	Maximum (mm)	6.80	3.40	3.50	4.10	4.50	4.30	4.38	4.20
Oct	Minimum (mm)	−6.20	−3.50	−3.90	−4.27	−4.13	−6.15	−4.65	−5.94
	Maximum (mm)	5.30	3.04	3.68	4.30	4.30	3.90	4.06	3.80
Nov	Minimum (mm)	−5.12	−3.16	−3.76	−3.84	−4.56	−5.20	−3.52	−5.12
	Maximum (mm)	4.79	2.37	2.87	3.35	4.13	3.60	3.17	2.96

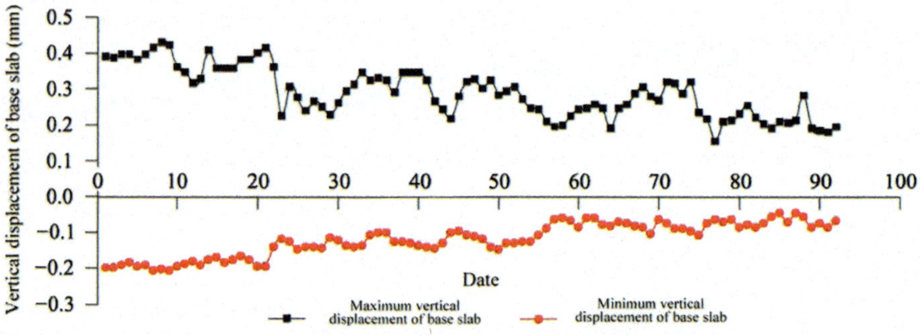

Fig. 7.58 Variations in maximum/minimum vertical displacement of base slab

girder surface at the shear notch positions without concreting within the temporary end thorn zone, the friction force at the temporary end thorn zone can still ensure stability within this zone. Hence, the safety at the critical part of the temporary end thorn zone is verified.

(2) Vertical relative displacement between the base slab and the bridge within temporary end thorn zone at Jianqiao Block Post

The measuring points for vertical relative displacements between the base slabs and the bridge were located at the center of the No. 172~173, No. 174~ 175, No. 176~ 177, and No. 178~179 girders on up and down tracks, respectively.

The maximum and minimum vertical displacements of the base slab in daily monitoring from September to November are shown in Fig. 7.58.

As shown in the figure, the variation amplitude of the vertical relative displacement between the base slab and the bridge at the girder center gradually narrows with decreasing temperature differences of the base slab.

See Table 7.15 for the maximum and minimum vertical displacements of the base slabs in daily monitoring from September to November.

As shown in Table 7.15, the maximum and minimum vertical displacements of the base slabs are 0.412 mm and -0.223 mm, respectively. The base slabs on the simply-supported girder are unit slabs under the action of temperature forces and are exposed to the influence of dead weight and longitudinal reinforcement. When their upper surface temperature is higher than the lower surface temperature, they exhibit a concave downward shape in the middle and an upwarp at both ends. When their upper surface temperature is lower than the lower surface temperature, they revert to their initial state, displaying a concave upward shape in the middle and contraction at both ends. Based on the monitoring results, the vertical displacements between the base slabs and the bridge deck within the temporary end thorn zone do not exceed 0.5 mm, thus verifying the stability of the base slab within the temporary end thorn zone.

(3) Longitudinal displacement at the pier top within temporary end thorn zone at Jianqiao Block Post

Table 7.15 Vertical displacement statistics of base slabs

Monitoring period	Extreme Value	No. 23 and 27	No. 24 and 28	No. 25 and 29	No. 26 and 30
Sep	Minimum (mm)	−0.112	−0.093	0.036	−0.223
	Maximum (mm)	0.392	0.294	0.412	0.331
Oct	Minimum (mm)	−0.108	−0.083	0.042	−0.203
	Maximum (mm)	0.368	0.273	0.345	0.305
Nov	Minimum (mm)	−0.090	−0.074	0.029	−0.178
	Maximum (mm)	0.306	0.229	0.321	0.258

The pier inclinometers within the temporary end thorn zone were located on the No. 172, 173, 174, 175, 176, and 180 piers, respectively.

The longitudinal displacement of the pier (height: 20.5 m) measured with the inclinometer on the No. 173 pier from September to November is shown in Fig. 7.59.

As shown in Fig. 7.59, the longitudinal displacement of the pier/abutment varies with changing temperature. Specifically, when the temperature changes at a rising rate, the longitudinal displacement increases. When the temperature changes at a lowering rate, the longitudinal displacement decreases. The longitudinal displacement of the No. 173 pier varies between 2.11 mm and -3.50 mm.

See Table 7.16 for the maximum and minimum longitudinal displacements at the pier top monitored with inclinometers from September to November.

The longitudinal displacements at the top of the fixed piers under the continuous girder and simple-supported girder vary from -11.63 mm to 8.3 mm and from −4.8

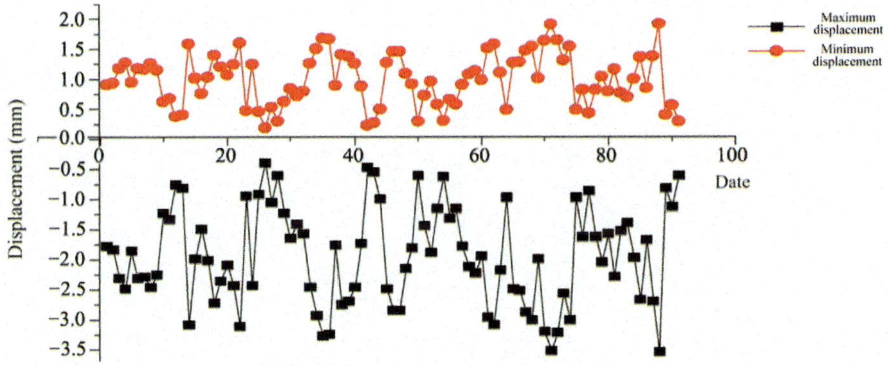

Fig. 7.59 Longitudinal displacement of No. 173 Pier

Table 7.16 Longitudinal displacement statistics of Piers/abutments

Monitoring Period	Pier No											
	172#		173#		174#		175#		176#		180#	
	Maximum /mm	Minimum /mm	Maximum /mm	Minimum /mm	Maximum /mm	Minimum /mm	Maximum /mm	Minimum /mm	Maximum /mm	Minimum /mm	Maximum /mm	Minimum /mm
Sep	1.45	−2.6	1.6	−3.11	2.86	−3.54	2.91	−4.80	4.84	−5.22	8.3	−11.5
Oct	1.35	−2.55	1.68	−3.26	2.3	−3	3.08	−3.94	4.24	−4.41	8.11	−11.1
Nov	1.50	−2.53	2.11	−3.5	2.998	−2.832	3.50	−3.424	4.472	−4.312	7.84	−11.63

mm to 3.5 mm respectively, both falling within the permissible ranges of design, verifying the design rationality and safety of the bridge piers/abutments.

7.2.5 Comparative Analysis Between Monitoring Results and Theoretical Results

1. No. 42 turnout section on the bridge
(1) Longitudinal relative displacement between base slab and bridge within turnout section

Based on the monitoring results of the relative displacement between the base slab and the bridge at the movable bearing within the No. 42 turnout (No. 11, 12, 13, 14, and 15 measuring points) on the down track, as well as the relative displacement between the base slab and the bridge at the movable bearing within the No. 42 turnout (No. 16, 17, 18, 19, and 20 measuring points) on the down track, the longitudinal relative displacement between the base slab and the bridge varies with temperature, The comparison between the calculated and measured values is shown in Table 7.17.

The No. 11/16 and No. 13/18 measuring points indicate the relative displacements between the base and the bridge at the movable end of the turnout girder. The comparison between the calculated and measured values indicates they are approximate to each other and the latter is less than the former. The No. 12/17, 14/19, and 15/20 measuring points indicate the relative displacements between the bases and the bridge at the fixed end of the turnout girder. Given that the bases are fixed onto the bridge through shear notches, thus maintaining mutual deformation, the measured data demonstrate no relative displacement between the bases and the bridge, thus verifying the stability of the base slabs.

(2) Vertical relative displacement between base slab and bridge

The No. 21~24 and No. 31~ 34 measuring points were placed to respectively monitor vertical relative displacements between the base slabs and the bridge preceding (over the movable bearing of 32.6 m girder), within (at the center of 43.6 m girder), and

Table 7.17 Statistics of relative displacement between base and bridge deck within turnout section

Measuring point	Calculated value (mm)		Measured value (mm)	
	Maximum	Minimum	Maximum	Minimum
11#/16#	13.58	−22.08	8.88	−13.57
12#/17#	0.65	−0.76	0.59	−0.63
13#/18#	21.8	−18.81	14.62	−24.76
14#/19#	0.83	−0.74	0.61	−0.62
15#/20#	0.71	−0.74	0.65	−0.65

following (at the center of 43.6 m girder and 32.6 m girder) of the turnout. The measured data at respective measuring points are summarized in Table 7.18.

The vertical displacements of the base slabs vary with temperature. Specifically, when the temperature rises, the vertical displacements of the base slabs gradually decrease. The maximum and minimum vertical displacements are 0.17 mm and -0.12 mm respectively. The minimum vertical displacement occurring at the base slab is caused by compressing the sliding layer. Vertical displacements at the base slabs varying in association with changing temperatures are restricted by the shear notch and lateral stop block on the girder surface. Consequently, the vertical displacements that occur are minimal.

(3) Base slab strain

The No. 41/51, 42/52, 43/53, 44/54, 45/55, and 46/56 measuring points were placed to respectively monitor turnout strains at the left girder end of the turnout girder, and over the No. 149, No. 151, No. 152, No. 153, and No. 154 movable bearings. See Table 7.19 for the comparison between the monitoring data and simulation results through calculations.

The above comparison between the calculated and measured values indicates they are approximate to each other and the latter is less than the former. The absence of cracking at the critical parts of the base slabs verifies the safety of the base slabs.

(4) Longitudinal displacement of pier/abutment within turnout section

See Table 7.20 for the longitudinal displacements at the piers/abutments within the turnout section.

Table 7.18 Measured maximum vertical displacements of base slabs

Measuring point No	21#/31#	22#/32#	23#/33#	24#/34#
Maximum vertical displacement (mm)	0.15	0.14	0.17	0.16
Minimum vertical displacement (mm)	−0.10	−0.12	−0.08	−0.07

Table 7.19 Strain statistics of base within turnout section

Measuring point	Calculated value (με)		Measured value (με)	
	Maximum tensile strain	Maximum compressive strain	Maximum tensile strain	Maximum compressive strain
41#/51#	98	−120	70	−132
42#/52#	127	−161	75	−130
43#/53#	35	−71	69	−134
44#/54#	74	−101	58	−134
45#/55#	50	−77	69	−125
46#/56#	48	−76	71	−130

Table 7.20 Statistics of longitudinal displacement at pier top within turnout section

Abutment/Pier No	Calculated value (mm)		Measured value (mm)	
	Maximum	Minimum	Maximum	Minimum
150#	7.8	−9.9	4.0	−5.0
153#	1.5	−2.2	1.32	−1.74
154#	1.6	−1.9	1.58	−1.88

The comparison between the calculated and measured values shown in Table 7.20 indicates they are approximate to each other and the latter is less than the former, thus verifying the design rationality and safety of the bridge piers/abutments.

2. Temporary end thorn zone
(1) Longitudinal relative displacement between base slab and bridge

Based on the monitoring and calculations of relative displacements between the base slabs and the bridge at the positions of shear notches without concreting on No 171, 172, 174, 175, 176, 177, and 178 piers along the simply-supported girder, and the shear notch position on No. 180 pier along the continuous girder, the monitoring data and numerical simulation results are shown in Table 7.21.

The comparison between the calculated and measured relative displacements between the base slabs and the bridge at the measuring points (No. 1, 2, 3, 4, 5, 6, 7, 8, 9, 10, 11, 12, 15, and 16) without concreting of the shear notches along the simply-supported girder and at the measuring points (No. 13 and 14) along the continuous girder indicates they are approximate to each other and the latter is less than the former. Despite relative displacements between the base slab and girder surface at the shear notch positions without concreting within the temporary end thorn zone, the friction force at the temporary end thorn zone can still ensure stability within this zone. Hence, the safety at the critical part of the temporary end thorn zone is verified.

Table 7.21 Calculated and measured relative displacements of base slabs

Measuring point	Calculated value (mm)		Measured value (mm)	
	Maximum	Minimum	Maximum	Minimum
No. 13 and 14	9.50	−11.65	6.9	−7.2
No. 1 and 2	3.70	−3.96	3.4	−3.9
No. 3 and 4	3.92	−4.21	3.68	−4.2
No. 5 and 6	4.28	−4.66	4.30	−4.5
No. 7 and 8	4.52	−5.24	4.5	−5.0
No. 9 and 10	4.91	−7.59	4.3	−6.5
No. 11 and 12	8.97	−13.4	4.38	−5.4
No. 15 and 16	5.88	−9.6	4.2	−6.4

(2) Vertical relative displacement between base slab and bridge

The measuring points within the temporary end thorn zones were respectively located at the center of the No. 172~173, 174~ 175, 176~177, and 178~179 girders on the up and down tracks. The measured data at respective measuring points are summarized in Table 7.22.

The vertical displacements of the base slabs vary with temperature. Specifically, when the temperature rises, the vertical displacements of the base slabs gradually decrease. The maximum and minimum vertical displacements are 0.412 mm and -0.223 mm, respectively. Since the measuring points are arranged in the middle of the simply-supported girder within the temporary end thorn zones and the post-cast strip of the shear notches of the simply-supported girder has not been concreted, the base slabs on the simply-supported girder can be considered unit slabs. When the temperature of the upper surface of the base slabs is higher than that of their lower surface during a temperature rise process of the base slabs, the base slabs exhibit a concave downward shape in the middle and an upwarp at both ends. Conversely, when the temperature of the upper surface is lower than that of the lower surface, the base slabs display a convex upward shape in the middle and contraction at both ends. The base slabs without the constraints from the shear notches result in minor vertical displacements due to temperature effects.

(3) Longitudinal displacement of pier/abutment within temporary end thorn zone

Refer to Table 7.23 for the longitudinal displacements at the piers/abutments within the temporary end thorn zone.

Table 7.22 Measured maximum vertical displacements of base slabs

Measuring point No	No. 23 and 27	No. 24 and 28	No. 25 and 29	No. 26 and 30
Maximum vertical displacement (mm)	0.392	0.294	0.412	0.331
Minimum vertical displacement (mm)	−0.112	−0.093	0.036	−0.223

Table 7.23 Statistics of displacements of piers/abutments within temporary end thorn zone

Pier/abutment	Measured value (mm)		Calculated value (mm)	
	Maximum	Minimum	Maximum	Minimum
172#	1.50	−2.6	1.82	−2.90
173#	2.10	−3.5	2.41	−3.74
174#	2.99	−3.54	3.19	−4.91
175#	3.5	−4.8	4.08	−5.94
176#	4.8	−5.2	5.02	−5.51
180#	8.3	−11.6	9.11	−13.10

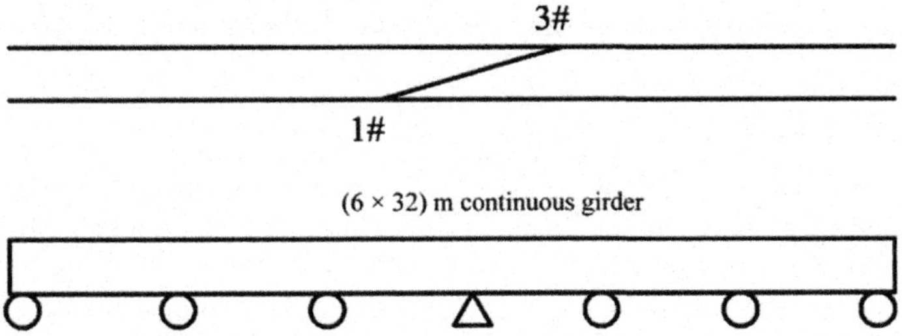

Fig. 7.60 Schematic arrangement of No. 18 Single-crossover turnout bridge within the throat area on the south side of Xuzhoudong station

The No. 180 pier features a fixed bearing beneath the continuous girder, and the No. 172~176 piers feature fixed bearings beneath the simply-supported girder. Due to the influence of friction force and longitudinal reinforcement of the base slabs under temperature effects, the temperature forces at the fixed bearings beneath the continuous girder are evident, resulting in noticeable longitudinal displacements at the piers/abutments. On the contrary, the displacements at the No. 172~176 piers are minor. In addition, as shown in Table 7.23, the measured values are less than the calculated values, and the measured displacement values fall within the safety range, thus verifying the design rationality and safety of the piers/abutments.

7.3 Operational Status Monitoring of no. 18 Slab Turnout on Xuzhoudong Station Bridge of Beijing–Shanghai HSR

7.3.1 Brief Introduction to Monitoring Site

The No. 18 single crossover within the throat area on the south side of Xuzhoudong Station of Beijing–Shanghai HSR consists of the CWR turnouts with the slab ballastless tracks, which are laid on a (6 × 32) m continuous girder in a north–south strike, with No. 1 turnout at the south and No. 3 turnout at the north, as shown in Fig. 7.60.

7.3.2 Monitoring Items and Methods

1. Monitoring Purpose

This turnout monitoring is aimed at verifying the correctness and rationality of the computational theory, design methodology, and design parameters of CWR turnouts on bridges, and providing theoretical guidance and accumulating practical experience for the design, construction, and maintenance of CWR turnouts with ballastless tracks on bridges.

2. Monitoring Item

Real-time monitoring of rail temperature forces (including additional force) preceding and following the turnouts, at the heel end of the switch rails within the turnouts, and at the switches;

Real-time monitoring of expansion and contraction displacements at the stock rails, switch rails, and point rails of the turnouts;

Real-time monitoring of longitudinal cumulative displacements of turnout slabs relative to longitudinally coupled slabs and longitudinally coupled slabs relative to the girder surface;

Real-time monitoring of expansion and contraction displacements at the ends of the turnout girder;

Real-time monitoring of rail temperatures and air temperatures within the turnout range;

Real-time image acquisition of the turnouts, turnout slabs, girder surface, and sensor status.

3. Measurement Method

Power supply from solar energy sources is utilized to enable the measurement and image acquisition for monitoring temperature forces, displacements, and temperatures of the CWR turnouts on the bridge. Field data (including images) is transmitted in real-time to designated servers using wireless network technology. Users can access and analyze data anytime and anywhere through network access.

4. Arrangement of measuring points
(1) Turnout temperature force

Measuring points for temperature forces on stock rails: A total of 22 temperature force measuring points are arranged, including 14 on straight stock rails, 3 near stock rail stoppers, and 5 on curved closure rails on the turnouts. The arrangement of the measuring points is shown in Fig. 7.61.

(2) Turnout-bridge relative displacement

Measuring points for girder-rail relative displacements: A total of 8 displacement measuring points are arranged, including 4 on stock rails, 1 respectively at the front end of a turnout, at a stopper, at the position corresponding to the actual point of a

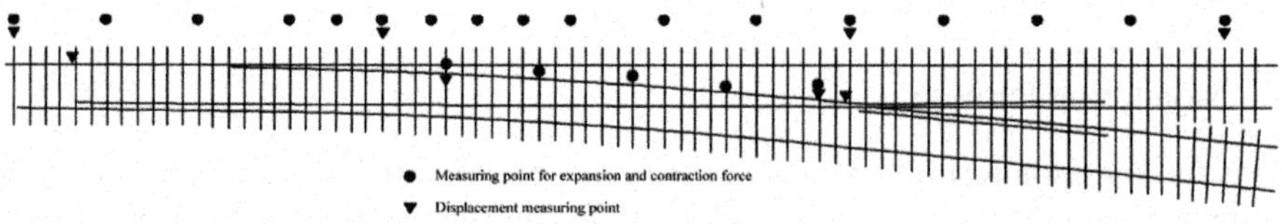

Fig. 7.61 Arrangement of measuring points for temperature forces and displacements of turnout

point rail, and at the rear end of a turnout, 1 respectively at the front and rear ends of a curved closure rail, and 1 respectively at the actual point of a switch rail and the actual point of a frog rail, as shown in Fig. 7.61.

7.3.3 Analysis of Monitoring Data

1. Rail temperature force

Under the conditions of a rail temperature difference of 20 °C, a turnout slab temperature difference of 13 °C, and a girder temperature difference of 4 °C, the measured temperature forces on the straight stock rails are shown in Fig. 7.62.

As shown in the figure, the maximum additional force on rails occurred at the continuous girder end. Due to the combination blocks of the stopper without contact, the temperature forces on the rails did not peak at the heel end of the switch rails. Due to a sliding layer laid between the bridge deck and the track bed slabs/base slabs, the impact of bridge expansion and contraction on the forces endured by the rails is minor, and the maximum additional force on the rails was only 11.6 kN, indicating that the sliding layer works well. After a gradual transition from a breathing zone to a non-breathing zone, the rails over the end thorn no longer bear additional forces.

2. Longitudinal displacement
(1) Longitudinal displacement of rail within turnout

The variation laws of displacements of the switch rail and point rail within the No. 1 turnout with temperature are shown in Figs. 7.63 and 7.64, and the measured displacements are shown in Table 7.24.

As shown in Table 7.24, the displacements of the switch rails and point rails are directly related to the rail temperature changes. The girder-rail interactions are minor, attributed to a "two-geotextile one-membrane" sliding layer laid between the bridge deck and base slab, and the influence of bridge temperature changes on

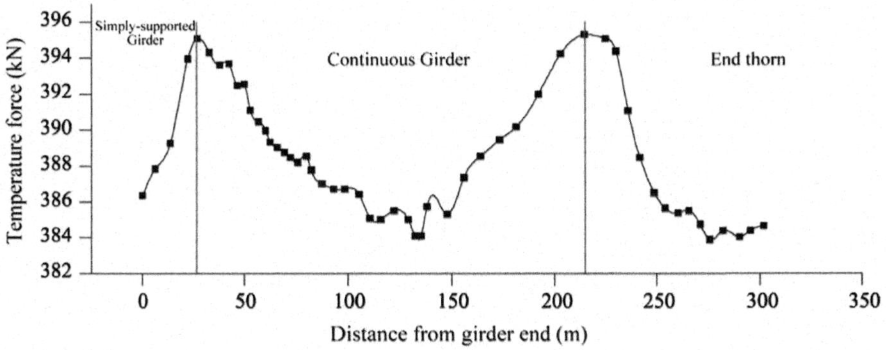

Fig. 7.62 Measured temperature forces on stock rails

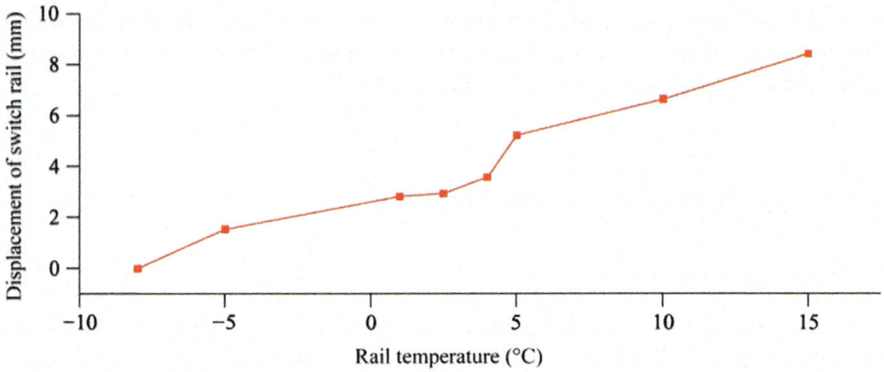

Fig. 7.63 Variation law of displacements of switch rail with rail temperature

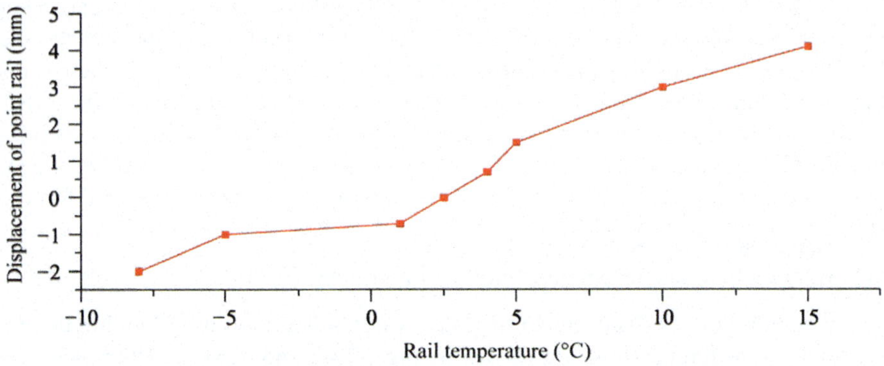

Fig. 7.64 Variation law of displacements of point rail with rail temperature

Table 7.24 Measured displacements

S/N	Rail temperature (°C)	Bridge temperature (°C)	Displacement of switch rail (mm)	Displacement of point rail (mm)
1	−8	−2	0.00	−2
2	−5	−1	1.54	−1
3	1	−0.7	2.82	−0.7
4	2.5	0	2.94	0
5	4	0.7	3.58	0.7
6	5	1.5	5.22	1.5
7	10	3	6.64	3
8	15	4.1	8.42	4.1

the displacements of the inner rails within the turnout is inapparent. When the rail temperature rises, the switch rails and point rails extend. On the contrary, when the rail temperature falls, the switch rails and point rails retract.

(2) Longitudinal displacement of bridge

The variation laws of expansion and contraction displacements at the continuous girder ends with temperature are shown in Figs. 7.65 and 7.66, and the measured displacements are shown in Table 7.25.

As shown in Table 7.25, the bridge displacements are directly related to the girder temperature changes, and the displacements at the girder ends increase linearly with the girder temperature increases. The measured values closely align with the theoretical analysis results. On the whole, the rail temperature changes do not impose

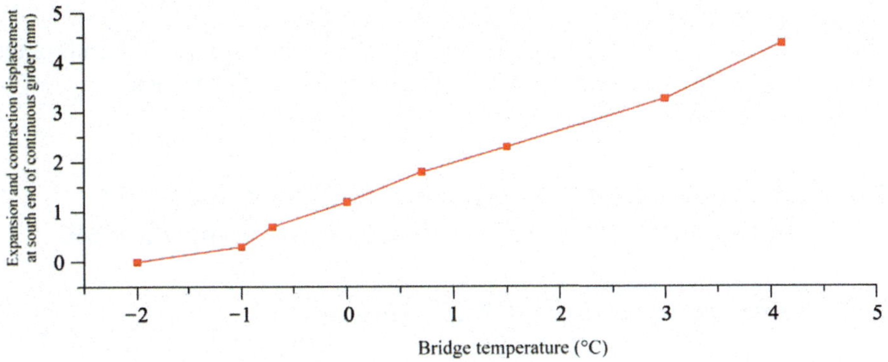

Fig. 7.65 Variation law of expansion and contraction displacements at the south end of continuous girder with bridge temperature

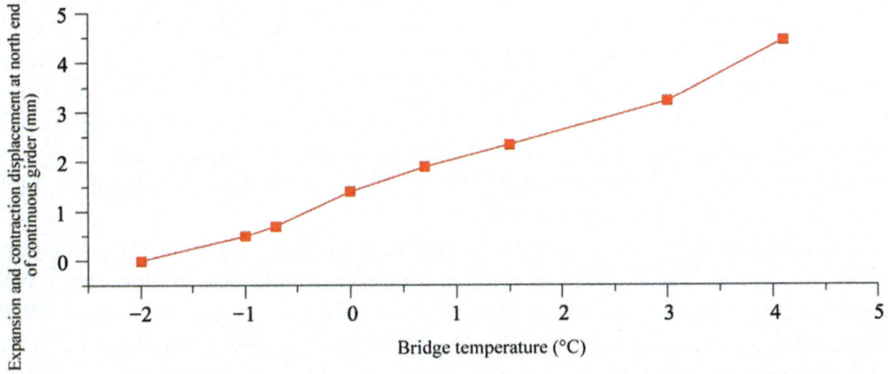

Fig. 7.66 Variation law of expansion and contraction displacements at the north end of continuous girder with bridge temperature

Table 7.25 Measured displacements

S/N	Rail temperature (°C)	Bridge temperature (°C)	Displacement at south end of continuous girder (mm)	Displacement at north end of continuous girder (mm)
1	−8	−2	0.00	0.00
2	−5	−1	0.30	0.50
3	1	−0.7	0.70	0.70
4	2.5	0	1.20	1.40
5	4	0.7	1.80	1.90
6	5	1.5	2.30	2.34
7	10	3	3.26	3.22
8	15	4.1	4.37	4.45

apparent impacts on the bridge displacements. The displacements at both ends of the continuous girder with a fixed bearing in its middle are nearly identical.

7.4 Operational Status Monitoring of Turnout on 102 M Continuous Girder-Arch of Hangzhou–Changsha HSR

7.4.1 Brief Introduction to Monitoring Site

Hangzhou–Changsha High-Speed Railway is designed for operation at 350 km/h, and accommodates bridges accounting for more than 72% of the whole line. The bridge span distributed at the monitoring site is a (3 × 32 + 102) m continuous girder + arch combination.

The span arrangement across the monitoring range and adjacent sections at both ends consist of 27–32 m simply-supported box girder + (6 × 32) m crossover continuous girder + 1–24 m simply-supported box girder + (3 × 32 + 102) m crossover continuous girder and arch + abutment. There is a sleeper-embedded ballastless track within the turnout section. A No. 18 high-speed turnout with a distance between centers of tracks of 5 m is used on the main line, and a bi-block ballastless track extends from the turnout at both ends.

The continuous girder + arch long-span turnout girder integrates the technical characteristics and challenges of long-span bridges, ballastless tracks, and CWR turnouts, representing one of the more difficult subjects in the ballastless CWR track field so far. Long-span bridges, ballastless tracks, and CWR turnouts have been extensively researched and widely applied in China and abroad. However, this turnout selected for this study represents the inaugural instance in China where CWR turnouts are installed on ballastless tracks atop a (3 × 32 + 102) m continuous girder + arch structure, marking the longest span to support such turnouts. There

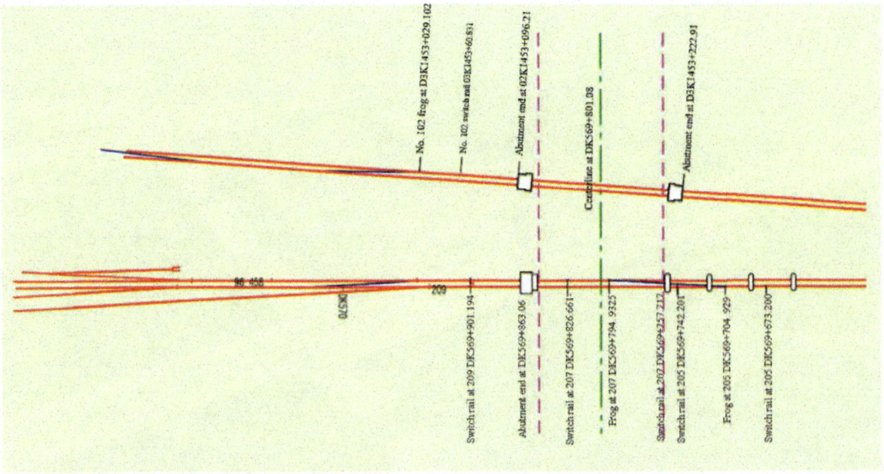

Fig. 7.67 Positional relationship between (3 × 32 + 102) m continuous Girder-arch bridge and turnout

is a deficiency in studies on laying CWR turnouts on long-span continuous girder and arch structures in China and abroad. Long-term comprehensive monitoring has been also absent for long-span continuous girder-arch bridges, ballastless tracks, and CWR turnouts. Therefore, it is necessary to carry out comprehensive monitoring of the girders, ballastless tracks, and CWRs for this (3 × 32 + 102) m continuous girder and arch structure. Figure 7.67 shows the positional relationship between the (3 × 32 + 102) m continuous girder-arch bridge and the turnout, and Fig. 7.68 shows the effect of the (3 × 32 + 102) m continuous girder and arch structure.

7.4.2 Monitoring Items and Measuring Points

1. Monitoring items

(1) Stress

Rail stress; track bed reinforcement stress.

(2) Temperature

Air temperature on bridge; rail temperature; track bed slab temperature; base slab temperature; bridge temperature.

(3) Displacement

Relative displacement between rail and track bed slab;
Relative displacement between track bed slab and base;
Relative displacement of base and bridge;
Longitudinal relative displacement at girder end.

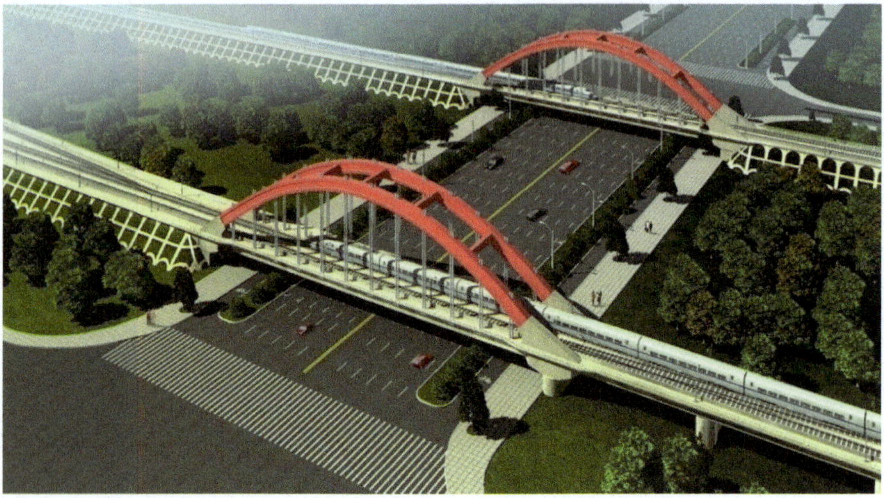

Fig. 7.68 Design sketch of (3 × 32 + 102) m continuous girder and arch

2. Arrangement of measuring points

Optical fiber sensors are used to monitor the temperatures, displacements, and stresses of the track structure. The monitoring of the CWR turnout on the continuous girder-arch bridge focused on the temperatures, stresses, strains, and displacements related to the track structure. The arrangement of measuring points is as follows:

(1) Temperature monitoring (16 measuring points in total)

1 measuring point for air temperature on the bridge: No. 1–1, located at the frog of the single crossover.

1 measuring point for girder surface temperature: No. 1–2, located at the end of the turnout girder.

1 measuring point for rail temperature: No. 1–3, located preceding the turnout.

7 measuring points for track bed slab temperature: No. 1–4 ~ No. 1–10, located at a girder joint, the actual point of a switch rail, a point rail, and the rear end of the turnout, respectively.

6 measuring points for base slab temperature: No. 1–11 ~ No. 1–16 are located at a girder joint, the actual point of a switch rail, and a point rail, respectively.

(2) Strain monitoring (24 measuring points in total)

Rail strain monitoring (16 measuring points in total).

14 Measuring points for rail strain: No. 2–1 ~ No. 2–11 within the turnout section; two measuring points for continuous girder: No. 2–12 ~ No. 2–14.

Reinforcement stress monitoring of track bed slab (10 measuring points in total).

10 measuring points for reinforcement stress of track bed slab: No. 3–1 ~ No. 3–10 respectively located at girder joints, the actual point of switch rails, the heel end of switch rails, point rails, and the rear end of the turnout.

(3) Displacement monitoring (28 measuring points in total)

Six measuring points for track bed slab warping: No. 4–1 ~ No. 4–6.

14 Measuring points for relative displacement between rail and track bed slab (or Track Slab): no. 5–1 ~ No. 5–14, Located Identically to Those for Rail Stress.

Three measuring points for relative displacement between track bed slab and base slab: No. 6–1 ~ No. 6–3.

Three measuring points for relative displacement between base slab and girder surface: No. 7–1 ~ No. 7–3.

Two measuring points for spacing between continuous girder joints: No. 8–1 ~ No. 8–2, respectively located at each end of the continuous girder.

7.4.3 Analysis of Monitoring Data

1. Analysis of temperature field monitoring data

(1) Air temperature and track bed slab temperature

Figure 7.69 shows the air temperature changes and track bed slab (inside) temperature changes with time. Both air temperature and track bed slab temperature vary periodically with time. See Table 7.26 for the monthly maximum and minimum temperatures in the monitoring period. As shown in the figure and table, the maximum air temperature and slab temperature were 40.3 °C and 51.3°C, and the minimum air temperature and slab temperature were −4 °C and −3.7 °C respectively during the monitoring period.

Table 7.27 shows the statistics of the monthly maximum air temperatures and maximum slab temperatures during the monitoring period. The differences between the monthly maximum track bed slab temperatures (inside) and the monthly maximum air temperatures were around 11 °C during the summer months.

(2) Air temperature and base slab temperature

Figure 7.70 shows the air temperature changes and base slab temperature changes with time. Both air temperature and base slab temperature vary periodically with time. See Table 7.28 for the monthly maximum and minimum temperatures during the monitoring period. As shown in the figure and table, the maximum air temperature and slab temperature were 40.3°C and 41.1°C, and the minimum air temperature and slab temperature were -4°C and -5.9°C respectively during the monitoring period.

Table 7.29 shows the statistics of the monthly maximum air temperature and maximum slab temperature during the monitoring period. The monthly maximum base slab temperatures (inside) were close to the monthly maximum air temperatures in the summer months and lower than the monthly maximum air temperatures in the remaining months.

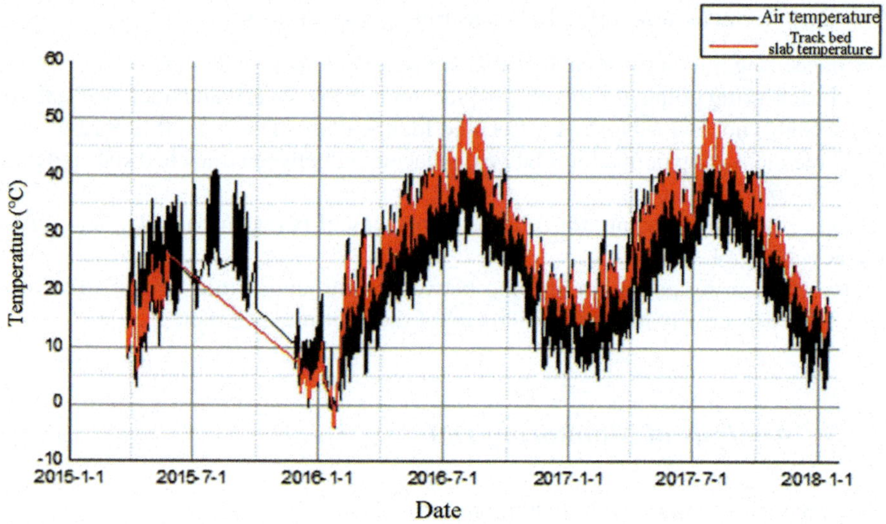

Fig. 7.69 Variations in air temperatures and track bed slab temperatures with time

Table 7.26 Statistics of annual maximum and minimum air temperatures and track bed slab temperatures

Year	Maximum air temperature (°C)	Minimum temperature (°C)	Maximum slab temperature (°C)	Minimum slab temperature (°C)
2015	40.1	−0.6	-	0.9
2016	40.3	-4.0	50.6	−3.7
2017	40.3	3.0	51.3	11.7

2. Analysis of temperature deformation monitoring data

(1) Rail-track bed slab longitudinal displacement

Figure 7.71 shows the rail-track bed slab longitudinal displacement variations and air temperature changes. The rail-track bed slab longitudinal displacements (preceding the turnout) varied in opposition to air temperatures. In 2016, the annual air temperature change range was 44.3 °C, and the displacement variation range was 12.2 mm, resulting in a displacement-air temperature ratio of 0.275 mm/°C. In 2017, the annual air temperature change range was 37.3°C, the displacement variation range was 10.7 mm, leading to a displacement-air temperature ratio of 0.287 mm/°C. Therefore, the ratio of rail-track bed slab longitudinal displacements (preceding turnout) to air temperature changes is preliminarily concluded at 0.28 mm/ °C.

Table 7.27 Statistics of monthly maximum air temperatures and track bed slab temperatures

Date	Monthly maximum air temperature (°C)			Monthly maximum slab temperature (°C)			Difference between monthly maximum track bed slab temperatures and monthly maximum air temperatures (°C)	
	2015	2016	2017	2015	2016	2017	2016	2017
Jan	–	19.6	26.9	–	11.1	23.8	−8.5	−3.1
Feb	–	29.0	32.2	–	26.0	26.3	−3.0	−6.1
Mar	–	32.6	30.2	–	29.5	25.6	−3.1	−5.4
Apr	35.8	37.5	38.2	–	36.1	37.1	−1.4	−1.1
May	35.4	39.6	39.2	–	38.2	44.4	−1.4	5.2
Jun	36.6	40.3	40.3	–	46.4	41.4	6.1	1.1
Jul	39.8	40.3	40.3	–	50.6	51.3	10.3	11.0
Aug	40.1	40.3	40.2	–	49.9	49.1	9.6	8.9
Sep	38.9	40.3	40.2	–	40.6	42.9	0.3	2.7
Oct	–	36.1	40.2	–	35.6	39.8	3.7	−0.3
Nov	–	31.5	32.2	–	29.2	30.8	−2.3	−3.0
Dec	16.7	24.7	21.1	8.9	23.2	22.6	−1.5	1.5

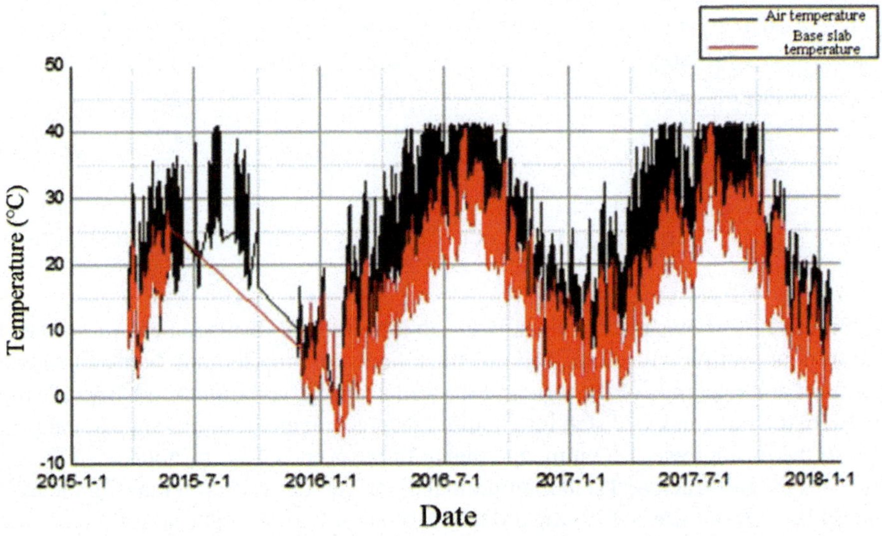

Fig. 7.70 Variations in air temperatures and base slab temperatures with time

Table 7.28 Statistics of annual maximum and minimum air temperatures and base slab temperatures

Year	Maximum air temperature (°C)	Minimum temperature (°C)	Maximum slab temperature (°C)	Minimum slab temperature (°C)
2015	40.1	−0.6	–	−0.3
2016	40.3	−4.0	40.4	−5.9
2017	40.3	3.0	41.1	−2.5

Table 7.29 Statistics of monthly maximum air temperatures and base slab temperatures

Date	Monthly maximum air temperature (°C)			Monthly maximum slab temperature (°C)			Difference between monthly maximum track bed slab temperatures and monthly maximum air temperatures (°C)	
	2015	2016	2017	2015	2016	2017	2016	2017
Jan	–	19.6	26.9	–	16.0	17.2	−3.6	−9.7
Feb	–	29.0	32.2	–	20.2	19.9	−8.8	−12.3
Mar	–	32.6	30.2	–	21.0	20.8	−11.6	−9.4
Apr	35.8	37.5	38.2	25.9	21.8	22.7	−15.7	−15.5
May	35.4	39.6	39.2	28.0	27.4	34.0	−12.2	−5.2
Jun	36.6	40.3	40.3	–	36.0	30.1	−4.3	−10.2
Jul	39.8	40.3	40.3	–	40.4	41.1	0.1	0.8
Aug	40.1	40.3	40.2	–	40.2	36.3	−0.1	−3.9
Sep	38.9	40.3	40.2	–	35.6	36.9	−4.7	−3.3
Oct	–	36.1	40.2	–	28.8	34.8	−7.3	−5.4
Nov	–	31.5	32.2	–	25.1	27.3	−6.4	−4.9
Dec	16.7	24.7	21.1	15.1	18.0	17.6	−6.7	−3.5

Figure 7.72 shows the rail-track bed slab longitudinal displacement variations in the turnout section with time. The rail-track bed slab longitudinal displacements at the front and rear ends of the turnout followed a similar variation trend. Specifically, in 2016, the rail-track bed slab longitudinal displacement variation range at the front end of the turnout was 12.2 mm, and that at the rear end of the turnout was 4.3 mm. In 2017, the rail-track bed slab longitudinal displacement variation range at the front end of the turnout was 10.7 mm, and that at the rear end of the turnout was 5.7 mm.

(2) Track bed slab-base slab longitudinal displacement

Figure 7.73 shows the longitudinal relative displacements between the track bed slab and base slab with time. The variation ranges of longitudinal relative displacements between the track bed slab and base slab were 8.8 mm in 2015, 10.1 mm in 2016, and 9.3 mm in 2017.

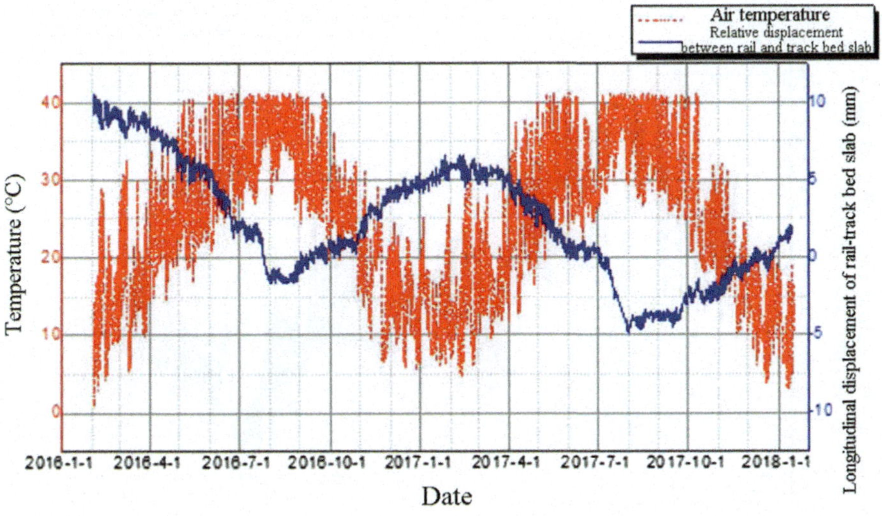

Fig. 7.71 Rail-Track bed slab longitudinal displacement variations and air temperature changes

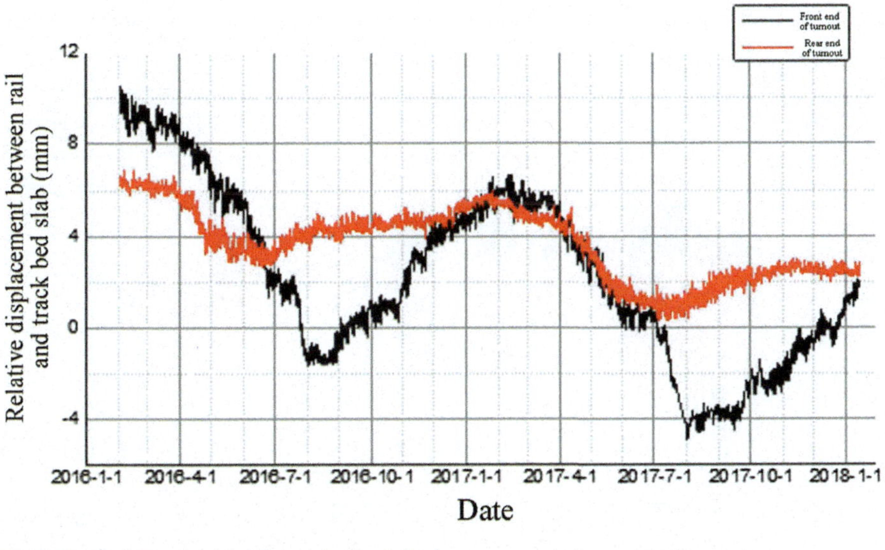

Fig. 7.72 Rail-Track bed slab longitudinal displacement variations with time

3. Analysis of temperature stress monitoring data

(1) Rail stress

Figure 7.74 shows the rail stresses at different positions on the bridge with time. The annual rail stress at the turnout front end ranges from −56.9 to 24.4 Mpa, and the stress variation range is 81.3 Mpa; the annual rail stress at the middle of turnout

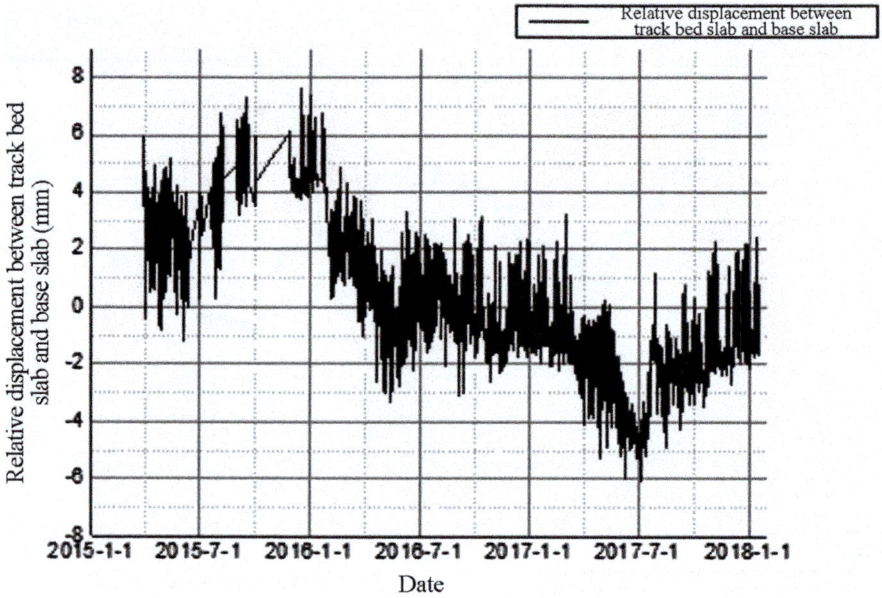

Fig. 7.73 Track bed slab and track slab longitudinal displacement variations with time

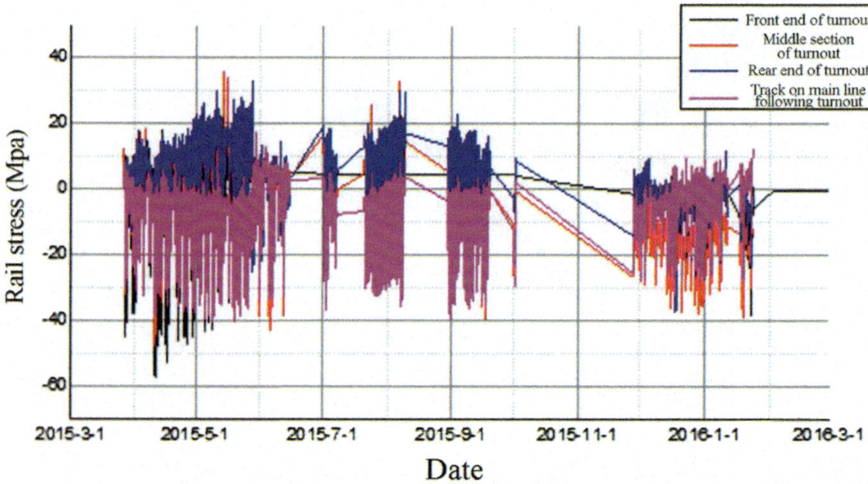

Fig. 7.74 Rail stress variations with time

ranges from −47.6 to 35.5 Mpa, and the stress variation range is 83.1 Mpa; the annual rail stress at the turnout tail varies from −36.9 to 33.0 Mpa, and the stress variation range is 69.9 Mpa; the annual rail stress of the normal track section preceding the turnout varies from −40.6 to 17.6 Mpa, and the stress variation range is 58.2 Mpa.

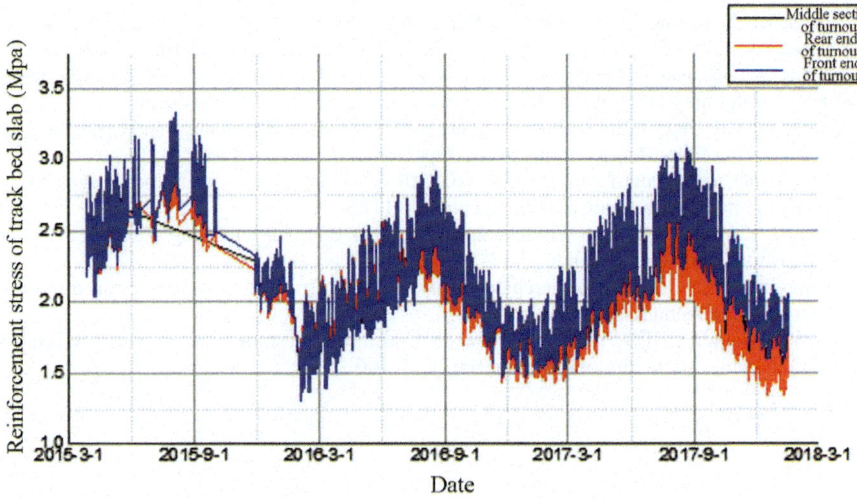

Fig. 7.75 Reinforcement stress variations of track bed slab with time

(2) Reinforcement stress of track bed slab

Figure 7.75 shows the stresses in track bed slabs at different positions on the bridge with time. The reinforcement stress of the track bed slab at the front end of the turnout annually varied from 1.3 to 3.3 Mpa, and the resulting stress variation range was 2.0 Mpa. The reinforcement stress of the track bed slab at the middle position of the turnout annually varied from 1.3 to 2.9 Mpa, and the resulting stress variation range was 1.6 Mpa. The reinforcement stress of the track bed slab at the rear end of the turnout annually varied from 1.3 to 3.1 Mpa, and the resulting stress variation range was 1.8 Mpa. The analysis indicates that the reinforcement stress in the track bed slabs is relatively small under the action of temperature load.

References

1. Shijie Zhang, Li Sun, Yanli Yang et al. (2010) Research on design of slab ballastless track in turnout section on lei bridge of Wuhan-Guangzhou passenger dedicated line. Railw Eng, (1): 13–16.
2. Weibin LI (2022) Distributed optical fiber sensors for comprehensive monitoring of railway infrastructure. China Munic Eng 1:81–84
3. Yuxin CHEN, Lixia WU (2022) Strain sensing characteristics of fiber grating and field experimental study. Highway 67(6):332–335
4. Zilin WANG, Wei FAN, Hong GAO et al (2022) Three-component fiber bragg grating geophone. Opt Commun Technol 46(3):47–51
5. Pengcheng NIE, Xianlun YU, Jiawei LIU (2022) Research on spectral characteristics of few-mode fiber bragg grating. Opt Commun Technol 46(1):16–19
6. Xinkui XING, Fanfan LIU, Lanri LIN et al (2021) Research on tension cantilever-mounted fiber bragg grating displacement sensor. Opt Commun Technol 45(12):1–4

7. Bowen LI, Yinggang LIU, Xiaoya SONG et al (2020) Study on Low-temperature sensing characteristics of fiber bragg grating. Piezoelectrics & Acoustooptics 42(6):787–790
8. Fugang Wu, Qingshan Zhang, Desheng Jiang, et al. (2007) Method of gaussian curve fitting for measuring fiber bragg wavelength. J Wuhan Univ Technol, (12): 116–118
9. Quanqiang LIU, Na CHEN, Congcong LI et al (2019) Study on Fabrication and Characteristics of Micro-nano Fiber Bragg Grating Probe. Opt Commun Technol 43(9):27–30
10. Feng ZHOU, Yongqian LI, Shaolong WANG (2017) Research Progress of Fiber Bragg Grating Sensor and Packaging. Opt Commun Technol 41(11):8–11
11. Yong XU (2016) Research on Wireless Sensor Network Based on Vibrating Wire Sensor. South China Univ Technol, Guangzhou
12. Chengyuan WANG (2021) Research and analysis of conversion circuit for resistance sensors. Electron Meas Technol 44(8):161–167
13. Jun WU, Weimin CHEN, Yuejie SHU et al (2015) Embedded strain homogenized FBG sensor for smart cables. Acta Photonica Sinica 44(7):95–100

Chapter 8
Field Test for Mechanical Properties of CWR Turnout on Bridge

To enhance the understanding of the transmission pattern of longitudinal forces on CWR turnouts on bridges, further study the computation theories, improve the design methods, optimize the design schemes, verify the correctness and rationality of design parameters, and provide theoretical guidance and accumulate practical experience for the design and construction of CWR turnouts on high-speed railway bridges in the future, the author team undertook field tests for the ballastless turnouts on Changshan Super Major Bridge at the right throat of Lechangdong Station of Wuhan–Guangzhou HSR and dynamic and static tests for the No. 42 slab turnout on the bridge of Yangzhou Connecting Track of Beijing–Shanghai HSR.

8.1 Field Testing for CWR Turnouts on Bridge at Lechangdong Station of Wuhan–Guangzhou High-Speed Railway

8.1.1 Test Overview

The turnout at the right throat of Lechangdong Station of Wuhan–Guangzhou HSR is placed on the (5 × 32) m continuous girder, and the single crossover is mounted on the (6 × 32) m continuous girder. These two continuous girders are connected by a 32 m simply-supported girder. This bridge carries four CNTT No. 18 turnouts with 350 km/h ballastless tracks, and the turnout sections accommodate long-sleeper embedded ballastless tracks [1]. The specific turnout bridge arrangement is shown in Fig. 8.1.

From August 29 to September 10, 2009, the test team carried out static tests for No. 4 and 8 CWR turnouts on Changshan Super Major Bridge at Lechangdong

Q. Li et al., *Continuously Welded Turnouts on High-Speed Railway Bridge*, Advances in High-speed Rail Technology, https://doi.org/10.1007/978-981-97-6524-9_8

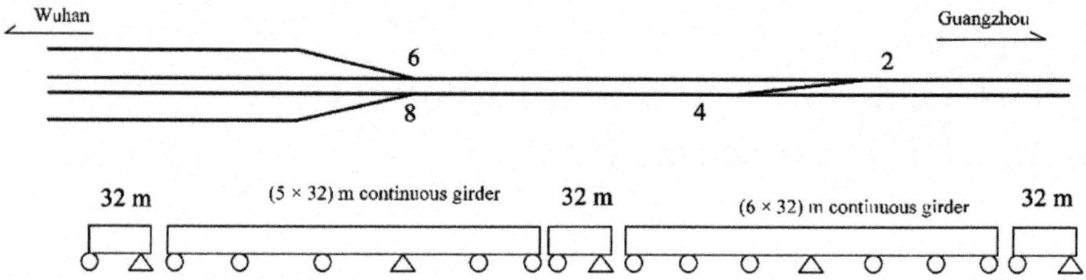

Fig. 8.1 Arrangement of bridge beneath CWR turnout group on Changshan super major bridge at Lechangdong station

Station of Wuhan–Guangzhou HSR. From October 12 to 29, 2009, they performed comprehensive dynamic testing for No. 2 and No. 6 turnouts.

8.1.2 Test Items

(1) Static tests: mainly involving the additional temperature forces on rails, relative displacements between the switch rails and stock rails, and displacements at girder ends.
(2) Dynamic tests: mainly involving the wheel-rail vertical and lateral forces at stock rails and closure rails, and the vertical force at switch rails and point rails within the CWR turnouts. The results from the above tests were utilized to calculate the derailment coefficient of inner and outer wheels, wheel unloading rates, and wheel-rail lateral force of rolling stock. These calculation results were further leveraged to deduce the operational safety indexes [2].
(3) Dynamic displacement tests: mainly including the vertical and lateral displacements of the stock rail in the section proceeding the switch rail, the vertical and lateral displacement of the switch rail and stock rail at the lead curve, the vertical and lateral displacement of the stock rail at the frog, the vertical displacement of the actual point of the point rail, and the opening at the actual point of the point rail within the turnout section.
(4) Strength of turnout components: including the rail bottom stress at the point rail cross section with a rail width of 20 mm, the rail bottom stress at the switch rail cross section with a rail width of 20 mm, and the rail bottom stress at the planed section of the wing rail.
(5) Vibration characteristics of the track structure at the CWR turnouts: focusing on the vertical vibration accelerations of rails in the main parts within the turnout sections and the vertical vibration accelerations of the turnout sleeper or track slab.
(6) Vibration characteristics of the bridge girder: including the vertical and lateral vibration accelerations at characteristic cross sections of the girder body.
(7) Deformation of the bridge structure: including the vertical deflection at characteristic cross sections of the girder body.

8.1.3 Arrangement of Measuring Points

1. Arrangement of Measuring Points for Static Tests

(1) Turnout temperature force

Considering the mutual influence of the CWR turnouts on the bridge, the rails within No. 4 and 8 turnouts on Changshan Super Major Bridge at Lechangdong Station were numbered 1, 2, 3, and 4 respectively, as shown in Fig. 8.2.

The layout plan of the No. 18 turnout and the sleeper numbers corresponding to its key positions are shown in Fig. 8.3. All the measuring points were mounted on the rail web.

The No. 1 rail includes three parts: the curved stock rail of the No. 4 turnout, the tangent between the No. 4 and 8 turnouts, and the straight stock rail of the No. 8 turnout. A total of 15 measuring points are placed, as shown in Table 8.1.

The No. 2 rail includes three parts: the curved stock rail of the No. 8 turnout, the tangent between the No. 8 and 4 turnouts, and the straight stock rail of the No. 4 turnout. A total of 17 measuring points are placed, as shown in Table 8.2.

The No. 3 rail mainly includes the straight closure rail within the No. 4 turnout. A total of 5 measuring points are arranged, as shown in Table 8.3.

The No. 4 rail mainly includes the straight closure rail within the No. 8 turnout. A total of 5 measuring points are arranged, as shown in Table 8.4.

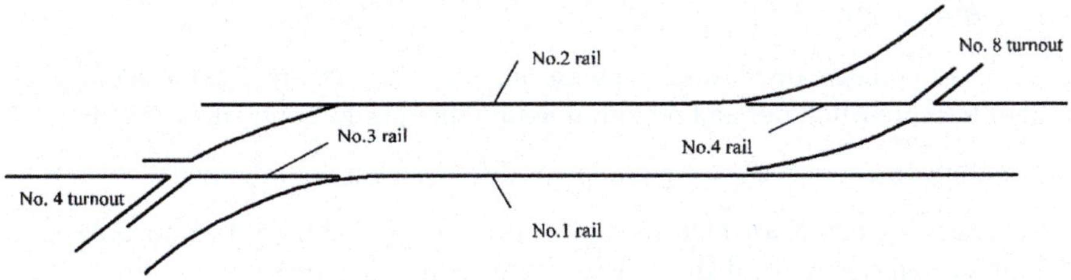

Fig. 8.2 Numbered test rails in CWR turnout group area on the bridge at Lechangdong station

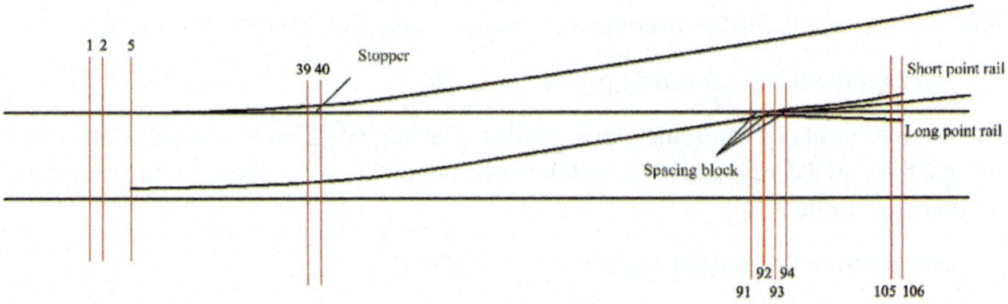

Fig. 8.3 Numbers of key turnout sleepers within No. 18 ballastless turnout

Table 8.1 Positions of measuring points for temperature force on No. 1 rail

Measuring point No.	Position	Sleeper No. at measuring point position	Distance to the actual point of switch rail within No. 4 turnout (m)
1	Curved stock rail within No. 4 turnout	90–91	135.9
2		79–80	129.3
3		61–62	118.5
4		40–41	105.9
5		2–3	83.1
6	Intermediate tangent	4–5	79.5
7		17–18	71.7
8		55–56	48.9
9		74–75	37.5
10		104–105	19.5
11	Straight stock rail within No. 8 turnout	132(0)–1	2.7
12		21–22	−9.3
13		40–41	−20.7
14		46–47	−36.3
15		91–92	−63.3

(2) Arrangement of measuring points for relative displacement between switch rail and stock rail

A girder-rail relative displacement measuring point was arranged at the actual points of the straight switch rail and point rail within the four turnouts, respectively.

(3) Arrangement of measuring points for displacement at girder end

Two measuring points are placed on the inner and outer sides of the movable end of the girder to measure the displacements at the end of the girder.

2. Arrangement of Measuring Points for Dynamic Tests

Measuring points are arranged at different positions along the turnouts to test the dynamic responses of the turnouts and bridge when an EMU passes through.

(1) Arrangement of measuring points for track

The arrangements of track measuring points for the No. 2 and 6 turnouts are shown in Figs. 8.4 and 8.5 respectively, and the specific test items of these measuring points are listed in Table 8.5.

(2) Arrangement of measuring points for bridge

The arrangement of acceleration and vertical dynamic displacement measuring points at the midspan of the bridge girders carrying the No. 2 and 6 turnouts is shown in Fig. 8.6, and the specific test items of these measuring points are listed in Table 8.6.

Table 8.2 Positions of measuring points for temperature force on No. 2 rail

Measuring point No.	Position	Sleeper No. at measuring point position	Distance to the actual point of switch rail within No. 4 turnout (m)
1	Straight stock rail within No. 4 turnout	56–57	115.5
2		40–41	105.9
3		38–39	104.7
4		27–28	98.1
5		12–13	89.1
6		7–8	86.1
7	Intermediate tangent	17–18	71.7
8		36–37	60.3
9		55–56	48.9
10		74–75	37.5
11		92–93	26.7
12		111–112	15.3
13	Curved stock rail within No. 8 turnout	6–7	~0.9
14		27–28	−13.5
15		40–41	−21.3
16		68–69	−38.1
17		95–96	−54.3

Table 8.3 Positions of measuring points for temperature force on No. 3 rail

Measuring point No.	Position	Sleeper No. at measuring point position	Distance to the actual point of switch rail within No. 4 turnout (m)
1	Straight closure rail within No. 4 turnout	87–88	134.1
2		73–74	125.7
3		62–63	119.1
4		51–52	112.5
5	Stopper within No. 4 turnout	40–41	105.9

8.1.4 Measurement Method

1. Static Test Methods

(1) Turnout temperature force

At different rail temperatures, a TS-3 rail temperature stress gauge is used to measure relative longitudinal displacements between the two measuring points at the neutral

Table 8.4 Positions of measuring points for temperature force on No. 4 rail

Measuring point No.	Position	Sleeper No. at measuring point position	Distance to the actual point of switch rail within No. 4 turnout (m)
1	Stopper within No. 8 turnout	40–41	−0.6
2	Straight closure rail within No. 8 turnout	51–52	−7.2
3		62–63	−13.8
4		73–74	−20.4
5		84–85	−27

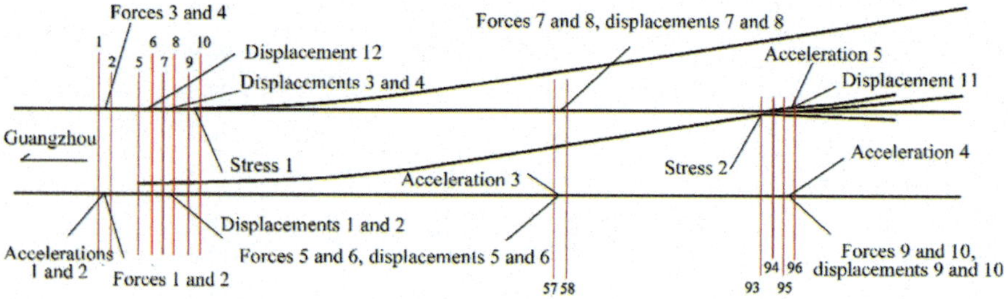

Fig. 8.4 Arrangement of measuring points for No. 2 turnout

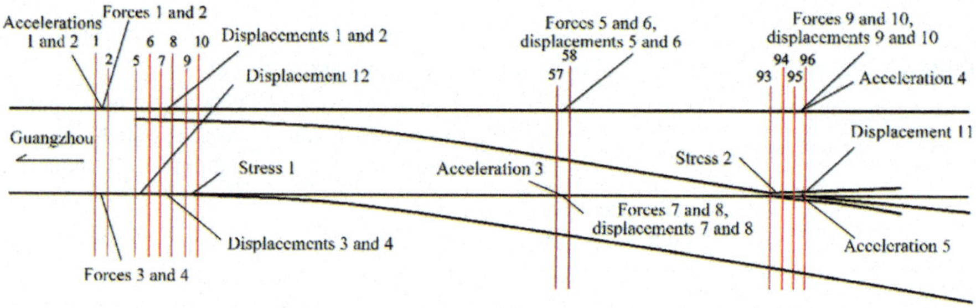

Fig. 8.5 Arrangement of measuring points for No. 6 turnout

axis (Fig. 8.7). The results are utilized to calculate the longitudinal distribution of rail temperature stress along the rail. The measurement accuracy is ±20 kN.

(2) Test of displacement of actual point of switch rail and point rail

At different temperatures, a Vernier calipers is used to measure the elongation and shortening of the actual point of the switch rail (point rail) relative to the stock rail, as shown in Fig. 8.8.

Table 8.5 Correspondence of measuring points and items

S/N	Measuring point	Monitoring item
1	Force 1	Vertical force on straight stock rail in section preceding switch rail
2	Force 2	Lateral force on straight stock rail in section preceding switch rail
3	Force 3	Vertical force on curved stock rail in section preceding switch rail
4	Force 4	Lateral force on curved stock rail in section preceding switch rail
5	Force 5	Vertical force on straight stock rail at lead curve
6	Force 6	Lateral force on straight stock rail at lead curve
7	Force 7	Vertical force on straight switch rail at lead curve
8	Force 8	Lateral force on straight switch rail at lead curve
9	Force 9	Vertical force of straight stock rail at frog
10	Force 10	Lateral force of straight stock rail at frog
11	Displacement 1	Vertical displacement of straight stock rail at the front end of switch
12	Displacement 2	Lateral displacement of straight stock rail at the front end of switch
13	Displacement 3	Vertical displacement of curved stock rail at front end of switch
14	Displacement 4	Lateral displacement of curved stock rail at front end of switch
15	Displacement 5	Vertical displacement of straight stock rail at lead curve
16	Displacement 6	Lateral displacement of straight stock rail at lead curve
17	Displacement 7	Vertical displacement of straight switch rail at lead curve
18	Displacement 8	Lateral displacement of straight switch rail at lead curve
19	Displacement 9	Vertical displacement of straight stock rail at frog
20	Displacement 10	Lateral displacement of straight stock rail at frog
21	Displacement 11	Vertical displacement at point rail section with 50 mm top width
22	Displacement 12	Opening at actual point of switch rail
23	Acceleration 1	Vertical acceleration of straight stock rail in section preceding switch rail
24	Acceleration 2	Vertical acceleration of sleeper under straight stock rail in section preceding switch rail
25	Acceleration 3	Vertical acceleration of straight stock rail at lead curve
26	Acceleration 4	Vertical acceleration of straight stock rail at frog
27	Acceleration 5	Vertical acceleration of sleeper under straight stock rail at frog
28	Stress 1	Rail bottom stress at switch rail with 20 mm top width
29	Stress 2	Wing rail stress preceding frog

(3) Test of displacement at girder end

A dial indicator is mounted on the pier top using a gauge stand, with its probe pressed against the movable end of the girder. The mounted dial indicator is used to measure displacement at the girder end, as shown in Fig. 8.9.

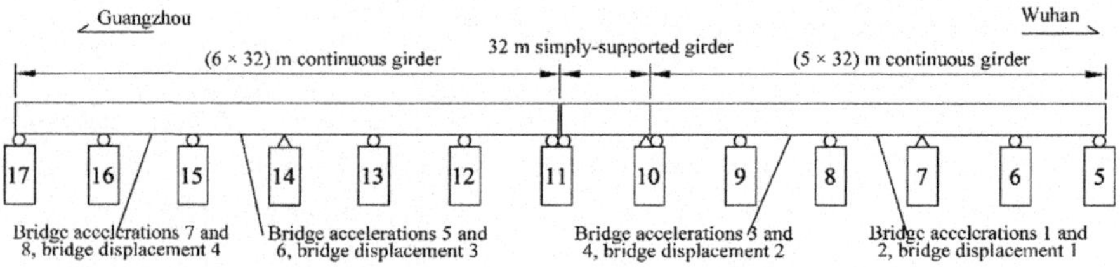

Fig. 8.6 Arrangement of measuring points

Table 8.6 Arrangement of measuring points on bridge

S/N	Measuring point	Monitoring item
1	Bridge acceleration 1	Vertical acceleration at the midspan of the girder carrying the point rail within No. 6 turnout
2	Bridge acceleration 2	Lateral acceleration at the midspan of the girder carrying the point rail within No. 6 turnout
3	Bridge acceleration 3	Vertical acceleration at the midspan of the girder carrying the switch rail within No. 6 turnout
4	Bridge acceleration 4	Lateral acceleration at the midspan of the girder carrying the switch rail within No. 6 turnout
5	Bridge acceleration 1	Vertical acceleration at the midspan of the girder carrying the point rail within No. 2 turnout
6	Bridge acceleration 2	Lateral acceleration at the midspan of the girder carrying the point rail within No. 2 turnout
7	Bridge acceleration 3	Vertical acceleration at the midspan of the girder carrying the switch rail within No. 2 turnout
8	Bridge acceleration 4	Lateral acceleration at the midspan of the girder carrying the switch rail within No. 2 turnout
9	Bridge displacement 1	Vertical dynamic displacement at the midspan of the girder carrying the point rail within No. 6 turnout
10	Bridge displacement 2	Vertical dynamic displacement at the midspan of the girder carrying the switch rail within No. 6 turnout
11	Bridge displacement 3	Vertical dynamic displacement at the midspan of the girder carrying the point rail within No. 2 turnout
12	Bridge displacement 4	Vertical dynamic displacement at the midspan of the girder carrying the switch rail within No. 2 turnout

Fig. 8.7 Test of temperature forces on stock rail

Fig. 8.8 Displacement of actual point of switch rail relative to stock rail

Fig. 8.9 Test of displacement at girder end

Fig. 8.10 Arrangement of strain gage rosette for wheel-rail force test

2. Dynamic Test Methods

(1) Wheel-rail lateral force (*Q*) and vertical force (*P*).

According to the *Track Side Test Methods of Vertical and Lateral Wheel-Rail Forces* (TB/T 2489-94) [3], the shear stress method is applied to measure the vertical wheel-rail force (*P*) and lateral wheel-rail force (*Q*) when a train passes through each cross-section measuring point within the turnout sections at different speeds. The measurement results are used to calculate the derailment coefficient and wheel unloading rate. The calculation results are utilized to determine the safety and stability of the test train operation. The arrangement of strain gauges is shown in Fig. 8.10.

(2) Lateral and vertical elastic displacements of rail

A spring-type displacement meter is used to measure the elastic displacement of rails at measuring points on the track structure, the opening of the switch rails and point rails within the turnout sections, and the dynamic gauge variations. The measurement results are used to evaluate the stability of the CWR turnouts, in conjunction with the test results of wheel-rail lateral force [4–6]. The arrangement of displacement meters is shown in Fig. 8.11.

(3) Strength of turnout components

The stress method is applied to measure the dynamic flexural stress of rails at turnout components. The measurement results are used to determine the strength and dynamic performance of turnout components during train operation. The arrangement of strain gauges is shown in Fig. 8.12.

(4) Vibration acceleration of track

A piezoelectric acceleration sensor is used to measure the track vibration acceleration. The rail acceleration sensor is attached to the upper surface of the rail base, the sleeper

Fig. 8.11 Arrangement of rail elastic displacement meter

Fig. 8.12 Arrangement of strain gauge for track strength test

acceleration sensor is fixed on the upper surface of the sleeper end, and the track bed slab acceleration sensor is mounted on the track bed slab under the rail, as shown in Fig. 8.13. The surface of the rails, sleepers, and track bed slabs shall be ground flat. In addition, the acceleration sensors shall be insulated from the rails.

(5) Vibration acceleration of bridge

A vibration pickup instrument is used to measure the vertical and lateral vibration accelerations of the bridge, and a deflection tester is used to measure the midspan deflection, as shown in Figs. 8.14 and 8.15.

(6) Test calibration

The wheel-rail vertical force (*P*) and lateral force (*Q*) are calibrated on site using a calibration frame, and all the displacement results are calibrated on site using a feeler gauge.

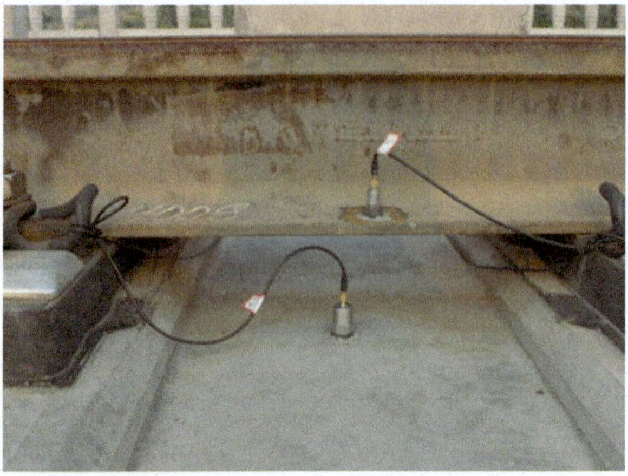

Fig. 8.13 Arrangement of acceleration sensor

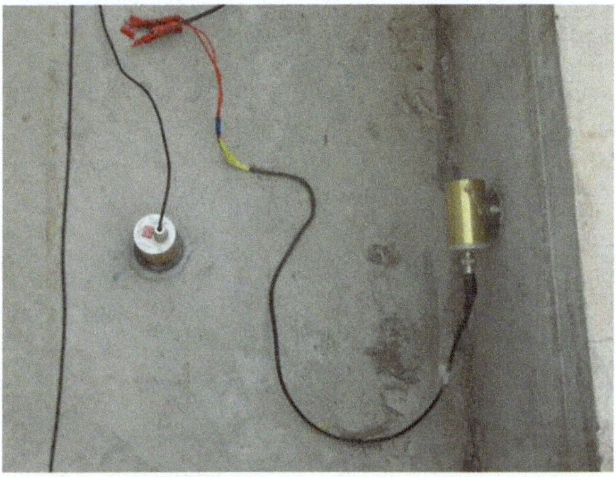

Fig. 8.14 Lateral and vertical vibration acceleration test of bridge

Fig. 8.15 Midspan deflection test of bridge

8.1.5 *Static Test of Turnout Group*

A total of 32 sets of valid data were acquired for the static tests. To verify the computational theories of CWR turnouts on bridge, the author team established a computational model for the CWR turnout group on Changshan Super Major Bridge, and compared the theoretical values with the measured ones.

1. Temperature Force on Rail

Figures 8.16 and 8.17 present the comparison between the measured temperature forces and theoretical values of Rails 1 and 2 formed by the No. 4 and No. 8 turnouts and the intermediate tangent on Changshan Super Major Bridge at Lechangdong Station, in which positive values represent compression and negative values indicate tension.

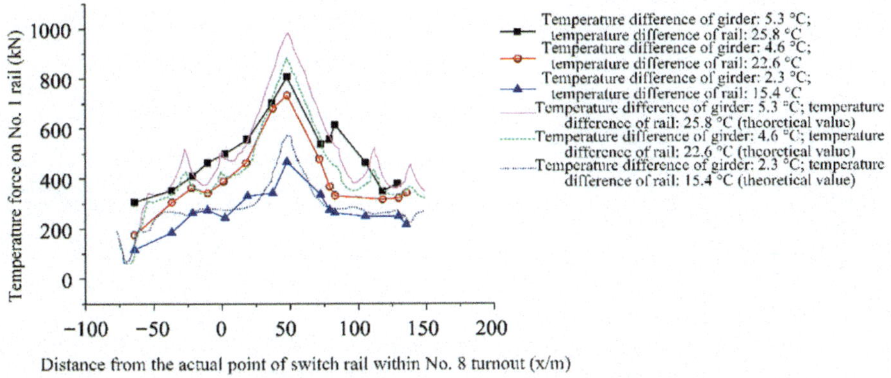

Fig. 8.16 Comparison between measured temperature forces and theoretical values of Rail 1

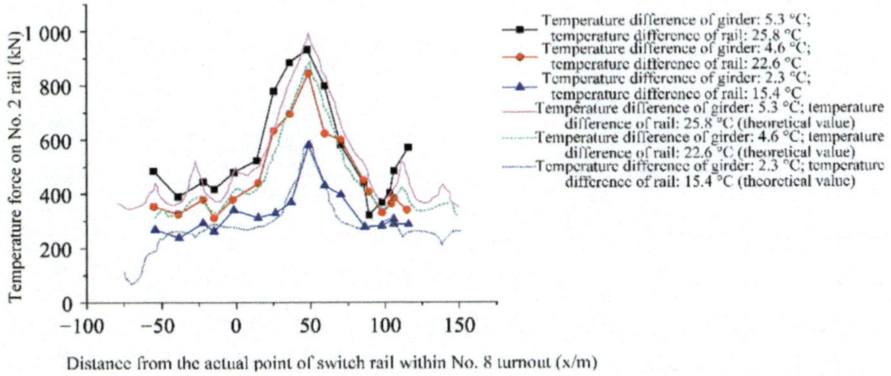

Fig. 8.17 Comparison between measured temperature forces and theoretical values of Rail 2

The comparison between measured temperature forces and theoretical values of Rail 3 formed by the straight closure rail within the No. 8 turnout and of Rail 4 formed by the straight closure rail within the No. 4 turnout on the bridge is shown in Figs. 8.18 and 8.19.

2. Expansion and Contraction Displacement of Turnout Components

Tables 8.7 and 8.8 present the displacements of the actual points of the switch rail and point rail within No. 4 and 8 turnouts on the Changshan Super Major Bridge at Lechangdong Station. In these tables, positive values signify that the actual points of the switch rail and long point rail are elongated relative to the stock rail due to rail temperature changes, while negative values indicate they are shortened.

The test results indicate that the expansion and contraction displacements at the actual points of the point rail and switch rail are closely aligned with the temperature variations. Due to the presence of a spacing block, the expansion and contraction

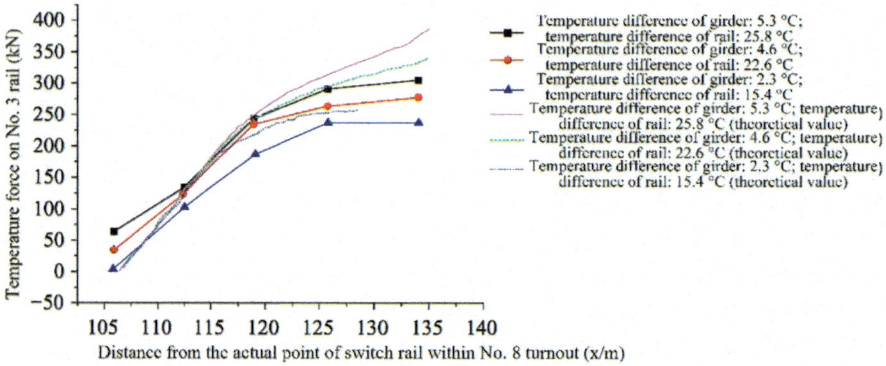

Fig. 8.18 Comparison between measured temperature forces and theoretical values of Rail 3

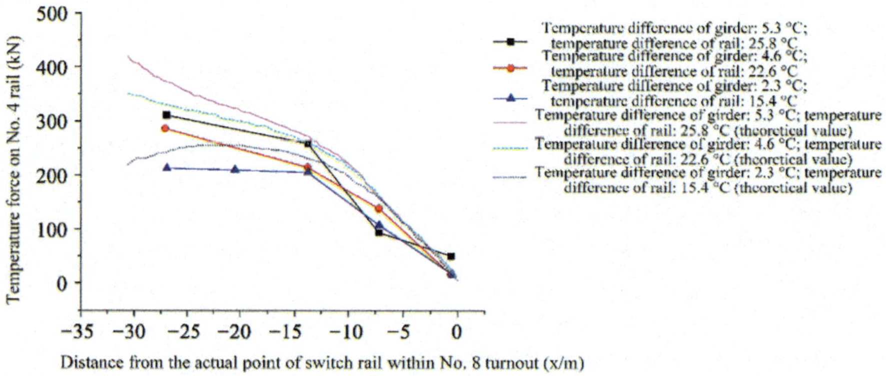

Fig. 8.19 Comparison between measured temperature forces and theoretical values of Rail 4

Table 8.7 Displacements of No. 4 turnout components on Changshan super major bridge at Lechangdong station

Ranges of girder temperature and rail temperature variations (°C)	Displacement of actual point of straight switch rail (mm)		Displacement at actual point of point rail (mm)	
	Measured value	Theoretical value	Measured value	Theoretical value
Temperature difference of girder: 5.0 °C; temperature difference of rail: 25.0 °C	7.0	7.85	2.8	3.13
Temperature difference of girder: 4.6 °C; temperature difference of rail: 22.6 °C	6.1	6.70	2.1	2.79
Temperature difference of girder: 2.3 °C; temperature difference of rail: 15.4 °C	4.1	4.51	1.9	2.12

Table 8.8 Displacements of No. 8 turnout components on Changshan super major bridge at Lechangdong station

Ranges of girder temperature and rail temperature variations (°C)	Displacement of actual point of straight switch rail (mm)		Displacement at actual point of point rail (mm)	
	Measured value	Theoretical value	Measured value	Theoretical value
Temperature difference of girder: 5.3 °C; temperature difference of rail: 25.8 °C	7.1	7.69	2.7	3.08
Temperature difference of girder: 3.9 °C; temperature difference of rail: 18.5 °C	5.9	6.62	2.4	2.76
Temperature difference of girder: 0.1 °C; temperature difference of rail: 2.0 °C	4.0	4.50	1.8	2.08

displacements of the heel end of the point rails were smaller than those of the switch rails. The test results were closely aligned with the theoretical calculation results.

3. Displacement at Girder End

Table 8.9 and Fig. 8.20 present the displacements at the girder end.

The test results of the expansion and contraction displacements at the girder end under temperature changes were closely aligned with the theoretical calculation values.

4. Summary

To verify the computational theories of CWR turnouts on bridges, a series of tests were performed on Changshan Super Major Bridge at Lechangdong Station of

Table 8.9 Analysis of test results of displacements at girder end

Temperature difference of bridge (°C)	Measured displacement at girder end (mm)	Theoretical calculation value (mm)
1.2	0.395	0.391
1.4	0.491	0.456
4.8	1.997	1.565
6.2	2.488	2.021
6.8	2.019	2.217
7.5	2.384	2.445
8.7	2.779	2.836

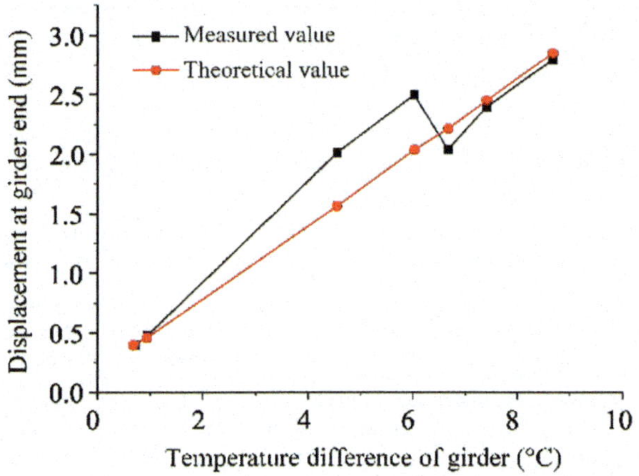

Fig. 8.20 Temperature displacements at simply-supported girder end

Wuhan–Guangzhou HSR, focusing on rail temperature forces, girder-rail relative displacements, and displacements at the girder end. The following conclusions are drawn from these tests:

(1) The temperature forces on the rails within the turnouts differed between the CWR turnout group area and individual turnouts on the bridge, owing to the mutual impacts among turnouts within the turnout group area.
(2) The comparisons between the measured values and the theoretical calculation results indicated the close alignment between the simulation results and test results regarding turnout stress, validating the correctness of the computational theories.
(3) Based on the measured values and simulation results, the stress in the CWR turnouts on Changshan Super Major Bridge complies with the design requirements.

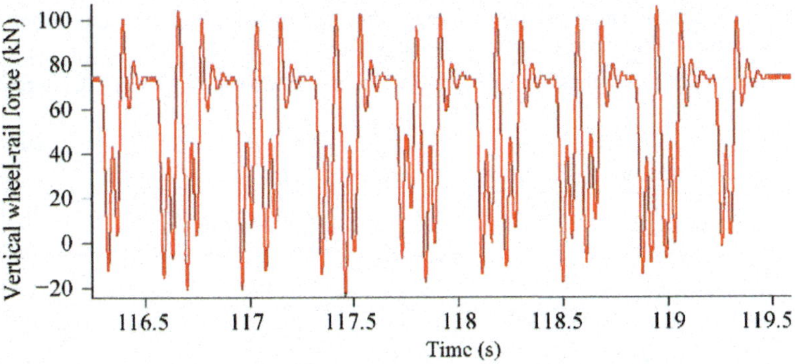

Fig. 8.21 Typical waveform of vertical wheel-rail forces

8.1.6 Dynamic Test of Turnout Group

Valid data were acquired from the dynamic tests on the turnouts with an EMU passing through on the main route 20 times. The basic information is shown as follows.

An inspection train consisting of a DF_{11} + a track inspection car (999,287) + a DF_{11} ran an up and down round trip between Guangzhoubei Station and Chenzhouxi Station, with a maximum running speed of 160 km/h.

A CRH_2-068C EMU ran from K1925 to Guangzhoudong Station for the level-by-level speed increase test. The EMU passed through turnout on the main route at the following speed levels: 240, 260, 280, 300, 310, 320, 330, 340, and 350 km/h.

1. Wheel-Rail Force, Derailment Coefficient, and Wheel Unloading Rate

The typical waveform of vertical wheel-rail forces when the EMU passes through the turnout on the main route is shown in Fig. 8.21.

Table 8.10 presents the maximum wheel unloading rate and derailment coefficient of the cross section in the section preceding the switch rail (between No. 1 and 2 turnout sleepers), and Figs. 8.22 and 8.23 present their relations with speeds, respectively.

As illustrated in Table 8.10, and Figs. 8.23 and 8.24, when the EMU passed through the No. 2 turnout on the main route, the maximum wheel unloading rate was 0.55, the maximum derailment coefficient was 0.51, and the maximum lateral wheel-rail force was 14 kN in the section preceding the switch rail. When the EMU passed through the No. 6 turnout on the main route, the maximum wheel unloading rate was 0.55, the maximum derailment coefficient was 0.55, and the maximum wheel-rail lateral force was 13.25 kN in the section preceding the switch rail. All the test results fell within the limits of safety indexes.

Table 8.10 Maximum values of safety indexes in the section preceding switch rail when passing through ballastless turnout on bridge on the main route

Speed (km/h)	Wheel unloading rate		Derailment coefficient		Wheel-rail lateral force (kN)	
	No. 2 turnout	No. 6 turnout	No. 2 turnout	No. 6 turnout	No. 2 turnout	No. 6 turnout
240	0.38	0.43	0.23	0.12	8.91	9.35
260	0.42	0.47	0.22	0.25	9.32	10.17
280	0.51	0.54	0.18	0.40	10.25	10.55
300	0.46	0.33	0.36	0.26	10.63	9.83
310	0.48	0.52	0.18	0.36	11.05	10.82
320	0.55	0.47	0.37	0.45	10.89	11.25
330	0.45	0.52	0.41	0.51	12.91	11.93
340	0.51	0.55	0.44	0.51	13.59	13.25
350	0.53	0.51	0.51	0.55	14.00	12.67

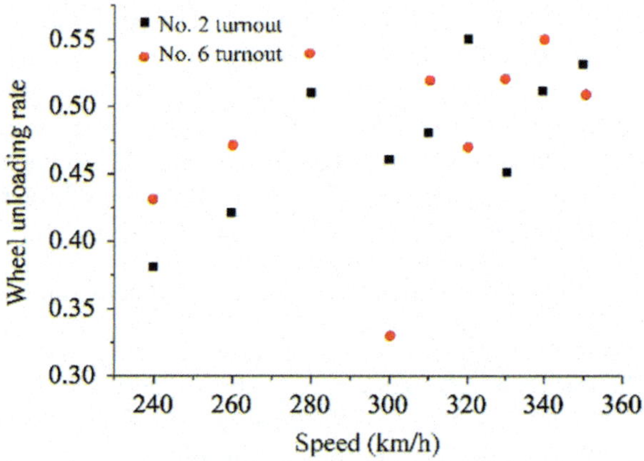

Fig. 8.22 Unloading rate variations preceding turnout with EMU speeds

Table 8.11 presents the maximum wheel unloading rate and derailment coefficient at the straight stock rail position of the frog (between No. 95 and 96 turnout sleepers), and Figs. 8.24 and 8.25 display their relations with speeds, respectively.

As illustrated in Table 8.11, and Figs. 8.24 and 8.25, when the EMU passed through the No. 2 turnout on the main route, the maximum wheel unloading rate was 0.58, the maximum derailment coefficient was 0.53, and the maximum wheel-rail lateral force was 13.14 kN at the straight stock rail position of the frog. When the EMU passed through the No. 6 turnout on the main route, the maximum wheel unloading rate was 0.55, the maximum derailment coefficient was 0.54, and the maximum wheel-rail lateral force was 16.77 kN at the straight stock rail position of the frog. All the test results fell within the limits of safety indexes.

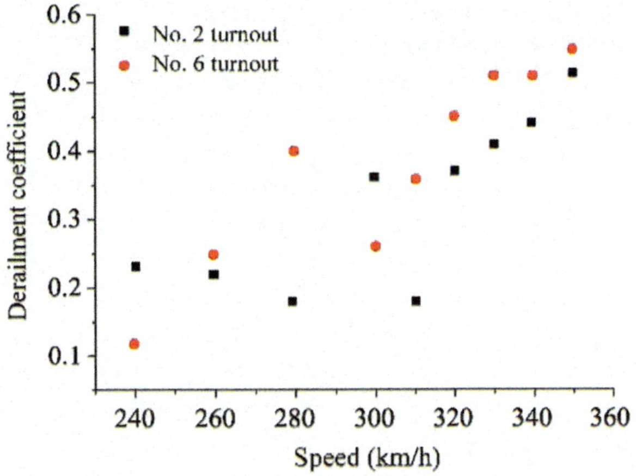

Fig. 8.23 Derailment coefficient variations preceding turnout with EMU speeds

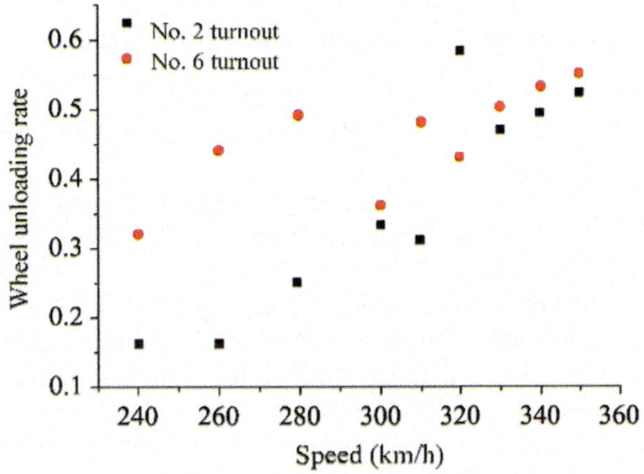

Fig. 8.24 Unloading rate variations at frog with EMU speeds

Table 8.11 Safety indexes of straight stock rail at frog when passing on the main route

Speed (km/h)	Wheel unloading rate		Derailment coefficient		Wheel-rail lateral force (kN)	
	No. 2 turnout	No. 6 turnout	No. 2 turnout	No. 6 turnout	No. 2 turnout	No. 6 turnout
240	0.16	0.32	0.12	0.09	9.66	11.68
260	0.16	0.44	0.10	0.19	9.83	8.99
280	0.25	0.49	0.34	0.38	10.36	13.15
300	0.33	0.36	0.28	0.30	9.87	12.81
310	0.31	0.48	0.36	0.38	11.05	12.97
320	0.58	0.43	0.45	0.47	11.91	13.41
330	0.47	0.50	0.48	0.50	12.29	11.56
340	0.49	0.53	0.51	0.54	13.14	14.28
350	0.52	0.55	0.53	0.52	12.79	16.77

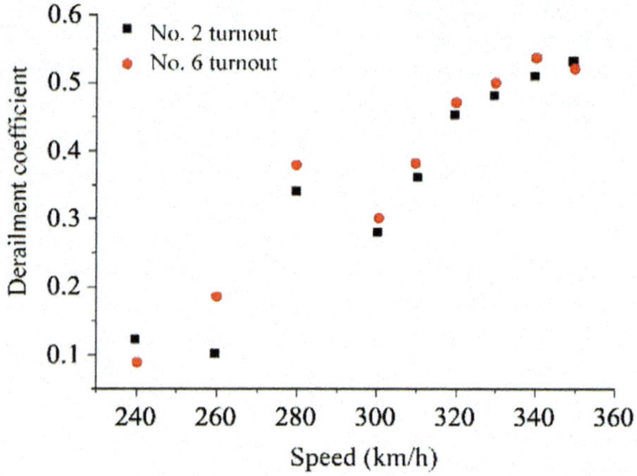

Fig. 8.25 Derailment coefficient variations at frog with EMU speeds

2. Lateral and Vertical Dynamic Displacement of Rail

Figures 8.26 and 8.27 present the waveform of vertical and lateral displacements of the rail when the EMU passed through the turnout section on the main route, respectively.

Table 8.12 presents the maximum vertical and lateral displacements of the rail at the front end of the switch. Figures 8.28, 8.29 and 8.30 present their relations with the EMU speeds.

As illustrated in Table 8.12 and Figs. 8.28, 8.29, 8.30, 8.31, when the EMU passed through the No. 2 turnout on the main route, the measured maximum vertical displacement of the straight stock rail at the front end of the switch was 0.460 mm and the

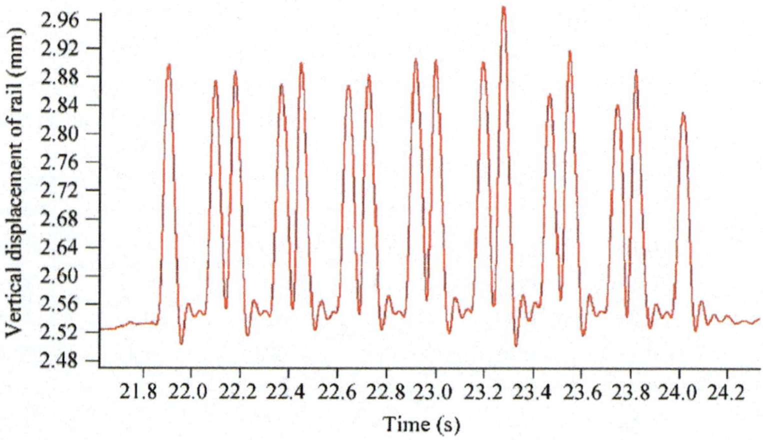

Fig. 8.26 Waveform of vertical displacements of rail in turnout section

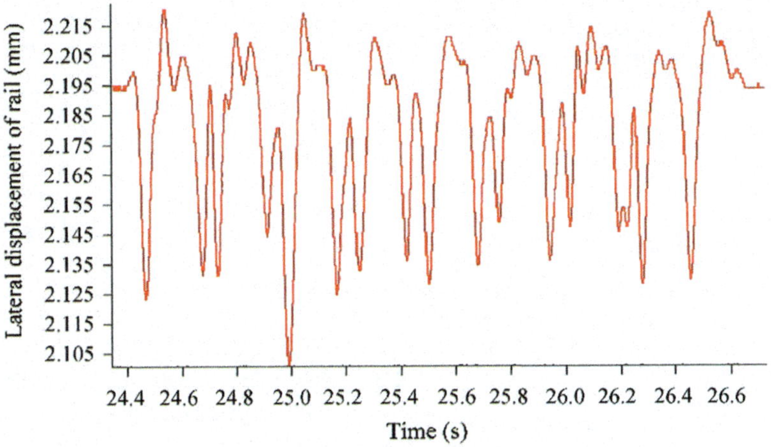

Fig. 8.27 Waveform of lateral displacements of rail in turnout section

Table 8.12 Maximum vertical and lateral displacements of rail at front end of switch

EMU speed (km/h)	Vertical displacement of straight stock rail (mm)		Lateral displacement of straight stock rail (mm)		Vertical displacement of curved stock rail (mm)		Lateral displacement of curved stock rail (mm)	
	No. 2 turnout	No. 6 turnout	No. 2 turnout	No. 6 turnout	No. 2 turnout	No. 6 turnout	No. 2 turnout	No. 6 turnout
240	0.420	0.563	0.193	0.162	0.469	0.270	0.140	0.241
260	0.427	0.560	0.215	0.169	0.524	0.313	0.134	0.216
280	0.437	0.497	0.141	0.162	0.463	0.384	0.160	0.185
300	0.427	0.583	0.169	0.138	0.876	0.313	0.159	0.135
310	0.424	0.380	0.108	0.139	0.467	0.379	0.195	0.162
320	0.399	0.320	0.112	0.162	0.512	0.323	0.187	0.129
330	0.460	0.445	0.207	0.162	0.430	0.331	0.201	0.188
340	0.444	0.751	0.178	0.359	0.544	0.284	0.176	0.157
350	0.432	0.297	0.113	0.164	0.441	0.283	0.157	0.171

measured maximum lateral displacement was 0.215 mm; the measured maximum vertical displacement of the curved stock rail was 0.876 mm and the measured maximum lateral displacement was 0.201 mm. When the EMU passed through the No. 6 turnout on the main route, the measured maximum vertical displacement of the straight stock rail at the front end of the switch was 0.751 mm and the measured maximum lateral displacement was 0.359 mm; the measured maximum vertical displacement of the curved stock rail was 0.384 mm and the measured maximum lateral displacement was 0.241 mm. All the test results fell within the permissible

Fig. 8.28 Vertical displacements of straight stock rail at front end of switch

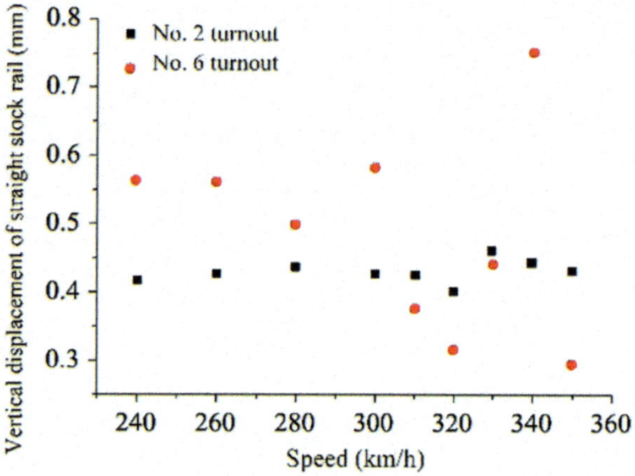

Fig. 8.29 Lateral displacements of straight stock rail at front end of switch

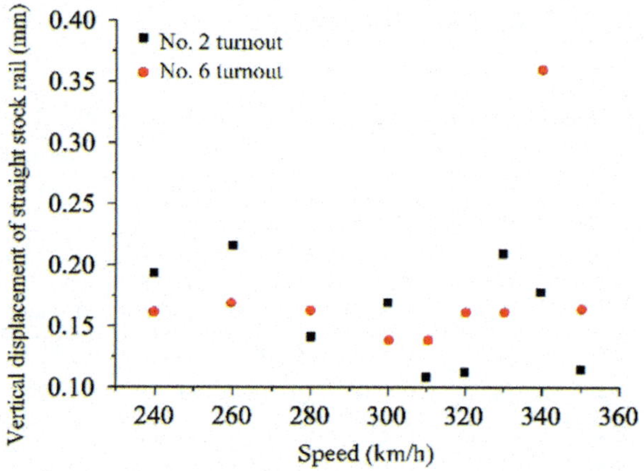

Fig. 8.30 Vertical displacements of curved stock rail at front end of switch

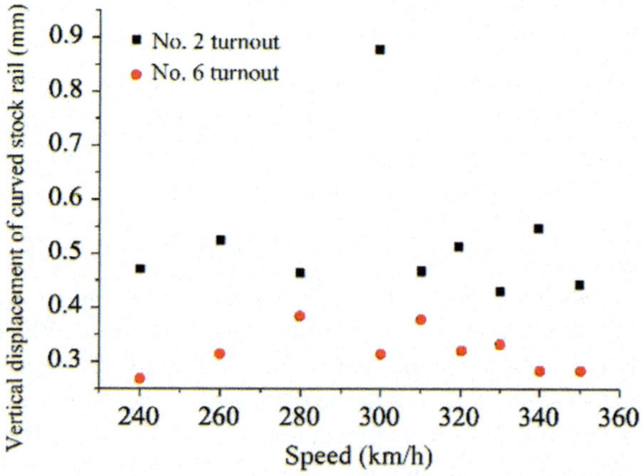

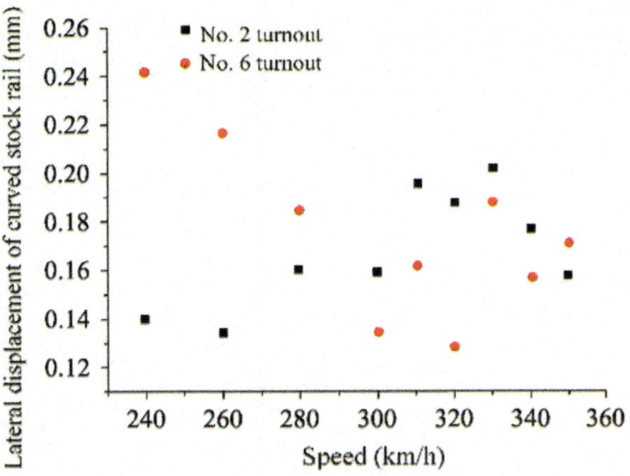

Fig. 8.31 Lateral displacements of curved stock rail at front end of switch

limits. As shown in the figures, the wheel-rail vertical and lateral displacements were not significantly influenced by EMU speeds.

Table 8.13 presents the maximum vertical and lateral displacements of the rail at the lead curve. Figures 8.32, 8.33, 8.34 and 8.35 display their relations with the EMU speeds.

As illustrated in Table 8.13 and Figs. 8.32, 8.33, 8.34, 8.35, when the EMU passed through the No. 2 turnout on the main route, the measured maximum vertical displacement was 0.683 mm and the measured maximum lateral displacement was 0.315 mm for the straight stock rail at the lead curve; the measured maximum vertical displacement was 0.379 mm and the measured maximum lateral displacement was

Table 8.13 Maximum vertical and lateral displacements of rail at lead curve

EMU speed (km/h)	Vertical displacement of straight stock rail (mm)		Lateral displacement of straight stock rail (mm)		Vertical displacement of straight switch rail (mm)		Lateral displacement of straight switch rail (mm)	
	No. 2 turnout	No. 6 turnout	No. 2 turnout	No. 6 turnout	No. 2 turnout	No. 6 turnout	No. 2 turnout	No. 6 turnout
240	0.296	0.358	0.165	0.173	0.255	0.271	0.210	0.253
260	0.472	0.513	0.178	0.167	0.374	0.306	0.294	0.281
280	0.395	0.427	0.166	0.158	0.352	0.391	0.246	0.193
300	0.463	0.536	0.162	0.181	0.364	0.353	0.175	0.146
310	0.456	0.423	0.179	0.156	0.365	0.365	0.160	0.165
320	0.399	0.382	0.168	0.167	0.379	0.336	0.211	0.193
330	0.492	0.461	0.213	0.171	0.299	0.293	0.232	0.243
340	0.641	0.674	0.289	0.362	0.311	0.368	0.212	0.251
350	0.683	0.695	0.315	0.357	0.374	0.312	0.288	0.296

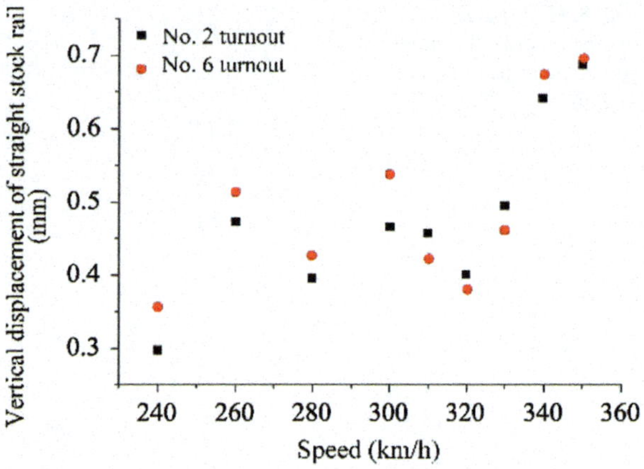

Fig. 8.32 Vertical displacements of straight stock rail at lead curve

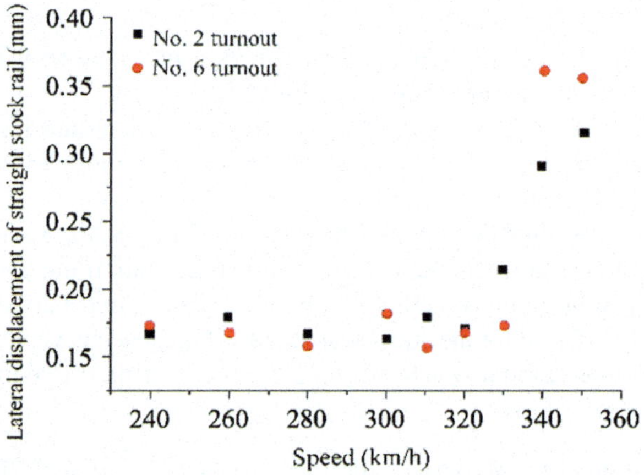

Fig. 8.33 Lateral displacements of straight stock rail at lead curve

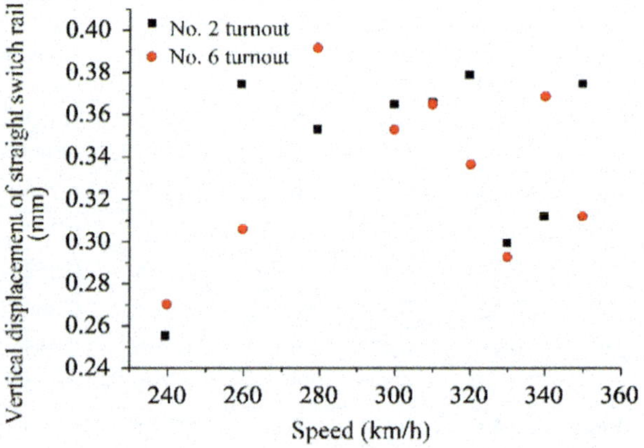

Fig. 8.34 Vertical displacements of straight switch rail at lead curve

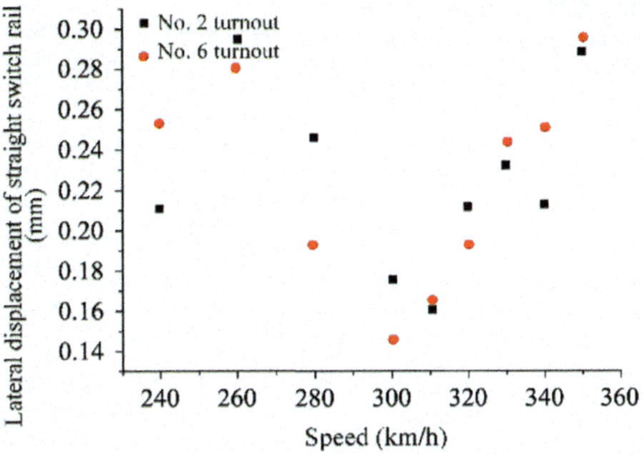

Fig. 8.35 Lateral displacements of straight switch rail at lead curve

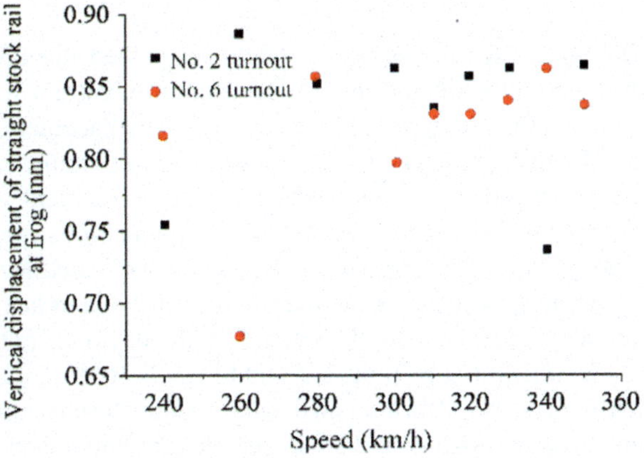

Fig. 8.36 Vertical displacements of straight stock rail at frog

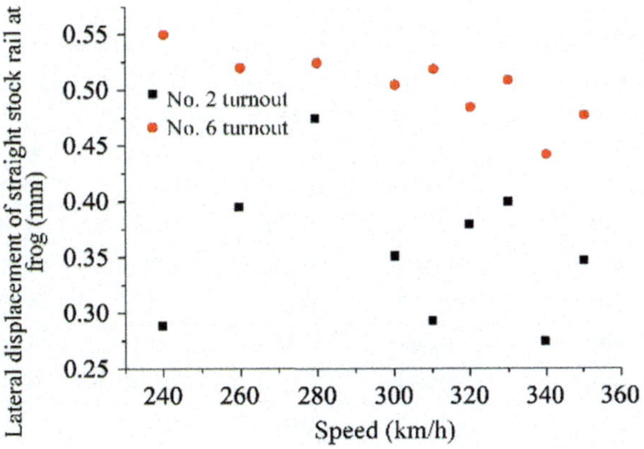

Fig. 8.37 Lateral displacements of straight stock rail at frog

Fig. 8.38 Vertical displacements at actual point of point rail

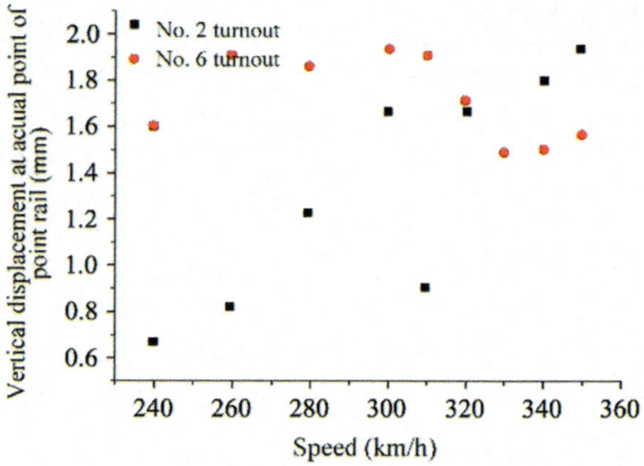

0.294 mm for the straight switch rail. When the EMU passed through the No. 6 turnout on the main route, the measured maximum vertical displacement was 0.695 mm and the measured maximum lateral displacement was 0.362 mm for the straight stock rail at the lead curve; the measured maximum vertical displacement was 0.391 mm, and measured maximum lateral displacement was 0.296 mm for the straight switch rail. All the test results fell within the limits specified in *Provisional Technical Specification for Turnout on Passenger Dedicated Railway.*

Table 8.14 presents the maximum vertical and lateral displacements of the rail at the frog. Figures 8.36, 8.37 and 8.38 display their relations with EMU speeds.

As illustrated in Table 8.14 and Figs. 8.36, 8.37 and 8.38, when the EMU passed through the No. 2 turnout on the main route, the measured maximum vertical displacement was 0.886 mm and the measured maximum lateral displacement was 0.473 mm

Table 8.14 Maximum vertical and lateral displacements of rail at frog

EMU Speed (km/h)	Vertical displacement of straight stock rail (mm)		Lateral displacement of straight stock rail (mm)		Vertical displacement of point rail (mm)	
	No. 2 turnout	No. 6 turnout	No. 2 turnout	No. 6 turnout	No. 2 turnout	No. 6 turnout
240	0.754	0.816	0.289	0.549	0.668	1.596
260	0.886	0.677	0.393	0.519	0.825	1.907
280	0.852	0.857	0.473	0.525	1.222	1.858
300	0.861	0.797	0.351	0.505	1.668	1.941
310	0.834	0.830	0.291	0.517	0.908	1.908
320	0.857	0.830	0.379	0.483	1.668	1.707
330	0.861	0.841	0.397	0.508	1.487	1.486
340	0.736	0.862	0.273	0.441	1.796	1.502
350	0.864	0.837	0.346	0.476	1.942	1.568

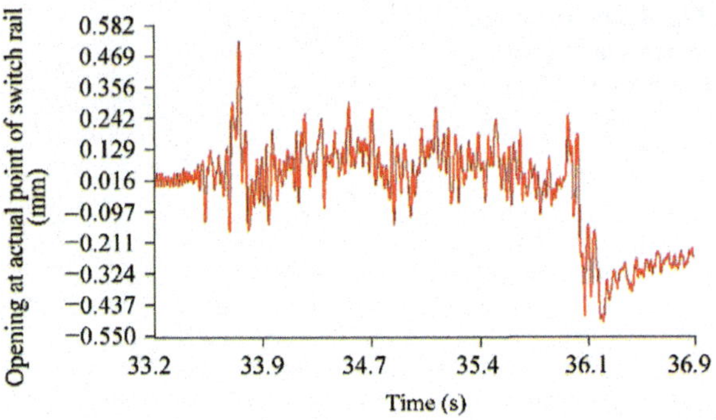

Fig. 8.39 Waveform of opening at actual point of switch rail

for the straight stock rail at the frog; the measured maximum vertical displacement at the actual end of the point rail was 1.942 mm. When the EMU passed through the No. 6 turnout on the main route, the measured maximum vertical displacement was 0.862 mm and the measured maximum lateral displacement was 0.549 mm for the straight stock rail at the frog; the measured maximum vertical displacement at the actual point of the point rail was 1.941 mm. All the test results fell within the specification limits.

3. Opening at Actual Point of Switch Rail

The opening of switch rails serves to measure their separation from the stock rails. When closed switch rails open beyond 4 mm, there is a risk that the wheel flange might collide with their actual points, potentially compromising the operational safety of trains. Figure 8.39 shows the waveform in the test of opening at the actual point of the switch rail when the EMU passed through the turnout on the main route.

Table 8.15 presents the test results at the actual point of the straight switch rail within the No. 6 turnout.

As illustrated in Table 8.15 and Fig. 8.39, the opening at the actual point of the switch rail was not significantly influenced by EMU speeds, and the maximum opening was 2.569 mm, falling within the limit specified in the specification and meeting the requirements.

4. Strength of Turnout Components

The rail bottom stress at the switch rail with a top width of 20 mm, when the EMU passed through the turnout on the main route, is shown in Fig. 8.40.

Table 8.15 Opening at actual point of switch rail

Speed (km/h)	240	260	280	300	310	320	330	340	350
Opening at actual point of switch rail (mm)	0.383	0.313	0.546	0.474	0.211	0.300	2.569	0.970	1.140

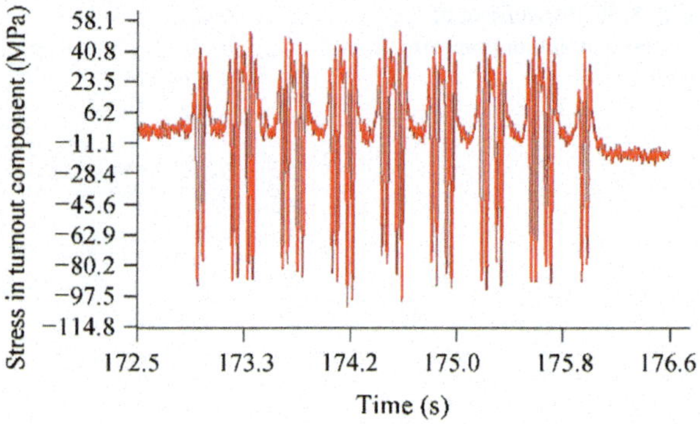

Fig. 8.40 Waveform of stresses in turnout components

Table 8.16 presents the rail bottom stresses in the switch rails and stress in the wing rails. Figures 8.41 and 8.42 display the relations between the rail bottom stress in the switch rails and wing rails and the EMU speeds, respectively.

As illustrated in Table 8.16, and Figs. 8.41 and 8.42, with the EMU speed increase, the collision impact between the wheelset and switch rail/wing rail was intensified, and the stresses in the switch rail bottom and wing rail rose. When the EMU passed through the No. 2 turnout on the main route, the measured maximum rail bottom stress at the switch rail section with a top width of 20 mm was 93.064 MPa and the measured maximum wing rail stress was 90.279 MPa. When the EMU passed through the No. 6 turnout on the main route, the measured maximum rail bottom stress at the switch rail section with a top width of 20 mm was 96.243 MPa and the measured maximum wing rail stress was 97.644 MPa. All the test values fell within

Table 8.16 Strength of turnout components

EMU speed (km/h)	Rail bottom stress at switch rail with 20 mm top width (MPa)		Stress in wing rail (MPa)	
	No. 2 turnout	No. 6 turnout	No. 2 turnout	No. 6 turnout
240	52.100	72.172	67.582	73.636
260	69.340	76.540	71.467	78.483
280	73.216	74.448	83.400	81.718
300	79.518	79.598	77.988	80.105
310	75.063	79.516	88.563	91.500
320	85.681	96.243	90.279	93.442
330	89.367	90.982	87.291	92.753
340	84.615	94.620	88.401	92.894
350	93.064	90.772	90.010	97.644

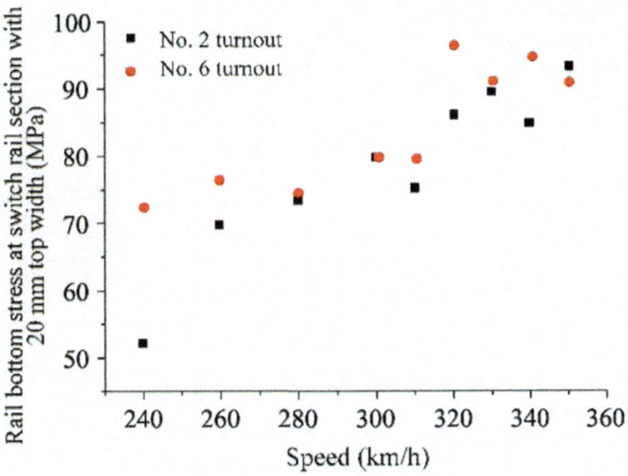

Fig. 8.41 Switch rail bottom stress variations with EMU speeds

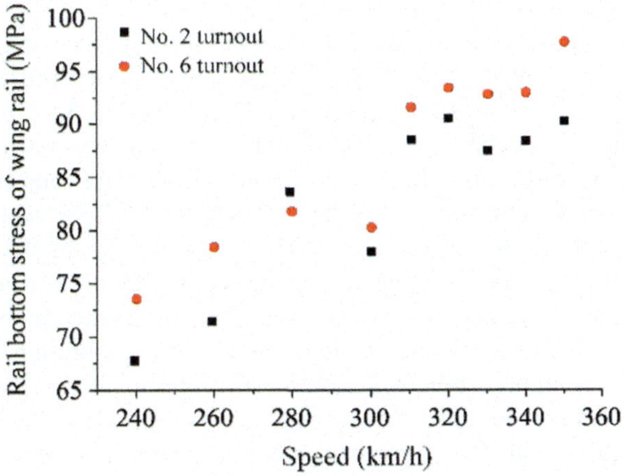

Fig. 8.42 Wing rail bottom stress variations with EMU speeds

the limits. The turnout components at susceptive positions have sufficient strength margins to meet the requirements for the safe operation of EMUs.

5. Vibration Acceleration of Track

Figure 8.43 shows the waveform of vertical vibration accelerations of the rail when the EMU passed through the turnout on the main route.

Table 8.17 displays the test results of vertical vibration accelerations of the rails and sleepers (or track slabs) within the turnout on the bridge. Figures 8.44, 8.45, 8.46, 8.47 and 8.48 illustrate the relations between vertical accelerations and EMU speeds.

As illustrated in Table 8.17 and Figs. 8.44, 8.45, 8.46, 8.47, 8.48, when the EMU passed through the No. 2 turnout on the main route, the maximum acceleration of the straight stock rail was 319.1 *g* and the maximum acceleration of the sleeper under the

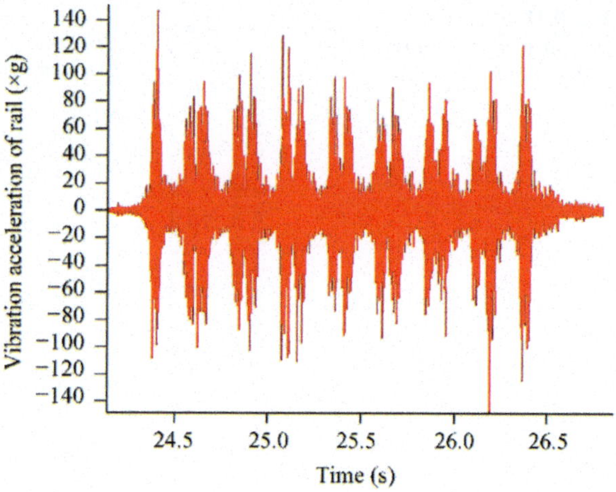

Fig. 8.43 Waveform of vertical vibration accelerations of rail

straight stock rail was 0.27 *g* in the section preceding the switch rail; the maximum acceleration of the straight stock rail at the lead curve was 667.8 *g*; the maximum acceleration of the straight stock rail was 351.32 *g*, and the maximum acceleration of the sleeper under the straight stock rail was 6.0 *g* at the frog. When the EMU passed through the No. 6 turnout on the main route, the maximum acceleration of the straight stock rail at the switch was 278.764 *g*; that at the lead curve was 603.780 *g*; that at the frog was 332.046 *g*. The maximum sleeper acceleration at the switch was 0.263 *g*; and that at the frog was 6.328 *g*. The vertical accelerations of the rails and sleepers increased with the EMU speed increment.

As illustrated in the diagrams of vertical vibration accelerations of the rails in the section preceding the switch rail, the accelerations within the No. 6 turnout were higher than those within the No. 2 turnout. This is mainly because the EMU passed through the No. 6 turnout in a facing-point manner from the No. 2 turnout, and the vibration responses of the No. 6 turnout were affected by the No. 2 turnout, i.e. the vibration responses intensified as the EMU passed through these turnouts.

6. Vibration Characteristics of Bridge

Figure 8.49 shows the waveform of vertical vibration accelerations of the bridge when the EMU passed through the turnout on the main route.

Table 8.18 presents the measured results of vertical and lateral vibration accelerations of the bridge. Figures 8.50, 8.51, 8.52 and 8.53 illustrate their relations with EMU speeds.

As illustrated in Table 8.18 and Figs. 8.50, 8.51, 8.52, 8.53, the measured maximum vertical vibration acceleration at the midspan was 3.952 and 2.958 m/s^2, 2.51 and 2.42 m/s^2 respectively when the EMU passed through the No. 15 and 16 girders, and the No. 8 and 9 girders, all meeting the specification requirements; the measured maximum lateral vibration acceleration at the midspan of the No. 15 and

Table 8.17 Maximum vertical vibration accelerations of turnout rails and sleepers ($\times g$)

Speed of EMU (km/h)	Straight stock rail in section preceding switch rail		Straight stock rail at lead curve		Straight stock rail at frog		Sleeper under straight stock rail in section preceding switch rail		Sleeper under straight stock rail at frog	
	No. 2 Turnout	No. 6 Turnout	No. 2 Turnout	No. 6 Turnout	No. 2 Turnout	No. 6 Turnout	No. 2 Turnout	No. 6 Turnout	No. 2 Turnout	No. 6 Turnout
240	147.10	159.94	233.27	183.62	216.61	162.51	0.14	0.13	0.16	0.51
260	143.99	152.26	193.25	193.43	269.32	159.35	0.11	0.12	6.00	1.30
280	154.32	176.69	205.10	205.10	245.78	173.22	0.18	0.18	5.98	1.00
300	160.67	177.67	235.41	232.56	227.71	189.34	0.20	0.19	0.31	0.92
310	319.10	194.73	245.04	216.73	268.55	219.28	0.19	0.18	6.01	3.01
320	148.64	207.64	352.29	373.16	272.11	285.19	0.27	0.23	0.28	4.99
330	167.07	235.24	365.24	423.62	239.33	312.36	0.22	0.22	5.99	5.02
340	140.86	276.25	574.56	591.76	296.05	308.74	0.27	0.26	6.00	5.79
350	148.25	278.76	667.80	603.78	351.32	332.05	0.22	0.23	5.64	6.33

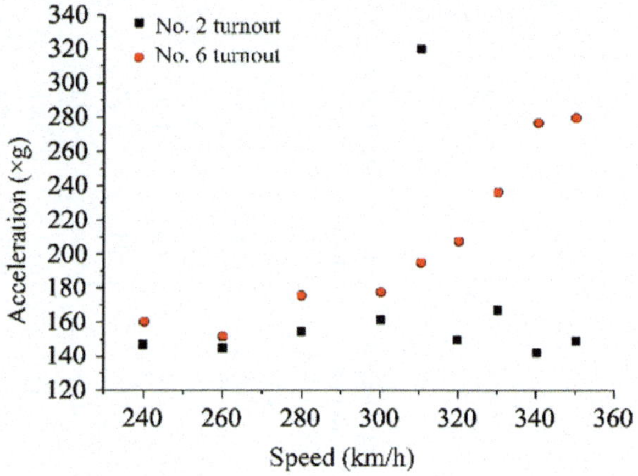

Fig. 8.44 Vertical accelerations of straight stock rail in section preceding switch rail

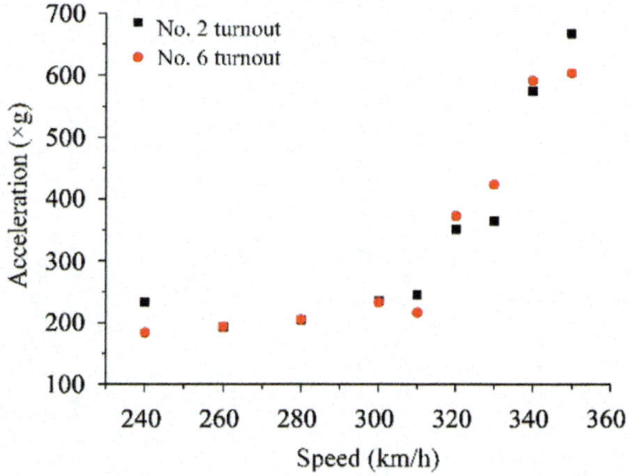

Fig. 8.45 Vertical accelerations of straight stock rail at lead curve

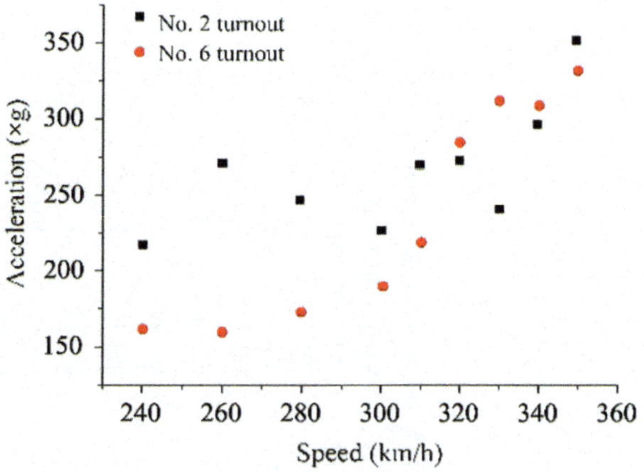

Fig. 8.46 Vertical accelerations of straight stock rail at frog

Fig. 8.47 Accelerations of sleeper under straight stock rail in section preceding switch rail

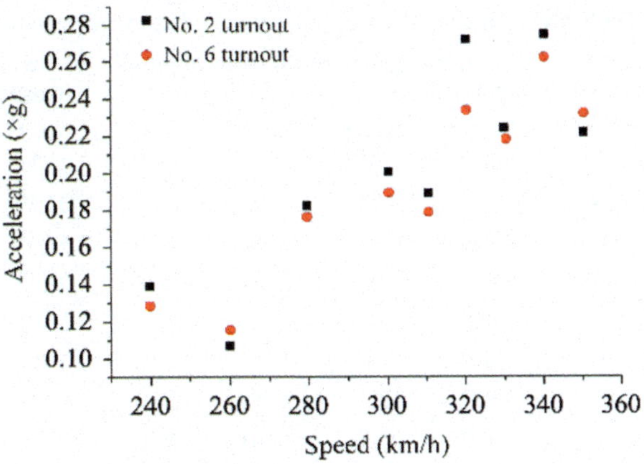

Fig. 8.48 Vertical accelerations of sleeper under straight stock rail at frog

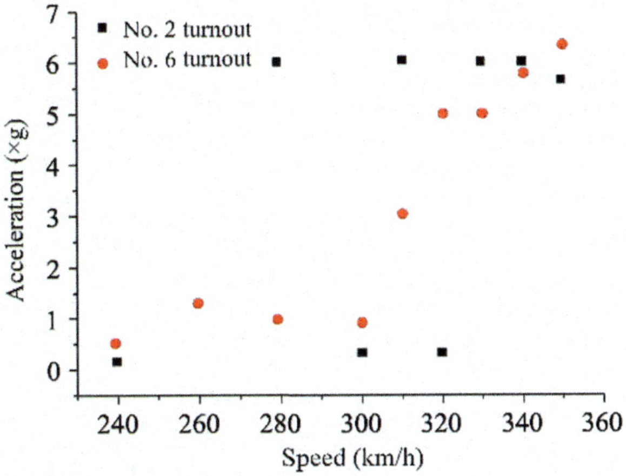

Fig. 8.49 Waveform of vertical vibration accelerations of bridge

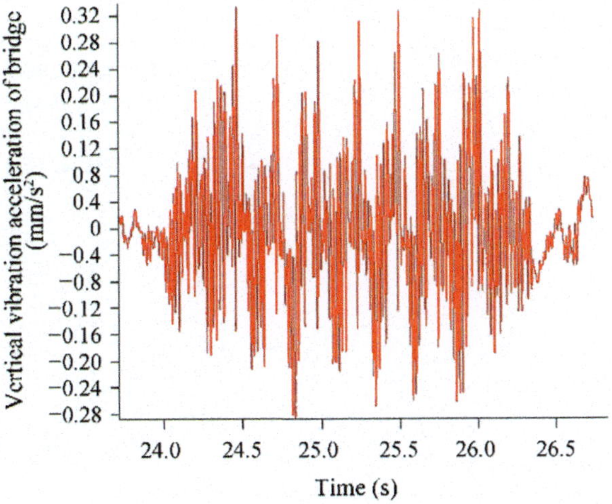

Table 8.18 Measured vertical and lateral vibration accelerations of bridge (Unit: m/s^2)

EMU speed (km/h)	Vertical vibration acceleration at girder midspan				Lateral vibration acceleration at girder midspan			
	No. 15 girder	No. 9 girder	No. 16 girder	No. 8 girder	No. 15 girder	No. 9 girder	No. 16 girder	No. 8 girder
240	2.365	0.73	1.857	0.84	0.229	0.08	0.249	0.23
260	2.960	1.21	2.380	1.50	0.165	0.25	0.365	0.31
280	3.266	1.05	2.544	0.94	0.262	0.32	0.284	0.33
300	3.952	1.69	1.740	1.78	0.371	0.46	0.683	0.49
310	3.372	1.30	2.320	1.10	0.317	0.51	0.324	0.46
320	3.875	1.86	2.958	2.10	0.333	0.48	0.345	0.41
330	2.272	2.15	2.161	2.26	0.213	0.46	0.260	0.44
340	3.297	2.20	2.575	2.35	0.210	0.57	0.229	0.65
350	3.353	2.42	2.798	2.51	0.246	0.83	0.256	0.92

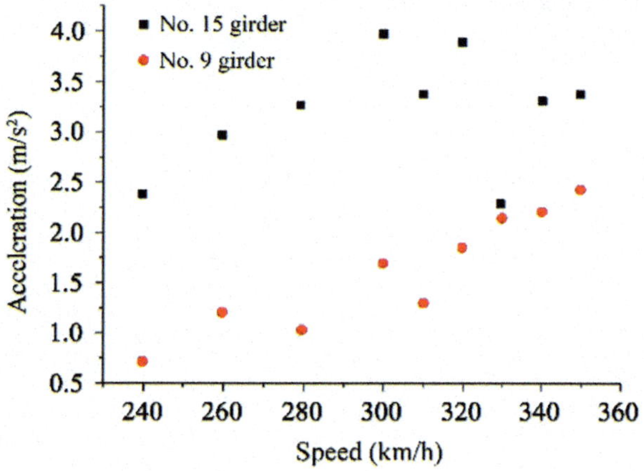

Fig. 8.50 Vertical accelerations of No. 15 and 9 girders

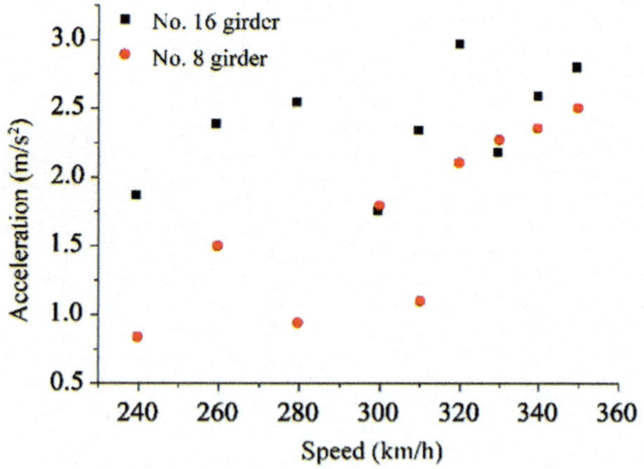

Fig. 8.51 Vertical accelerations of No. 16 and 8 girders

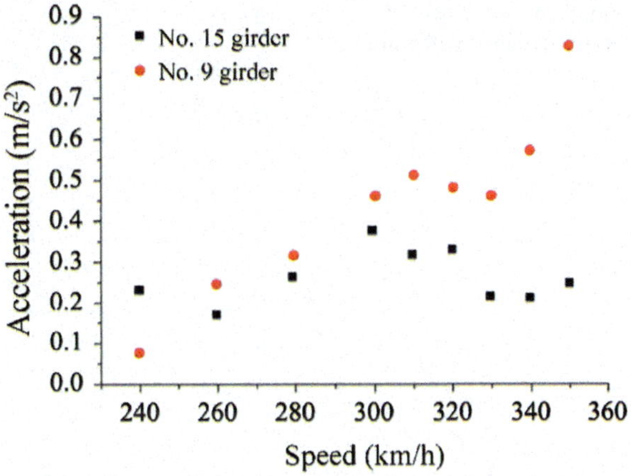

Fig. 8.52 Lateral accelerations of No. 15 and 9 girders

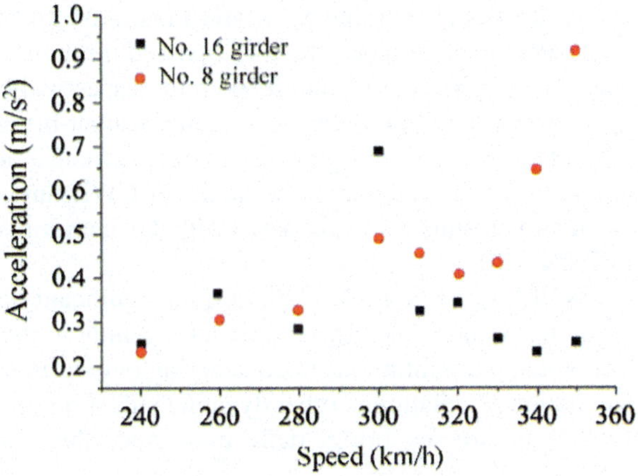

Fig. 8.53 Lateral accelerations of No. 16 and 8 girders

16 girders, and No. 8 and 9 girders were 0.371 and 0.683 m/s^2, and 0.92 and 0.83 m/s^2 respectively, all falling within the allowable limits.

Moreover, the vertical accelerations of the No. 8 and 9 girders were generally lower than those of the No. 15 and 16 girders. This difference is mainly attributed to the bridge structure at the throat area containing the No. 6 turnout, which features a variable cross-section girder with high rigidity. This contributes to the reduced vertical vibration accelerations. The lateral accelerations exhibited an opposite trend, primarily due to the relatively lower lateral rigidity of the No. 8 and 9 piers, stemming from their greater heights in comparison to the No. 15 and 16 piers. This factor exacerbates the vibrations experienced in these sections.

7. Dynamic Deflection of Bridge

Figure 8.54 illustrates the dynamic deflections of the bridge when the EMU passed through the turnout on the main route.

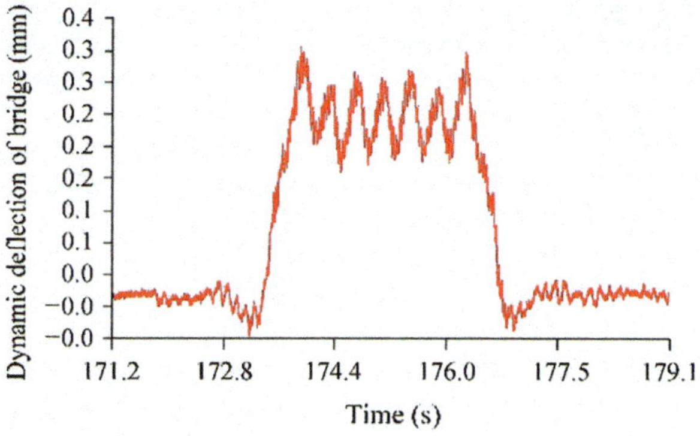

Fig. 8.54 Waveform of bridge dynamic deflections

Table 8.19 displays the measured dynamic deflections of the bridge. Figures 8.55 and 8.56 illustrate their relations with EMU speeds.

As illustrated in Table 8.19, and Figs. 8.55 and 8.56, the measured maximum midspan deflection when the EMU passed through the No. 8, 9, 15, and 16 girders was 0.680, 0.764, 0.730, and 0.768 mm, respectively.

To verify the scientificity of the train-turnout-bridge dynamic simulation computation theories, further optimize the train-turnout-bridge coupling vibration model, and improve the calculation methods for CWR turnouts on bridges, the simulation calculation results were compared with the measurements in the train tests, as shown in Table 8.20.

As illustrated in Table 8.20, despite significant disparities in the unloading rate at the front end of the switch and the opening at the exact point of the switch rail (attributable to limitations in considering only turnout structural irregularities in the simulation calculations), other dynamic indicators such as wheel-rail vertical forces, switch rail stresses, bridge deflections, and vibration accelerations exhibited close

Table 8.19 Measured deflections at bridge midspan

EMU speed (km/h)	Midspan deflection of No. 15 girder (mm)	Midspan deflection of No. 9 girder (mm)	Midspan deflection of No. 16 girder (mm)	Midspan deflection of No. 8 girder (mm)
240	0.465	0.666	0.616	0.660
260	0.454	0.684	0.724	0.604
280	0.447	0.680	0.679	0.608
300	0.467	0.764	0.768	0.680
310	0.441	0.646	0.731	0.636
320	0.476	0.616	0.484	0.548
330	0.730	0.684	0.645	0.614
340	0.576	0.654	0.618	0.630
350	0.557	0.670	0.674	0.614

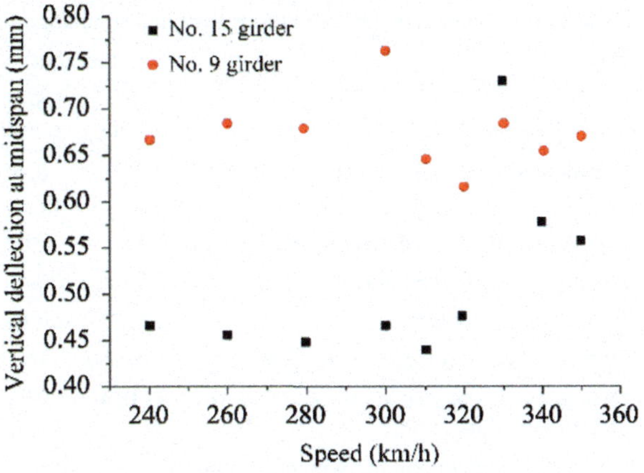

Fig. 8.55 Midspan deflections of No. 15 and 9 girders

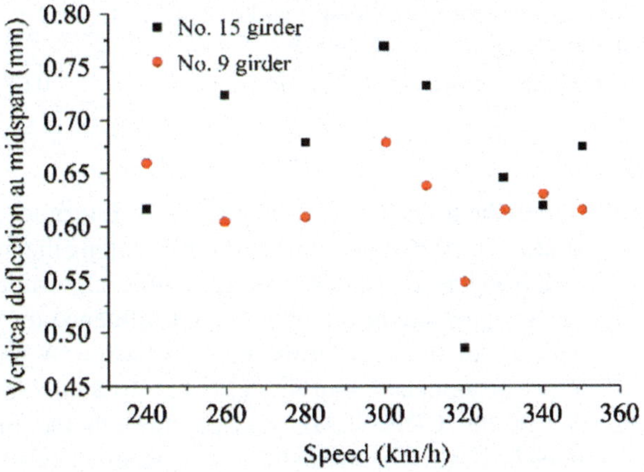

Fig. 8.56 Midspan deflections of No. 16 and 8 girders

alignment. These findings suggest that the simulation results of dynamic indexes for turnouts on bridges can effectively portray the dynamic characteristics of the train-turnout-bridge system under the dynamic loads exerted by high-speed trains.

8. Summary

To verify the computational theories for dynamic simulations of the train-turnout-bridge system, dynamic testing was conducted on the CWR turnouts on Changshan Super Major Bridge, focusing on the wheel-rail vertical (lateral) forces, vertical (lateral) displacements of rails, opening at the actual point of switch rails, strength of turnout components, and vibration characteristics of the bridge. The following conclusions and recommendations are drawn from the study results:

(1) The dynamic testing results were closely aligned with the theoretical simulation results, validating the correctness of the dynamic theory for CWR turnouts on bridges.

Table 8.20 Comparison between test results and simulation results at 350 km/h

Item	No. 2 turnout		No. 6 turnout	
	Test result	Simulation result	Test result	Simulation result
Derailment coefficient at front end of switch	0.51	0.40	0.55	0.40
Unloading rate at front end of switch	0.53	0.04	0.51	0.04
Derailment coefficient at frog	0.53	0.40	0.52	0.36
Unloading rate at frog	0.52	0.88	0.55	0.81
Vertical wheel-rail force at frog (kN)	108.260	160.9	120.814	159.3
Dynamic stress at switch rail section with 20 mm top width (MPa)	93.064	93.1	90.772	95.2
Opening at actual point of switch rail (mm)	–	–	2.569	0.43
Midspan deflection (mm)	0.674	0.63	0.768	0.58
Vertical vibration acceleration at girder midspan (m/s^2)	3.353	2.8	2.51	2.1
Lateral vibration acceleration at girder midspan (m/s^2)	0.256	0.61	0.92	0.46

(2) When the train passed through the CWR turnout group on the bridge, due to the mutual impact of dynamic responses among the turnouts, the dynamic responses amplified as the train passed through each turnout in sequence.
(3) EMU speeds exhibited significant influence on various dynamic performance indexes of the train, turnouts, and bridge. A higher speed corresponded to a more pronounced coupling effect between the turnouts and the bridge.
(4) When the CRH_2 EMU passed through the turnout on the bridge at high speeds, due to light axle loads and small wheel-rail vertical forces, the dynamic responses within the CWR turnout group area on Changshan Super Major Bridge fell within the allowable limits.
(5) As a result of the inadequate post-construction track geometry in the turnout section, the test EMU experienced lurching at the switches and frogs, highlighting the rigorous demands placed on geometry for high-speed turnouts. Therefore, it is recommended to take the following enhancing measures in the design of turnouts on bridges in the future: (1) Increase the vertical rigidity of bridges; (2) Rationally position turnouts, especially frogs; (3) Appropriately improve the elasticity of substructure; (4) Further optimize the wheel-rail interactions within turnouts and reduce structural irregularities.

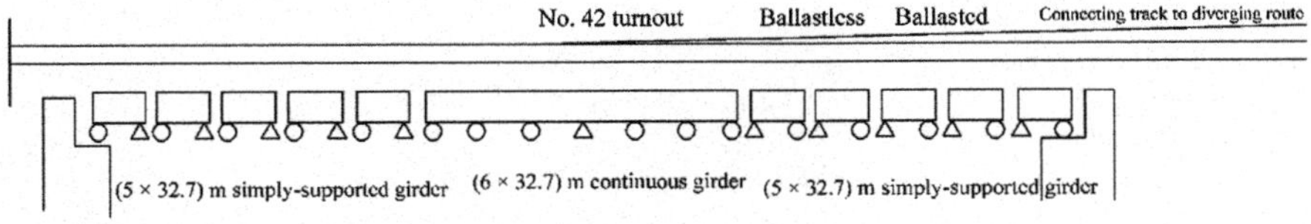

Fig. 8.57 Arrangement of test site

Fig. 8.58 Test scene

8.2 Dynamic and Static Testing of No. 42 Slab Turnout on Bridge of Yangzhou Connecting Track of Beijing–Shanghai High-Speed Railway

8.2.1 Test Overview

The No. 42 turnout on the up track of Yangzhou Connecting Track of Beijing–Shanghai HSR is installed on a (6 × 32.7) m prestressed concrete continuous girder, with a front length of 60.573 m and a rear length of 96.627 m, and a total length of 157.2 m. The permissible negotiation speed across the turnout is 350 km/h on the main route and 160 km/h on the diverging route. The arrangement of the turnout and bridge is shown in Fig. 8.57, and the testing scene is shown in Fig. 8.58.

A new-generation CRH-380A high-speed train was employed for testing [7]. It was operated at escalating speed levels as follows: 60, 80, 90, 110, 150, 160, 180, 190, 200, 210, 220, 230, 300, 320, 340, 350, 360, 370, and 380 km/h.

8.2.2 Test Items

(1) Safety: wheel-rail vertical forces (*P*) and lateral force (*Q*) at the switch, lead curve, and frog. The test results are utilized to calculate the derailment coefficient of inner and outer wheels, wheel unloading rates, and wheel-rail lateral forces

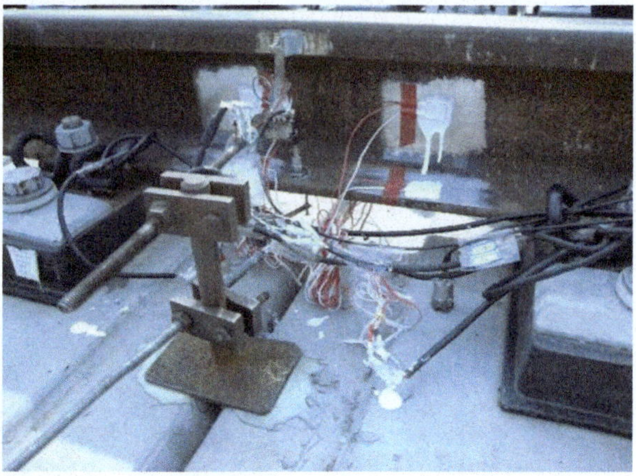

Fig. 8.59 Measuring points for wheel-rail force, rail displacement, rail acceleration, and track slab acceleration

of rolling stock. These calculation results were further leveraged to determine the operational safety of trains.

(2) Deformation of turnout components: vertical and lateral displacements of rails within the switch and frog.
(3) Strength of turnout components: stress of switch rail at 30 mm cross section, stress of switch rail at 50 mm cross section, and rail bottom stress of point rail at 35 mm cross section.
(4) Vibration characteristics of ballastless track and rail: vertical vibration accelerations of turnout sleeper and rail at point rail, vertical vibration accelerations of turnout sleeper and rail at lead curve.
(5) Vibration characteristics of bridge structure: vertical and lateral vibration accelerations at midspan in the turnout section.
(6) Rail temperature force.
(7) Temperature gradient and warping stress of turnout slab.

The test items are shown in Figs. 8.59, 8.60, 8.61 and 8.62.

8.2.3 Test Method

(1) Wheel-rail vertical force (P) and lateral force (Q): The "shear stress method" [8] is used for measurements according to the requirements of *Track Side Test Methods of Vertical and Lateral Wheel-Rail Forces* (TB/T 2489-94) when a train passes through the measuring points at varying speeds. The test results are used to calculate the derailment coefficient (*Q/P*) and wheel unloading rate.
(2) Vertical and lateral elastic displacement of stock rail: The vertical and lateral elastic displacements of the switch rail, switch, and frog in the turnout section are measured using spring-type displacement meters.

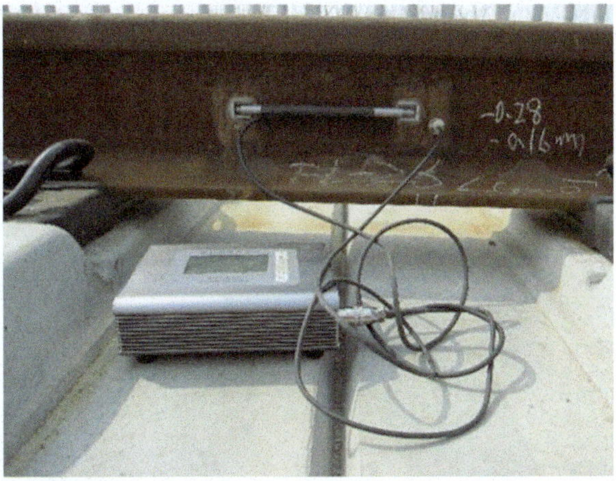

Fig. 8.60 Measuring points for rail temperature force

Fig. 8.61 Measuring points for vertical and lateral acceleration of bridge

(3) Vibration acceleration of rail and track slab at lead curve and frog: The vibration accelerations of the track slab and rail at the lead curve and traction point of the frog within the turnout using accelerometers. The test results are used to determine the dynamic influence of train operation on the turnout structure.

(4) Rail bottom stress at switch rail sections with a top width of 30 and 50 mm and at point rail section with a top width of 35 mm: The "stress method" is used for measurements. The test results are used to determine the strength and dynamic performance of the key turnout components.

(5) Vibration characteristics of bridge: The vibration accelerations at the midspan of the continuous girder are measured using vibration pickup instruments to determine the dynamic impact of train operation on the bridge structure.

Fig. 8.62 Measuring points for temperature gradient and warping stress of turnout slab

(6) Temperature gradient and warping stress of ballastless turnout slab: JMZX-212 intelligent digital vibrating wire strain gauges and JMZX-300X general-purpose tester are used for testing.
(7) Rail temperature force: TS-5 rail temperature strain gauges developed by Beijing Jiaotong University are used for testing.
(8) Displacement of switch rail and point rail relative to stock rail and wing rail: Vernier calipers are used for testing.

8.2.4 Static Test of Single Turnout

1. Temperature Force on Rail

A total of 33 measuring points were arranged for testing, including 26 on the straight stock rail and 7 on the curved closure rail within the turnout. All the measuring points were installed at the rail web. To facilitate data recording using TS-5 test instruments, the straight stock rail was numbered No. 1 rail and the curved closure rail was numbered No. 2 rail.

From February 15 to March 10, 2011, the test team performed tests on the rail temperature forces within the No. 42 turnout on the up track of Yangzhou Connecting Track of Beijing–Shanghai HSR, as shown in Fig. 8.63. A total of 15 sets of valid test data were acquired.

Through comparison, 4 sets of rail temperature force test data were selected for analysis. Table 8.21 displays the temperature forces on the No. 1 rail under different ranges of track slab and rail temperature variations.

Table 8.22 displays the temperature forces on the No. 2 rail under different ranges of track slab and rail temperature variations.

Fig. 8.63 Rail temperature force test

Table 8.23 displays the displacements at the actual points of the point rail and switch rail relative to the stock rail and wing rail under different ranges of temperature variations of the track slab and rail.

Using an overall model established for the No. 42 CWR turnout on the bridge of the up track of the Yangzhou Connecting Track along Beijing–Shanghai HSR, theoretical simulations were carried out for the above three sets of test results, and the simulation results were subsequently compared with the measured results, as shown in Figs. 8.64, 8.65 and Table 8.24.

As illustrated in the above figures and tables, under temperature changes, the temperature forces near the actual point of the switch rail were significant, mainly attributed to the expansion/contraction zone formed due to the track slab break preceding the turnout, leading to notable additional forces on the stock rail. The measured values were closely aligned with the theoretical calculation results, validating the correctness of the computational theories of longitudinal forces on CWR turnouts on bridges.

2. Stress of Turnout Slab

From February 15 to 20, 2011, The No. 14 and 41 turnout slabs of the No. 42 turnout on the up track were selected for temperature and stress tests. The turnout slabs, constructed as irregular quadrangles, were made of prestressed reinforced concrete, with a C55 compressive strength. The layout of the measuring points is shown in Figs. 8.66 and 8.67.

One JMZX-215 embedded concrete strain gauge was pre-embedded respectively on the top surface of the base slab at the corresponding turnout slab positions, and one JMZX-212 intelligent digital vibrating wire strain gauge was installed transversely and longitudinally at each measuring point on the upper surface of the turnout slabs. 1-Z indicates the longitudinal concrete strain gauge at the No. 1 measuring point, and 1-H indicates the transverse concrete strain gauge at the No. 1 measuring point.

Table 8.21 Measured temperature forces on rail 1 (Unit: kN)

Measuring point	Temperature difference of rail: 4.1 °C; temperature difference of track slab: 2.9 °C	Temperature difference of rail: 10.2 °C; temperature difference of track slab: 7.5 °C	Temperature difference of rail: 12.3 °C; temperature difference of track slab: 9.1 °C
1-1	−63.70	−192.66	−221.06
1-2	−105.50	−205.01	−224.65
1-3	−98.80	−203.70	−236.56
1-4	−89.60	−185.38	−240.24
1-5	−93.00	−172.94	−235.73
1-6	−89.50	−198.85	−225.01
1-7	−76.20	−173.04	−217.01
1-8	−70.00	−165.96	−230.39
1-9	−87.00	−167.74	−207.19
1-10	−67.20	−172.02	−215.33
1-11	−43.10	−192.37	−232.94
1-12	−73.20	−181.3	−195.39
1-13	−56.20	−165.68	−189.26
1-14	−45.30	−165.24	−205.36
1-15	−43.50	−172.09	−218.60
1-16	−55.40	−205.55	−245.64
1-17	−64.30	−210.85	−240.94
1-18	−85.30	−1 927	−252.84
1-19	−89.50	−211.3	−273.77
1-20	−66.40	−200.05	−267.95
1-21	−63.00	−213.59	−260.85
1-22	−89.60	−252.39	−272.53
1-23	−83.90	−285.95	−310.62
1-24	−115.00	−289.01	−327.79
1-25	−97.01	−177.71	−245.63
1-26	−63.00	−189.30	−244.89

(1) Temperature test

The test results of daily temperature variations at the center of the turnout slab are shown in Fig. 8.68.

As illustrated in Fig. 8.68, the temperature reached its lowest point at 5 a.m., recorded at 4 °C, and peaked at 2 p.m., recorded at 15.5 °C. The surface temperature of the turnout slab varied periodically, with a minimum of 7.7 °C at 7 a.m. and a maximum of 13 °C at 1 p.m.

Table 8.22 Measured temperature forces on rail 2 (Unit: kN)

Measuring point	Temperature difference of rail: 4.1 °C; temperature difference of track slab: 2.9 °C	Temperature difference of rail: 10.2 °C; temperature difference of track slab: 7.5 °C	Temperature difference of rail: 12.3 °C; temperature difference of track slab: 9.1 °C
2-1	−52.46	−110.76	−142.53
2-2	−47.45	−139.89	−202.12
2-3	−67.94	−180.95	−207.66
2-4	−63.46	−167.65	−228.46
2-5	−49.05	−174.57	−245.72
2-6	−52.83	−192.31	−202.64
2-7	−78.97	−187.46	−233.72

Table 8.23 Measured relative displacement of switch rail and point rail (unit: mm)

Measuring point	Temperature difference of rail: 4.1 °C; temperature difference of track slab: 2.9 °C	Temperature difference of rail: 10.2 °C; temperature difference of track slab: 7.5 °C	Temperature difference of rail: 12.3 °C; temperature difference of track slab: 9.1 °C
Actual point of switch rail	2.90	5.80	6.98
Actual point of point rail	0.84	1.50	2.94

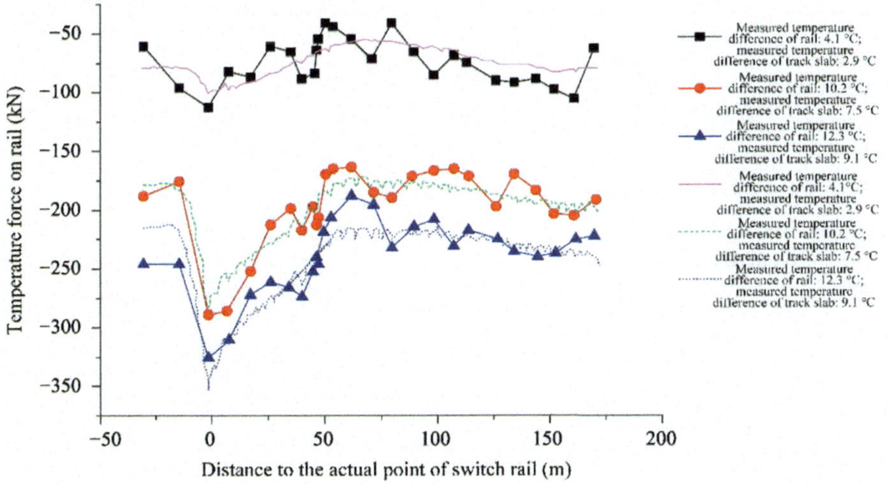

Fig. 8.64 Comparison between theoretical values and measured temperature forces on No. 1 rail

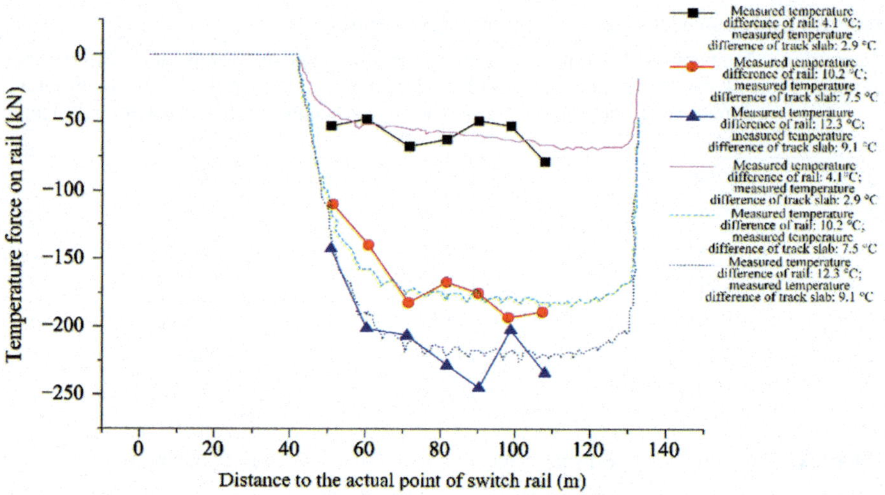

Fig. 8.65 Comparison between theoretical values and measured temperature forces on No. 2 rail

Table 8.24 Displacements of switch rail and point rail (unit: mm)

Measuring point	Temperature difference of rail: 4.1 °C; temperature difference of track slab: 2.9 °C		Temperature difference of rail: 10.2 °C; temperature difference of track slab: 7.5 °C		Temperature difference of rail: 12.3 °C; temperature difference of track slab: 9.1 °C	
	Theoretical value	Measured value	Theoretical value	Measured value	Theoretical value	Measured value
Actual point of switch rail	2.42	2.90	6.71	5.80	28	6.98
Actual point of point rail	0.95	0.84	2.46	1.50	3.34	2.94

Fig. 8.66 No. 14 turnout slab and layout of measuring points

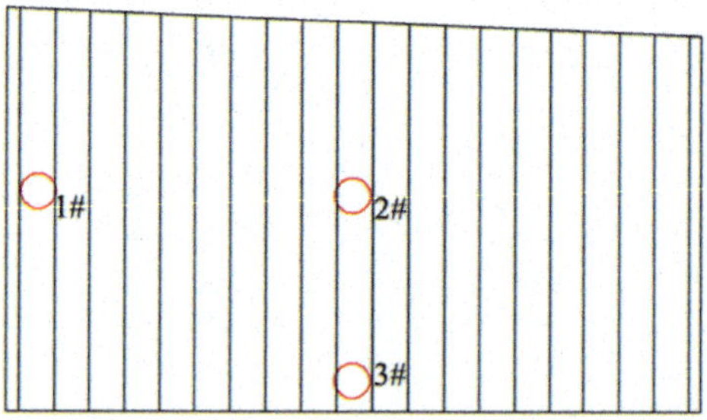

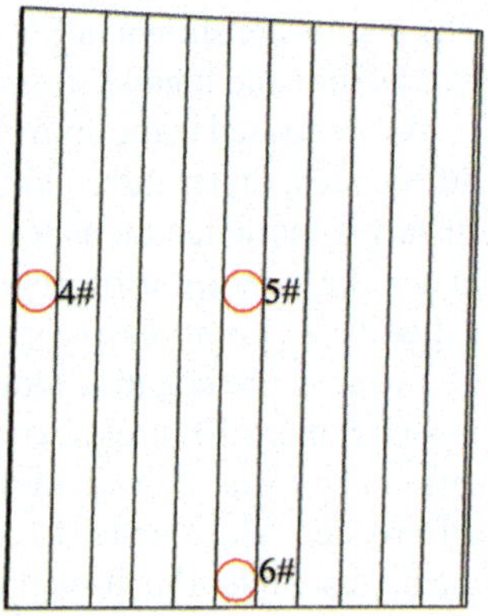

Fig. 8.67 No. 41 turnout slab and layout of measuring points

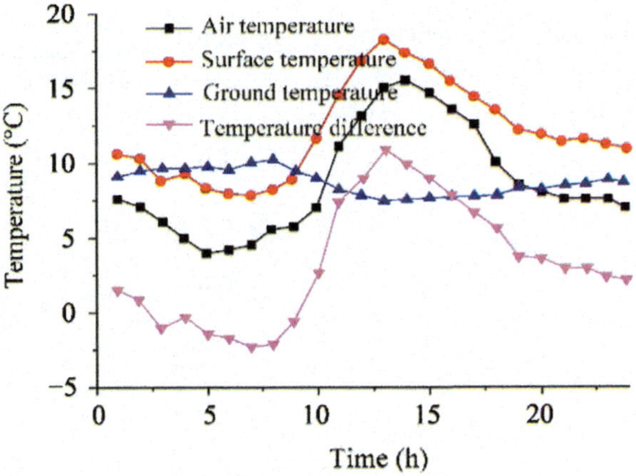

Fig. 8.68 Daily temperature variation curve of turnout slab

During the test in winter, heat was gradually transferred downward in the turnout slab through heat conduction, and the bottom surface temperature of the turnout slab varied inversely to the alterations in air temperature. Under high air temperatures, the bottom surface temperature of the turnout slab reached the lowest, recorded at 7.4 °C; when the air temperature dropped, the bottom surface temperature of the turnout slab reached the highest, recorded at 10.2 °C.

The daily temperature differences closely followed the variations in air temperature, with the negative temperature differences notably smaller than the positive ones. The maximum positive temperature difference was 10.9 °C at 1 p.m., and the maximum negative temperature difference was −2.3 °C at 7 a.m.

(2) Stress of turnout slab

The strain of the turnout slab was read on the general-purpose tester, and the longitudinal and transverse stresses of the turnout slab were obtained through calculations.

The results are shown in Figs. 8.69, 8.70, 8.71, 8.72, 8.73 and 8.74, in which positive results indicate tensile stress while negative results indicate compressive stress.

As illustrated in the above figures, warping stress amplified with rising temperature differences; under the condition of positive temperature differences, the turnout slab surface experienced tensile stress while the bottom experienced compressive stress; under the condition of negative temperature differences, the turnout slab surface experienced compressive stress while the bottom experienced tensile stress.

Because the negative temperature differences were significantly smaller than the positive ones, the stress experienced by the turnout slab under negative temperature differences was considerably lower compared to that under positive temperature differences. The tensile stress at each point on the surface of the turnout slab reached the largest under the maximum positive temperature difference. The maximum value of the No. 14 slab occurred at the No. 2 measuring point in the center of the turnout

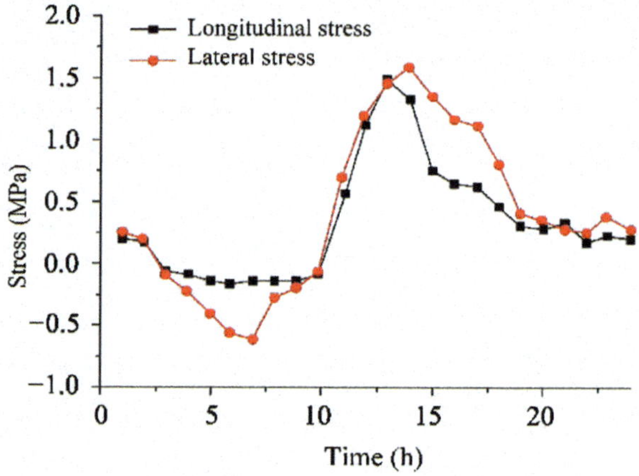

Fig. 8.69 Stress variation curve at No. 1 measuring point

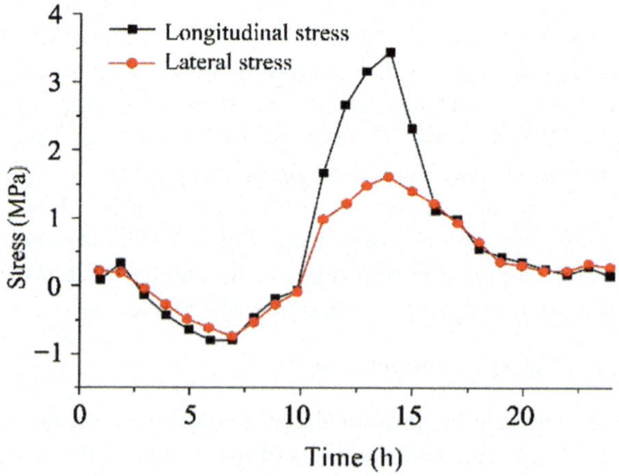

Fig. 8.70 Stress variation curve at No. 2 measuring point

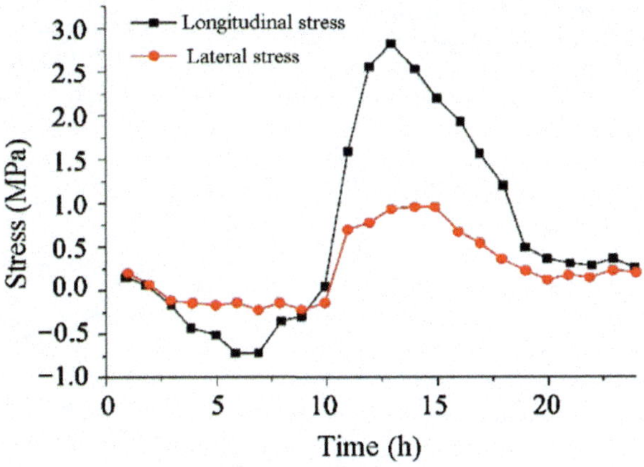

Fig. 8.71 Stress variation curve at No. 3 measuring point

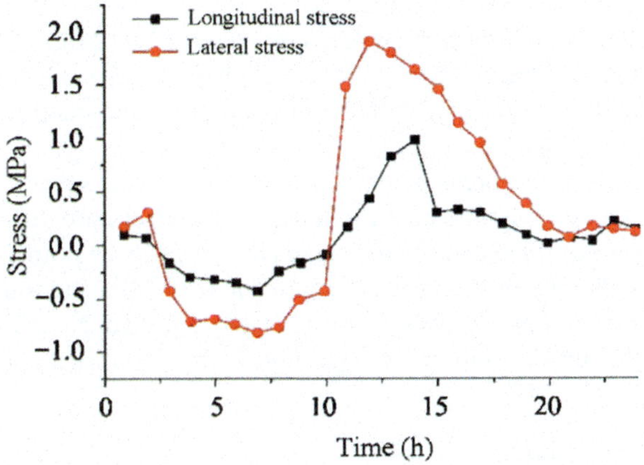

Fig. 8.72 Stress variation curve at No. 4 measuring point

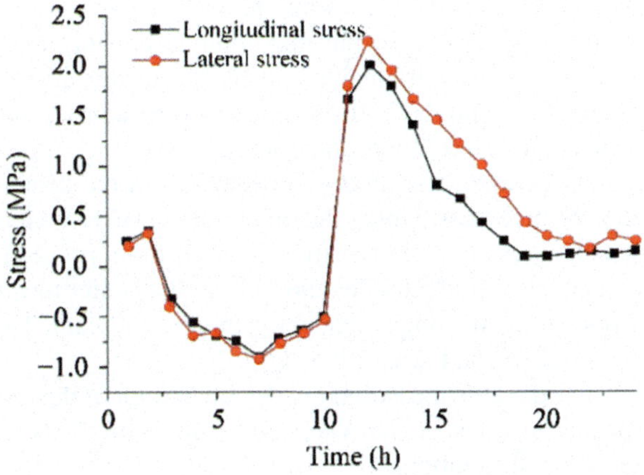

Fig. 8.73 Stress variation curve at No. 5 measuring point

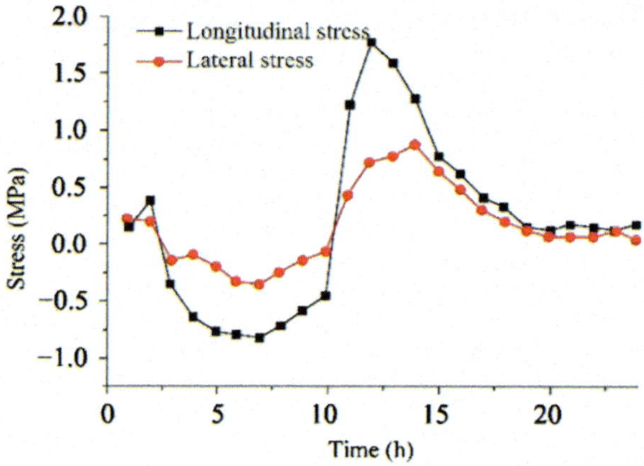

Fig. 8.74 Stress variation curve at No. 6 measuring point

slab, with longitudinal stress recorded at 3.41 MPa and transverse stress at 1.59 MPa; the maximum value of the No. 41 slab occurred at No. 5 measuring point in the center of the turnout slab, with longitudinal stress recorded at 1.97 MPa and transverse stress at 2.22 MPa.

The compressive stress at each point on the surface of the turnout slab reached the largest under the maximum negative temperature difference. The maximum value of the No. 14 slab occurred at the No. 2 measuring point in the center of the turnout slab, with longitudinal stress recorded at 0.81 MPa and transverse stress at 0.75 MPa; the maximum value of the No. 41 slab occurred at No. 5 measuring point in the center of the turnout slab, with longitudinal stress recorded at 0.89 MPa and transverse stress at 0.91 MPa.

3. Summary

The daily temperature variations on the surface of the turnout slabs exhibited a periodic pattern, with the maximum occurring around 2 p.m. During the test in winter, heat was gradually transferred downward in the turnout slab through heat conduction, and the bottom surface temperature of the turnout slab varied inversely to the alterations in air temperature.

The positive temperature differences of the turnout slab were much larger than the negative ones. Under the maximum positive temperature gradient, the maximum tensile stress occurred at the center of the turnout slab. The warping stress of the turnout slab gradually decreased from the center toward the edges, with the warping stress along the longer side being more significant. These variations were influenced by the shape and size of the slab.

Based on the measured results, the design of the turnout slab at the No. 42 turnout on the up track of the Yangzhou Connecting Track of Beijing–Shanghai HSR is considered as rational.

8.2.5 Dynamic Test of Single Turnout

A total of 34 measuring points were placed for the dynamic tests of the No. 42 turnout, and the detailed distribution of these measuring points is shown in Table 8.25.

1. Wheel-Rail Force Test

The maximum values of safety indexes measured when the CRH-380A EMU passed through the section preceding the turnout are shown in Table 8.26.

As illustrated in the Table 8.26, when the EMU passed through the section preceding the turnout, the maximum derailment coefficient was 0.717, the maximum unloading rate was 0.519, and the maximum wheel-rail lateral force was 21.418 kN, all falling within the EMU limits of 0.8, 0.8, and 35 kN.

The maximum values of safety indexes measured when the CRH-380A EMU passed through the lead curve are shown in Table 8.27.

As illustrated in the Table 8.27, when the EMU passed through the lead curve, the maximum derailment coefficient was 0.756, the maximum unloading rate was 0.129, and the maximum wheel-rail lateral force was 21.897 kN, all falling within the EMU limits of 0.8, 0.8, and 35 kN.

2. Stress of Turnout Components

Table 8.28 displays the measured stress at the switch rail with a top width of 30 and 50 mm respectively, and at the point rail with a top width of 35 mm.

Table 8.25 Distribution of measuring points on No. 42 CWR turnout

S/N	No.	Number	Locations of measuring point	Remarks
1	Wheel-rail vertical force	5	A, B, C, D, E	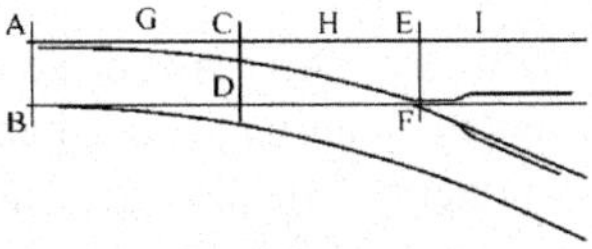
2	Wheel-rail lateral force	5	A, B, C, D, E	
3	Dynamic flexural stress at rail bottom	2	G, I	
4	Vertical dynamic displacement of rail	5	A, B, C, D, E	At the same measuring point position, the measuring points for wheel-rail vertical forces, wheel-rail lateral forces, vertical dynamic displacements of rails, lateral dynamic displacements of rails, and rail bottom dynamic flexural stresses are arranged on the same rail cross section in principle
5	Lateral dynamic displacement of rail	5	A, B, C, D, E	
6	Vertical acceleration of rail	3	A, C, E	
7	Vertical acceleration of track slab	3	A, C, E	
8	Vertical acceleration of girder	3	G, H, I	
9	Lateral acceleration of girder	3	G, H, I	

Table 8.26 Calculated values of safety indexes at front end of switch when passing through ballastless turnout on bridge on the main route

Speed (km/h)	Wheel unloading rate	Derailment coefficient	Wheel-rail lateral force (kN)
60	0.086	0.125	16.405
80	0.062	0.450	15.78
90	0.056	0.717	21.418
110	0.065	0.087	6.118
150	0.099	0.172	8.919
160	0.140	0.259	16.672
180	0.106	0.246	15.153
190	0.061	0.211	14.457
200	0.060	0.279	12.225
210	0.187	0.481	17.137
220	0.205	0.425	20.053
230	0.519	0.072	2.952
300	0.329	0.042	3.094
320	0.042	0.134	6.493
340	0.395	0.222	11.204
350	0.086	0.209	9.497
360	0.051	0.198	10.081
370	0.118	0.157	6.885
380	0.083	0.175	10.081

Figure 8.75 illustrates the typical waveform of stresses on the turnout components, and Figs. 8.76, 8.77 and 8.78 respectively display the stress variations on them with EMU speeds.

With the EMU speed increase, the impact between the wheelset and switch rail escalated, leading to a notable intensification of rail bottom stress at the turnout components. The measured maximum rail bottom stress at the switch rail section with a top width of 30 mm was 114.992 MPa, and the corresponding average maximum value was 86.378 MPa; the measured maximum rail bottom stress at the switch rail section with a top width of 50 mm was 104.977 MPa, and the corresponding average maximum value was 84.327 MPa; the measured maximum rail bottom stress at the point rail section with a top width of 35 mm was 25.818 MPa, and the corresponding average maximum value was 18.428 MPa. The measured values were far less than the permissible stress limit of 335 MPa, indicating that the strength of the turnout components at susceptive positions is sufficient to meet the requirements for the safe operation of EMUs.

3. Rail Displacement

Table 8.29 shows the measured maximum vertical and lateral displacements of the rails when CRH-380A EMU passed through the CWR turnout on the bridge.

Table 8.27 Maximum values of safety indexes at lead curve when passing through ballastless turnout on bridge on the main route

Speed (km/h)	Wheel unloading rate	Derailment coefficient	Wheel-rail lateral force (kN)
60	0.044	0.245	12.745
80	0.071	0.664	16.894
90	0.057	0.756	20.29
110	0.065	0.074	3.599
150	0.034	0.141	6.864
160	0.090	0.510	13.532
180	0.085	0.437	11.868
190	0.046	0.613	13.41
200	0.058	0.719	14.671
210	0.101	0.706	21.404
220	0.129	0.756	21.295
230	0.025	0.174	5.878
300	0.033	0.194	6.606
320	0.099	0.348	14.117
340	0.077	0.452	14.867
350	0.039	0.150	21.897
360	0.047	0.407	15.762
370	0.061	0.114	14.626
380	0.023	0.294	12.367

As illustrated in the above figures and tables, when the EMU passed through the turnout on the bridge, the measured maximum vertical displacement of the straight stock rail at the front end of the switch was 0.998 mm and the measured maximum lateral displacement was 0.998 mm; the measured maximum vertical displacement of the curved stock rail was 0.941 mm and the measured maximum lateral displacement was 0.974 mm. All the test results fell within the permissible limit of 1.5 mm.

4. Vibration of Rail and Turnout Slab

Table 8.30 displays the test results of vertical vibration accelerations of the rails and turnout slabs within the single turnout on the bridge. Figures 8.79 and 8.80 show the waveforms of the vibration accelerations of the rail and turnout slab, respectively.

When the EMU passed through the turnout on the bridge, the measured maximum vertical vibration acceleration of the straight stock rail at the front end of the turnout switch was 283.839 g and the measured maximum vertical vibration acceleration of the turnout slab was 4.846 g; the measured maximum vertical vibration acceleration of the straight stock rail at the frog was 425.014 g, and the measured maximum vertical vibration acceleration of the turnout slab was 1.840 g. EMU speeds exhibited significant influence on turnout vibrations.

Table 8.28 Rail bottom stress of turnout components

Speed (km/h)	Stress at switch rail with a top width of 30 mm (MPa)	Stress at switch rail with a top width of 50 mm (MPa)	Stress at point rail with a top width of 35 mm (MPa)
60	58.742	55.37	12.351
80	69.384	64.357	11.563
90	66.813	65.342	12.227
110	68.952	64.328	13.956
150	71.445	63.954	15.678
160	79.354	70.356	14.365
180	73.49	71.289	17.259
190	77.669	67.877	17.334
200	92.641	80.287	16.998
210	87.574	78.54	15.563
220	90.276	81.669	17.718
230	94.353	83.584	14.73
300	94.578	79.549	18.596
320	96.331	80.69	19.612
340	105.346	93.674	11.67
350	108.946	95.467	16.953
360	106.354	97.364	20.174
370	110.558	100.387	19.116
380	114.992	104.977	25.818

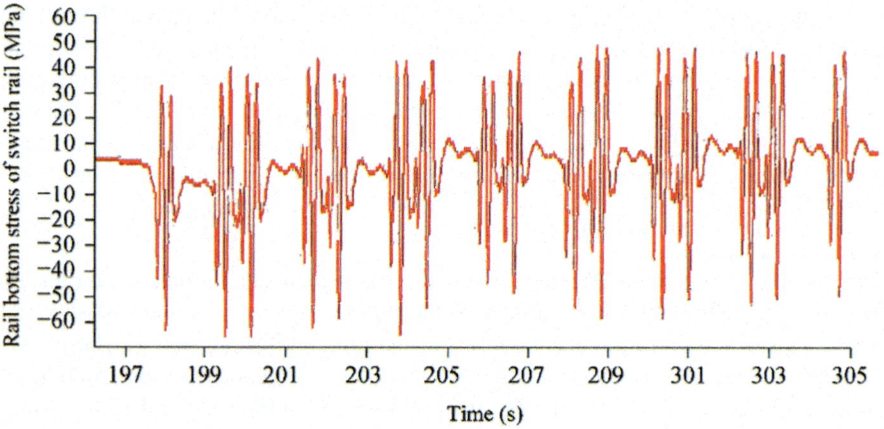

Fig. 8.75 Waveform of rail bottom stress at switch rail

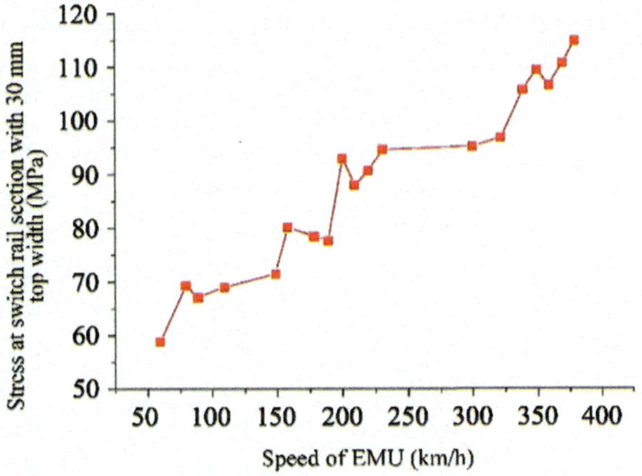

Fig. 8.76 Stress variations at switch rail with a top width of 30 mm with EMU speeds

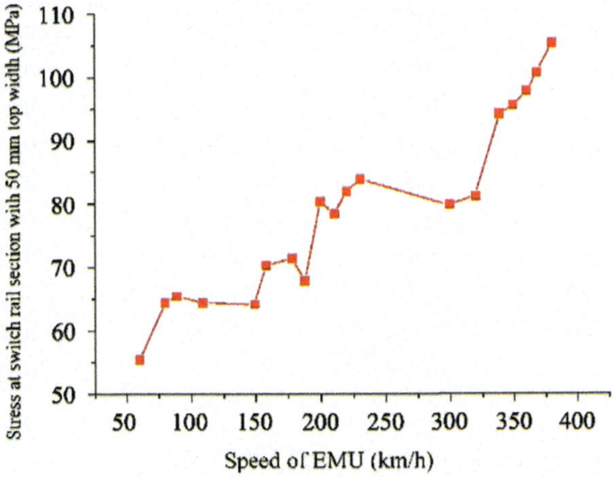

Fig. 8.77 Stress variations at switch rail with a top width of 50 mm with EMU speeds

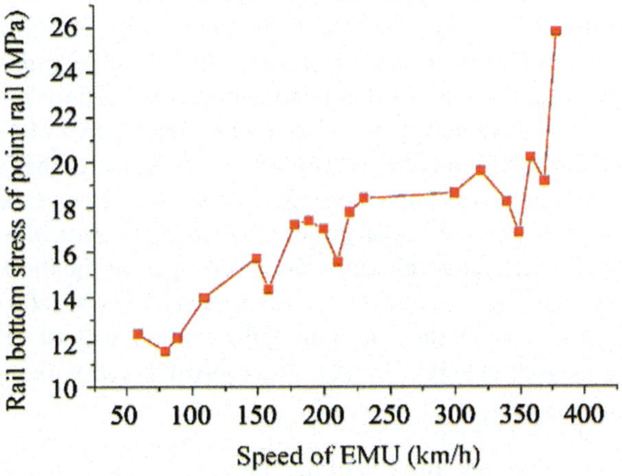

Fig. 8.78 Stress variations at main parts of point rail with EMU speeds

Table 8.29 Maximum vertical and lateral displacements of rail at front end of switch

Speed (km/h)	Vertical displacement of straight stock rail (mm)	Lateral displacement of straight stock rail (mm)	Vertical displacement of curved stock rail (mm)	Lateral displacement of curved stock rail (mm)
60	0.700	0.897	0.768	0.569
80	0.667	0.925	0.742	0.322
90	0.811	0.826	0.738	0.515
110	0.551	0.426	0.735	0.348
150	0.628	0.315	0.639	0.103
160	0.659	0.400	0.631	0.468
180	0.569	0.534	0.550	0.287
190	0.510	0.341	0.581	0.395
200	0.736	0.874	0.941	0.947
210	0.776	0.745	0.872	0.296
220	0.998	0.784	0.665	0.811
230	0.514	0.748	0.653	0.351
300	0.498	0.712	0.549	0.569
320	0.563	0.998	0.804	0.793
340	0.718	0.968	0.838	0.611
350	0.630	0.642	0.610	0.634
360	0.457	0.877	0.755	0.974
370	0.454	0.248	0.702	0.580
380	0.528	0.570	0.813	0.800

5. Bridge Deformation and Vibration

The vibration acceleration measuring points were arranged in the second, third, and fifth spans of the continuous girder, respectively. Table 8.31 displays the measured maximum vibration accelerations of the bridge. Figures 8.81 and 8.82 illustrate the waveforms of the vertical and lateral accelerations of the bridge, respectively.

As illustrated in the above table and figures, the vibration accelerations at the midspan of the continuous girder escalated with the EMU speed increase; when the EMU passed through the bridge, the vertical acceleration at the midspan of the second span of the continuous girder reached 0.212 *g* and the lateral acceleration was 0.111 *g*; the vertical acceleration at the midspan of the third span of the continuous girder reached 0.029 *g*, and the lateral acceleration was 0.007 *g*; the vertical acceleration at the midspan of the fifth span of the continuous girder reached 0.267 *g*, and the lateral acceleration is 0.021 *g*. All the test results fell with the specification limits.

6. Summary

(1) As a consequence of the inherent structural irregularities inevitable in turnouts, significant vertical and lateral vibrations are induced at key turnout components when trains transit through them on the main and diverging routes. This

Table 8.30 Maximum vertical vibration accelerations of turnout rails and turnout sleepers (×*g*)

Speed (km/h)	Straight stock rail at front end of switch	Turnout sleeper at front end of switch	Straight stock rail at frog	Turnout sleeper at frog
60	78.044	2.406	79.421	0.432
80	85.343	2.027	81.862	0.893
90	84.177	2.678	83.927	1.047
110	89.397	1.588	87.73	1.238
150	92.106	2.123	95.321	1.034
160	96.9	2.207	9 332	1.542
180	104.803	1.87	102.106	1.029
190	129.291	1.86	147.877	1.259
200	147.174	1.872	144.999	1.996
210	168.965	2.777	425.014	1.571
220	182.857	3.536	362.449	1.421
230	178.964	2.320	167.225	1.455
300	218.791	1.854	211.008	1.663
320	243.847	4.846	236.379	1.497
340	272.148	2.904	249.136	1.03
350	280.948	2.486	257.595	1.02
360	271.304	3.215	264.36	0.917
370	279.839	2.96	269.36	1.243
380	283.839	2.96	27 397	1.840

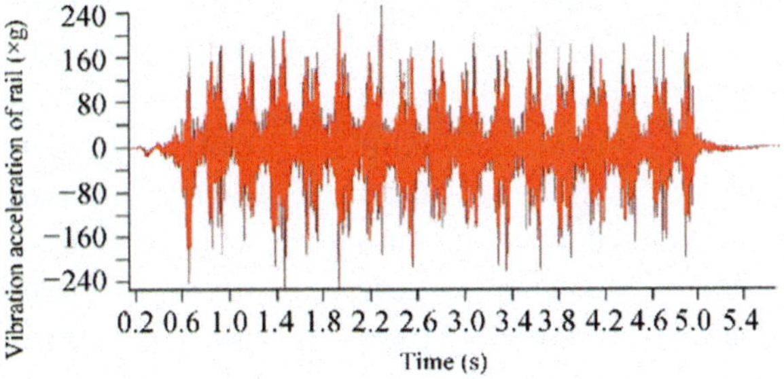

Fig. 8.79 Waveform of vibration accelerations of rail

leads to heightened dynamic responses in the turnout section compared to the general section. When trains traverse turnouts on bridges, the train-turnout-bridge coupling effect induces amplified vibrations, leading to elevated dynamic indexes of the trains, turnouts, and bridges.

(2) The measured maximum values of all the test items exhibited an adequate safety margin, ensuring the secure and stable operation of trains.

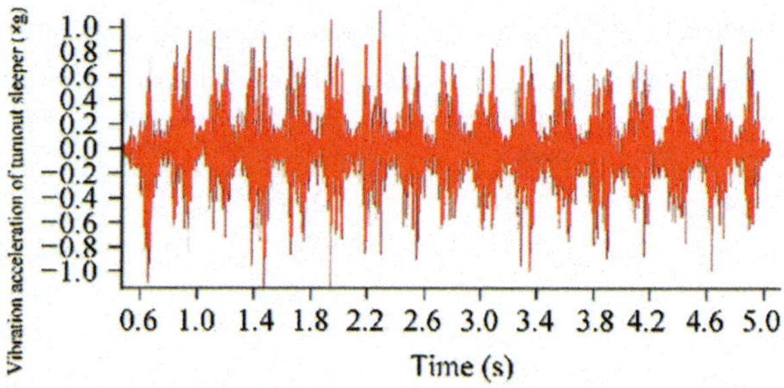

Fig. 8.80 Waveform of vibration accelerations of turnout sleeper

Table 8.31 Maximum vertical and lateral vibration accelerations at midspan of continuous girder ($\times g$)

Speed (km/h)	Vertical vibration acceleration at midspan of second span	Lateral vibration acceleration at midspan of second span	Vertical vibration acceleration at midspan of third span	Lateral vibration acceleration at midspan of third span	Vertical vibration acceleration at midspan of fifth span	Lateral vibration acceleration at midspan of fifth span
60	0.153	0.055	0.013	0.004	0.152	0.009
80	0.145	0.057	0.019	0.001	0.129	0.011
90	0.159	0.088	0.018	0.004	0.158	0.010
110	0.127	0.02	0.006	0.003	0.049	0.007
150	0.101	0.040	0.009	0.003	0.053	0.009
160	0.118	0.069	0.006	0.004	0.091	0.021
180	0.105	0.061	0.006	0.002	0.081	0.009
190	0.126	0.076	0.005	0.002	0.097	0.014
200	0.166	0.068	0.019	0.003	0.148	0.010
210	0.186	0.111	0.012	0.002	0.161	0.010
220	0.160	0.083	0.023	0.004	0.200	0.010
230	0.161	0.101	0.011	0.002	0.113	0.009
300	0.111	0.073	0.008	0.003	0.146	0.010
320	0.201	0.056	0.013	0.007	0.173	0.006
340	0.170	0.091	0.005	0.007	0.217	0.010
350	0.154	0.072	0.013	0.004	0.204	0.009
360	0.170	0.068	0.020	0.007	0.223	0.011
370	0.212	0.096	0.029	0.007	0.267	0.009
380	0.193	0.076	0.029	0.007	0.267	0.009

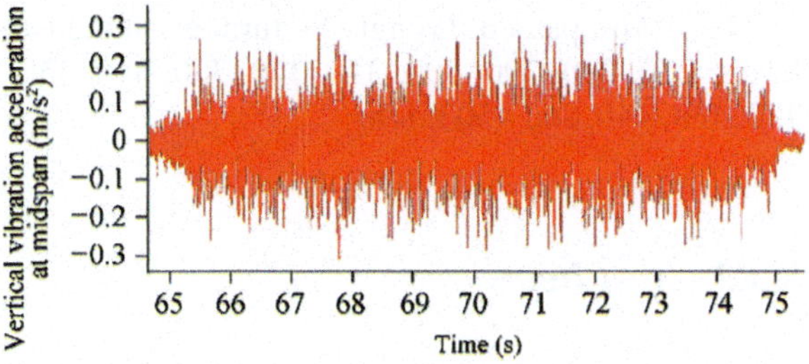

Fig. 8.81 Waveform of vertical vibration accelerations at midspan of continuous girder

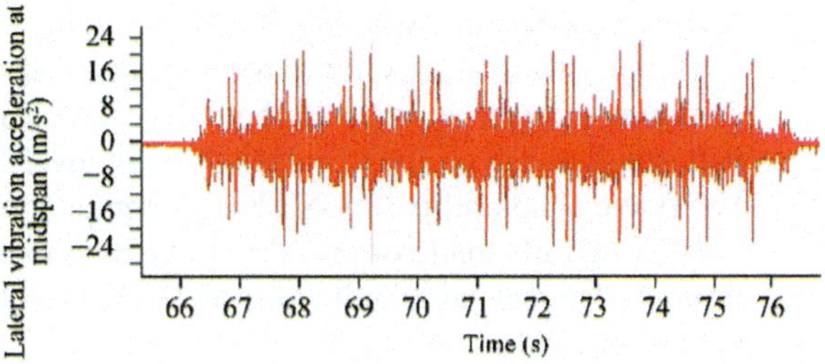

Fig. 8.82 Waveform of lateral vibration accelerations at midspan of continuous girder

(3) Trains, turnouts, and bridges make up an integral interaction system. When trains pass through turnouts, their interactions induce coupling vibrations, which escalate in intensity with higher train speeds. Additionally, train-turnout-bridge coupling dynamic interactions are directly associated with train axle load, track conditions, and bridge structures.

8.3 Dynamic Performance Test of No. 18 Turnout on Bridge at Xuzhoudong Station

8.3.1 Test Overview

The test trains ran from Zaozhuangxi Station to Bengbunan Station, across the test object of the No. 18 single turnout on the bridge.

CRH2-068C, CRH-380AL, and CRH-380BL high-speed EMUs were employed as the test trains. The CRH2-068C EMU consists of 6 motor cars and 2 trailers, with an axle load of less than 14 t and a fixed wheelbase of 2.5 m. The CRH-380AL and CRH-380BL EMUs respectively consist of 14 motor cars and 2 trailers, with an axle load of less than 14 t and a fixed wheelbase of 2.5 m.

The EMUs passed through the turnout on the main route at the following speed levels: 230, 280, 290, 300, 310, 320, 330, 340, 350, 360, 370, 380, 390, 400, 410, 420, 430, 440, and 450 km/h.

8.3.2 Test Items

1. Test Purpose

(1) Assess the safety and stability of EMUs passing through movable-point turnouts with ballastless tracks on passenger-dedicated railways on the main and diverging routes, as well as the safety and stability of CWR turnouts with ballastless tracks on long bridges through level-by-level speed increase tests.
(2) Verify the adaptability of CNTT key turnout components such as Zul60 switch rails, wing rails made of ordinary rails, monobloc forged point rails, fasteners, turnout sleepers, and switching and locking devices under high-speed train operation conditions.
(3) Evaluate the dynamic performance of slab ballastless track structures within the turnout section and their adaptability to the turnouts through dynamic tests.
(4) Verify the rationality of turnout track rigidity, the uniformity of track rigidity within the turnout section, and the smoothness of track rigidity transition preceding and following turnouts.
(5) Enhance the understanding of the transmission pattern of longitudinal forces between ballastless CWR tracks and bridges and train-track-bridge coupling vibration pattern; further study the computational theories, improve the design methods, and optimize the design schemes for CWR tracks on bridges.

2. Test Items

(1) Safety

Wheel-rail vertical force (P) and lateral force (Q). The test results are used to calculate the derailment coefficient (Q/P) of inner and outer wheels, wheel unloading rate ($\Delta P/P$), and wheelset lateral force ($Q1–Q2$) of rolling stock and further determine the operational safety of test trains.

(2) Turnout regularity

The Sperling index of trains passing through the turnout on the main and diverging routes is calculated according to the accelerations of the trains, and the results are used to evaluate the turnout regularities.

(3) Lateral stability of track structure

① Opening at the actual point of the switch rail for facing-point movement into the turnout;

② Lateral displacement of wing rail head;
③ Lateral displacement of stock rail head preceding actual point of switch rail;
④ Lateral displacement of rail head in closure section of switch rail.

(4) Strength of key turnout components

① Rail bottom stress of Zul-60 switch rail at rail head width of 40 and 50 mm;
② Rail bottom stress at planed section of wing rail.

(5) Rigidity and uniformity index of track in turnout section

One point is chosen respectively at the transition section into and from the turnout section, the front part, middle part, and heel end of the switch rail, offset base plate of the turnout closure, the wing rail, the middle part of the point rail, and the heel end of the point rail, to measure the vertical displacements of the rails, verify the rationality of base plate rigidity value of 17.5 kN/mm, and assess the actual application effect of the track rigidity homogenization technology.

(6) Transition range and magnitude of vertical wheel-rail force

Measuring points are placed continuously at the straight switch rail positions with a top width of 10 mm, 20 mm, 30 mm, 40 mm, and 50 mm, to measure the changes in vertical forces acting on them, and to determine the transition scopes and proportions of wheel-rail vertical forces on the straight switch rail and curved stock rail.

(7) Vibration of ballastless track structural components in turnout section

① Vibration acceleration of rail;
② Vertical acceleration of turnout slab;
③ Vertical acceleration of concrete base slab.

8.3.3 *Test Method and Arrangement of Measuring Points*

1. Test Method

(1) According to the *Track Side Test Methods of Vertical and Lateral Wheel-Rail Forces* (TB/T 2489-1994), the shear stress method is applied to measure the vertical wheel-rail force (P) and lateral wheel-rail force (Q). The measurement results are used to calculate the safety parameters for EMU operation: derailment coefficient (Q/P), wheel unloading rate ($\Delta P/P$), and wheel-rail lateral force ($Q1$–$Q2$).

(2) Lateral elastic displacement of switch rail and stock rail.

The lateral elastic displacements of the switch rail and stock rail within the turnout section are measured using spring-type displacement meters. The test results are used to calculate the opening of the curved switch rail and gauge expansion at the actual point of the switch rail, and evaluate the stability of the turnout structure in conjunction with the test results of wheel-rail lateral force.

(3) Vertical displacement of rail

The vertical displacements of the rail within the turnout section are measured using spring-type displacement meters. The test results are used to evaluate the rationality and uniformity of track rigidity within the turnout area and the rigidity consistency with the sectional tracks in conjunction with the designed static rigidity at fastener nodes within the turnout section.

(4) The vibration accelerations of the rails, track slabs, and bases are measured using piezoelectric or resistance strain type accelerometers.
(5) Transition range and magnitude of vertical wheel-rail force

The vertical force acting on the curved stock rail is used to calculate the proportions of vertical force acting on the curved stock rail and straight switch rail and to further evaluate the transition ranges and magnitudes of wheel-rail vertical forces.

The wheel-rail vertical force (*P*) and lateral force (*Q*) are calibrated using a calibration frame on site, and subsequent quasi-static calibration and verification are conducted by operating inspection trains passing through the measuring points at a speed of 5 km/h; displacements are calibrated using a feeler gauge on site; the dynamic strain gauges are checked and conductor errors are corrected using a standard strain source.

2. Arrangement of Measuring Points

The arrangement of measuring points for dynamic tests of the CWR turnout on the bridge is shown in Fig. 8.83 and Table 8.32.

(1) Vertical and lateral wheel-rail forces on CWR turnout

Two measuring points for vertical and lateral forces on stock rail respectively;

Two measuring points for vertical and lateral forces on closure rail respectively;

Two measuring points for vertical forces on point rail.

(2) Vertical and lateral dynamic wheel-rail displacement of CWR turnout

Two measuring points for vertical and lateral displacement of stock rail respectively;

Two measuring points for vertical and lateral displacement of switch rail respectively;

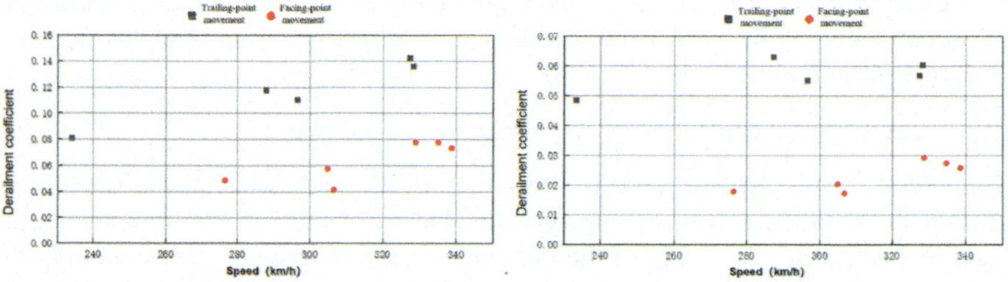

Fig. 8.83 Arrangement of measuring points for dynamic tests

Table 8.32 Distribution of measuring points

S/N	No	Number	Locations of measuring point	Remarks
1	Wheel-rail vertical force	8	A, B, C, D, E, F, G, H	At the same measuring point position, the measuring points for wheel-rail vertical forces, wheel-rail lateral forces, vertical dynamic displacements of rails, lateral dynamic displacements of rails, and rail bottom dynamic flexural stresses are arranged on the same rail cross section in principle
2	Wheel-rail lateral force	4	A, B, E, F	
3	Dynamic flexural stress at rail bottom	4	A, C, D, E	
4	Vertical dynamic displacement of rail	6	A, C, D, E, G, H	
5	Lateral dynamic displacement of rail	6	A, B, E, F, G, H	
6	Vertical acceleration of rail	5	A, C, E, G, H	
7	Vertical acceleration of sleeper	5	A, C, E, G, H	
8	Girder acceleration	5	A, C, E, G, H	

Two measuring points for vertical and lateral displacement of closure rail respectively;

Two measuring points for vertical forces on point rail.

(3) Vibration acceleration of CWR turnout

Six measuring points for vertical acceleration of rail;

Six measuring points for vertical acceleration of sleeper;

Three measuring points for girder acceleration;

One measuring point for dynamic deflection at midspan;

Three measuring points for longitudinal girder-rail relative displacement.

A total of 45 measuring points are arranged for dynamic tests of the CWR turnout on the bridge.

8.3.4 Analysis of Test Data

From November 16, 2010, to January 30, 2011, the CRH2-068C, CRH-380AL, and CRH-380BL comprehensive inspection trains were operated to cross the No. 18 turnout on the main route at the following speed levels: 230, 280, 290, 300, 310, 320, 330, 340, 350, 360, 370, 380, 390, 400, and 410 km/h.

1. Parameters of Train Operation Safety

Table 8.33 shows the measured maximum value, average value, standard difference, and maximum possible value of the derailment coefficient, wheel unloading rate, and wheel-rail lateral force when the CRH2-068C comprehensive inspection train passed through the turnout on the main route. Figures 8.84, 8.85 and 8.86 illustrate their relations with train speeds.

As illustrated in Table 8.33 and Figs. 8.84, 8.85 and 8.86, when the EMU passed through the turnout on the main route:

The maximum derailment coefficient ranged from 0.04 to 0.14, and the average value ranged from 0.02 to 0.06, which were relatively small and did not show significant dependency on EMU speeds.

The maximum wheel unloading rate ranged from 0.11 to 0.32, and the average value ranged from 0.05 to 0.15, which were relatively small and did not show significant dependency on EMU speeds.

The maximum wheel-rail lateral force ranged from 2.7 to 10.8 kN, and the average value ranged from 1.4 to 7.5 kN, which were relatively small and did not show significant dependency on EMU speeds.

The measured maximum derailment coefficient, unloading rate, and wheel-rail lateral force were respectively less than the limits of 0.8, 0.8, and 46.8 kN for EMUs.

Table 8.34 displays the measured maximum derailment coefficient, wheel unloading rate, and wheel-rail lateral force when the CRH-380AL comprehensive inspection train passed through the turnout on the main route. Figures 8.87, 8.88 and 8.89 illustrate their relations with EMU speeds.

As illustrated in Table 8.34 and Figs. 8.86, 8.87, 8.88, when the EMU passed through the turnout on the diverging route:

The maximum derailment coefficient ranged from 0.07 to 0.47, and the average value ranged from 0.02 to 0.08.

The maximum unloading rate ranged from 0.15 to 0.78, and the average value ranged from 0.05 to 0.22.

The maximum wheel-rail lateral force ranged from 2.9 to 24.5 kN, and the average value ranged from 0.9 to 12.2 kN.

The measured maximum derailment coefficient, unloading rate, and wheel-rail lateral force were respectively less than the limits of 0.8, 0.8, and 46.8 kN for EMUs.

2. Lateral Stability of Track Structure Within Turnout Section

The measuring points for deformation of the rails within the turnout section are arranged to measure the lateral displacement of the wing rail head, opening at the

Table 8.33 Safety parameters for EMU passing through turnout on the main route

Item	Direction	Speed (km/h)	Maximum	Average value	Standard difference	Maximum possible value
Derailment coefficient	Facing-point movement	276.5	0.05	0.02	0.01	0.05
		305.1	0.06	0.02	0.01	0.05
		306.6	0.04	0.02	0.01	0.04
		329.1	0.08	0.03	0.02	0.07
		335.2	0.08	0.03	0.02	0.07
		339.0	0.07	0.03	0.02	0.07
	Trailing-point movement	233.5	0.08	0.05	0.02	0.09
		287.5	0.12	0.06	0.02	0.12
		296.5	0.11	0.06	0.02	0.10
		327.3	0.14	0.06	0.02	0.11
		327.9	0.14	0.06	0.02	0.11
Wheel unloading rate	Facing-point movement	276.5	0.22	0.08	0.05	0.20
		305.1	0.23	0.11	0.06	0.26
		306.6	0.25	0.11	0.07	0.27
		329.1	0.29	0.15	0.08	0.33
		335.2	0.32	0.13	0.09	0.36
		339.0	0.32	0.12	0.09	0.33
Wheel unloading rate	Trailing-point movement	233.5	0.19	0.09	0.05	0.22
		287.5	0.15	0.05	0.04	0.14
		296.5	0.14	0.05	0.03	0.12
		327.3	0.24	0.11	0.06	0.26
		327.9	0.23	0.10	0.06	0.27
Wheel-rail lateral force (kN)	Facing-point movement	276.5	3.2	1.6	1	4.3
		305.1	2.7	1.5	0.9	3.7
		306.6	2.7	1.4	0.9	3.7
		329.1	3.1	1.7	1.0	4.3
		335.2	3.2	1.8	0.6	3.2
		339.0	3.7	2.1	1	4.8
	Trailing-point movement	233.5	8.4	6.2	1.2	9.1
		287.5	10.8	7.4	2.3	13.3
		296.5	1	6.5	1.4	9.9
		327.3	9.1	6.2	1.4	9.7
		327.9	9.0	6.4	1.3	9.6

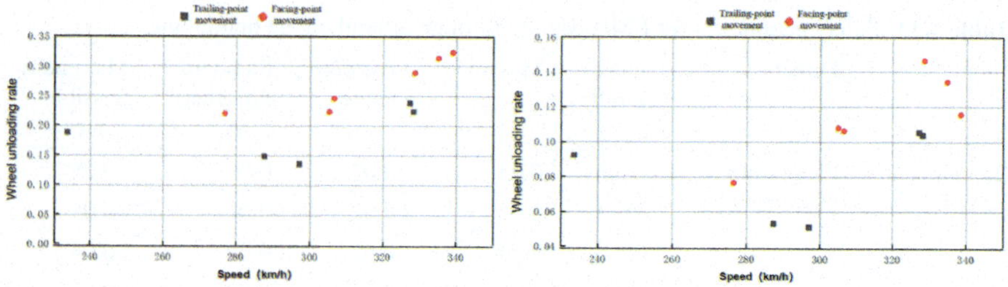

Fig. 8.84 Relation between derailment coefficient and speed of EMU passing on the main route

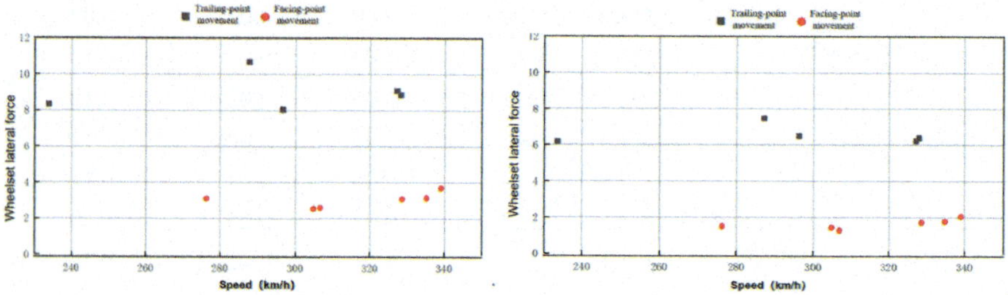

Fig. 8.85 Relation between unloading rate and speed of EMU passing on the main route

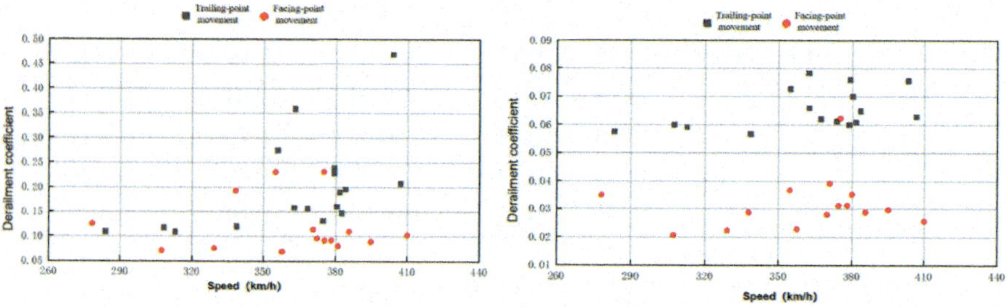

Fig. 8.86 Relation between wheel-rail lateral force and speed for EMU passing on the main route

actual point of the switch rail, and lateral displacements of the stock rail preceding the actual point of the switch rail, and at the straight switch rail sections with a top width of 30 and 40 mm, respectively.

Table 8.34 Safety parameters for EMU passing through turnout on the main route

Item	Direction	Speed (km/h)	Maximum	Average value	Standard difference	Maximum possible value
Derailment coefficient	Facing-point movement	278.6	0.13	0.03	0.03	0.11
		307.2	0.08	0.02	0.01	0.05
		329.7	0.08	0.02	0.02	0.06
		333	0.20	0.03	0.03	0.10
		354.7	0.23	0.04	0.03	0.12
		357.5	0.07	0.02	0.01	0.06
		370.4	0.12	0.03	0.02	0.08
		371.9	0.10	0.04	0.02	0.10
		374.6	0.09	0.03	0.02	0.08
		375.0	0.23	0.06	0.06	0.20
		377.8	0.09	0.03	0.02	0.08
		381.0	0.08	0.03	0.02	0.08
		385.4	0.11	0.03	0.02	0.08
		394.7	0.09	0.03	0.02	0.08
		410.0	0.11	0.03	0.02	0.07
	Trailing-point movement	283.9	0.11	0.06	0.02	0.12
		302	0.12	0.06	0.02	0.12
		313.0	0.11	0.06	0.03	0.12
		339.0	0.12	0.06	0.02	0.11
		355.7	0.28	0.07	0.04	0.18
		362.5	0.16	0.07	0.03	0.14
		362.9	0.36	0.08	0.06	0.22
		361	0.16	0.06	0.03	0.13
		374.2	0.13	0.06	0.03	0.13
		378.9	0.23	0.08	0.04	0.18
		378.9	0.24	0.06	0.04	0.16
		380.6	0.16	0.07	0.04	0.16
		381.4	0.19	0.06	0.03	0.15
		382.2	0.15	0.06	0.03	0.14
		383.4	0.20	0.07	0.04	0.17
		404.0	0.47	0.08	0.06	0.23
		407.2	0.21	0.06	0.04	0.17
Wheel unloading rate	Facing-point movement	278.6	0.15	0.06	0.04	0.16
		307.2	0.27	0.11	0.08	0.30
		329.7	0.34	0.14	0.10	0.39

(continued)

Table 8.34 (continued)

Item	Direction	Speed (km/h)	Maximum	Average value	Standard difference	Maximum possible value
		333	0.26	0.12	0.07	0.30
		354.7	0.27	0.11	0.07	0.29
		357.5	0.23	0.11	0.07	0.29
		370.4	0.26	0.10	0.06	0.25
		371.9	0.30	0.12	0.08	0.32
		374.6	0.23	0.08	0.06	0.24
		375.0	0.30	0.10	0.07	0.27
		377.8	0.27	0.10	0.07	0.28
		381.0	0.32	0.09	0.07	0.26
		385.4	0.24	0.09	0.05	0.22
		394.7	0.24	0.09	0.06	0.24
		410.0	0.23	0.09	0.05	0.23
	Trailing-point movement	283.9	0.16	0.06	0.04	0.16
		302	0.22	0.05	0.05	0.17
		313.0	0.18	0.06	0.04	0.15
		339.0	0.22	0.08	0.04	0.19
		355.7	0.21	0.09	0.05	0.23
		362.5	0.25	0.13	0.06	0.28
		362.9	0.40	0.15	0.08	0.33
		361	0.39	0.14	0.07	0.31
		374.2	0.39	0.20	0.09	0.42
		378.9	0.47	0.22	0.11	0.50
		378.9	0.49	0.21	0.11	0.50
		380.6	0.40	0.17	0.09	0.40
		381.4	0.40	0.21	0.09	0.44
		382.2	0.45	0.22	0.11	0.49
		383.4	0.46	0.20	0.12	0.51
		404.0	0.78	0.20	0.15	0.58
		407.2	0.59	0.21	0.16	0.61
Wheel-rail lateral force (kN)	Facing-point movement	278.6	5.5	2.0	1.5	5.7
		307.2	3.6	0.9	0.8	3.0
		329.7	3.0	1.0	0.6	2.6
		333	7.1	1.4	1.6	5.3
		354.7	3	2.4	1.6	6.4
		357.5	5.9	1.3	0.9	3.5

(continued)

Table 8.34 (continued)

Item	Direction	Speed (km/h)	Maximum	Average value	Standard difference	Maximum possible value
		370.4	4.0	1.0	1.0	3.4
		371.9	5.8	2.7	1.6	6.8
		374.6	2.9	1.4	0.8	3.5
		375.0	9.2	3.0	2.7	9.8
		377.8	5.4	2.0	1.2	4.9
		381.0	4.4	1.5	1.0	4.1
		385.4	5.7	1	1.0	3.5
		394.7	3.8	1.2	1.0	3.6
		410.0	3.6	0.9	0.9	3.1
	Trailing-point movement	283.9	11.3	7.7	2.0	12.7
		302	12.1	7.6	2.2	13.2
		313.0	12.5	7.5	2.7	14.2
		339.0	11.9	6.8	2.1	11.9
		355.7	14.7	7.5	2.8	14.4
		362.5	17.7	7.5	2.8	14.5
		362.9	24.5	12.2	7.8	31.8
		361	10.9	7.2	2.0	12.2
		374.2	10.3	6.7	1.9	11.6
		378.9	11.3	4.9	2.8	11.9
		378.9	11.4	4.7	2.4	10.7
		380.6	12.7	7.9	3.3	16.1
		381.4	11.7	5.6	2.3	11.3
		382.2	11.5	5.6	2.4	11.6
		383.4	12.2	6.4	3.9	16.1
		404.0	12.2	4.8	2.2	10.4
		407.2	14.9	6.3	5.3	19.6

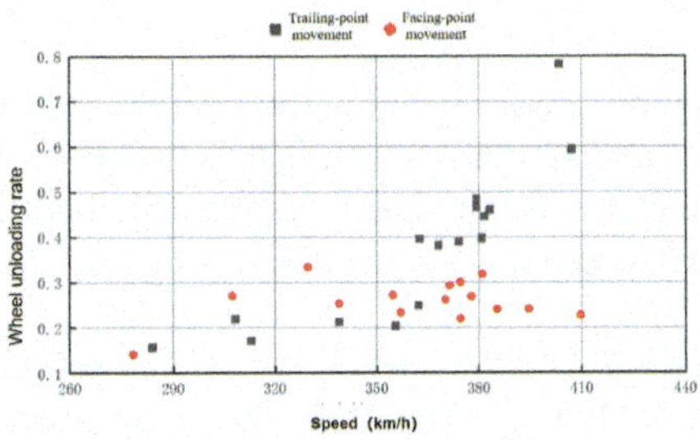

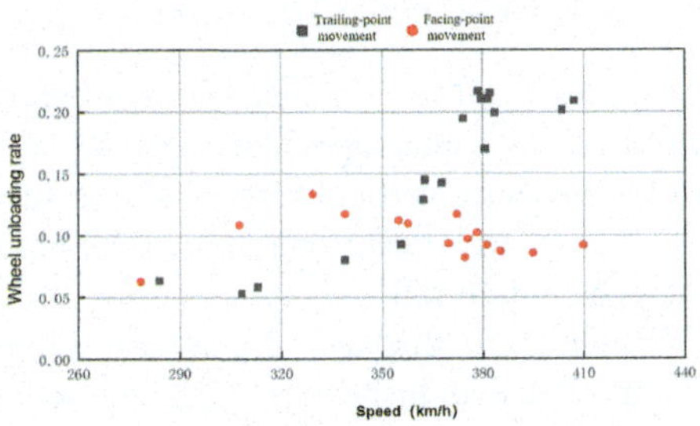

Fig. 8.87 Relation between derailment coefficient and speed of EMU passing on the main route

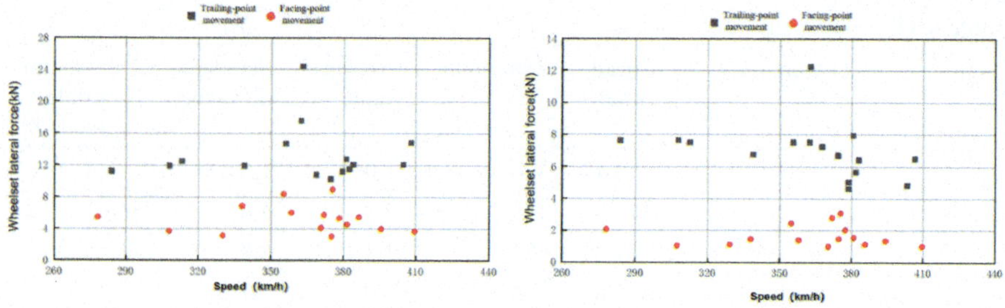

Fig. 8.88 Relation between unloading rate and speed of EMU passing on the main route

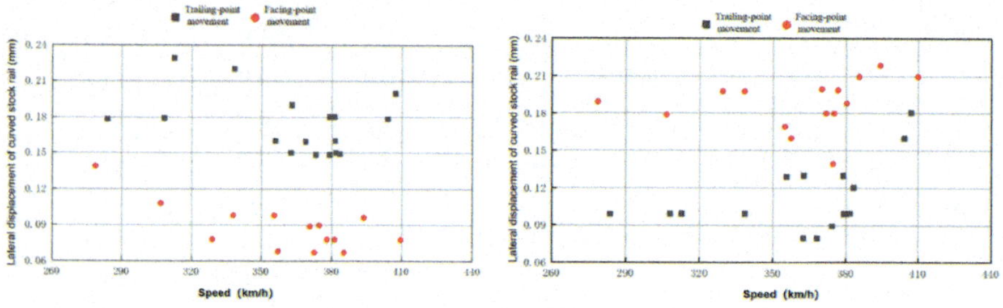

Fig. 8.89 Relation between wheel-rail lateral force and speed for EMU passing on the main route

Table 8.35 displays the measured maximum deformation parameters of the turnout components when the CRH-380AL comprehensive inspection train passed through the turnout on the main route. Figures 8.90, 8.91, 8.92 and 8.93 illustrate their relations with EMU speeds.

As illustrated in Table 8.35 and Figs. 8.90, 8.91, 8.92 and 8.93, when the EMU passed through the turnout on the main route:

The lateral displacement of the stock rail head ranged from 0.07 to 0.29 mm, and that of the wing rail ranged from 0.16 to 0.42 mm, all less than the limit of 1.5 mm; the opening of the switch rail ranged from 0.19 to 0.40 mm, less than the limit of 3 mm.

3. Wheel-Rail Force Transition

The wheel-rail force transition measuring points are arranged at the straight switch rail sections with a top width of 10, 20, 30, 40, and 50 mm, as well as the curved stock rail to measure vertical forces. The test results are used to calculate the distribution coefficient and transition range of the vertical forces on the curved stock rail and the straight switch rail.

Table 8.36 displays the measured maximum values when the CRH-380AL comprehensive inspection train passed through the turnout on the main route.

Table 8.35 Measured maximum deformation parameters of rails for EMU passing turnout on the main route (mm)

Direction	Speed	Lateral displacement of curved stock rail	Lateral displacement of straight stock rail	Lateral displacement of curved stock rail at 30 mm	Lateral displacement of straight stock rail at 30 mm	Lateral displacement of curved stock rail at 40 mm	Lateral displacement of straight stock rail at 40 mm	Lateral displacement of wing rail	Opening of switch rail
Facing-point movement	278.6	0.14	0.19	0.14	0.15	0.18	0.22	0.16	0.28
	307.2	0.11	0.18	0.12	0.20	0.21	0.22	0.23	0.24
	329.7	0.08	0.20	0.11	0.16	0.19	0.21	0.28	0.27
	338.4	0.10	0.20	0.12	0.16	0.19	0.18	0.31	0.30
	354.9	0.10	0.17	0.14	0.14	0.14	0.20	0.23	0.34
	357.5	0.07	0.16	0.12	0.14	0.15	0.21	0.31	0.25
	370.4	0.09	0.20	0.14	0.19	0.17	0.17	0.42	0.24
	371.9	0.07	0.18	0.18	0.14	0.15	0.18	0.28	0.30
	374.6	0.09	0.14	0.14	0.12	0.15	0.14	0.19	0.27
	375.0	0.09	0.18	0.14	0.13	0.22	0.17	0.32	0.40
	377.7	0.08	0.20	0.11	0.21	0.12	0.15	0.33	0.22
	380.9	0.08	0.19	0.13	0.10	0.10	0.17	0.29	0.20
	385.4	0.07	0.21	0.12	0.10	0.15	0.16	0.38	0.26
	394.7	0.10	0.22	0.13	0.12	0.16	0.15	0.22	0.35
	410.0	0.08	0.21	0.13	0.13	0.17	0.17	0.16	0.27
Trailing-point movement	283.9	0.18	0.10	0.18	0.15	0.12	0.23	0.31	0.37
	302	0.18	0.10	0.19	0.17	0.11	0.17	0.34	0.31
	313.0	0.23	0.10	0.16	0.14	0.13	0.26	0.27	0.25
	339.0	0.22	0.10	0.16	0.16	0.12	0.28	0.32	0.40

(continued)

Table 8.35 (continued)

Direction	Speed	Lateral displacement of curved stock rail	Lateral displacement of straight stock rail	Lateral displacement of curved stock rail at 30 mm	Lateral displacement of straight stock rail at 30 mm	Lateral displacement of curved stock rail at 40 mm	Lateral displacement of straight stock rail at 40 mm	Lateral displacement of wing rail	Opening of switch rail
	355.7	0.16	0.13	0.19	0.15	0.16	0.25	0.29	0.32
	362.5	0.15	0.08	0.16	0.13	0.18	0.23	0.25	0.27
	362.9	0.19	0.13	0.17	0.17	0.21	0.22	0.30	0.29
	361	0.16	0.08	0.15	0.14	0.20	0.23	0.22	0.19
Trailing-point movement	374.2	0.15	0.09	0.19	0.16	0.20	0.24	0.35	0.23
	378.9	0.18	0.13	0.18	0.18	0.16	0.26	0.24	0.35
	378.9	0.15	0.10	0.14	0.13	0.17	0.24	0.33	0.29
	380.5	0.18	0.10	0.17	0.15	0.16	0.26	0.41	0.27
	381.4	0.16	0.10	0.13	0.15	0.18	0.26	0.32	0.29
	382.2	0.15	0.10	0.17	0.15	0.20	0.22	0.26	0.29
	383.4	0.15	0.12	0.15	0.15	0.16	0.24	0.31	0.32
	404.0	0.18	0.16	0.15	0.21	0.19	0.23	0.37	0.30
	407.2	0.20	0.18	0.17	0.22	0.17	0.29	0.29	0.28

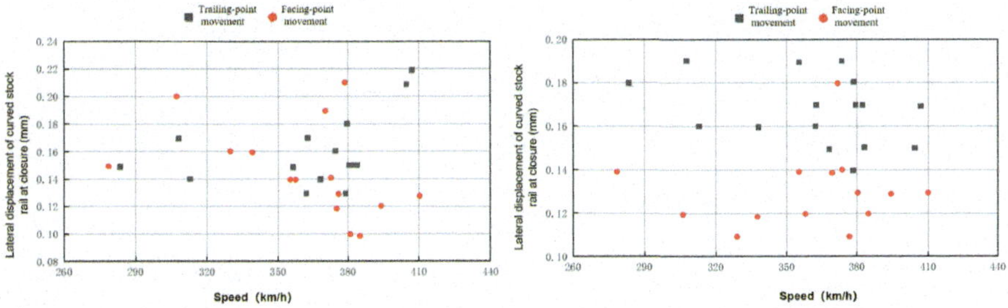

Fig. 8.89 Relation between lateral displacement of stock rail and speed of EMU passing on the main route

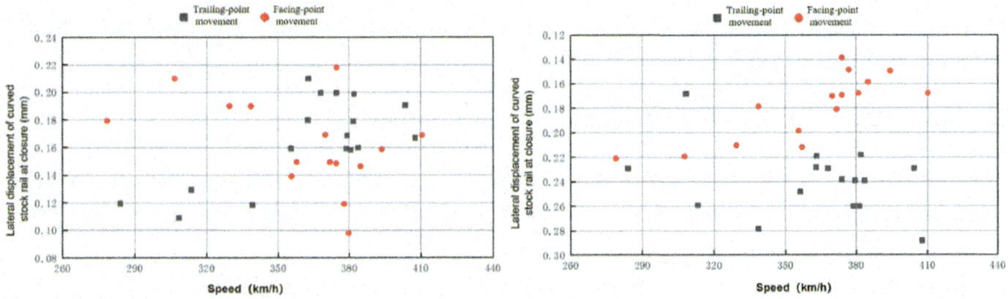

Fig. 8.90 Relation between lateral displacement of stock rail at straight switch rail section with 30 mm top width and speed of EMU passing on the main route

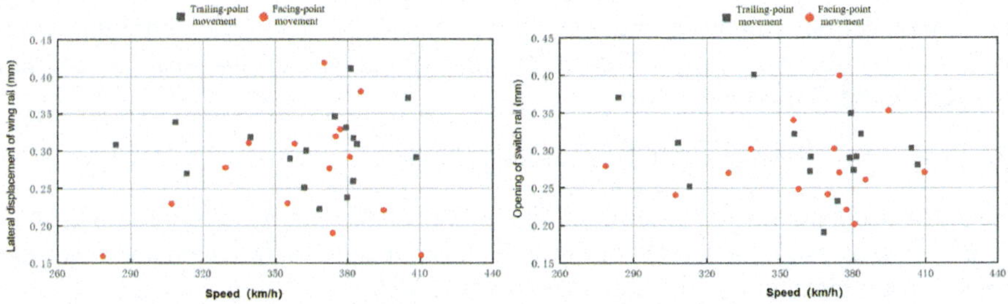

Fig. 8.91 Relation between lateral displacement of stock rail at straight switch rail section with 40 mm top width and speed of EMU passing on the main route

Figure 8.93 illustrates the transition proportions and ranges of wheel-rail vertical forces.

When the EMU passed through the turnout on the main route, the straight switch rail cross section with a top width of 20 mm began to experience vertical forces, that with a top width of 30 mm experienced about 52% of the vertical forces, that with a top width of 40 mm experienced about 88% of the vertical forces, and that with a top width of 50 mm experienced 100% of the wheel-rail vertical forces. The transfer

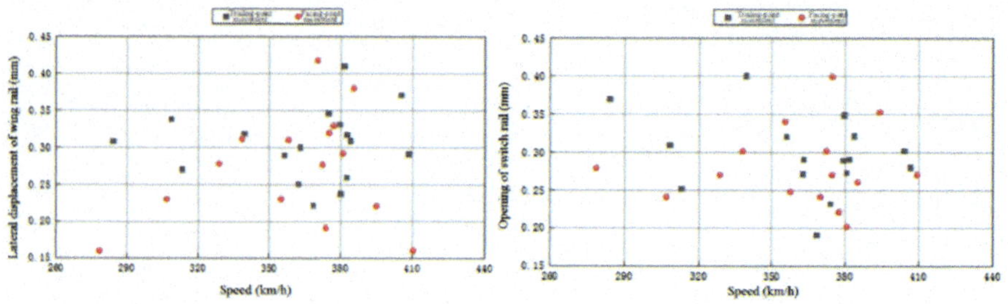

Fig. 8.92 Relation between lateral displacement of wing rail/opening of switch rail and speed of EMU passing on the main route

of wheel-rail loads met the design requirements and can meet the requirements for the safe and stable operation of trains.

4. Strength of Turnout Components

The measuring points for component strength are arranged at the rail bottom of the straight switch rail sections with a top width of 40 and 50 mm respectively for stress tests.

Table 8.37 displays the measured maximum and average rail bottom stress of the straight switch rail when the CRH-380AL comprehensive inspection train passed through the turnout on the main route. Figures 8.95 and 8.96 illustrate their relations with EMU speeds.

The measured maximum rail bottom stress was 88.49 MPa at the straight switch rail cross section with a top width of 40 mm and 72.54 MPa at the straight switch rail cross section with a top width of 50 mm. All the test results were far less than the permissible limit of 335 MPa.

5. Test Data and Analysis of Turnout Slab Strain

The measuring points for the strain of the turnout slabs within the No. 18 turnout are arranged to measure the lateral strain of the turnout slabs on the main and diverging routes. They were all positioned on the No. 117 slab.

Table 8.38 displays the measured maximum strain of the turnout slabs when the CRH-380AL comprehensive inspection train passed through the turnout on the main route. Figure 8.97 illustrates their relations with speeds.

As shown in Table 8.38 and Fig. 8.96, the maximum lateral strain of the straight turnout slab on the diverging route was 11.54 $\mu\varepsilon$, and the curved turnout slab on the diverging route was 7.11 $\mu\varepsilon$ when the CRH-380AL comprehensive inspection train passed on the main route.

6. Displacement of Turnout Slab Within Turnout Section

The measuring points for displacement of the turnout slab within the turnout section are arranged to measure the vertical, lateral, and longitudinal displacement of the

Table 8.36 Proportions of vertical forces on straight switch rail for EMU passing the turnout on the main route

Direction	Speed (km/h)	Straight switch rail section with 10 mm top width	Straight switch rail section with 20 mm top width	Straight switch rail section with 30 mm top width	Straight switch rail section with 40 mm top width	Straight switch rail section with 50 mm top width
Facing-point movement	278.6	0	0.10	0.54	0.86	1
	307.2	0	0.21	0.53	0.95	1
	329.7	0	0.15	0.49	0.93	1
	338.4	0	0.02	0.42	0.92	1
	354.9	0	0.20	0.55	0.93	1
	357.5	0	0.15	0.54	0.94	1
	370.4	0	0.16	0.53	0.91	1
	371.9	0	0.03	0.35	0.78	1
	374.6	0	0.09	0.50	0.89	1
	375.0	0	0.15	0.43	0.89	1
	377.7	0	0.01	0.43	0.87	1
	380.9	0	0.14	0.51	0.88	1
	385.4	0	0.09	0.49	0.88	1
	394.7	0	0.07	0.49	0.89	1
	410.0	0	0.03	0.45	0.96	1
Trailing-point movement	283.9	0	0.24	0.51	0.92	1
	302	0	0.25	0.59	0.88	1
	313.0	0	0.42	0.60	0.89	1
	339.0	0	0.47	0.53	0.89	1
	355.7	0	0.48	0.50	0.92	1
	362.5	0	0.49	0.55	0.83	1
	362.9	0	0.44	0.54	0.88	1
	361	0	0.47	0.53	0.85	1
	374.2	0	0.39	0.58	0.89	1
	378.9	0	0.39	0.68	0.90	1
	378.9	0	0.46	0.54	0.73	1
	380.5	0	0.45	0.60	0.86	1
	381.4	0	0.49	0.51	0.72	1
	382.2	0	0.50	0.57	0.79	1
	383.4	0	0.42	0.48	0.75	1
	404.0	0	0.47	0.54	0.87	1
	407.2	0	0.33	0.57	0.89	1

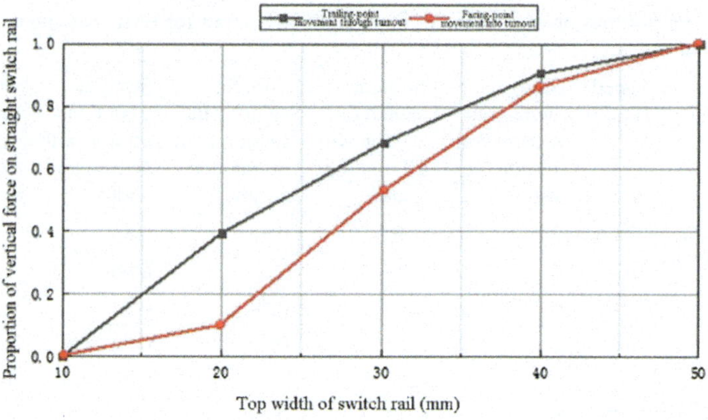

Fig. 8.94 Transition proportions and ranges of wheel-rail vertical forces for EMU passing on the main route

turnout slab relative to the base. They were positioned on both sides of the turnout respectively and at No. 51 turnout slab.

Table 8.39 displays the measured maximum displacement of the turnout slab within the turnout section when the CRH-380AL comprehensive inspection train passed through the turnout on the main route.

As shown in Table 8.39, when the CRH-380AL comprehensive inspection train passed on the main route.

The lateral displacement of the turnout slab ranged from 0.003 to 0.010 mm, the vertical displacement ranged from 0.003 to 0.022 mm, and the longitudinal displacement ranged from 0.002 to 0.050 mm.

7. Vibration Acceleration of Turnout Section

The measuring points for vibration within turnout section are arranged to measure the rail vibration acceleration, turnout slab vibration acceleration, and base vibration acceleration.

Table 8.40 displays the maximum measured vibration acceleration within the turnout section when the CRH-380AL comprehensive inspection train passed on the main route. Figures 8.98, 8.99 and 8.100 illustrate their relations with train speeds.

Table 8.37 Statistics of measured maximum and average strengths of turnout components for EMU passing the turnout on the main route (MPa)

Direction	Speed (km/h)	Straight switch rail section with 40 mm top width		Straight switch rail section with 50 mm top width	
		Maximum stress	Average stress	Maximum stress	Average stress
Facing-point movement into turnout	278.6	68.47	51.60	62.70	50.69
	307.2	60.88	50.97	61.67	53.89
	329.7	62.28	52.64	62.21	54.05
	338.4	63.31	54.84	60.23	51.4
	354.9	54.56	45.72	72.54	50.03
	357.5	61.83	48.78	59.77	49.37
	370.4	63.14	52.93	54.83	48.58
	371.9	66.60	52.90	57.20	46.52
	374.6	65.87	51.82	57.40	45.00
	375.0	64.00	51.88	56.45	47.24
	377.7	64.49	52.77	523	50.62
	380.9	66.13	41.1	68.99	54.81
	385.4	62.14	47.16	66.34	55.40
	394.7	57.93	46.82	63.33	52.26
	410.0	58.46	46.71	58.87	51.73
Trailing-point movement through turnout	283.9	80.39	67.17	27.33	17.19
	302	88.49	67.15	35.70	26.22
	313.0	86.93	66.15	38.60	21.98
	339.0	80.18	70.78	29.42	17.21
	355.7	75.64	66.63	20.14	14.99
	362.5	55.74	45.59	34.82	22.18
	362.9	68.63	59.48	31.87	18.60
	361	76.41	67.39	24.94	15.23
	374.2	79.27	65.56	31.83	19.26
	378.9	75.85	64.54	27.13	17.25
	378.9	55.74	43.36	50.50	27.30
	380.5	632	58.58	33.86	19.65
	381.4	76.03	66.49	25.67	19.00
	382.2	70.05	56.40	29.20	18.97
	383.4	64.88	55.09	36.80	22.42
	404.0	66.61	58.81	27.72	17.36
	407.2	72.74	62.52	28.09	17.20

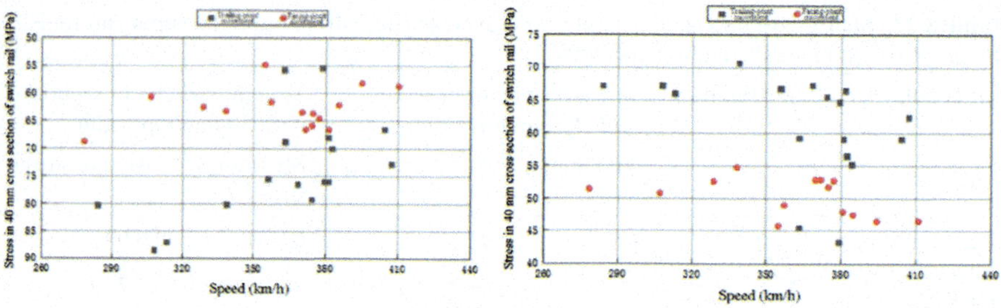

Fig. 8.95 Relation between maximum and average stress at straight switch rail cross section with 40 mm top width and speed of EMU passing on the main route

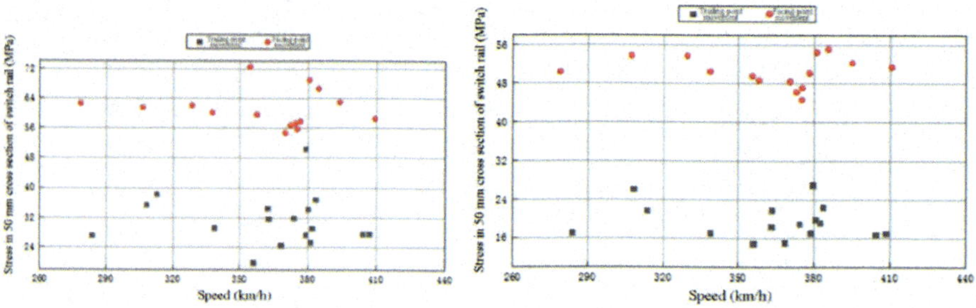

Fig. 8.96 Relation between maximum and average stress at straight switch rail cross section with 50 mm top width and speed of EMU passing on the main route

Table 8.38 Measured maximum strain of turnout slabs for EMU passing through No. 3 turnout location on main and diverging routes (με)

Direction	Speed (km/h)	Lateral strain of turnout slab on main route	Lateral strain of turnout slab on diverging route
Facing-point movement	278.6	9.86	2.01
	307.2	5.95	6.01
	329.7	6.46	5.39
	338.4	4.32	3.49
	354.9	4.80	2.49
	357.5	4.81	2.64
	370.4	9.11	4.30
	371.9	11.06	4.35
	374.6	10.07	2.58
	375.0	1.0	4.55
	377.7	9.48	5.06

(continued)

Table 8.38 (continued)

Direction	Speed (km/h)	Lateral strain of turnout slab on main route	Lateral strain of turnout slab on diverging route
	380.9	8.75	3.98
	385.4	11.54	5.71
	394.7	7.12	5.90
	410.0	9.44	4.48
Trailing-point movement	283.9	6.81	5.61
	302	3.72	3.76
	313.0	5.86	5.85
	339.0	6.28	7.11
	355.7	6.60	5.60
	362.5	4.61	4.02
	362.9	8.52	5.80
	361	4.83	3.83
	374.2	10.76	5.12
	378.9	6.51	4.20
	378.9	5.11	4.53
	380.5	7.93	3.66
	381.4	5.22	4.45
	382.2	4.89	3.02
	383.4	6.93	6.69
	404.0	8.99	6.10
	407.2	9.27	5.90

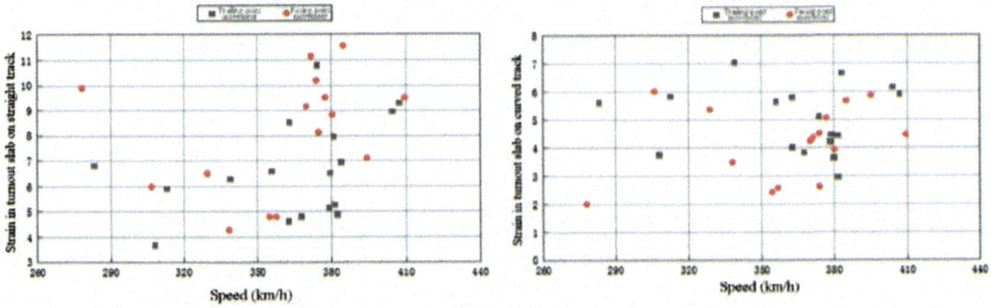

Fig. 8.96 Relation between strain of turnout slab and speed of EMU passing on the main route

As shown in Table 8.40, when the CRH-380AL comprehensive inspection train passed on the main route.

Table 8.39 Measured maximum displacement of turnout slab for EMU passing through turnout on the main route (mm)

Direction	Speed	Vertical displacement of straight slab on diverging route	Lateral displacement of straight slab on diverging route	Longitudinal displacement of straight slab on diverging route	Vertical displacement of curved slab on diverging route	Lateral displacement of curved slab on diverging route	Longitudinal displacement of curved slab on diverging route
Facing-point movement	278.6	0.003	0.005	0.006	0.004	0.008	0.004
	307.2	0.005	0.005	0.004	0.005	0.008	0.003
	329.7	0.005	0.009	0.020	0.006	0.012	0.004
	338.4	0.004	0.008	0.023	0.003	0.009	0.004
	354.9	0.005	0.010	0.020	0.005	0.012	0.003
	357.5	0.005	0.007	0.012	0.004	0.012	0.003
	370.4	0.006	0.004	0.006	0.004	0.008	0.004
	371.9	0.003	0.010	0.049	0.004	0.021	0.004
	374.6	0.004	0.010	0.040	0.005	0.018	0.003
	375.0	0.006	0.003	0.007	0.003	0.009	0.005
	377.7	0.009	0.008	0.009	0.005	0.012	0.005
Facing-point movement	380.9	0.007	0.013	0.043	0.005	0.022	0.004
	385.4	0.008	0.008	0.012	0.003	0.008	0.005
	394.7	0.008	0.004	0.006	0.004	0.009	0.013
	410.0	0.010	0.010	0.011	0.004	0.009	0.005
Trailing-point movement	283.9	0.004	0.006	0.018	0.003	0.008	0.003
	302	0.004	0.008	0.047	0.004	0.009	0.004
	313.0	0.004	0.004	0.006	0.004	0.010	0.005
	339.0	0.006	0.008	0.043	0.004	0.017	0.003

(continued)

Table 8.39 (continued)

Direction	Speed	Vertical displacement of straight slab on diverging route	Lateral displacement of straight slab on diverging route	Longitudinal displacement of straight slab on diverging route	Vertical displacement of curved slab on diverging route	Lateral displacement of curved slab on diverging route	Longitudinal displacement of curved slab on diverging route
	355.7	0.003	0.007	0.023	0.005	0.006	0.003
	362.5	0.007	0.011	0.050	0.004	0.019	0.002
	362.9	0.005	0.007	0.009	0.005	0.006	0.003
	361	0.007	0.009	0.039	0.004	0.019	0.002
	374.2	0.005	0.008	0.032	0.004	0.017	0.002
	378.9	0.006	0.004	0.028	0.004	0.008	0.004
	378.9	0.008	0.004	0.009	0.004	0.011	0.002
	380.5	0.006	0.007	0.036	0.004	0.012	0.004
	381.4	0.006	0.007	0.013	0.004	0.013	0.002
	382.2	0.006	0.003	0.010	0.003	0.012	0.003
	383.4	0.007	0.006	0.004	0.004	0.013	0.004
	404.0	0.006	0.007	0.014	0.006	0.012	0.003
	407.2	0.005	0.008	0.031	0.005	0.006	0.004

Table 8.40 Maximum measured vibration acceleration for EMU passing turnout on the main route (m/s^2)

Direction	Speed	Vibration acceleration of rail	Vibration acceleration of turnout slab	Vibration acceleration of base
Facing-point movement	278.6	676.7	22.0	14.4
	307.2	652.3	17.1	20.0
	329.7	745.5	29.9	14.4
	338.4	890.2	24.5	12
	354.9	3376.7	26.9	20.2
	357.5	2490.1	41.3	18.8
	370.4	2474.5	39.0	22.5
	371.9	2622	46.6	17.0
	374.6	1094.7	59.2	20.4
	375.0	1648.9	41	21.3
	377.7	2075.5	35.7	22.0
	380.9	1537.8	25.9	17.1
	385.4	1368.8	28.6	19.2
	394.7	2152.3	32.2	23.5
	410.0	2584.5	29.9	25.0
Trailing-point movement	283.9	512.1	23.6	17.2
	302	693.3	22.3	16.8
	313.0	811.8	23.8	17.0
	339.0	902.6	26.9	14.9
	355.7	1032.3	28.7	15.7
	362.5	2157.1	29.0	17.0
	362.9	1639.2	30.3	17.2
	361	2504.3	32.3	17.2
	374.2	2991.5	25.7	16.2
	378.9	1484.5	26.8	17.5
	378.9	2079.9	29.1	17.4
	380.5	2084.5	29.2	20.4
	381.4	2027.8	47.4	16.6
	382.2	3348.6	31.8	19.5
	383.4	1687.3	29.2	20.4
	404.0	2563.3	32.2	18.5
	407.2	2679.3	30.6	20.3

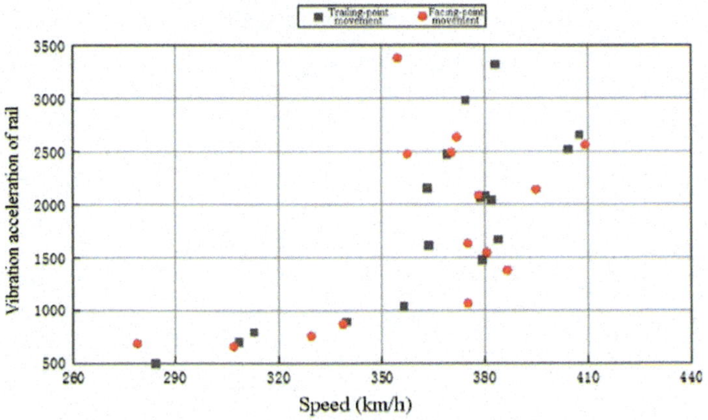

Fig. 8.98 Relation between vibration acceleration of rail and speed of EMU passing on the main route

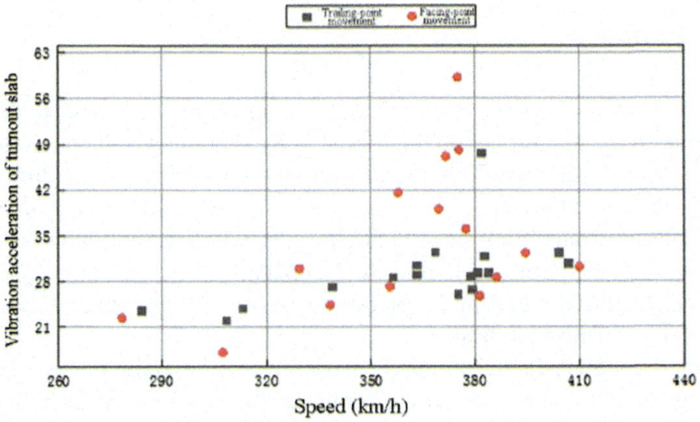

Fig. 8.99 Relation between vibration acceleration of turnout slab and speed of EMU passing on the main route

The vibration acceleration of the rail ranged from 512.1 to 3376.7 m/s^2, that of the turnout slab ranged from 17.1 to 59.2 m/s^2, and that of base ranged from 14.4 to 25.0 m/s^2, all meeting the turnout requirements for dynamic acceptance.

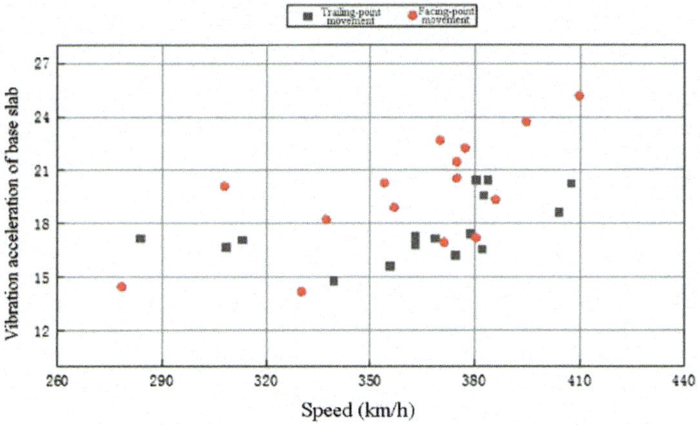

Fig. 8.100 Relation between vibration acceleration of base and speed of EMU passing on the diverging route

8.3.5 *Summary*

(1) When the CRH2-068C comprehensive inspection train passed through the turnout on the main route on the bridge at Xuzhoudong Station, all the requirements were met regarding the derailment coefficient, unloading rate, and wheel-rail lateral force.
(2) The measured maximum stress at the straight switch rail bottom was 72.54 MPa, far less than the permissible limit of 335 MPa.
(3) All the measured vibration accelerations of the rail, turnout slab, and base fell within the permissible limits.
(4) When the CRH2-068C comprehensive inspection train passed through the turnout on the main route on the bridge at Xuzhoudong Station, the dynamic acceptance requirements were met regarding all dynamic indexes, offering train operation guarantee in terms of safety, comfort, and stability.

References

1. Gao Z (2009) Analysis of dynamic characteristics of sleeper-embedded ballastless track on bridge. Southwest Jiaotong University, Chengdu
2. Lou P, Gong K, Zhao C (2020) Operational safety of CRTS III slab ballastless track under earthquake. J Railway Sci Eng 17(1):1–7
3. Railway Engineering Research Institute of China Academy of Railway Sciences (1994) Track side test methods of vertical and lateral wheel-rail forces: TB/T 2489-94. China Railway Publishing House, Beijing
4. Liu F (2009) Finite element analysis on stability of CWR track. Beijing Jiaotong University, Beijing

5. Yang R (2008) Computational theory and experimental study on longitudinal force on CWR turnout on bridge. Southwest Jiaotong University, Chengdu
6. Tian C, Wang P (2003) Stability analysis of CWR turnout based on uniform formula. J Southwest Jiaotong Univ 4:375–379
7. Li X, Yang J, Rong J et al (2017) Characteristic harmonic modeling for CRH380A high-speed EMU based on measured data. J China Railway Soc 39(9):32–39
8. Bi L, Zhao P, Wang P et al (2021) Research on the applicability of track side test methods for wheel-rail vertical impact force. J China Railway Soc 43(2):118–124

Chapter 9
Design Cases of CWR Turnout on Bridge

Through the process of summarizing, digesting, and absorbing Chinese and foreign design experience related to CWR turnouts on bridges, China has successfully overcome technical bottlenecks in the design of ballastless turnouts on bridges. This achievement led to the development of two types of ballastless track structures in turnout sections: slab ballastless track for turnouts on bridges [1–3] and long-sleeper embedded ballastless track for turnouts on bridges [4, 5]. With the intensive construction of high-speed railways in China, the technology of CWR turnouts on bridges has been rapidly developed and successfully applied to numerous railways, including Beijing–Shanghai, Wuhan–Guangzhou, Guangzhou–Shenzhen–Hong Kong, Hangzhou–Ningbo, Hangzhou–Changsha, and Wuhan–Shiyan HSRs. In addition, to meet the special requirements for overpassing urban roads or navigation, the inclusion of long-span continuous girders becomes inevitable. This leads to another major technical challenge in the development of CWR track systems: laying CWR turnouts on long-span bridges. Based on the design of CWR turnouts on conventional bridges, this chapter presents the study on the CWR turnout laying on large-span bridges, focusing on two design examples of the continuous girder with a 100-m main span at Keqiao Station of Hangzhou–Ningbo HSR and the continuous girder + arch with a 102-m main span crossing Jiulong Avenue of Hangzhou–Changsha HSR.

9.1 Design of CWR Turnout on the Bridge at Suzhoubei Station of Beijing–Shanghai HSR

Five elevated stations (Zhenjiang, Changzhou, Wuxi, Suzhou, and Kunshan) are distributed in the Xuzhou–Shanghai section of Beijing–Shanghai HSR. They all incorporate slab ballastless tracks featuring longitudinally coupled base slabs in their turnout sections on the bridges. Before the development of this railway, slab turnouts

Q. Li et al., *Continuously Welded Turnouts on High-Speed Railway Bridge*, Advances in High-speed Rail Technology, https://doi.org/10.1007/978-981-97-6524-9_9

had not been applied globally, without Chinese experience available for reference. China launched special studies on the stress and deformation patterns, computational theories, and design approaches of ballastless CWR turnouts on bridges featuring longitudinally coupled base slabs, aiming to fulfill CWR turnout laying on slab ballastless tracks on bridges. This section introduces the CWR turnouts with slab ballastless tracks on the bridge at Suzhoubei Station.

9.1.1 Design Overview

1. Design Criteria

As one north–south corridor within the HSR network of "Four North–South and Four East–West HSR Corridors" in China, the Beijing–Shanghai HSR connects the Bohai Economic Rim and Yangtze River Delta, passes through three municipalities directly under the Central Government (Beijing, Tianjin, and Shanghai) and four provinces (Hebei, Shandong, Anhui, and Jiangsu), with 24 stations distributed throughout a total length of 1,318 km, and a design speed of 350 km/h. This HSR marks the longest construction within a single phase, the highest investment, and the highest standards in China.

2. Station Arrangement

Suzhoubei Station is a large elevated passenger station, covering a station site of 3.2 km in total length, and located in the straight and flat slope section. The station is provided with 4 arrival and departure tracks, respectively set on both sides of the up- and down-traffic main lines, with an effective length of 650 m. There are 2 passenger platforms (450 m × 12 m × 1.25 m), as shown in Fig. 9.1. The main lines at the station are laid with CRTS II slab tracks generally and slab ballastless tracks in the turnout sections. The arrival and departure tracks are laid with CRTS I bi-block ballastless tracks generally and sleeper-embedded ballastless tracks in the turnout sections.

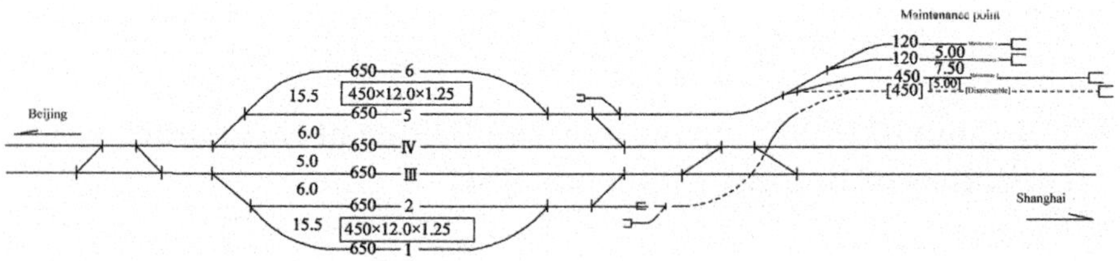

Fig. 9.1 Arrangement of Suzhoubei station (unit: m)

9.1.2 Design of Slab Ballastless Turnout on Bridge

The slab ballastless tracks for turnouts on the bridge consist of rails, fasteners, turnout slabs, mortar cushions, base, sliding layers, and lateral stops, as shown in Figs. 9.2 and 9.3.

1. Fastener System

The ERL fasteners matched with BWG [6, 7] turnouts are adopted, which consist of Vossloh SKL12 clips/washers, T-shaped screws, and nuts. The clip deformation allowance is 14.5 mm and the clamping force is 12 kN, as shown in Fig. 9.4.

2. Turnout slab

The turnout slabs are 240 mm thick, transversely covered with pile caps measuring 340 mm in width and spaced at 600 mm. The turnout slab surface between the pile

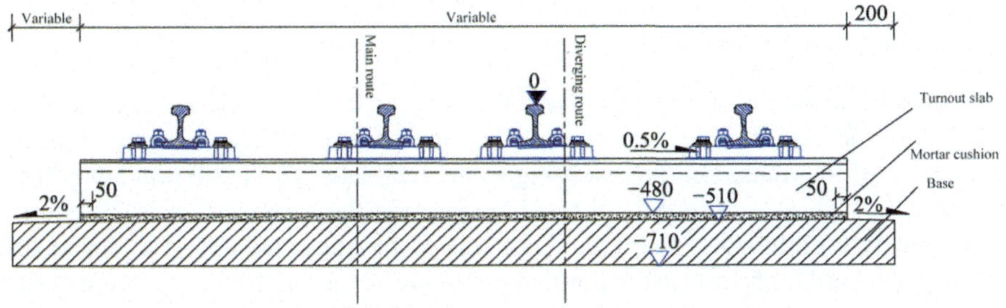

Fig. 9.2 Cross section of slab ballastless track in turnout section on bridge (unit: mm)

Fig. 9.3 No. 18 single turnout on the bridge

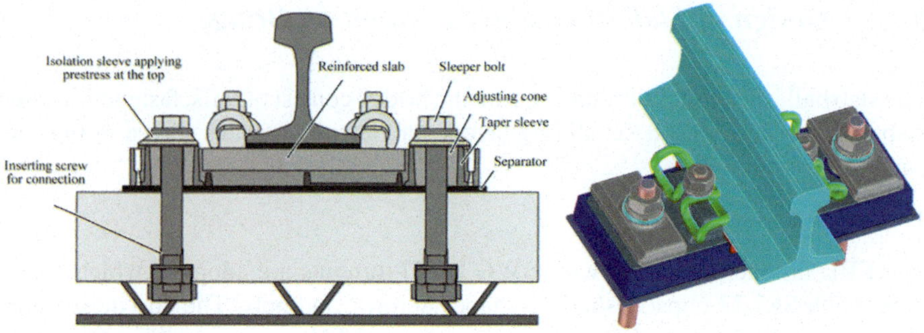

Fig. 9.4 BWG ERL fastener system

caps is 0.5% sloped transversely for drainage and transversely pre-cracked with a crack depth of 4 cm, built with C55 concrete and HRB335 reinforcement, as shown in Fig. 9.5.

3. Mortar Cushion and Base

The mortar cushion is 30 mm thick, and the base is 200 mm thick and is cast in situ with C30 concrete. On the main route, the base side is 1,475 mm away from the centerline of the track. On the diverging route, the base side is 200 mm away from the turnout slab side. The part of the base beyond the turnout slab is 2% sloped transversely for drainage toward the outer side of the track, and the turnout slab side is shifted inward for 50 mm at the point of gradient change. HRB500 rebars are placed inside the base throughout the CRTS II slab track within the whole turnout section and adjacent bridge sections. Insulation treatment with plastic clips is adopted between the longitudinal and transverse rebars inside the base within the turnout section up to the track circuit requirements, as shown in Fig. 9.6.

Fig. 9.5 Precast track bed slab

Fig. 9.6 Base construction

4. Sliding Layer and Shear Notch

A sliding layer is arranged between the base slab and the heightening layer of the bridge deck to reduce the influence of girder expansion on the longitudinal stress of the rails. The sliding layer consists of two layers of geotextile and one layer of membrane, that is, a sliding membrane is set between the two layers of geotextile, which is laid continuously in the longitudinal direction [8, 9]. The sliding membrane may be made of PE-HD high-density polyethylene film with a thickness of 1 mm, and the geotextile may be made of shear-resistant polypropylene (PP) fiber with a thickness of 2.2 mm.

The shear notch [10, 11] is set above the fixed bearing of the bridge for connection to the base, and HRB335 rebars are used for anchorage.

5. Lateral Stop

The lateral stop is 800 mm long and made of C35 concrete. The working face of the position-limiting plate on the main route/diverging route in contact with the base side shall be parallel to the centerline of the main route, as shown in Fig. 9.7, to ensure the relative sliding between the base and the position-limiting plate.

9.1.3 Design Checking Calculation of CWR Turnout Group on Bridge

1. Calculation Parameters

The arrangement of the turnout bridge at the left throat of Suzhoubei Station is shown in Fig. 9.8.

Fig. 9.7 Lateral stop

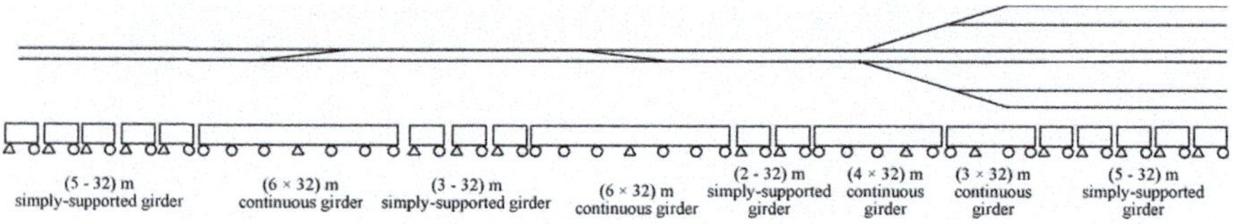

Fig. 9.8 Arrangement of turnout bridge at left throat of Suzhoubei station

No. 18 high-speed turnout with 60 kg/m rails is adopted, with a total length of 69 m, a front length of 31.729 m, and a rear length of 37.271 m. The longitudinal linear rigidity of the piers is shown in Table 9.1.

Table 9.1 Longitudinal linear rigidity

S/N	Span (m)	Longitudinal linear rigidity (kN/m)	Remarks
1	32.6	595	
2	32.6	595	
3	24.6	595	
4	24.6	595	
5	24.78	595	
6	6 × 32.6	900	Fixed pier
7	24.6	595	
8	24.6	595	
9	32.6	595	
10	6 × 32.6	900	Fixed pier
11	32.6	595	
12	32.6	595	
13	4 × 32.6	1,750	Fixed pier
14	3 × 32.6	820	Fixed pier

Temperature load: The temperature drop is 54 °C for the base slab and 30 °C for the bridge.

Braking load: The configuration of one-track traction and one-track braking is adopted. The traction load was calculated as per 33.3 × 0.8 kN/m and a loading distance of 30 m. The braking load was calculated as per 16 kN/m and a loading distance of 300 m.

2. Calculation Results

(1) Calculation results corresponding to various friction coefficients between base and substructure

The rigidity reduction coefficient of the base slab was taken as 0.1, and the friction coefficients between the base and the substructure were taken as 0.1, 0.3, and 1.0, respectively. In Condition II, the stress deformation of components within the ballastless turnout under temperature effects is shown in Figs. 9.9, 9.10, 9.11, 9.12, 9.13 and 9.14, respectively.

In Condition II, the stress deformation of components within the ballastless turnout under braking effects is shown in Figs. 9.15, 9.16, 9.17, 9.18, 9.19 and 9.20, respectively.

The stress and deformation results of the ballastless turnout under various temperature loads and braking loads in different conditions are shown in Tables 9.2 and 9.3.

Based on the above calculation results, the increases in the friction coefficient between the base slab and bridge led to gradual reductions in both the rail temperature force and the rail braking force. While minor fluctuations were observed for the rail

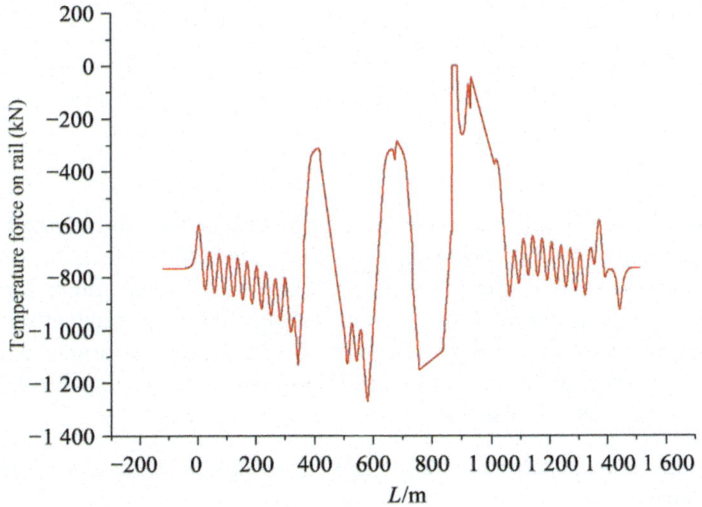

Fig. 9.9 Temperature force on rail

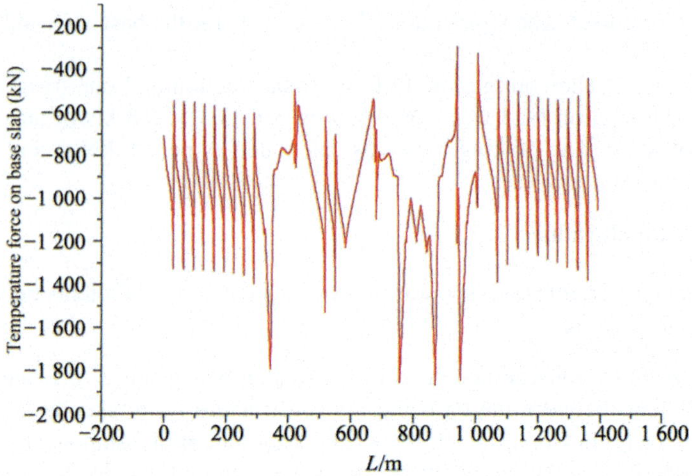

Fig. 9.10 Temperature force on base slab

Fig. 9.11 Longitudinal displacement of rail within No. 2 turnout

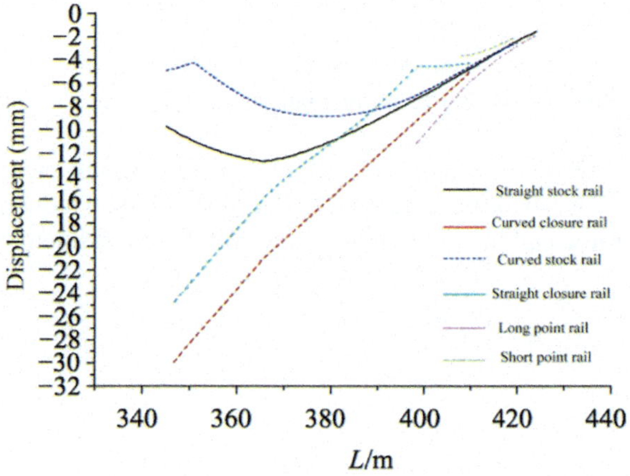

temperature force, significant variations were noted for the rail braking force. The force transfer device within the turnout was observed slightly affected by the friction coefficient changes. With the gradual increments in the temperature force on the bridge pier/abutment, the braking force on the pier/abutment gradually decreased. Under the action of temperature loads, the displacement of turnout components gradually increased; under the action of braking loads, the displacement of turnout components gradually decreased.

(2) Calculation results corresponding to various rigidity reduction coefficients of base slab

The friction coefficient between the base slab and the bridge was taken as 0.1, and the rigidity reduction coefficient of the base slab was taken as 0.1, 0.3, and 1.0,

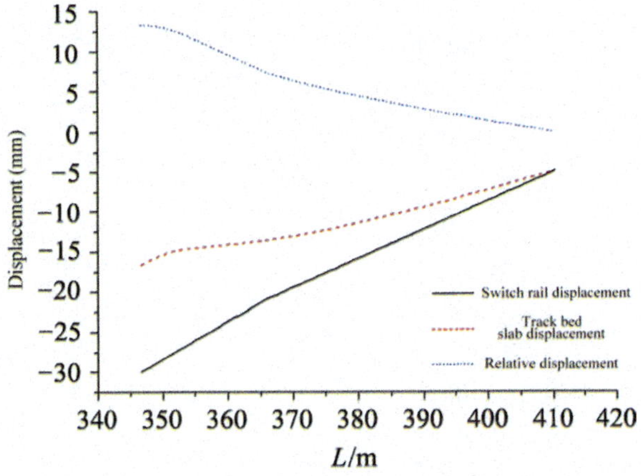

Fig. 9.12 Relative displacement between switch rail and track slab within No. 2 turnout

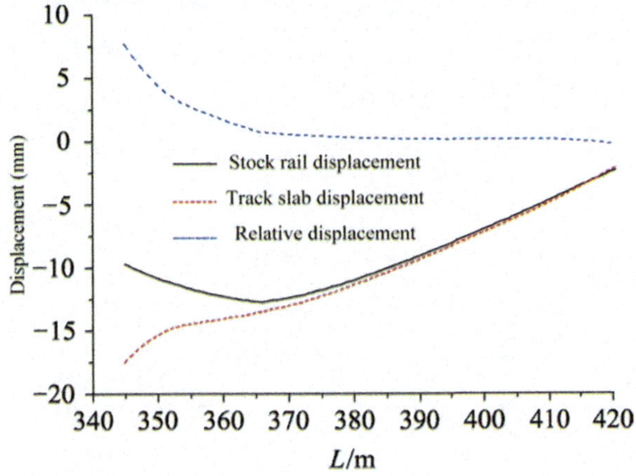

Fig. 9.13 Relative displacement between stock rail and track bed within No. 2 turnout

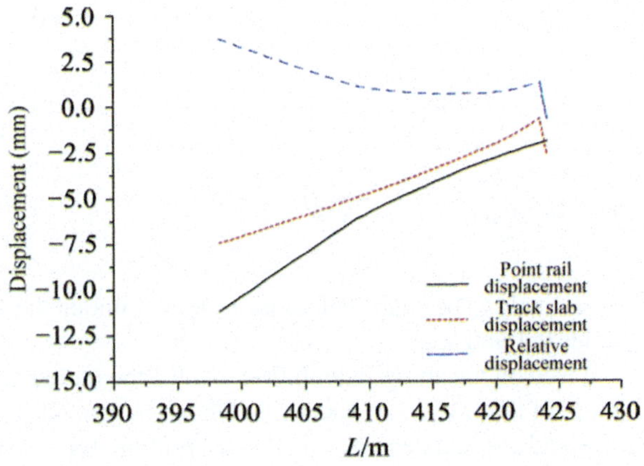

Fig. 9.14 Relative displacement between point rail and track slab within No. 2 turnout

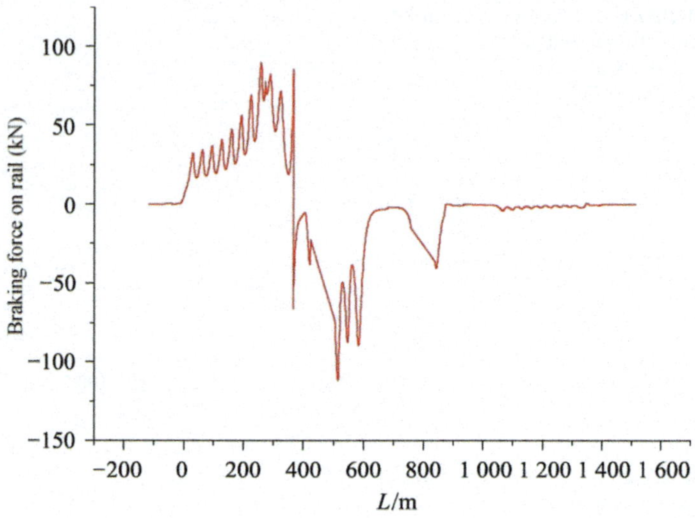

Fig. 9.15 Braking force on rail

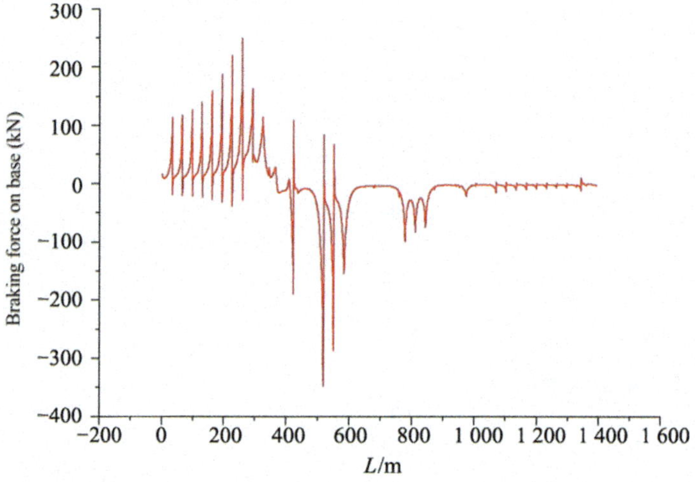

Fig. 9.16 Braking force on base

respectively. The main calculation results under the three conditions are shown in Tables 9.4 and 9.5.

Based on the above calculation results, the increases in the rigidity reduction coefficient of the track slab and base slab led to gradual increases in the rail temperature force and gradual decreases in the rail braking force. Under the action of temperature, the stress on the stopper and spacing block gradually increased; under the action of braking forces, the stress on the stopper gradually increased while that on the spacing

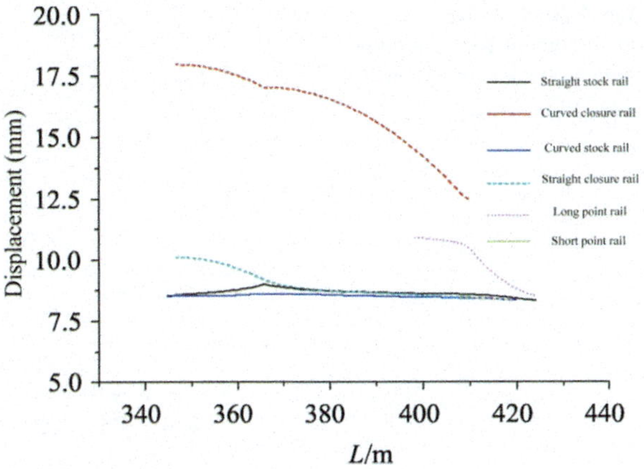

Fig. 9.17 Displacement of No. 2 turnout

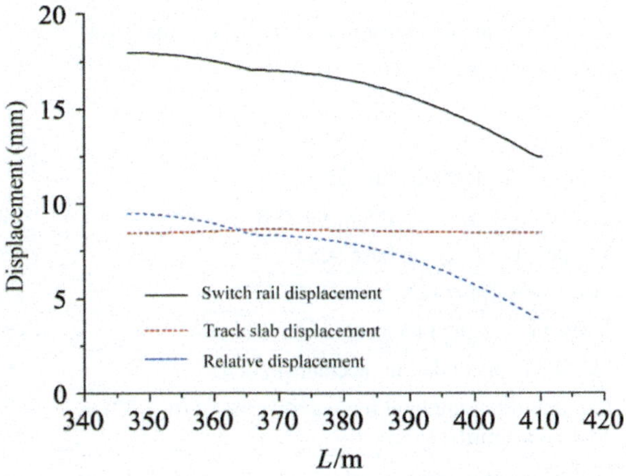

Fig. 9.18 Relative displacement between switch rail and track slab within No. 2 turnout

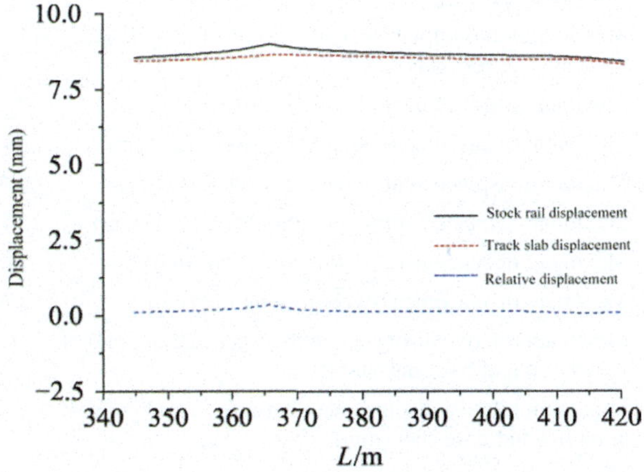

Fig. 9.19 Relative displacement between stock rail and track slab within No. 2 turnout

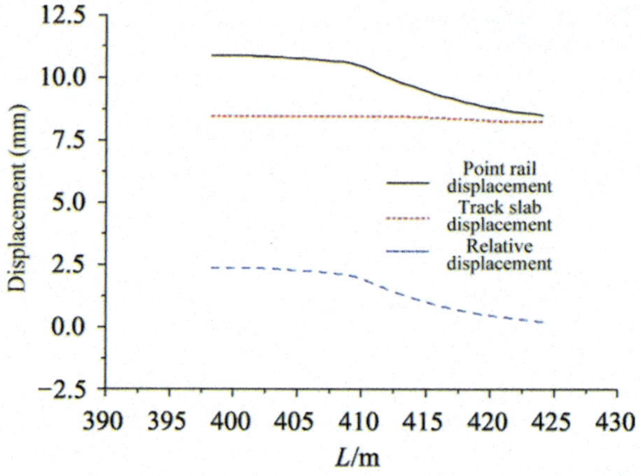

Fig. 9.20 Relative displacement between point rail and track slab within No. 2 turnout

Table 9.2 Calculation results under temperature loads

Item	Temperature force		
	Condition I	Condition II	Condition III
Temperature force on rail (kN)	1,130	1,183	1,268
Temperature force on track slab (kN)	1,326	1,302	1,875
Temperature force on base (kN)	1,848	1 866	2,710
Stress on stopper (kN)	176.5	162.5	166.5
Stress on spacing block (kN)	75	82	85
Stress on consolidation mechanism (kN)	751	973	2,400
Maximum longitudinal force at pier top of (6 × 32) m continuous girder (kN)	67	221	436
Maximum longitudinal force at pier top of (4 × 32) m continuous girder (kN)	888	532	246
Maximum longitudinal force at pier top of (3 × 32) m continuous girder (kN)	1,582	1,637	1 689
Maximum displacement of straight stock rail (mm)	7.76	9.7	13.37
Maximum displacement of curved closure rail (mm)	26.8	29.95	32.83
Maximum displacement of curved stock rail (mm)	0.72	4.91	8.63
Maximum displacement of straight closure rail (mm)	20.58	24.78	28.64
Maximum displacement of long point rail (mm)	11.4	11.17	12.77
Maximum displacement of short point rail (mm)	1.65	3.63	5.64
Maximum relative displacement between actual point of switch rail and base slab (mm)	16.63	13.38	8.04
Maximum relative displacement between actual point of point rail and base slab (mm)	7.49	3.73	2.82
Maximum relative displacement between stock rail and base slab (mm)	5.99	7.8	12.341

Table 9.3 Calculation results under braking loads

Item	Braking force		
	Condition I	Condition II	Condition III
Braking force on rail (kN)	127	112	83
Braking force on track slab (kN)	195	194	182
Braking force on base (kN)	354	348	391
Stress on stopper (kN)	140	159	163
Stress on spacing block (kN)	94	89	89
Stress on consolidation mechanism (kN)	629	488	544
Maximum longitudinal force at pier top of (6 × 32) m continuous girder (kN)	1,231	982	755
Maximum longitudinal force at pier top of (4 × 32) m continuous girder (kN)	1,416	1,237	1,016
Maximum longitudinal force at pier top of (3 × 32) m continuous girder (kN)	998	924	804
Maximum displacement of straight stock rail (mm)	12.17	9	6.85
Maximum displacement of curved closure rail (mm)	20.96	17.963	15.87
Maximum displacement of curved stock rail (mm)	11.83	8.54	6.5
Maximum displacement of straight closure rail (mm)	13.33	10.12	7.98
Maximum displacement of long point rail (mm)	13.35	10.89	8.92
Maximum displacement of short point rail (mm)	10.89	8.46	6.43
Maximum relative displacement between actual point of switch rail and base slab (mm)	11.43	9.52	9.5
Maximum relative displacement between actual point of point rail and base slab (mm)	2.1	2.4	2.5
Maximum relative displacement between stock rail and base slab (mm)	0.29	0.355	0.37

block gradually decreased with slight variations. With the gradual increments in the temperature force on the bridge pier/abutment, the braking force on the pier/abutment gradually decreased.

(3) Checking calculation for turnout featuring longitudinally coupled base slabs on bridge

According to the above calculation results, the strength of rails, displacement at the actual point of the switch rail, displacement at the actual point of the point rail, and displacement of the switch and frog relative to the base slab within the CWR turnout on the bridge at the left throat of Suzhoubei Station were checked. The results are shown in Table 9.6.

Based on the above checking calculation results, the strength of rails, the relative displacement at the actual point of the switch rail, the relative displacement at the

Table 9.4 Calculation results under temperature loads

Item	Temperature force		
	Condition I	Condition II	Condition III
Rail force (kN)	1,130	1,289	1,562
Track slab (kN)	1,326	3,687	11,892
Base (kN)	1,848	4,755	13,658
Stress on stopper (kN)	176.5	208.5	285
Stress on spacing block (kN)	75	105	144
Stress on consolidation mechanism (kN)	751	748	2,241
Maximum longitudinal force at pier top of (6 × 32) m continuous girder (kN)	67	121	1,148
Maximum longitudinal force at pier top of (4 × 32) m continuous girder (kN)	888	773	1,438
Maximum longitudinal force at pier top of (3 × 32) m continuous girder (kN)	1,582	1,752	2,705

Table 9.5 Calculation results under braking loads

Item	Braking force		
	Condition I	Condition II	Condition III
Rail force (kN)	127	86	80
Track slab (kN)	195	273	399
Base (kN)	354	443	687
Stress on stopper (kN)	140	153	160
Stress on spacing block (kN)	94	91	89
Stress on consolidation mechanism (kN)	629	686	695
Maximum longitudinal force at pier top of (6 × 32) m continuous girder (kN)	1,231	955	613
Maximum longitudinal force at pier top of (4 × 32) m continuous girder (kN)	1,416	1,180	852
Maximum longitudinal force at pier top of (3 × 32) m continuous girder (kN)	998	874	674

actual point of the point rail, and the displacement of the switch and frog relative to the base slab within the CWR turnout on the bridge at the left throat of Suzhoubei Station all meet the requirements.

Table 9.6 Checking calculation results of turnout

Item	Calculated maximum	Permissible value
Rail strength (MPa)	320	351
Relative displacement between switch rail and base slab (mm)	16.6	30
Relative displacement between point rail and base slab (mm)	8.5	10
Relative displacement between switch and base slab (mm)	0.5	3.7
Relative displacement between frog and base slab (mm)	7.8	20.7
Shear stress of stopper bolt (MPa)	194	265
Shear stress of spacing block bolt (MPa)	80	265

9.2 Design of CWR Turnout on the Bridge at Tung Chung Station of Guangzhou–Shenzhen–Hong Kong HSR

9.2.1 Design Overview

The Guangzhou–Shenzhen Section of Guangzhou–Shenzhen–Hong Kong HSR extends for a total length of 113.827 km and is designed for train operation up to 350 km/h. Five stations are distributed along the whole line: Tung Chung, Humen, Guangming, Shenzhenbei, and Futian, among which Tung Chung, Humen, and Guangming are elevated stations. This section focuses on Tung Chung Station to introduce the design of CWR turnouts with sleeper-embedded ballastless tracks. The sleeper-embedded ballastless track is adopted in the turnout section at Tung Chung Station. The specific layout plan is shown in Fig. 9.21, and the bridge span arrangement is shown in Fig. 9.22. No. 1 and 3 turnouts are located on the (32 + 2 × 50 + 32) m continuous girder, while No. 2 and 4 turnouts are located on the (5 × 32) m continuous girder.

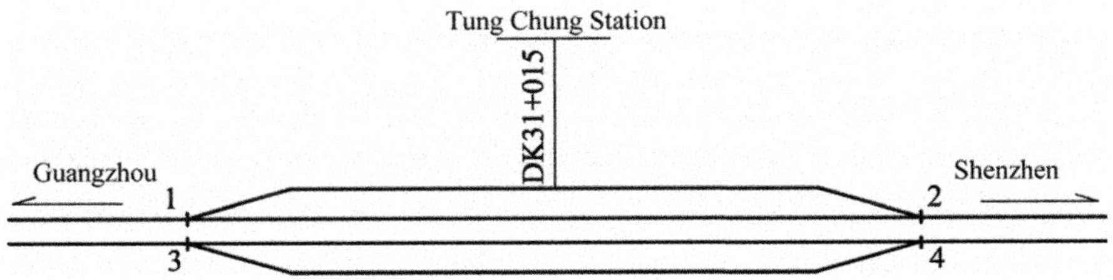

Fig. 9.21 Layout plan of Tung Chung station

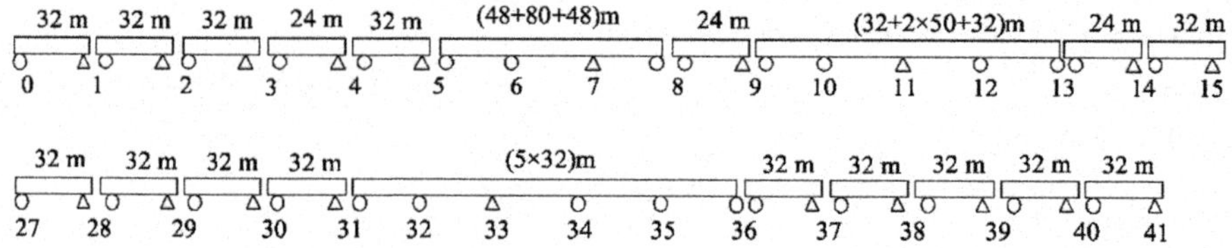

Fig. 9.22 Span arrangement at Tung Chung station

9.2.2 *Design of Sleeper-Embedded Ballastless Turnout on Bridge*

The sleeper-embedded ballastless track within the turnout section on the bridge consists of the turnout rails, fasteners, turnout sleepers, reinforced concrete track bed slabs, intermediate isolation layer, and deck protection layer from top to bottom. The track structure is 800 mm high, as shown in Figs. 9.23 and 9.24.

1. Fastener System

The fastener system is not reiterated here as it is the same as the one described in Sect. 9.1.2 of this chapter.

2. Turnout Sleeper

The turnout sleepers are made of concrete and truss reinforcement, through the cast in situ practice to produce good connection performance. The turnout sleepers are positioned perpendicular to the main route at the turnout and perpendicular to the centerline of the line following the diversion from the heel end of the frog. Long turnout sleepers are hinged, which is conducive to stress and deformation mitigation, as shown in Fig. 9.25.

3. Track Bed Slab

The track bed slab within the turnout section is made of C40 concrete by the cast in situ practice, with a thickness of 400 mm. The edge of the track bed slab is 1,600 mm away from the centerline of the outer rail. The top surface of the track bed slab is sloped transversely for drainage adaptive to specific conditions. The track bed slab within the turnout on the bridge is placed on the concrete protection layer of the bridge deck, with an intermediate separation layer set between them. The track bed slab is secured by the longitudinal and transverse bollards of the bridge deck protection layer as shown in Figs. 9.26 and 9.27.

4. Design of Deck Protection Layer

The concrete protection layer of the bridge deck is made of C40 concrete through the cast in situ practice, and the design thickness of the protection layer is 140 mm. An intermediate separation layer of "two-geotextile one-membrane" is set between the track bed slab and the protection layer within the turnout section. The intermediate

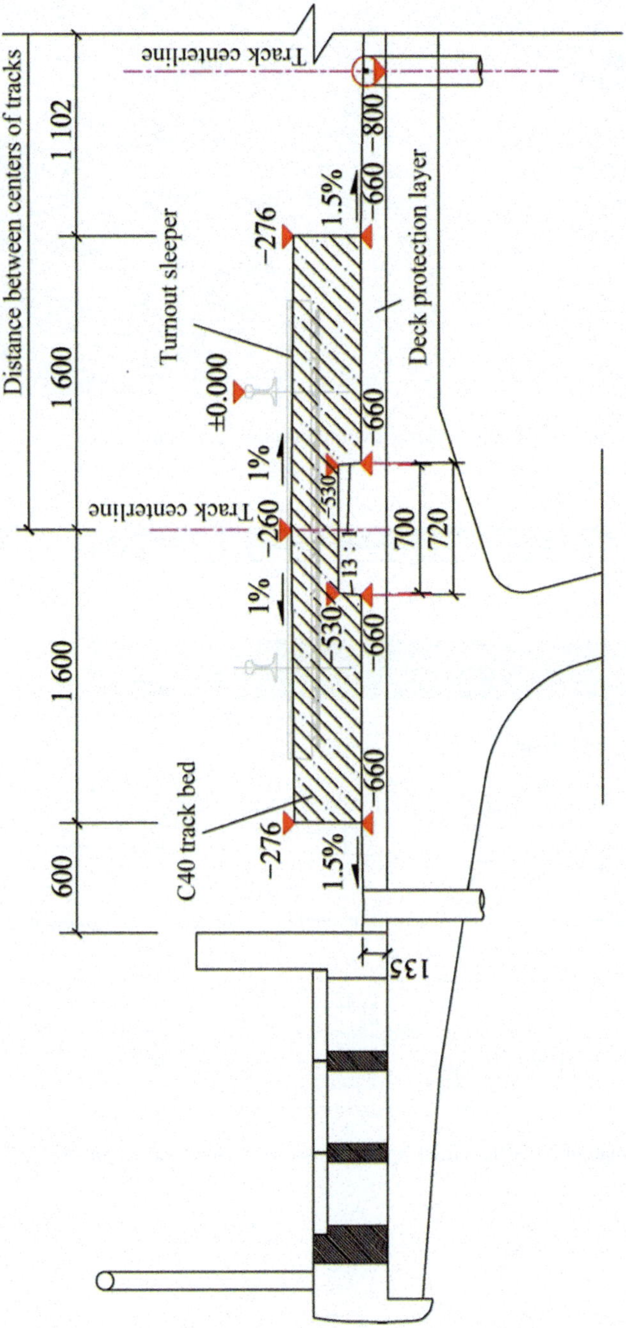

Fig. 9.23 Sleeper-embedded ballastless track within turnout section on bridge (unit: mm)

Fig. 9.24 Sleeper-embedded ballastless track within turnout section on bridge

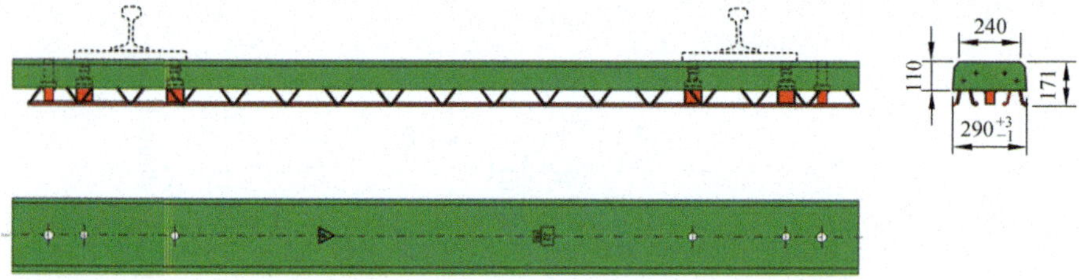

Fig. 9.25 BWG turnout sleeper (unit: mm)

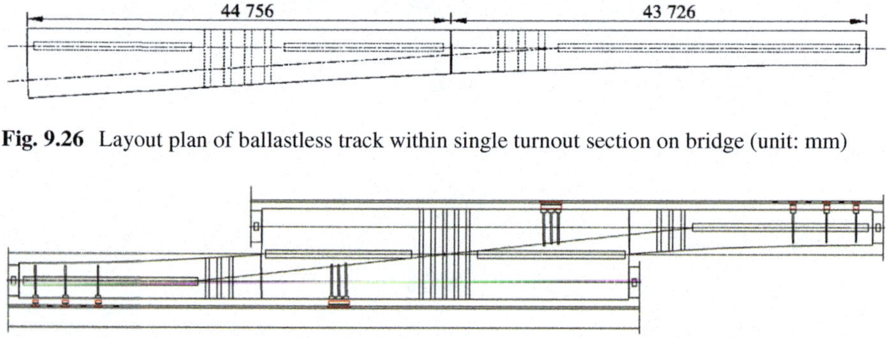

Fig. 9.26 Layout plan of ballastless track within single turnout section on bridge (unit: mm)

Fig. 9.27 Layout plan of ballastless track within crossover turnout section on bridge

Fig. 9.28 Isolation layer on deck protection layer and bollards

separation layer consists of upper and lower layers of 2.2 mm thick polypropylene geotextile and a middle layer of 1 mm thick high-density polyethylene film (PE film). There are expansion joints set longitudinally along the railway at the center of the bridge concrete protection layer and filled with polyurethane sealant, as shown in Fig. 9.28.

9.2.3 Design Checking Calculation of CWR Turnout Group on Bridge

1. Calculation Parameters

See Table 9.7 for the rigidity of middle piers/abutments in Fig. 9.22.

60 kg/m rails are laid on sleepers spaced at 0.6 m. The longitudinal resistances of the turnout track were calculated as per 10 kN/sleeper.

Table 9.7 Linear rigidity of piers/abutments

Pier No.	Linear rigidity (kN/cm)	Pier No.	Linear rigidity (kN/cm)	Pier No.	Linear rigidity (kN/cm)
0	767.9	11	1,945	33	1,842
1	797.7	12	5,513	34	1,871
2	823.8	13	1,950	35	1,424
3	691.9	14	1,950	36	1,394
4	724.8	15	1,950	37	720
5	1,945	27	1,950	38	748.3
6	5,757	28	1,950	39	721.7
7	5,513	29	1,950	40	744.6
8	1,950	30	1,950	41	897.2
9	1,437	31	1,749		
10	1,489	32	1,712		

2. Calculation of Dynamic Flexural Stress

The maximum axle load of high-speed EMUs is taken as 170 kN, and the rigid wheelbase as 2.5 m. The design speed is 350 km/h and the speed coefficient is $\alpha = 1$. This bridge is located on a straight section with an eccentric load coefficient of $\beta = 0$ and lateral force coefficient of $f = 1.25$. The rail support rigidity D of the ballastless track is taken as 23 kN/mm. The sleeper spacing is taken as $a = 625$ mm. The rail section parameter is taken as 9 mm for wear:

$$\sigma_{底d} = \frac{M_0(1+\alpha+\beta)}{W_{底}} \cdot f$$

where M_0—static bending moment of rail (N mm);
α—velocity coefficient;
β—eccentric load coefficient.
See Table 9.8 for the calculation results of the dynamic flexural stress of rails.
Based on the calculation, the dynamic flexural tensile stress of the rail base is $\sigma_d = 119.78$ MPa.

3. Analysis and Checking Calculation

(1) Strength of Rails within Turnout Section

The calculation results of longitudinal forces on rails at the Guangzhou and Shenzhen ends are shown in Figs. 9.29 and 9.30, respectively. The maximum longitudinal force

Table 9.8 Calculation of dynamic flexural stress of rail

Calculation of dynamic flexural stress			Calculation parameters	
Wheel position	I	II	E/MPa	210,000
P_0/N	85,000	85,000	I/mm^4	26,364,000
x/mm	0	2,500	$W_{头}$/mm^3	279,000
kx	0	2.838,4	$W_{底}$/mm^3	362,000
u_0	1.00	−0.0733	D/(kN/mm)	23
P_0u_0	85,000.00	−6,230.50	a/mm	625
$\sum P_0u_0$	78,769.500,000		α	0.72
M_j/(N mm)	17,344,456.00		β	0.18
M_d/(N mm)	34,688,912.00		f	2.0
$a_{底}$/MPa	119.78		k	0.001,135,37
$a_{头}$/MPa	155.42			

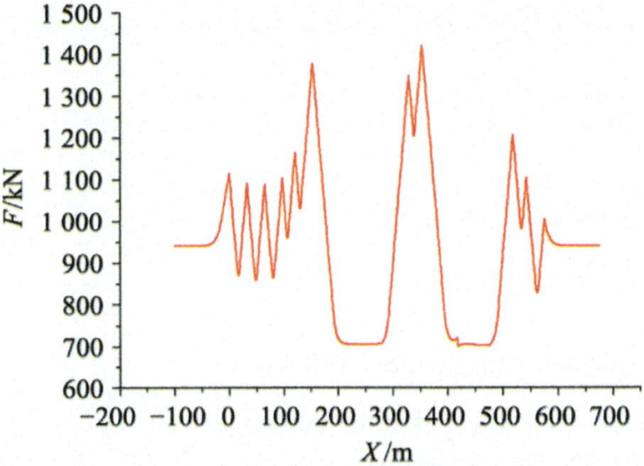

Fig. 9.29 Longitudinal forces on stock rails (Guangzhou end)

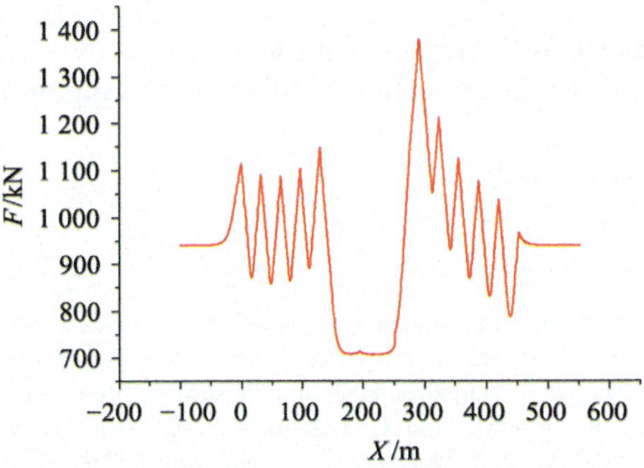

Fig. 9.30 Longitudinal forces on stock rails (Shenzhen end)

on the stock rail at the Guangzhou end is 1,439.9 kN (including temperature force), and that at the Shenzhen end is 1,497.5 kN (including temperature force).

The checking calculation results of turnout section strength are shown in Table 9.9, indicating compliance with design specifications regarding rail stress.

Table 9.9 Checking calculation of turnout rail strength (unit: MPa)

Item	Additional stress of rail	Dynamic flexural stress of rail	Temperature force on rail	Total stress of rail	Permissible stress of rail
Guangzhou end	62.3	126	123.6	311.9	352
Shenzhen end	56.9	126	123.6	306.5	352

Table 9.10 Expansion and contraction displacements at actual points of switch rail and point rail (unit: mm)

Turnout No.	Absolute displacement at actual point of switch rail	Displacement of switch rail relative to bridge	Absolute displacement at actual point of point rail	Displacement of point rail relative to bridge
1, 3	26.978	16.565	6.225	3.295
2, 4	28.857	16.315	8.205	3.125

(2) Checking calculation of turnout displacement

See Table 9.10 for calculations of turnout displacements.

According to the above checking calculation results, the rail strength and displacements of the switch rail and point rail meet the requirements.

9.3 Design of CWR Turnout on the Bridge at Guangzhounan Station of Wuhan–Guangzhou HSR

9.3.1 Design Overview

1. Station Arrangement

Guangzhounan Station is the main large passenger station of Guangzhou Railway Hub, one of China's four major railway passenger transportation hubs. It stands at the end point of Wuhan–Guangzhou HSR and the starting point of both Guangzhou–Shenzhen–Hong Kong HSR and Guangzhou–Zhuhai Intercity Railway. It also connects to the Guangzhou–Maoming Line westward. This station is equipped with two yards respectively for Wuhan–Guangzhou HSR and Guangzhou–Shenzhen Intercity Railway, and Guangzhou–Zhuhai Intercity Railway. These two yards are arranged in a transverse layout. It is provided with 28 arrival and departure tracks and 15 platforms, and all the track are ballasted, as shown in Fig. 9.31.

2. Turnout–Bridge Arrangement

The bridge span arrangement at the north throat of the yard for Wuhan–Guangzhou HSR is (4 × 31) m continuous girder + (3 × 32.7) m continuous girder + 2–32 m simply-supported girder + (4 × 32.7) m continuous girder + (29.3 + 4 × 32.7 + 29.3) m continuous girder + (28 + 2 × 32.7 + 28) m continuous girder. The bridge is laid with the ballasted track, and No. 18 high-speed turnout on the main line, as shown in Fig. 9.32.

The bridge span arrangement at the south throat of the yard for Wuhan–Guangzhou HSR is (21.5 + 2 × 32.7 + 21.5) m continuous girder + (27.7 + 4 × 30.0 + 27.7) m continuous girder + (32.7 + 35.64 + 32.7) m continuous girder + (2 × 32.7 + 2

Fig. 9.31 Aerial view of elevated Guangzhounan station

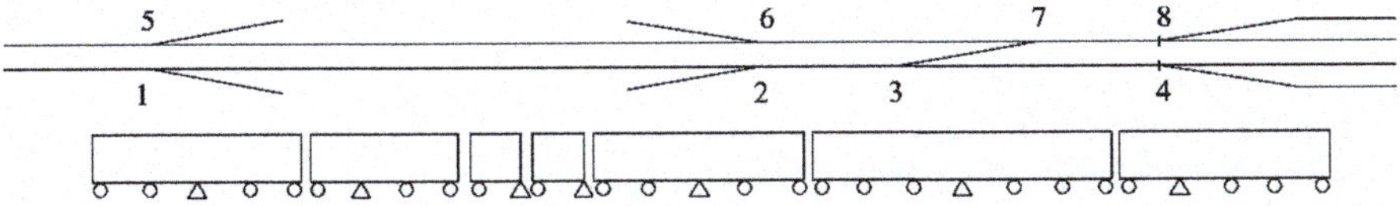

Fig. 9.32 Turnout-bridge arrangement at north throat of yard for Wuhan–Guangzhou HSR at Guangzhounan station

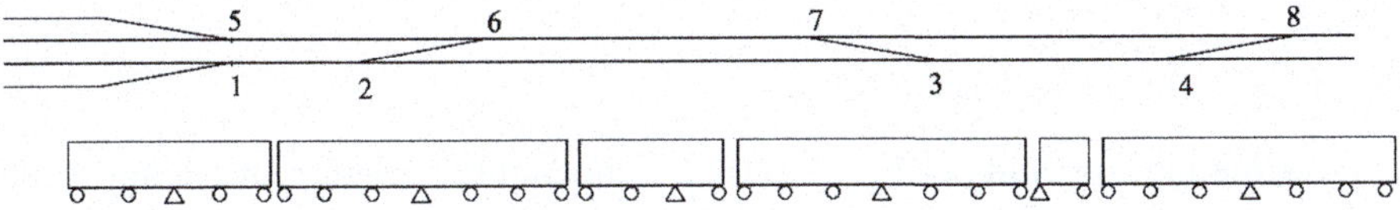

Fig. 9.33 Turnout-bridge arrangement at the south throat of yard for Wuhan–Guangzhou HSR at Guangzhounan station

$\times$ 36.6 + 2 $\times$ 32.7) m + 27.5 m simply-supported girder + (6 $\times$ 32.7) m continuous girder. The bridge is laid with ballasted track, and No. 18 high-speed turnout on the main line, as shown in Fig. 9.33.

9.3.2 Design of Ballasted Turnout on Bridge

The ballasted turnout on the bridge consists of the turnout rails, fasteners, turnout sleepers, and track bed. The track structure is 802 mm high, as shown in Fig. 9.34.

Fig. 9.34 Ballasted turnout on bridge

9.3.3 Design Checking Calculation of CWR Turnout Group on Bridge

1. Calculation Parameters

(1) Temperature difference between girder and rail

Temperature difference of girder: 15 °C.

The maximum rail temperature is 58 °C, and the minimum rail temperature is −1.5 °C at Guangzhounan Station. The stress-free rail temperature of the ballasted track is (35 ± 5) °C.

(2) Longitudinal resistance of track

For calculating longitudinal forces due to temperature variation, the longitudinal resistance is taken as 70 N/cm.

For calculating longitudinal forces due to deflection in the case of no load on the rail surface, the longitudinal resistance is taken as 70 N/cm. In the case of the rail surface under loading, the longitudinal resistance under locomotives is taken as 110 N/cm and that under cars is taken as 70 N/cm.

For calculating forces from rail breakage, the longitudinal resistance is taken as 110 N/cm.

(3) Turnout parameters

CNTT No. 18 high-speed turnouts are laid at Guangzhounan Station, with a total length of 69 m, a front length of 31.729 m, and a rear length of 37.271 m.

2. North Throat of Yard for Wuhan–Guangzhou HSR

1) Calculation results of additional force and relative displacement of rail

(1) Longitudinal force due to temperature variation of rail and girder-track relative displacement

The maximum additional tension of rails was 343.8 kN, the maximum additional compression was 295.3 kN, and the maximum girder-rail relative displacement was 15.1 mm, as shown in Figs. 9.35 and 9.36.

(2) Longitudinal force due to deflection of rail and girder-rail relative displacement

The maximum additional longitudinal tension due to deflection of rails was 52.62 kN, the maximum additional longitudinal compression due to deflection was 52.0 kN, and the maximum girder-rail relative displacement was 0.35 mm, as shown in Figs. 9.37 and 9.38.

(3) Braking force on rail and girder-rail relative displacement

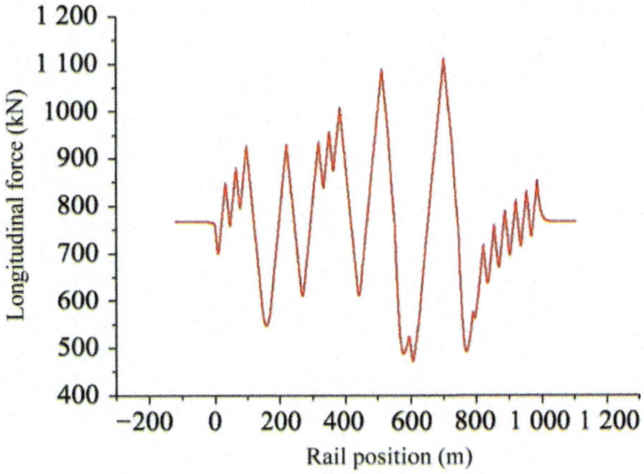

Fig. 9.35 Additional expansion and contraction force on rail

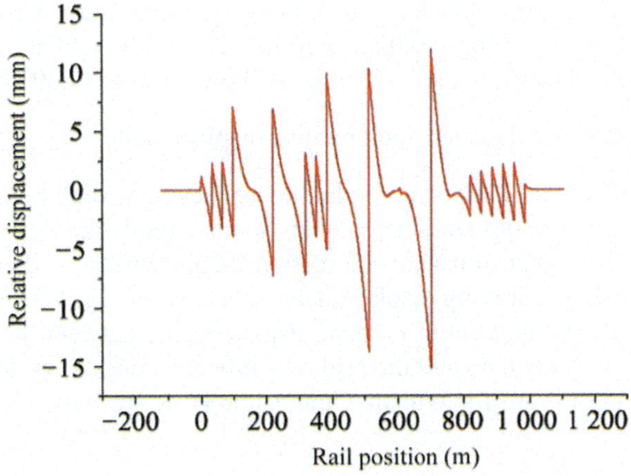

Fig. 9.36 Relative girder-rail expansion and contraction displacement

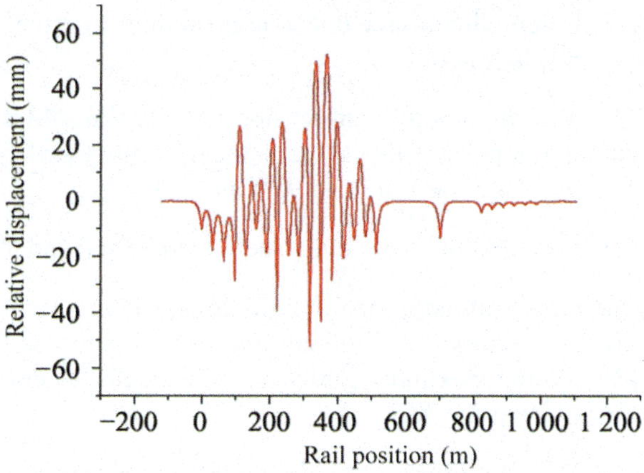

Fig. 9.37 Additional longitudinal force due to deflection on rail

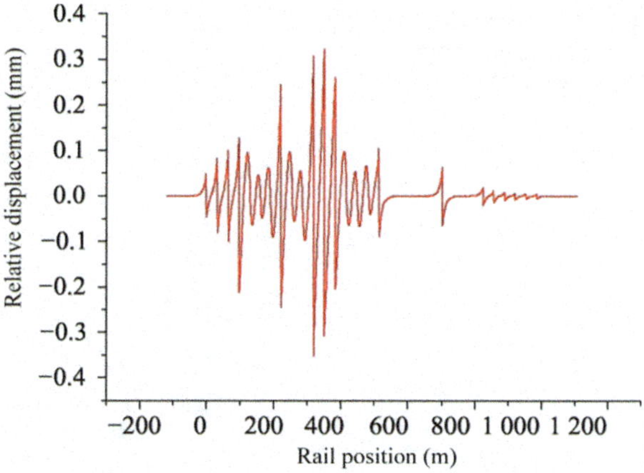

Fig. 9.38 Relative displacement between girder and rail due to deflection

The maximum additional braking tension of rails was 280.5 kN, the maximum additional braking compression was 288.3 kN, and the maximum girder-rail relative displacement was 3.75 mm, as shown in Figs. 9.39 and 9.40.

(4) Total girder-track relative displacement

The total girder-track relative displacement shall be calculated, taking into account the most unfavorable combination, i.e. total relative rail-bridge displacement = relative expansion and contraction displacement + relative flexural displacement + relative braking displacement, as shown in Fig. 9.41.

The maximum relative displacements between the sensitive components (frogs and switches) and the bridge within each turnout on the main line at the north throat of the yard for Wuhan–Guangzhou HSR are shown in Table 9.11.

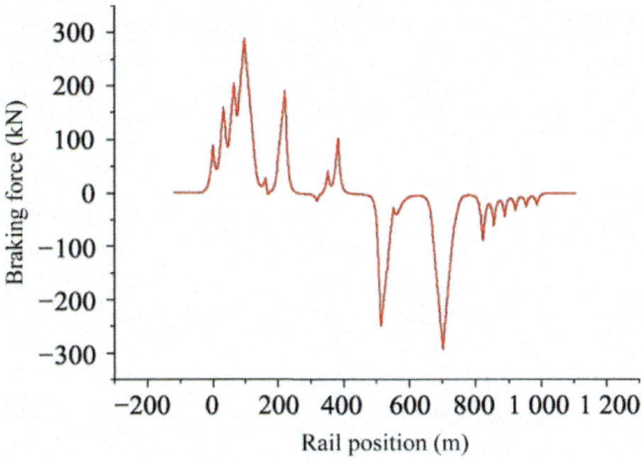

Fig. 9.39 Additional braking force on fail

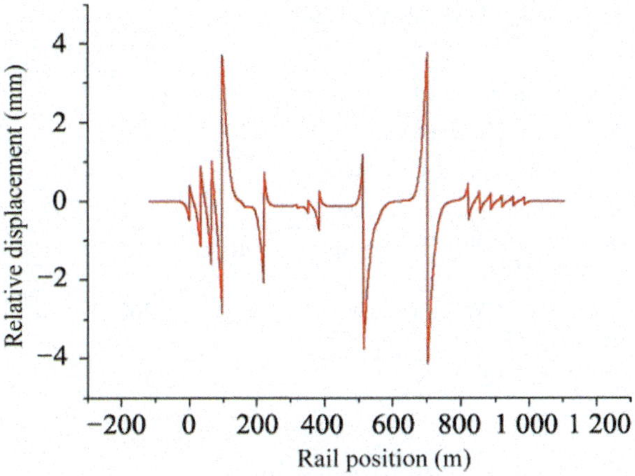

Fig. 9.40 Relative displacement between girder and rail due to braking

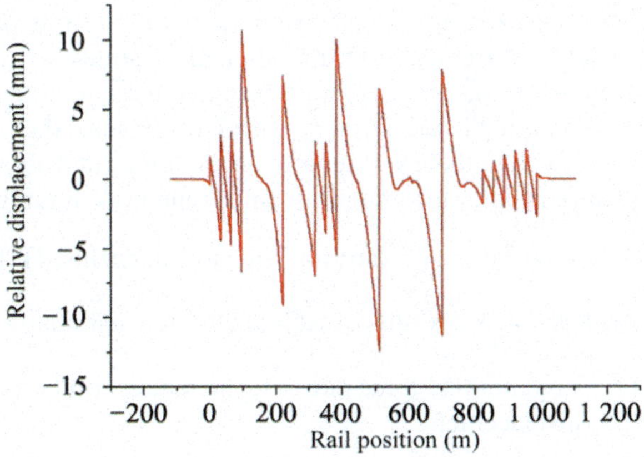

Fig. 9.41 Total girder-track relative displacement

Table 9.11 Statistics of relative displacements between turnout frogs/switches and bridge

Turnout S/N	Relative displacement between frog and bridge (mm)	Relative displacement between switch and bridge (mm)
1	1.8	1.8
2	2.2	5.0
3	0.75	3.1
4	0.45	3.3
5	1.76	1.78
6	2.1	4.9
7	0.5	8.3
8	0.58	3.3

Table 9.12 Checking calculation results

Item		Calculated value	Permissible value	Conclusion
Relative displacement at frog (mm)		2.2	5.2	Pass
Relative displacement at switch (mm)		8.3	22.2	Pass
Girder-rail rapid displacement (mm)		3.7	4	Pass
Calculation of rail stress within turnout section (MPa)	Tensile stress	223.2	352	Pass
	Compressive stress	264.56	352	Pass
Stability of CWR track within turnout section (kN)		2,208	3,410	Pass
Break joint (cm)		3.3	10	Pass

2) Checking calculation results

See Table 9.12 for the checking calculation results of relative displacements at the switch and frog, turnout rail strength, stability of CWR track within the turnout section, turnout stopper, bolt shear, and broken rail gap.

According to the checking calculation results, the relative displacement at the frog and switch of the CWR turnout on the bridge at the north throat, the girder-rail rapid displacement, rail strength, stability, and broken rail gap meet the requirements.

3. South Throat of Yard for Wuhan–Guangzhou HSR

1) Calculation results of additional force and relative displacement of rail

(1) Longitudinal force due to temperature variation of rail and girder-track relative displacement

The maximum additional tension of rails was 290.5 kN, the maximum additional compression was 441.611 kN, and the maximum girder-rail relative displacement was 15.9 mm, as shown in Figs. 9.42 and 9.43.

(2) Longitudinal force due to deflection of rail and girder-rail relative displacement

The maximum additional longitudinal tension due to deflection of rails was 28.0 kN, the maximum additional longitudinal compression due to deflection was 20.9 kN, and the maximum girder-rail relative displacement was 0.17 mm, as shown in Figs. 9.44 and 9.45.

(3) Braking force on rail and girder-rail relative displacement

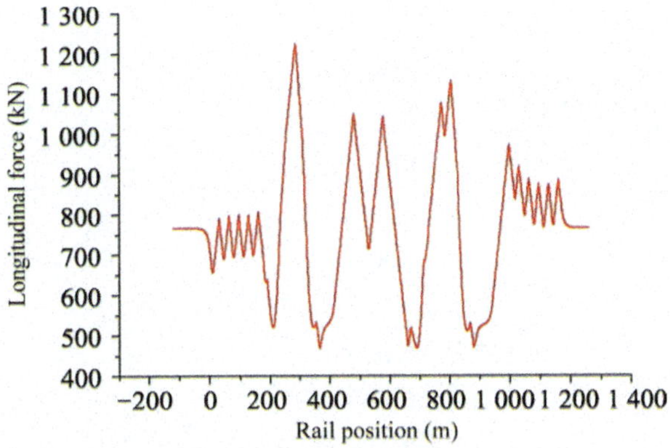

Fig. 9.42 Additional expansion and contraction force on rail

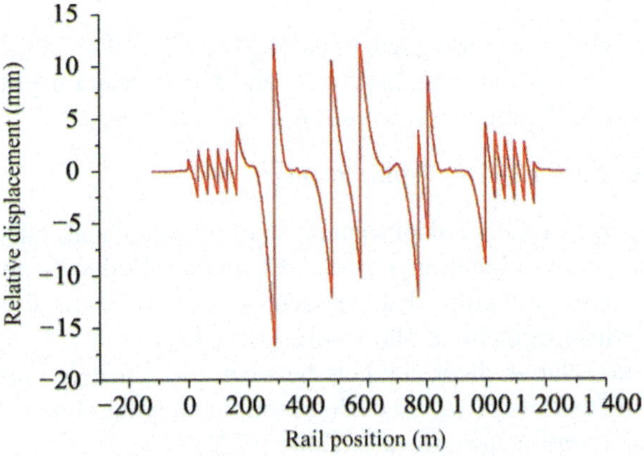

Fig. 9.43 Relative girder-rail expansion and contraction displacement

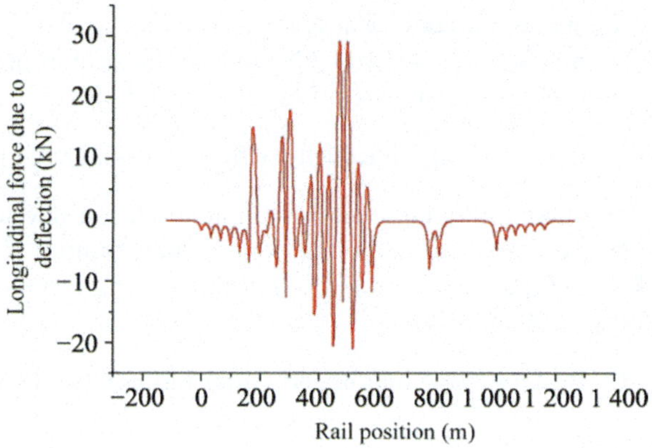

Fig. 9.44 Additional longitudinal force due to deflection on rail

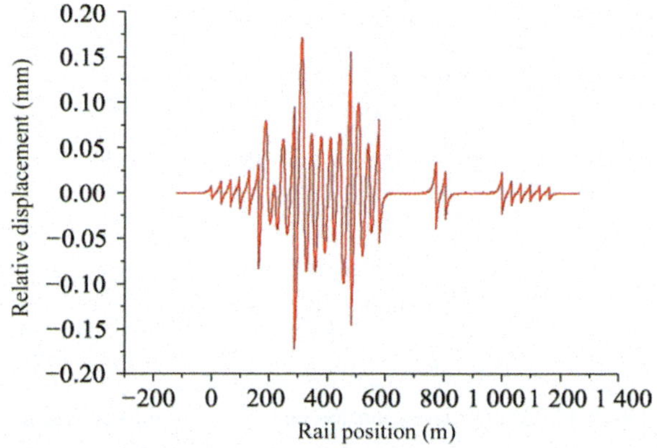

Fig. 9.45 Relative displacement between girder and rail due to deflection

The maximum additional braking tension of rails was 229.9 kN, the maximum additional braking compression was 234.5 kN, and the maximum girder-rail relative displacement was 3.73 mm, as shown in Figs. 9.46 and 9.47.

(4) Total girder-track relative displacement

The total girder-track relative displacement shall be calculated, taking into account the most unfavorable combination, i.e. total relative rail-bridge displacement = relative expansion and contraction displacement + relative flexural displacement + relative braking displacement, as shown in Fig. 9.48.

The maximum relative displacements between the sensitive components (frogs and switches) and the bridge within each turnout on the main line at the south throat of Guangzhounan Station are shown in Table 9.13.

2) Checking calculation results

See Table 9.14 for the checking calculation results of relative displacements at the switch and frog, turnout rail strength, stability of CWR track within the turnout

Fig. 9.46 Additional braking force on fail

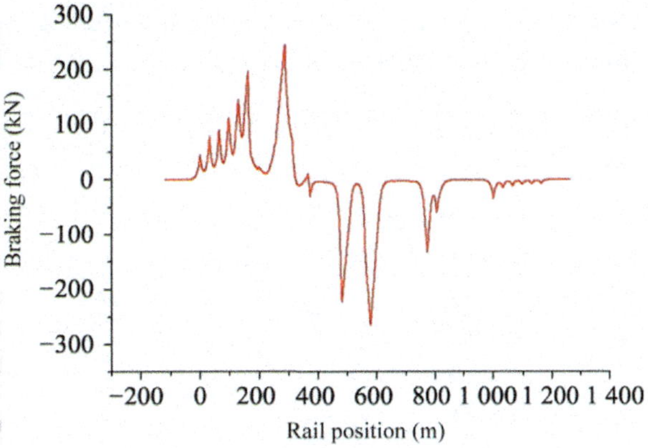

Fig. 9.47 Relative displacement between girder and rail due to braking

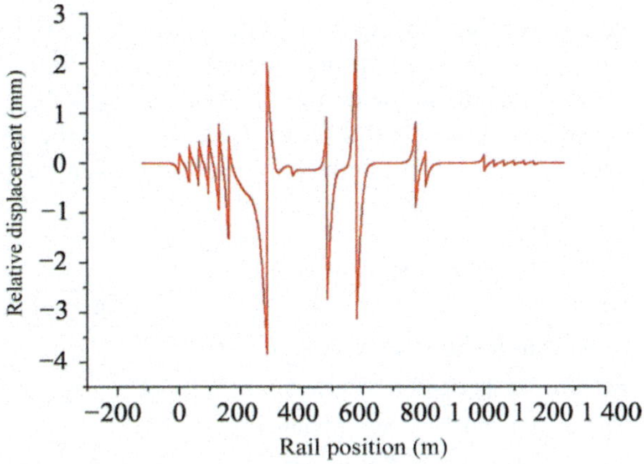

Fig. 9.48 Total girder-track relative displacement

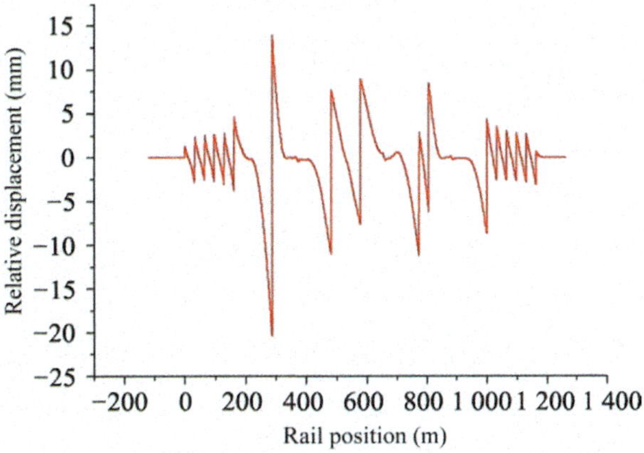

Table 9.13 Statistics of relative displacements between turnout frogs/switches and bridge

Turnout S/N	Relative displacement between frog and bridge (mm)	Relative displacement between switch and bridge (mm)
1	0.98	6.9
2	0.11	11.96
3	0.64	2.3
4	0.42	7.9
5	0.98	6.9
6	0.3	3.4
7	1.9	8.76
8	0.1	1.3

section, turnout stopper, bolt shear, and broken rail gap within the turnout on the main line at the south throat of Guangzhounan Station.

According to the checking calculation results, the relative displacement at the frog and switch of the CWR turnout on the bridge at the south throat, the girder-rail rapid displacement, rail strength, stability, and broken rail gap meet the requirements.

Table 9.14 Checking calculation results

Item		Calculated value	Permissible value	Conclusion
Relative displacement at frog (mm)		1.9	5.2	Pass
Relative displacement at switch (mm)		11.9	22.2	Pass
Girder-rail rapid displacement (mm)		3.73	4	Pass
Calculation of rail stress within turnout section (MPa)	Tensile stress	230.3	352	Pass
	Compressive stress	271.7	352	Pass
Stability of CWR track within turnout section (kN)		2,418	3,410	Pass
Break joint (cm)		3.3	10	Pass

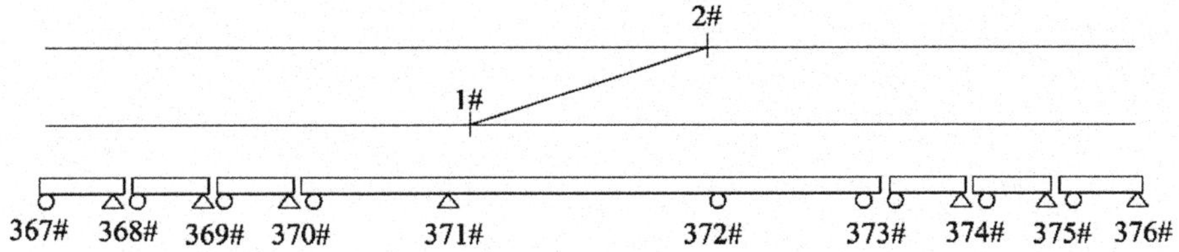

Fig. 9.49 Arrangement of CWR turnout on Keqiao super major bridge in Shaoxing

9.4 Design of CWR Turnout on the (60 + 100 + 60) m Continuous Girder at Keqiao Station of Hangzhou–Ningbo HSR

9.4.1 Overview of Keqiao Station on Hangzhou–Ningbo HSR

Hangzhou–Ningbo HSR is an important component of China's national high-speed railway network for passenger transport and the high-speed railway network across the Yangtze River Delta. This railway extends for a total length of 149.89 km and is designed for train operation up to 350 km/h. There are 7 stations distributed along the whole line: Hangzhoudong, Hangzhounan, Shaoxingbei, Shangyubei, Yuyaobei, Zhuangqiao, and Ningbo. It was commenced for traffic on July 1, 2013 [12, 13]. The Hangzhou–Ningbo HSR features the CRTS II slab ballastless track system along the whole line, and slab ballastless tracks within its turnout sections [14, 15].

Due to the restrictions of route conditions and urban planning, the turnout girder at Keqiao Station in Shaoxing is designed with a 100-m-long main span and a span arrangement of (3 × 32) m simply-supported girder + (60 + 100 + 60) m continuous girder + (3 × 32) m simply-supported girder. The arrangement of the No. 18 single crossover on the bridge is shown in Fig. 9.49. The distance between centers of tracks is 5 m.

9.4.2 Static Checking Calculation for Design of CWR Turnout on Bridge

The checking calculations on the rail strength and displacement of the CWR turnout on the (60 + 100 + 60) m continuous girder at Keqiao Station are detailed below.

1. Calculation Parameters

(1) Pier rigidity

See Table 9.15 for the rigidity of piers beneath the simply-supported girder and continuous girder for the turnout.

Table 9.15 Rigidity of piers/abutments

S/N	Pier No.	Longitudinal linear rigidity of pier/abutment (kN/cm)	Span length (m)
1	367#	461.7	32
2	368#	495.0	32
3	369#	518.0	32
4	370#	1,357.1	(60 + 100 + 60) m continuous girder for turnout
5	371#	3,980.2	
6	372#	4,134.3	
7	373#	1,210.0	
8	374#	518.0	32
9	375#	495.0	32
10	376#	461.7	32

(2) Resistance parameters

The clearance between the large and small stopper blocks is maintained at 7 mm. The same resistance parameter is taken for the stopper and the spacing block, and nonlinear resistance is taken for the spacing block. The resistance function is expressed as follows:

$$F = 40.49 + 5.8 \times 10^4 x + 7.93 \times 10^6 x^2 + 39 \times 10^8 x^3 + 1.02 \times 11 x^4 + 7.7 \times 10^{12} x^5$$

where x denotes the relative displacement of rails on both sides of the spacing block; F denotes the resistance of the spacing block.

(3) Train load

The train load is calculated using the standard live load of the Chinese railway. The train length is taken as 400 m, and the wheel-rail adhesion coefficient as 0.164. When calculating braking forces, the loading mode of single-track braking is adopted, and the loading conditions are shown in Fig. 9.50.

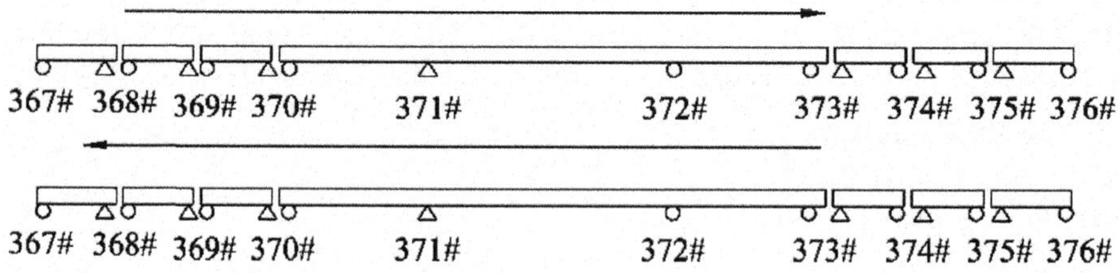

Fig. 9.50 Loading conditions of braking force

Condition 1: The car head under braking force is at 252.18 m away from the left abutment, and the force is applied from left to right.

Condition 2: The car tail under braking force is at 252.18 m away from the left abutment, and the force is applied from right to left.

(4) Rail temperature

According to the on-site meteorological data, the local maximum rail temperature is 62.1 °C, and the minimum rail temperature is −10.5 °C. The design stress-free rail temperature is 27 ± 5 °C.

2. Calculation of Temperature Stress

According to the rail and stress-free rail temperature data, the maximum temperature drop range of rails is calculated as follows:

$$\Delta T_{\mathrm{j}} = 27 + 5 - (-10.5) = 42.5\,^{\circ}\mathrm{C}$$

The maximum temperature rise range is calculated as follows:

$$\Delta T_{\mathrm{s}} = 62.1 - (27 - 5) = 40.1\,^{\circ}\mathrm{C}$$

The rail temperature stress is calculated as follows:

$$\sigma_{\mathrm{t}} = 2.1 \times 10^{11} \times 1.18 \times 10^{-5} \times 42.5 = 105.4\ \mathrm{MPa}$$

3. Calculation of Dynamic Flexural Stress

Since Keqiao Station has the same values as Tung Chung Station of Guangzhou–Shenzhen–Hong Kong HSR in terms of train axle load, speed coefficient, eccentric load coefficient, and lateral force coefficient, the deduced dynamic flexural stress of 119.78 MPa remains the same for both. Consequently, the specifics are not reiterated in this section.

4. Analysis and Checking Calculation

(1) Calculations for expansion and contraction

The additional longitudinal forces due to temperature variations of the stock rails, with the turnout temperature increase of 40.1 °C and decrease of 42.5 °C, are illustrated in Figs. 9.51, 9.52, 9.53 and 9.54. At a distance of 252.18 m from the left abutment, the maximum value recorded was 354.71 kN.

(2) Calculations for braking

The stress and displacement of the stock rails under different braking conditions are shown in Figs. 9.55 and 9.56, respectively. Under the two conditions, the extreme value of the braking force of 147.8 kN was recorded at 174 m away from the left abutment.

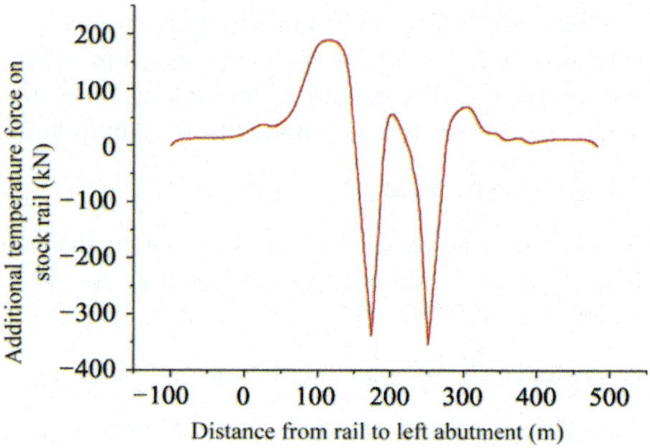

Fig. 9.51 Additional temperature force on stock rail upon temperature rise of 40.1 °C

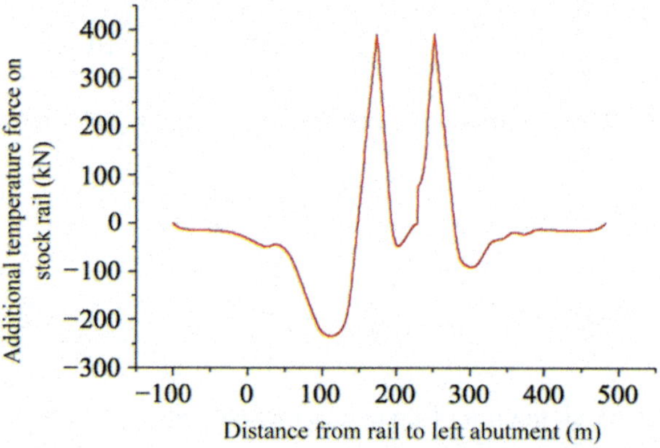

Fig. 9.52 Additional temperature force on stock rail upon temperature drop of 42.5 °C

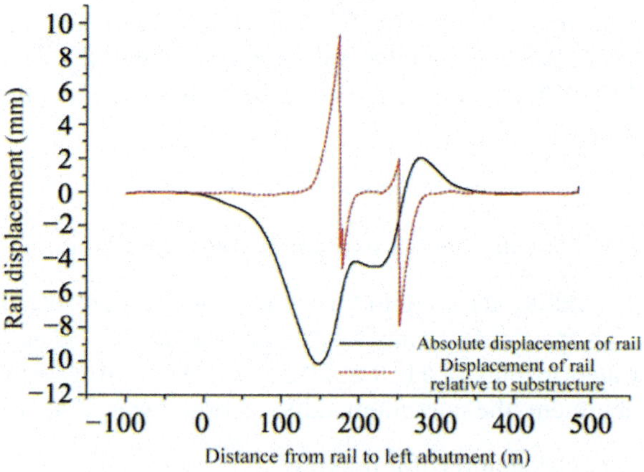

Fig. 9.53 Displacement of long rail upon temperature drop of 42.5 °C

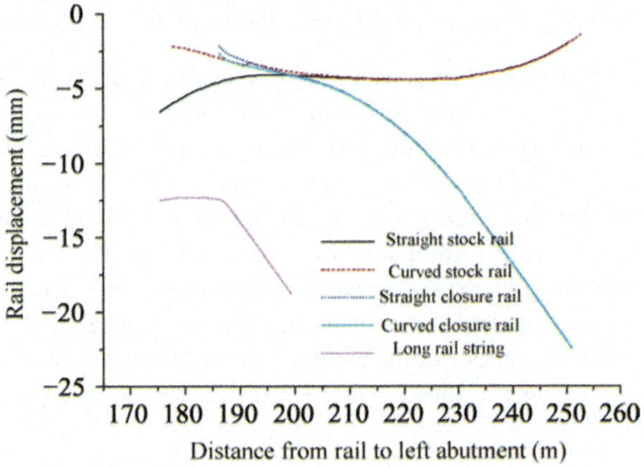

Fig. 9.54 Displacement of turnout rail upon temperature drop of 42.5 °C

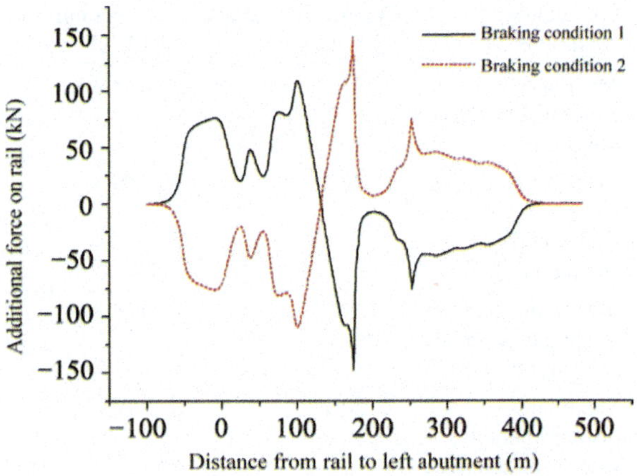

Fig. 9.55 Additional force on long rail under braking conditions

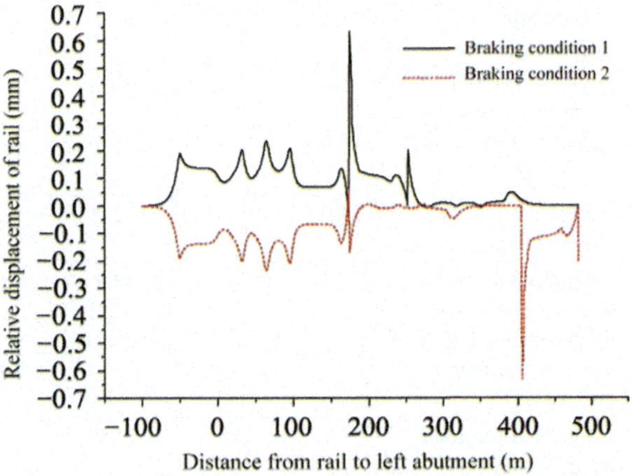

Fig. 9.56 Relative displacement between rail and substructure under braking conditions

(3) Checking calculations of strength and displacement

See Table 9.16 for the calculation results of stress and displacement of various turnout components under different conditions.

During expansion and contraction conditions, significant longitudinal relative displacements were observed between the movable parts within the turnout and the bridge. Conversely, the girder-track relative displacements induced by braking forces were minimal and could be disregarded. Since the expansion and contraction displacements of the movable parts within the turnout occurred in directions consistent with those of the bridge, the girder-track relative displacements at these movable parts within the turnout were negligible. Consequently, the longitudinal forces on the turnout force transfer device were also minor. Under the action of

Table 9.16 Calculation results under various conditions

Calculation conditions	Temperature drop of 42.5 °C	Temperature rise of 40.1 °C	Braking condition I	Braking condition II
Maximum additional force on rail (kN)	393.3	−354.7	−147.8	147.8
Maximum stress on force transfer device at heel end of switch rail (kN)	72.2	69.3	–	–
Maximum stress on spacing block at end of wing rail (kN)	129.2	115.7	–	–
Maximum displacement of switch rail relative to stock rail (mm)	20.7	19.2	–	–
Maximum displacement of point rail relative to wing rail (mm)	14.6	12.7	–	–
Additional force on No. 368 pier (kN)	46.9	31.2	149.2	149.2
Additional force on No. 369 pier (kN)	174.9	132.0	478.9	478.9
Additional force on No. 372 pier with fixed bearing beneath continuous girder (kN)	1,837.5	1,554.8	1,475.4	1,475.4
Additional force on No. 374 pier (kN)	58.1	21.7	89.1	89.1
Additional force on No. 375 pier (kN)	9.6	16.8	11.0	11.0

braking forces, significant longitudinal forces were observed on the fixed piers beneath the continuous girder. Additionally, substantial longitudinal forces occurred on the fixed bearings of the simply-supported girder located next to the continuous girder.

To sum up, the checking calculation of rail strength is expressed as follows:

$$\begin{aligned}&[\sigma_{\mathrm{d}}]+[\sigma_{\mathrm{t}}]+[\sigma_{\mathrm{f}}]+[\sigma_{\mathrm{z}}]\\&\quad=105.4+119.78+50.78+19.08\\&\quad=295.04\ \mathrm{MPa}\ <\ [\sigma]\end{aligned}$$

The calculation result indicates that the relative displacements at the frog and switch of the CWR turnout on the bridge and the strength of the turnout rails meet the requirements.

9.4.3 Dynamic Simulation Analysis of CWR Turnout on Bridge

When trains pass through turnouts on railway bridges at high speeds, the dynamic interactions between the wheels and rails intensify due to various factors, including track irregularities within the turnout section on bridges, vertical and lateral displacements of the bridges, the rotation angles at the girder ends, differences in rigidity at the subgrade–bridge transition, and foundation settlement, not only causing damages to the turnouts and bridge structures but also impairing the safety and comfort of train operation [16–24]. When the excitation frequencies of trains near or match the natural frequencies of bridges, resonance may occur between the trains and bridges. This situation can lead to significant impacts on driving safety [25]. Hence, dynamic checking calculations are necessitated before laying high-speed CWR turnouts on long-span bridges [26].

1. Calculation Parameters

(1) Computing parameters (train)

Relevant parameters of CRH_3 EMUs were used, and the length of the train is taken as an 8-vehicle high-speed train.

(2) Computing parameters (turnout)

The front length of 31.729 m and the rear length of 37.271 m were adopted for turnouts, and the vertical rigidity of 25 kN/mm for the fastener system, based on the No. 18 ballastless turnout for train operation up to 350km/h independently developed by China. The structural irregularities resulting from train operation are shown in Figs. 9.57 and 9.58.

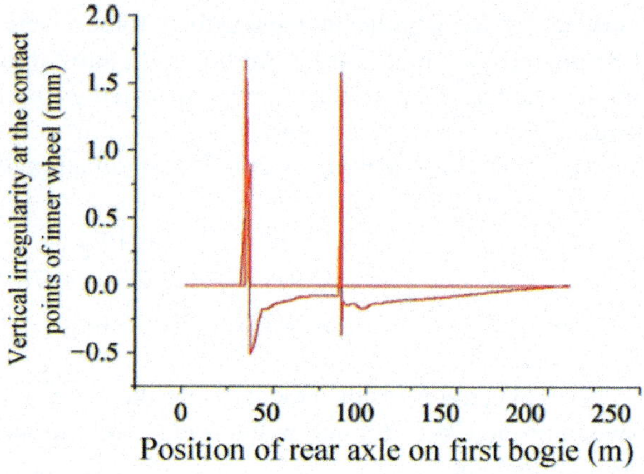

Fig. 9.57 Vertical irregularity of structure

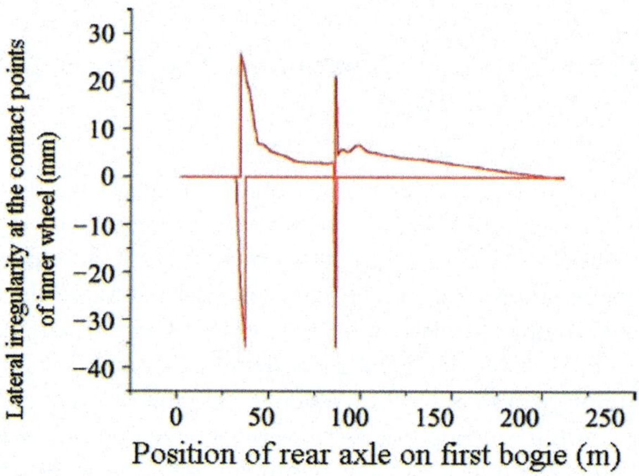

Fig. 9.58 Lateral irregularity of structure

(3) Calculation parameters (bridge)

The continuous girder beneath the turnout is an integral structure with a single-box single-cell variable cross section that accommodates two tracks. The girder is made of C50 concrete with an elastic modulus of 3.55×10^4 N/mm^2 and a unit weight of 26.5 kN/m^3. See Table 9.17 for typical section parameters.

2. Simulation Calculation and Analysis

To identify the rational layout of crossover turnouts on the turnout girders, calculations were conducted in three conditions respectively: the center position of the turnout girder, 15 m left shift from the center position, and 30 m left shift from the center position.

Table 9.17 Section parameters of (60 + 100 + 60) m continuous girder at Keqiao Station, Shaoxing

Characteristic section in design sketch	Sectional area (m^2)	Vertical moment of inertia of section (m^4)	Transverse moment of inertia of section (m^4)
I	18.453	60.918	138.417
II	13.153	44.613	117.951
III	26.522	214.104	196.774
IV	13.153	44.613	117.951
V	18.453	60.918	138.417

1) Calculations in the case of center position

The crossover is positioned at the center of the turnout girder in a symmetrical layout, as shown in Fig. 9.59.

When trains pass through the turnout, the dynamic response of the trains, turnout, and bridge structures is shown as follows:

(1) Dynamic wheel load distribution

With the left abutment as the origin of the coordinates, the dynamic wheel load distribution and vertical displacements of the wheelsets of the motor car's first bogie are shown in Figs. 9.60 and 9.61, respectively. Due to the effect of bridge vibration and turnout structural irregularities, the wheelsets experienced intensified vertical displacement variations, and the dynamic wheel loads fluctuated obviously. The maximum dynamic wheel load in the frog section was recorded at about 170.7 kN, falling within the permissible limit.

(2) Flange force distribution

The contact between the wheelset's inner wheel and the switch rail, as well as the point rail, occurs within a limited range. The flange force distribution is shown in Fig. 9.62. The maximum flange force in the switch section was 23.5 kN, and the maximum instantaneous lateral impact force in the frog section was about 42.8 kN, indicating the intensified wheel-rail lateral impact at the frog still fell within the permissible range.

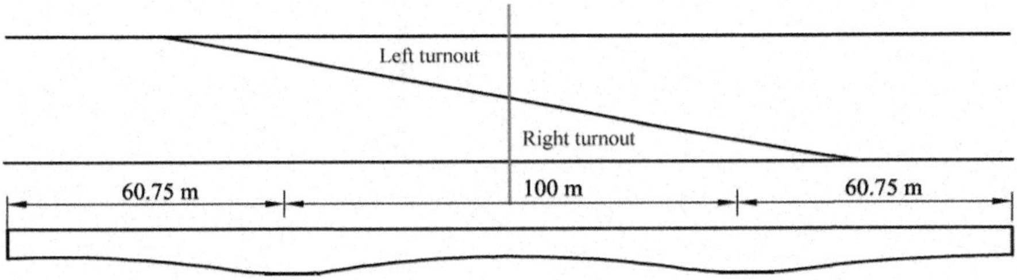

Fig. 9.59 Turnout–bridge arrangement

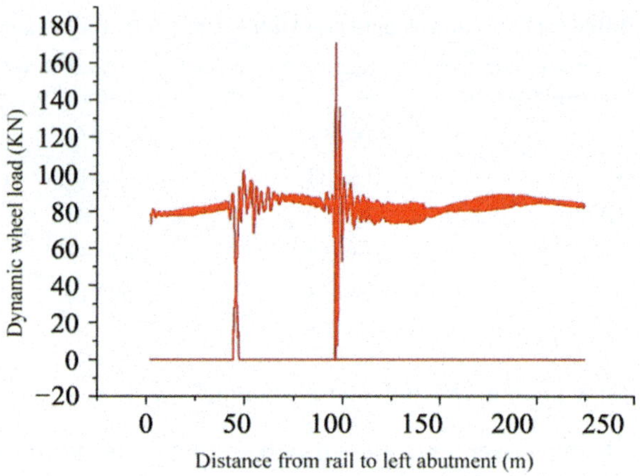

Fig. 9.60 Dynamic wheel load distribution under wheel

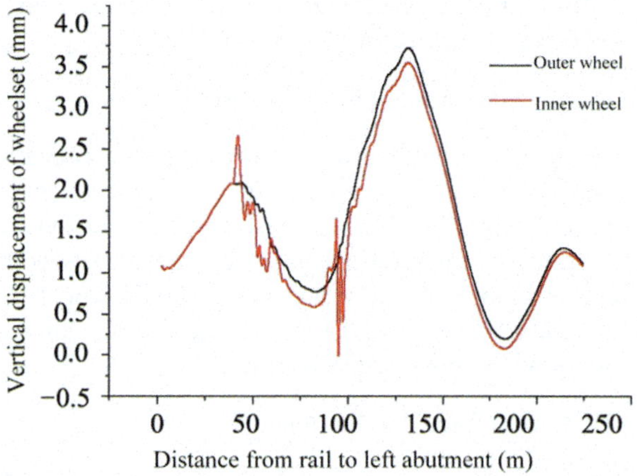

Fig. 9.61 Vertical displacement of wheelset

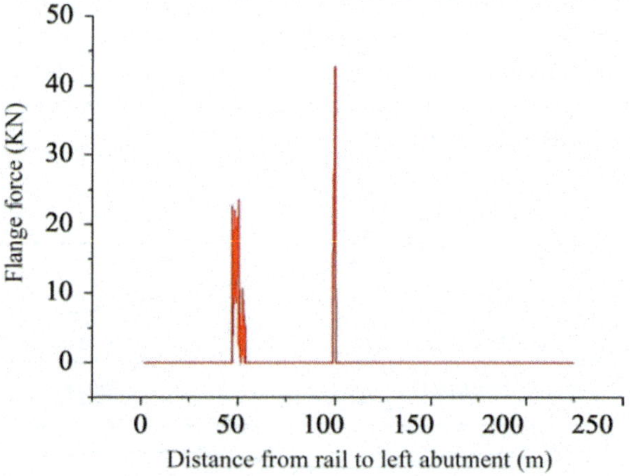

Fig. 9.62 Flange force distribution of inner wheel

(3) Lateral shift of wheelset and bogie

Figure 9.63 illustrates the lateral shift distribution of the front axle, rear axle, and frame of the leading bogie. As shown in Fig. 9.63, the maximum lateral shift of the front axle was recorded at about 7.4 mm.

(4) Derailment coefficient

Figure 9.64 illustrates the derailment coefficient distribution of wheels. As depicted in the figure, the maximum derailment coefficients of the front axle in the switch and frog sections were 0.40 and 0.21 respectively, while those of the rear axle in the switch and frog sections were 0.24 and 0.66 respectively, all falling within the safety limit.

(5) Wheel unloading rate

Figure 9.65 illustrates the wheel unloading rate distribution of wheels. As depicted in the figure, the unloading rates of the inner wheel on the front axle in the switch and

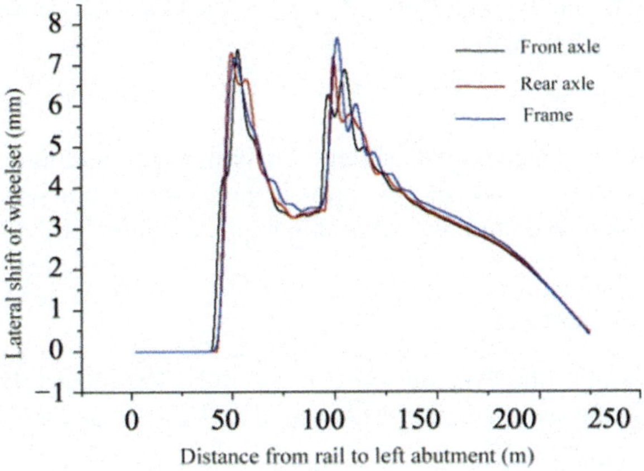

Fig. 9.63 Lateral shift of wheelset

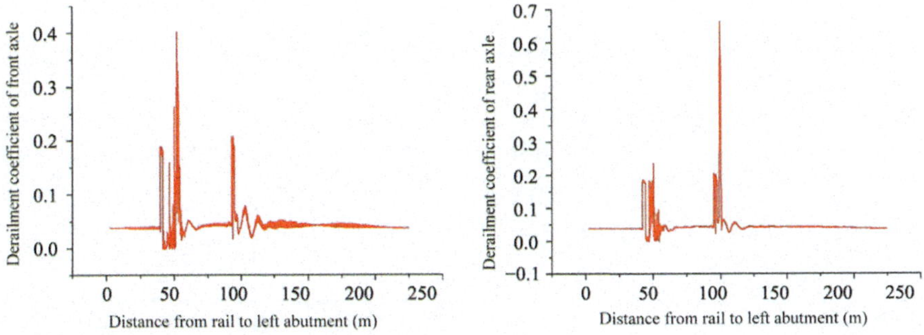

Fig. 9.64 Derailment coefficient distribution

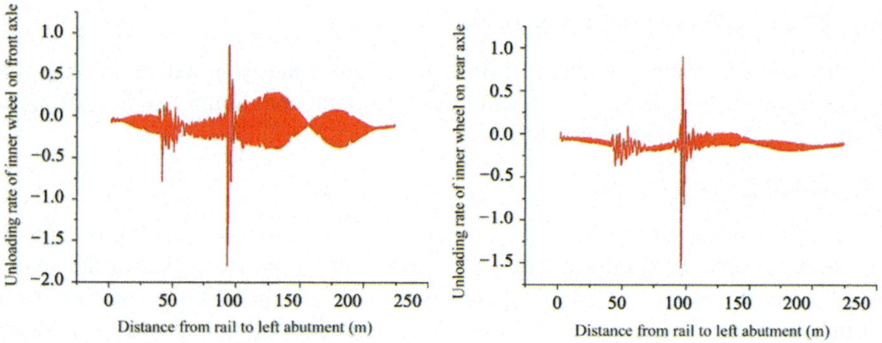

Fig. 9.65 Unloading rates of wheels

frog sections were 0.10 and 0.86 respectively, while those of the inner wheel on the rear axle in the switch and frog sections were 0.09 and 0.90, respectively. The duration with a wheel unloading rate greater than 0.8 was about 0.01s. The comprehensive evaluation reveals that the wheel unloading rates meet the safety index requirements, albeit with a low safety margin.

(6) Riding quality of car body

Figure 9.66 illustrates the vertical and lateral vibration accelerations of the car body. The maximum lateral vibration acceleration in the switch section was around 0.39 m/s^2, and that in the frog section was approximately 0.22 m/s^2, both falling within the permissible limit.

(7) Dynamic stress of rail

Figure 9.67 illustrates the dynamic stress of the rails. The impact dynamic stress of the switch rail in the switch section was around 209.2 MPa, and that of the point rail

Fig. 9.66 Lateral vibration acceleration of car body

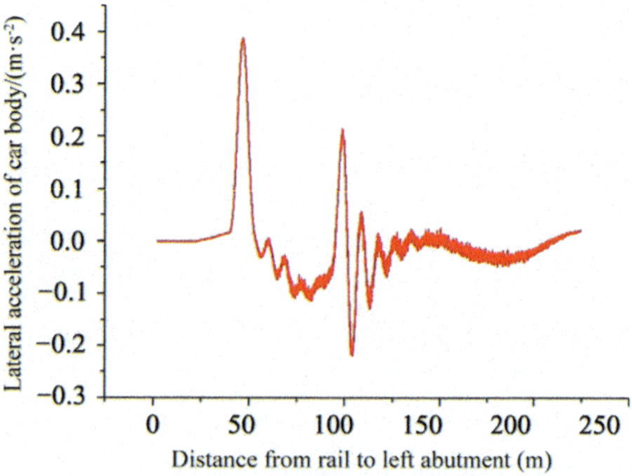

in the frog section was approximately 97.9 MPa, both falling within the permissible limit.

(8) Opening of switch rail and point rail

Figures 9.68 and 9.69 illustrate the opening of the switch rail and point rail, which were recorded around 0.43 and 0.02 mm respectively, both falling within the permissible limit.

(9) Bridge displacement

Figure 9.70 illustrates the corresponding relationship between the vertical dynamic displacement of the bridge under the front axle/deflection change rate of the bridge and the wheel unloading rate. As depicted in the figure, under the action of the EMU, the maximum dynamic displacement of the 100-m girder span was about 4.59 mm,

Fig. 9.67 Dynamic stress of rail

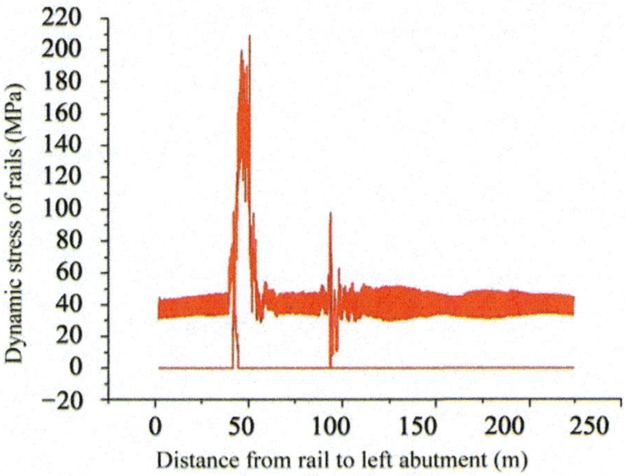

Fig. 9.68 Opening of switch rail

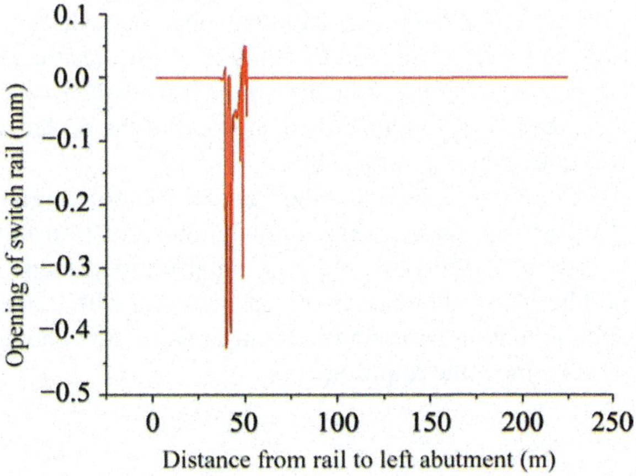

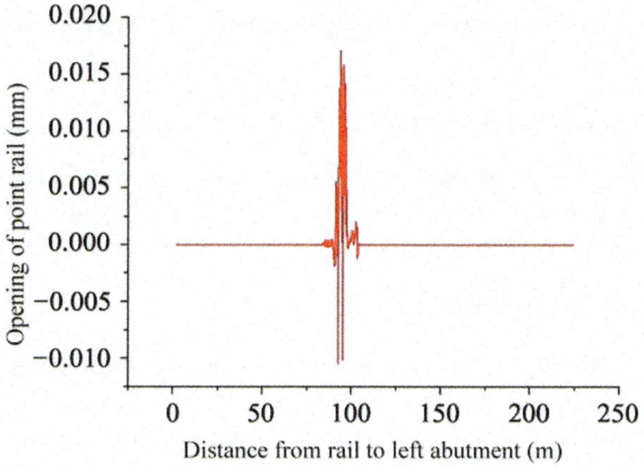

Fig. 9.69 Opening of point rail

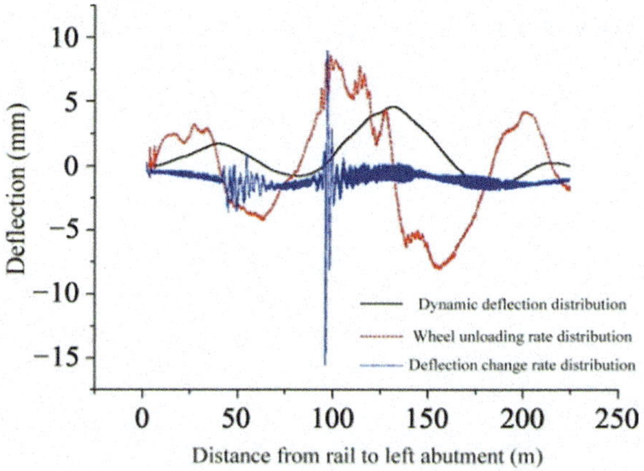

Fig. 9.70 Vertical displacement of bridge under wheel

while that of the 60-m side span was about 1.74 mm. The dynamic deflection at the frog was small but the deflection change rate was high, indicating non-ideal turnout-bridge relative positions. Figure 9.71 illustrates the lateral dynamic displacement of the bridge under the front axle. Under the sole action of turnout structural irregularities, the lateral dynamic displacement of the bridge did not exceed 0.15 mm, which was considered unnoticeable.

Figure 9.72 illustrates the vertical rotation angle distribution at the bearings in middle spans, and the maximum dynamic rotation angle was approximately 0.153‰. Figure 9.73 illustrates the horizontal rotation angle distribution at the bearings in middle spans, showing an increasing trend with more trains running onto the bridge, and the maximum dynamic rotation angle was about 0.003‰. All the displacement indexes meet the requirements.

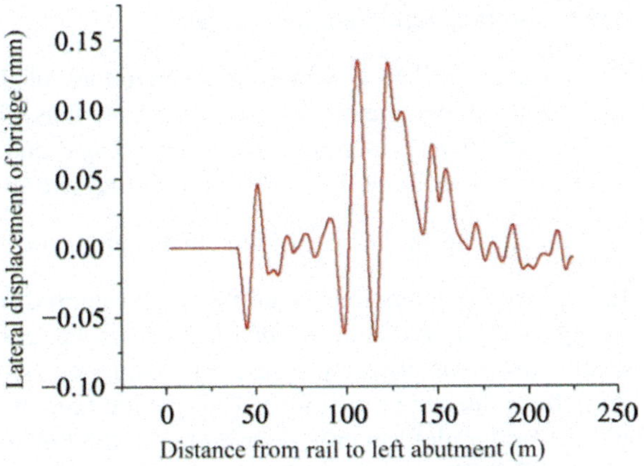

Fig. 9.71 Lateral displacement of bridge under wheel

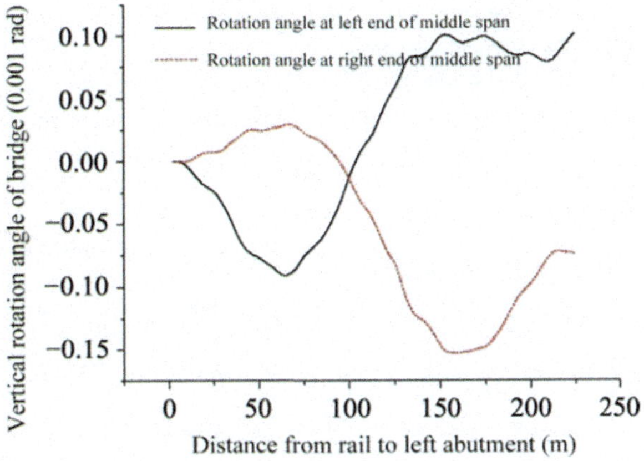

Fig. 9.72 Vertical rotation angle distribution at bearings in middle spans

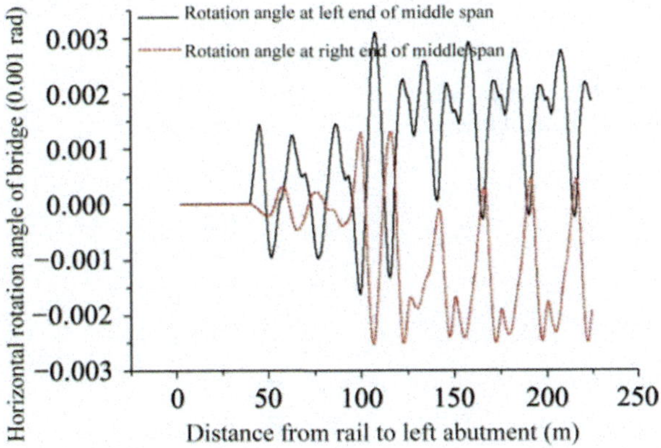

Fig. 9.73 Horizontal rotation angle distribution at bearings in middle spans

(10) Vibration acceleration of bridge

The vertical and lateral vibration accelerations of the middle span of the continuous girder under the wheels are shown in Figs. 9.74 and 9.75. The maximum vertical vibration acceleration was about 2.3 m/s^2 and the maximum lateral vibration acceleration was about 0.35 m/s^2, both falling within the permissible limit.

2) Calculations in the case of left shift from the center position

To improve the safety margin given by the centered crossover turnout on the continuous girder, the dynamic simulation analysis was carried out on the turnout-bridge relative positions properly adjusted (considering two conditions: 15 and 30 m left shifts from the center position). The specific layouts of the crossover on the bridge and the bridge deformation distribution at the left and right turnouts are shown in Figs. 9.76, 9.77, 9.78, 9.79, 9.80 and 9.81.

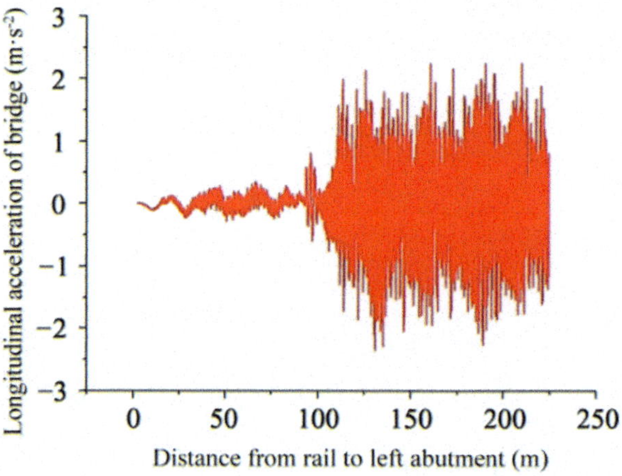

Fig. 9.74 Vertical vibration acceleration of bridge

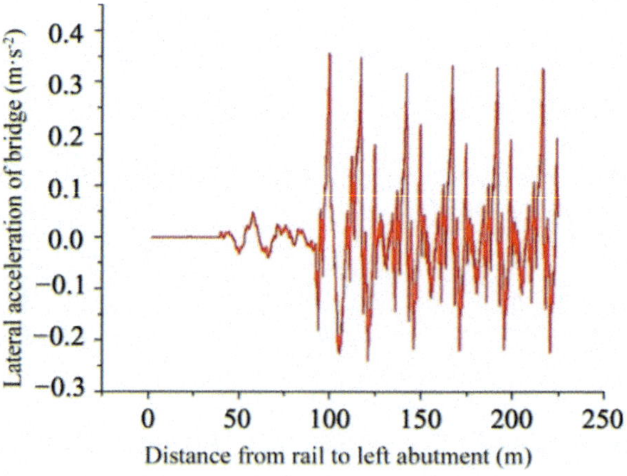

Fig. 9.75 Lateral vibration acceleration of bridge

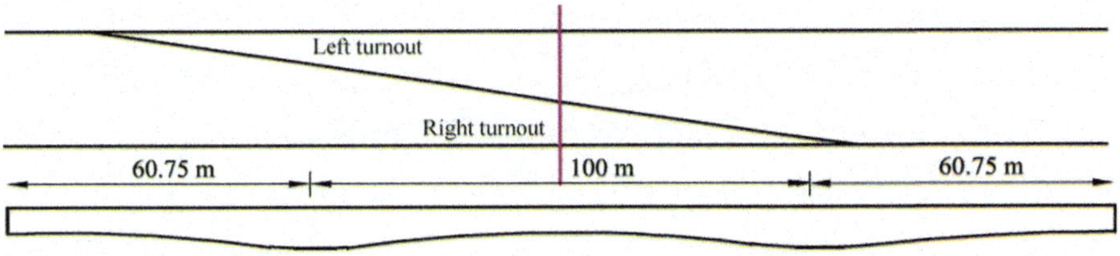

Fig. 9.76 Left shift of No. 18 crossover turnout for 15 m from the center position

Fig. 9.77 Bridge deformation distribution at left turnout

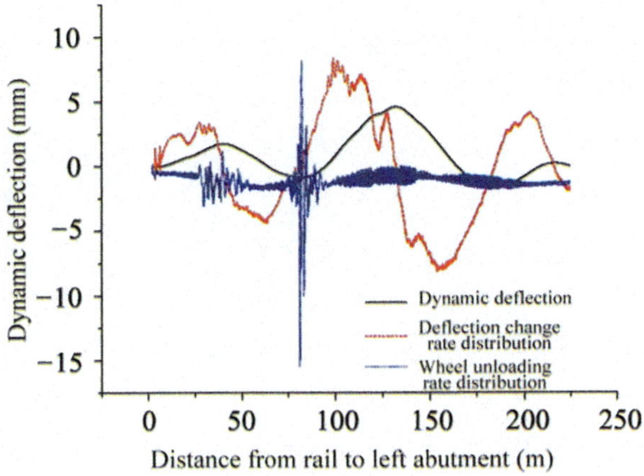

Fig. 9.78 Bridge deformation distribution at right turnout

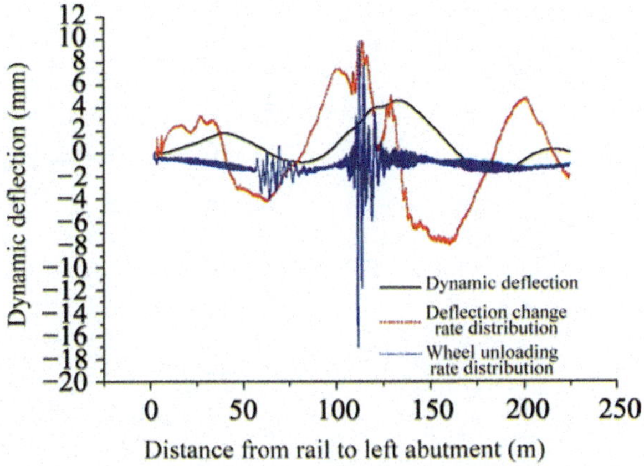

As depicted in the figure for two conditions, adjustments of the crossover turnout position on the continuous girder contributed to appropriate improvements in the interactions between the turnout and the bridge near the turnout at the fixed bearing

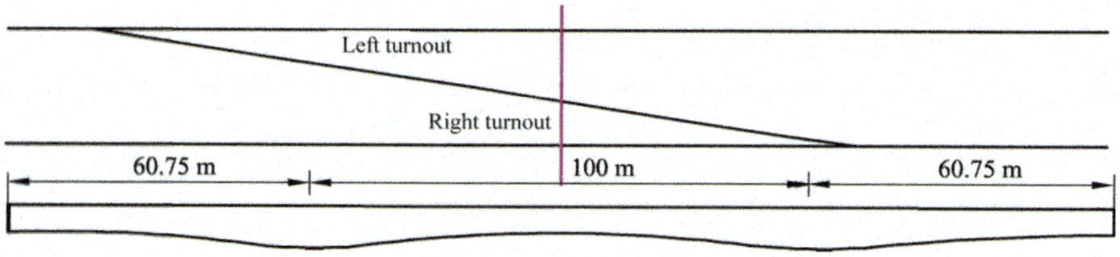

Fig. 9.79 Left shift of No. 18 crossover turnout for 30 m from the center position

Fig. 9.80 Bridge deformation distribution at left turnout

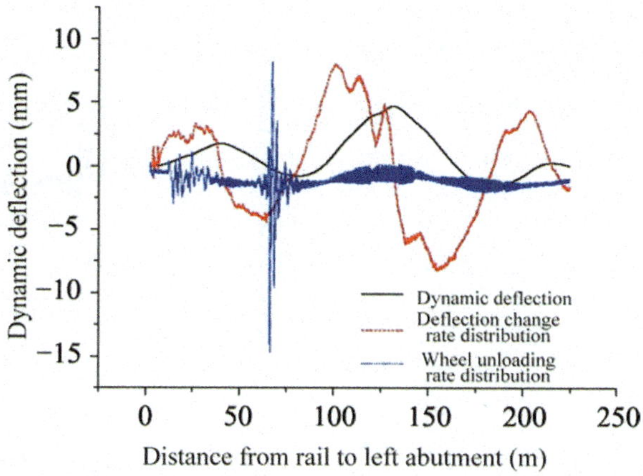

Fig. 9.81 Bridge deformation distribution at right turnout

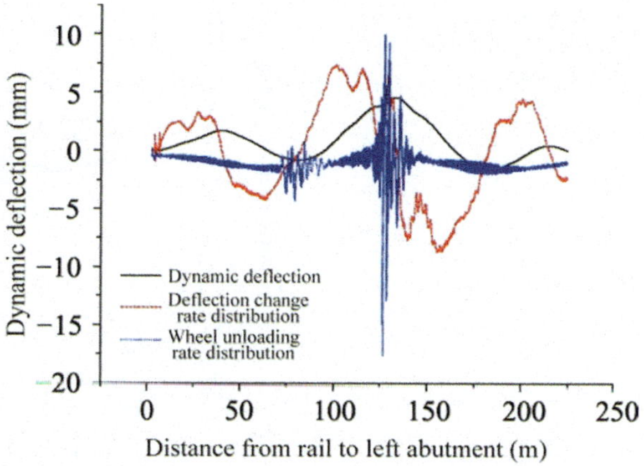

side. However, because of the span length and the number of spans in the continuous girder, achieving such enhancements for both the left and right turnouts simultaneously is impractical.

3. Analysis and Conclusion

See Table 9.18 for the summary of calculation results for the CWR turnout under the three working conditions.

By comparing the results among different positions of the crossover turnout on the bridge, the following conclusions can be drawn:

(1) The results of the dynamic simulation analysis on the No. 18 crossover turnout arranged at the center position at Keqiao Station in Shaoxing indicate that all the dynamic indexes are approaching the safety limits. Adjusting the position of the crossover turnout on (60 + 100 + 60) m continuous girder can appropriately improve the interactions between the turnout and bridge near the turnout at the fixed bearing side. However, because of the span arrangement and the number of spans in the continuous girder, achieving such enhancements for both the left and right turnouts simultaneously is impractical. In the actual design, the span length or number of spans of continuous girders shall be reduced as much as possible, or the vertical rigidity of bridges shall be appropriately improved if conditions are available.
(2) For applying CWR turnouts on long-span bridges, it is necessary to make a comprehensive analysis following calculations of dynamic and static indexes, integrating factors such as turnouts, bridge structures, and turnout–bridge arrangements.
(3) In general, measures should be taken to strengthen the performance of the No. 18 crossover turnout on the (60 + 100 + 60) m long-span continuous girder at Keqiao Station, Shaoxing. The specific measures are as follows: (1) Increase the vertical rigidity of the bridge; (2) Adjust the span length or number of spans of the continuous girder, and determine the rational turnout position, especially the frog position; (3) Appropriately improve the substructure elasticity; (4) Further optimize the wheel-rail interactions within the turnout section to mitigate structural irregularities.

9.5 Design of CWR Turnout on the 102-m Continuous Girder and Arch Crossing Jiulong Avenue of Hangzhou–Changsha HSR

9.5.1 Overview of Continuous Girder and Arch Crossing Jiulong Avenue on Hangzhou–Changsha HSR

The Hangzhou–Changsha HSR spans a total length of 927 km and encompasses 21 stations. Designed for train operation up to 350 km/h, it functions as a primary route within the "Eight North–South and Eight East–West Corridors" high-speed

Table 9.18 Summary of maximum calculation results

Turnout arrangement	Center position on continuous girder	15 m Left shift from center position		30 m Left shift from center position	
		Left turnout	Right turnout	Left turnout	Right turnout
Max. dynamic wheel load (kN)	170.7	172.6	213.6	164.0	243.6
Max. flange force in switch section (kN)	23.5	24.0	34.3	25.8	40.3
Max. flange force in frog section (kN)	42.8	45.4	35.8	40.8	29.6
Derailment coefficient of front axle in switch section	0.40	0.39	0.43	0.37	0.44
Derailment coefficient of front axle in frog section	0.21	0.21	0.20	0.25	0.23
Derailment coefficient of rear axle in switch section	0.24	0.24	0.36	0.27	0.41
Derailment coefficient of rear axle in frog section	0.66	0.70	0.79	0.66	0.83
Max. unloading rate of front axle in switch section	0.10	0.27	0.15	0.21	0.30
Max. unloading rate of front axle in frog section	0.86	0.79	0.93	0.79	0.95
Max. unloading rate of rear axle in switch section	0.09	0.12	0.07	0.11	0.08
Max. unloading rate of rear axle in frog section	0.90	0.82	0.92	0.83	0.94

(continued)

Table 9.18 (continued)

Turnout arrangement	Center position on continuous girder	15 m Left shift from center position		30 m Left shift from center position	
		Left turnout	Right turnout	Left turnout	Right turnout
Lateral vibration acceleration of car body (×*g*)	0.039	0.037	0.039	0.036	0.037
Max. dynamic stress of switch rail (MPa)	209.2	186.7	207.9	157.5	226.1
Max. dynamic stress of point rail (MPa)	97.9	99.3	105.9	96.3	111.2
Max. opening of switch rail (mm)	0.43	0.41	0.44	0.40	0.44
Max. opening of point rail (mm)	0.02	0.02	0.02	0.02	0.03
Max. vertical dynamic displacement of bridge (mm)	4.59	4.60	4.66	4.71	4.56
Max. lateral dynamic displacement of bridge (mm)	0.14	0.07	0.16	0.07	0.12
Vertical vibration acceleration of bridge (×g)	0.23	0.22	0.43	0.21	0.29
Lateral vibration acceleration of bridge (×g)	0.035	0.019	0.044	0.014	0.044

railway network. This railway links with the Shanghai–Hangzhou HSR and the Changsha–Kunming HSR, forming the Shanghai–Kunming High-Speed Corridor that traverses China from east to west. On December 22, 2009, the construction of Hangzhou–Changsha HSR was officially commenced. On December 10, 2014, the Hangzhou–Nanchang Section of this railway was officially inaugurated, signifying the commencement of operation along the entire line.

The interchange crossing Jiulong Avenue, the central axis of Honggutan District, Jiangxi, is located at the throat area of Nanchangxi Station. The People's Government of Jiangxi Province requested a clear width under the bridge of 96 m. Therefore, after

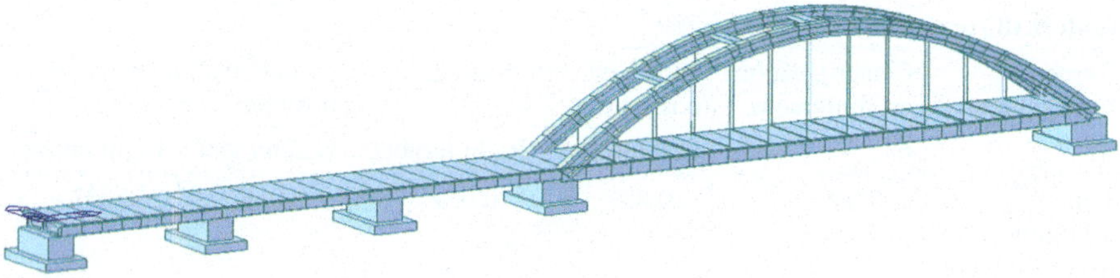

Fig. 9.82 Physical model of (3 × 32 + 102) m continuous girder and arch

consultation with provincial and ministerial high-level officials, a turnout girder plan with a single span crossing the 96 m road was chosen [27]. To address the challenges posed by 100-m span CWR turnout bridges with ballastless tracks, relevant entities have conducted specific studies to guarantee the safety of turnouts.

The design of the interchange crossing Jiulong Avenue includes a continuous girder and arch combination with a 102-m-long main span, which was the largest among continuous girders supporting turnouts in China and abroad at that time. The main span girder features a single-box twin-cell section, while the arch ribs a reinforced concrete section. The suspenders are arranged in a vertical system with parallel wires [28]. The central main pier is provided with a longitudinal fixed bearing, and the rest piers are provided with multi-directional movable bearings. The two main piers are provided with 3 bearings, with a transverse fixed bearing in the middle. All the piers in the side spans are provided with 2 multi-directional movable bearings. The arrangement is shown in Fig. 9.82.

9.5.2 *Static Checking Calculation for Design of CWR Turnout on Bridge*

It is necessary to perform checking calculations to ensure the rational stress state of Jiulong Avenue Interchange under the action of temperature and train loads, focusing on the strength and displacements of the turnout structure. In addition, broken rail gaps shall also be checked through calculations.

1. Calculation Parameters

(1) Basic track information

Hangzhou–Changsha HSR extends with trans-section CWR tracks, particularly with CRTS II slab ballastless tracks and No. 18 KZ (07) 009 high-speed turnouts on the main line.

(2) Bridge data

The bridge span is arranged as (27–32) m simply-supported box girder + (6 × 32) m turnout continuous girder + (1–24) m simply-supported box girder + (6 × 32) m turnout continuous girder (single crossover) + (1–24) m simply-supported box girder + (1–24) m simply-supported box girder + (32 + 32.7 + 32.7 + 102) m continuous girder and arch. The turnout arrangement is shown in Fig. 9.83.

(3) Rail temperature

According to the on-site meteorological data, the local maximum rail temperature is 60.1 °C, and the minimum rail temperature is −9.7 °C. The design stress-free rail temperature is 25 ± 5 °C.

(4) Train load

The ZK standard load is used for braking force calculations of this high-speed railway, as shown in Fig. 9.84.

2. Calculation of Temperature Stress

According to the local rail and stress-free rail temperature data, the maximum temperature drop range of rails on this bridge is as follows:

$$\Delta T_{\mathrm{j}} = 25 + 5 - (-9.7) = 39.7\,^{\circ}\mathrm{C}$$

The maximum temperature rise range is calculated as follows:

$$\Delta T_{\mathrm{s}} = 60.1 - (25 - 5) = 40.1\,^{\circ}\mathrm{C}$$

The rail temperature stress is calculated as follows:

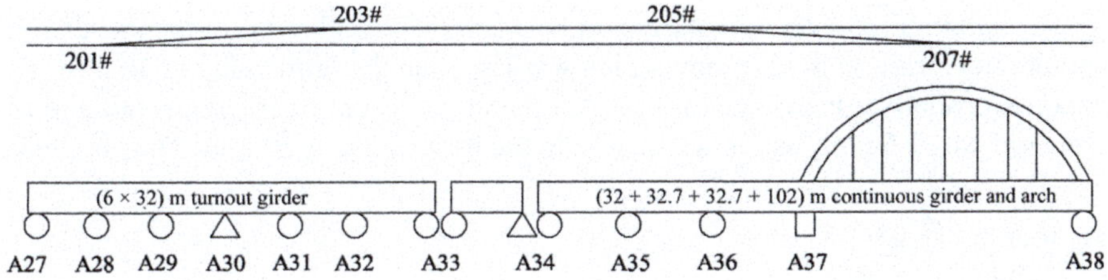

Fig. 9.83 Turnout arrangement

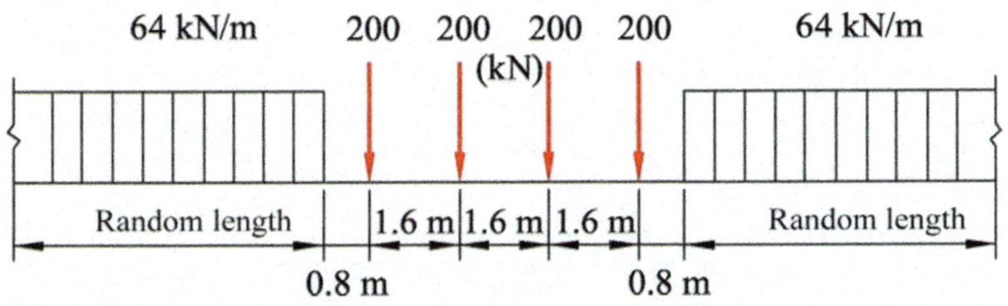

Fig. 9.84 Schematic diagram of ZK standard live load for high-speed railway

$$\sigma_t = 2.1 \times 10^{11} \times 1.18 \times 10^{-5} \times 39.7 = 98.46\ \text{MPa}$$

3. Calculation of Dynamic Flexural Stress

Since Jiulong Avenue Interchange has the same values as the (60 + 100 + 60) m continuous girder at Keqiao Station of Hangzhou–Ningbo HSR in terms of train axle load, speed coefficient, eccentric load coefficient, and lateral force coefficient, the calculated dynamic flexural stress of 119.78 MPa remains the same for both. Consequently, the specifics are not reiterated in this section.

4. Analysis and Checking Calculation of Option I

The (32 + 32.7 + 32.7 + 102) m continuous girder and arch arrangement is adopted, in which the fixed bearing of the 24 m simply-supported girder between the left single-crossover turnout girder and the continuous girder + arch is located at the large mileage end.

(1) Calculation of additional longitudinal forces due to temperature variations

The maximum additional longitudinal forces due to temperature variations on the stock rails are shown in Figs. 9.85 and 9.86. These maximums on Track 1 and Track 2 both occurred at the right end of the single-crossover turnout girder, recorded at 737 and 746 kN, respectively.

Under expansion and contraction conditions, the displacements of the switch rail and point rail of the turnout are shown in Table 9.19. (Note: In consideration of the small difference between the displacements of the straight and curved switch rails within the same turnout and the slightly larger displacements of the latter in general, the displacements of the curved switch rail are incorporated in the table.)

The maximum relative displacement between the actual point of the switch rails and the stock rails was 12.7 mm, which was less than the limit value of 40 mm. The maximum relative displacement between the actual point of the point rail and the wing rail was 3.5 mm, which was less than the limit value of 20 mm. The maximum

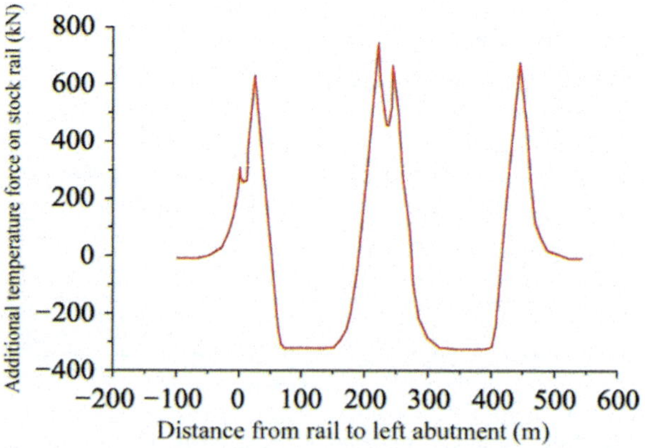

Fig. 9.85 Additional longitudinal forces due to temperature variations on Track 1

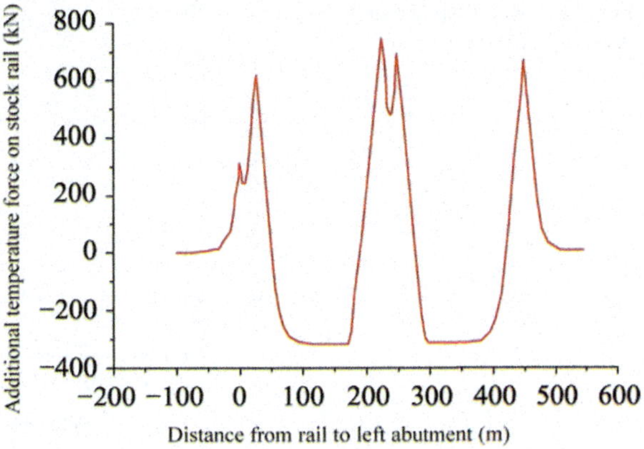

Fig. 9.86 Additional longitudinal forces due to temperature variations on Track 2

Table 9.19 Displacements of switch rail and point rail (unit: mm)

Turnout No.	201#	207#	203#	205#
Absolute displacement of switch rail	−23.2	22.8	22.4	−22.8
Expansion and contraction displacement of switch rail relative to stock rail	−9.9	9.9	12.7	−11.5
Expansion and contraction displacement of switch rail relative to turnout sleeper	−8.1	8.1	8.1	−8.1
Expansion and contraction displacement of switch rail relative to bridge	−8.2	8.1	8.0	−8.2
Displacement of stock rail relative to bridge	−1.7	1.8	4.7	−3.2
Absolute displacement of point rail	−8.4	8.0	7.6	−8.0
Expansion and contraction displacement of point rail relative to wing rail	−3.4	3.5	3.5	−3.5
Expansion and contraction displacement of point rail relative to turnout sleeper	−3.4	3.4	3.5	−3.4
Expansion and contraction displacement of point rail relative to bridge	−3.6	3.4	3.3	−3.4

Note In this table, "–" indicates the left direction, while values without "–" indicate rightward displacements

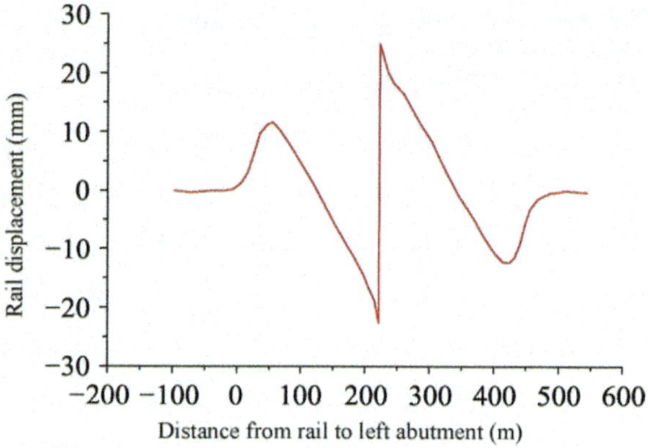

Fig. 9.87 Rail displacement from rail breakage

relative displacement between the stock rail at the turnout switch and the bridge was 4.7 mm, which was less than 5 mm specified in the specification. All the indicators meet the requirements.

(2) Rail fracture

The calculations of forces from rail breakage focused on a rail fracture on Track 2, at the position with the maximum additional longitudinal force due to temperature variation. This point is positioned at the right end of the (6 × 32) m single-crossover turnout girder, 220.2 m away from the left abutment. As shown in Fig. 9.87, the rail displacement due to rail breakage was 47.6 mm, remaining within the permissible value of 70 mm, which meets the requirements.

(3) Longitudinal forces on turnout components

See Table 9.20 for the longitudinal forces on turnout force transfer devices. Considering the limited temperature fluctuations in turnouts and the alignment of expansion and contraction directions of the movable parts with the bridge, the longitudinal forces on the turnout force transfer devices are small. Specifically, the stoppers were not subjected to longitudinal forces, and the maximum longitudinal force on the spacing blocks was 77 kN, falling within the limit of 373 kN.

5. Analysis and Checking Calculation of Option II

The (32 + 32.7 + 32.7 + 102) m continuous girder and arch arrangement is adopted, in which the fixed bearing of the 24 m simply-supported girder between the left single-crossover turnout girder and the continuous girder + arch is located at the small mileage end.

(1) Expansion and contraction

The maximum additional longitudinal forces due to temperature variations on the stock rails are shown in Figs. 9.88 and 9.89. These maximums on Track 1 and Track

Table 9.20 Longitudinal forces on turnout force transfer devices under various conditions (unit: kN)

Turnout No.	Track state	Stopper on main route	Stopper on diverging route	Spacing block on main route	Spacing block on diverging route
201#	Expansion and contraction	0	0	−77	−77
	Rail breakage	0	0	76	76
207#	Expansion and contraction	0	0	77	77
	Rail breakage	0	0	−76	−76
203#	Expansion and contraction	0	0	77	77
	Rail breakage	0	0	−76	−76
205#	Expansion and contraction	0	0	−77	−77
	Rail breakage	0	0	76	76

Note "+" indicates tension, and "−" indicates compression

2 both occurred at the left end of the (32 + 32.7 + 32.7 + 102) m continuous girder and arch, recorded at 735 and 743 kN, respectively.

See Table 9.21 for the displacements of the switch rail and point rail of the turnout under expansion and contraction conditions.

As shown in the above table, the maximum relative displacement between the actual point of the switch rails and the stock rails was 11.8 mm, which was less than the limit value of 40 mm. The maximum relative displacement between the actual point of the point rail and the wing rail was 3.5 mm, which was less than the limit value of 20 mm. The maximum relative displacement between the stock rail at the

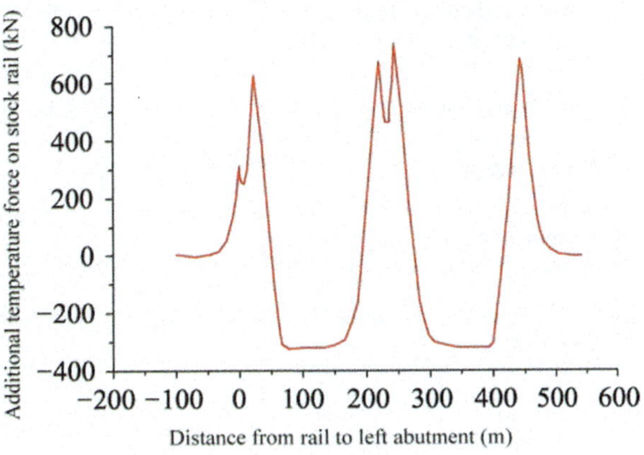

Fig. 9.88 Additional longitudinal forces due to temperature variations on Track 1

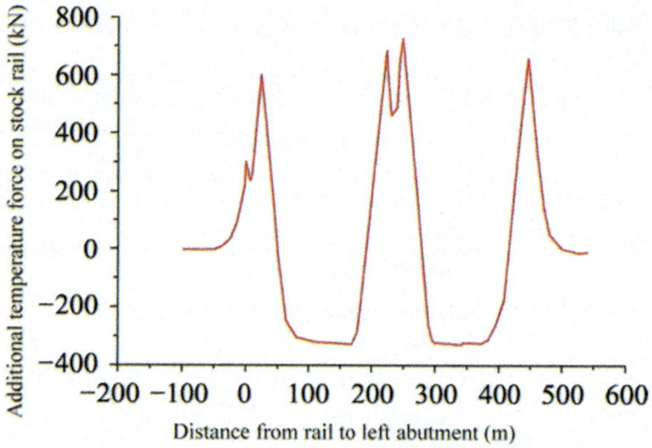

Fig. 9.89 Additional longitudinal forces due to temperature variations on Track 2

Table 9.21 Displacements of switch rail and point rail (unit: mm)

Turnout No.	201#	207#	203#	205#
Absolute displacement of switch rail	−22.9	23.2	22.7	−22.4
Expansion and contraction displacement of switch rail relative to stock rail	−9.9	10.0	11.8	−12.2
Expansion and contraction displacement of switch rail relative to turnout sleeper	−8.1	8.1	8.1	−8.1
Expansion and contraction displacement of switch rail relative to bridge	−8.2	8.1	8.0	−8.2
Displacement of stock rail relative to bridge	−1.7	1.9	3.9	−4.0
Absolute displacement of point rail	−8.1	8.4	7.9	−7.6
Expansion and contraction displacement of point rail relative to wing rail	−3.4	3.5	3.5	−3.5
Expansion and contraction displacement of point rail relative to turnout sleeper	−3.4	3.4	3.4	−3.5
Expansion and contraction displacement of point rail relative to bridge	−3.6	3.4	3.3	−3.5

Note In this table, "−" indicates the left direction, while values without "−" indicate rightward displacements

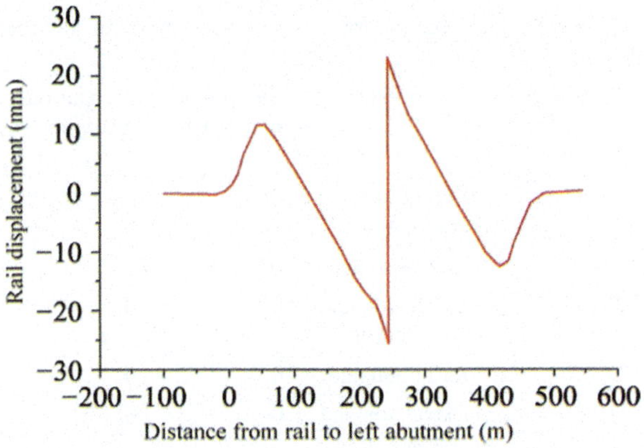

Fig. 9.90 Rail displacement from rail breakage

turnout switch and the bridge was 4.0 mm, which was less than 5 mm specified in the specification. All the indicators meet the requirements.

(2) Rail fracture

The calculations of forces from rail breakage focused on a rail fracture on Track 2, at the position with the maximum additional longitudinal force due to temperature variation. This point is positioned at the left end of the (32 + 32.7 + 32.7 + 102) m continuous girder and arch, 244.8 m away from the left abutment.

As shown in Fig. 9.90, the rail displacement due to rail breakage was 47.9 mm, remaining within the permissible value of 70 mm, which meets the requirements.

(3) Longitudinal forces on turnout force transfer devices

See Table 9.22 for the longitudinal forces on turnout force transfer devices. Considering the limited temperature fluctuations in turnouts and the alignment of expansion and contraction directions of the movable parts with the bridge, the longitudinal forces on the turnout force transfer devices are small. Specifically, the stoppers were not subjected to longitudinal forces, and the maximum longitudinal force on the spacing blocks was 77 kN, falling within the limit of 373 kN.

9.5.3 Dynamic Simulation Analysis of CWR Turnout on Bridge

In the previous section, static simulation calculation and analysis have been carried out for rail strength, turnout component displacement, and structural stress of the CWR turnouts on the continuous girder and arch crossing Jiulong Avenue. To ensure the driving safety and comfort of trains passing through this long-span bridge with the CWR turnouts, this section discusses the dynamic simulation calculations using a three-dimensional solid model of the (32 + 32.7 + 32.7 + 102) m continuous

Table 9.22 Longitudinal forces on turnout force transfer devices under various conditions (unit: kN)

Turnout No.	Track state	Stopper on main route	Stopper on diverging route	Spacing block on main route	Spacing block on diverging route
201#	Expansion and contraction	0	0	−77	−77
	Rail breakage	0	0	76	76
207#	Expansion and contraction	0	0	77	77
	Rail breakage	0	0	−76	−76
203#	Expansion and contraction	0	0	77	77
	Rail breakage	0	0	−76	−76
205#	Expansion and contraction	0	0	−77	−77
	Rail breakage	0	0	76	76

Note "+" indicates tension, and "−" indicates compression

turnout girder over Jiulong Avenue, and in the scenarios of a CRH_3 EMU running at speeds of 200, 220, 230, and 250 km/h.

1. Dynamic Simulation Analysis of No. 205 Turnout on Continuous Girder and Arch Bridge

(1) Dynamic wheel load distribution

With the left abutment as the origin of the coordinates, the dynamic wheel load distribution of the inner wheel on the front axle of the bogie is shown in Fig. 9.91. As a result of the parametric vibration of the bridge, the dynamic wheel load undergoes noticeable fluctuations. This is primarily due to the intensified vertical displacement change of the wheelsets resulting from the combined influence of bridge vibration and inherent structural irregularities present in the turnout, as shown in Fig. 9.92. The maximum dynamic wheel load in the frog section was about 128.3 kN, less than the permissible value.

(2) Flange force distribution

The contact between the wheelset's inner wheel and the switch rail, as well as the point rail, occurs within a limited range. The flange force distribution is shown in Fig. 9.93. The maximum flange force in the switch section was 51.5 kN, and the maximum instantaneous lateral impact force in the frog section was approximately 37.4 kN, both falling within the safety limit.

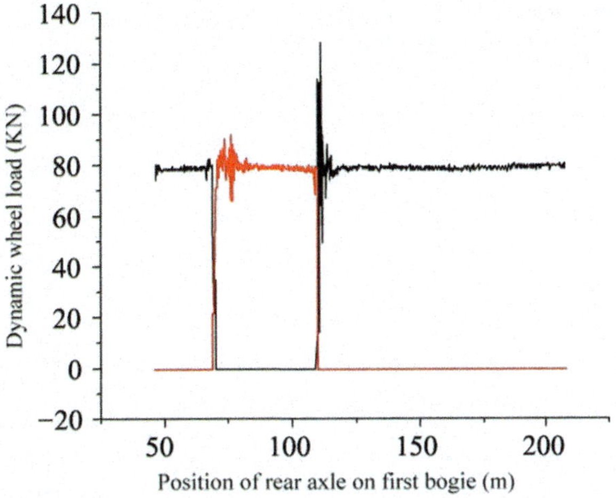

Fig. 9.91 Dynamic wheel load distribution under wheel

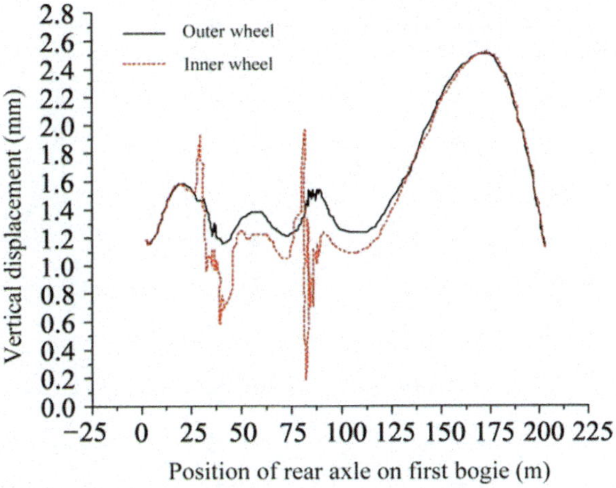

Fig. 9.92 Vertical displacement of wheelset

(3) Lateral Shift of Wheelset and Bogie

Figure 9.94 illustrates the lateral shift distribution of the front axle, rear axle, and frame of the leading bogie. As depicted in the figure, the maximum lateral shift of the front axle was recorded at about 7.36 mm.

(4) Derailment coefficient

Figure 9.95 illustrates the derailment coefficient distribution of wheels. As depicted in the figure, the maximum derailment coefficients of the front axle in the switch and frog sections were 0.22 and 0.24 respectively, while those of the rear axle in the switch and frog sections were 0.52 and 0.76 respectively, all falling within the safety limit.

Fig. 9.93 Flange force distribution of inner wheel

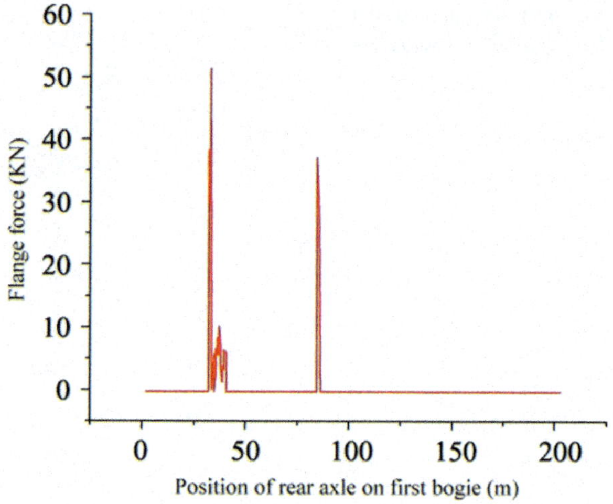

Fig. 9.94 Lateral shift of wheelset

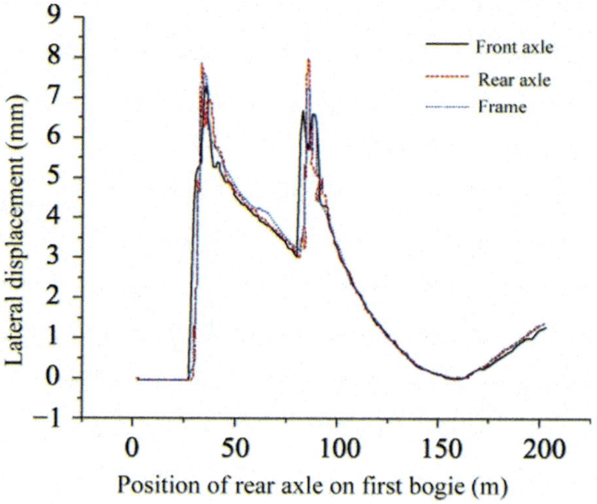

(5) Wheel unloading rate

Figure 9.96 illustrates the wheel unloading rate distribution of wheels. As depicted in the figure, the unloading rates of the inner wheel on the front axle in the switch and frog sections were 0.23 and 0.78 respectively, while those of the inner wheel on the rear axle in the switch and frog sections were 0.31 and 0.79 respectively, all falling within the safety limit.

(6) Riding quality of car body

The vertical and lateral vibration accelerations of the car body are shown in Fig. 9.97. The maximum lateral vibration acceleration in the switch section was about 0.36 m/ s^2, falling within the permissible limit.

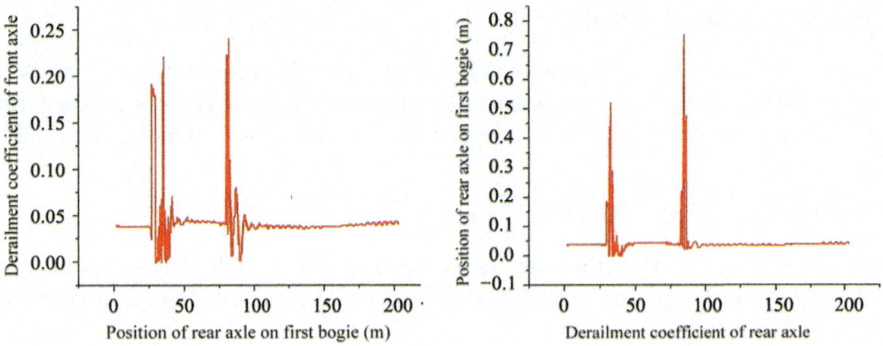

Fig. 9.95 Wheel derailment coefficient distribution

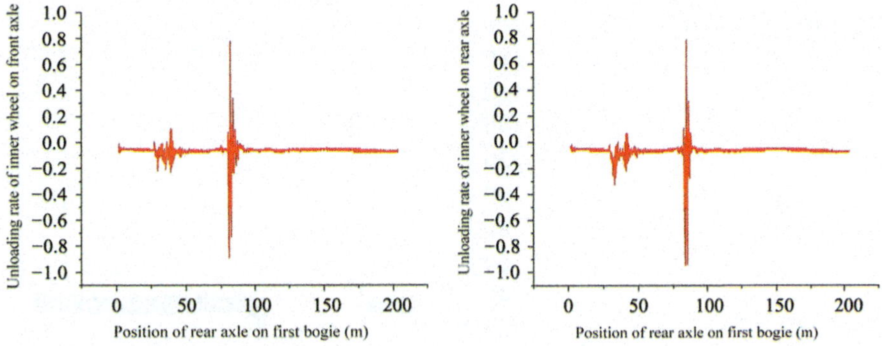

Fig. 9.96 Unloading rates of wheels

Fig. 9.97 Lateral vibration acceleration of car body

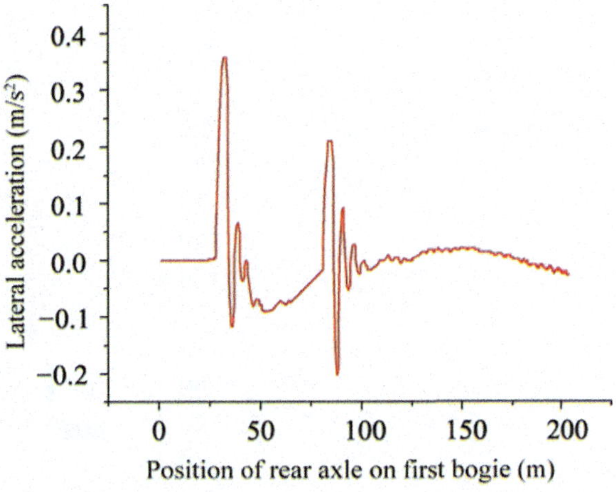

(7) Dynamic Stress of Rail

Figure 9.98 illustrates the dynamic stress of the rails. The impact dynamic stress of the switch rail in the switch section was around 180.5 MPa, and that of the point rail in the frog section was approximately 82.5 MPa, both falling within the permissible limit.

(8) Opening of switch rail and point rail

Figures 9.99 and 9.100 illustrate the opening of the switch rail and point rail, which were recorded around 0.49 and 0.018 mm respectively, both falling within the permissible limit.

(9) Bridge displacement

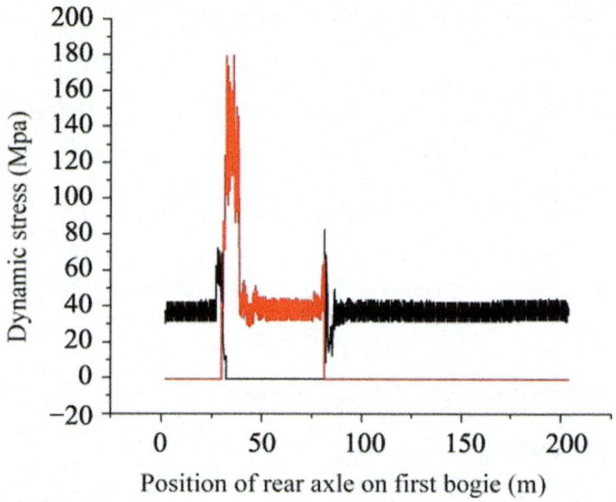

Fig. 9.98 Dynamic stress of rail

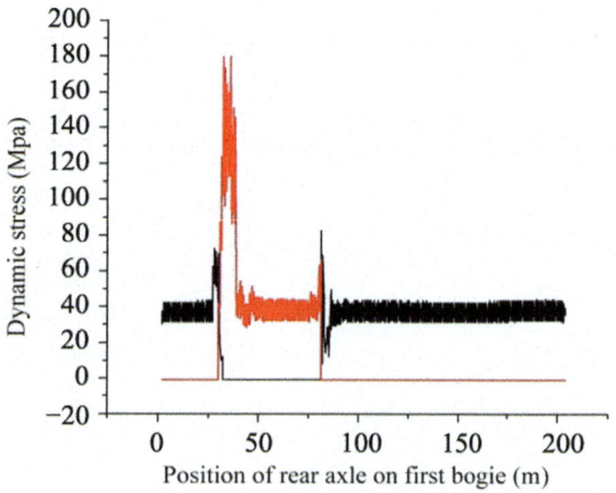

Fig. 9.99 Opening of switch rail

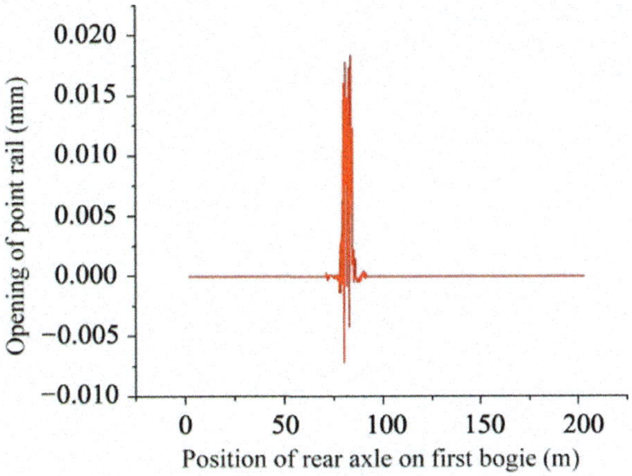

Fig. 9.100 Opening of point rail

Figure 9.101 illustrates the corresponding relationship between the vertical dynamic displacement of the bridge under the front axle/deflection change rate of the bridge and the unloading rate. As depicted in the figure, under the action of EMU loads, the maximum dynamic displacement of the 102-m-long girder span was about 2.53 mm, while that of the 32-m-long span on the left side was about 0.76 mm. The dynamic deflection at the frog was small but the deflection change rate was high, indicating non-ideal turnout-bridge relative positions. Figure 9.102 illustrates the lateral dynamic displacement of the bridge under the front axle. Under the sole action of turnout structural irregularities, the lateral dynamic displacement of the bridge did not exceed 0.11 mm, which was considered unnoticeable.

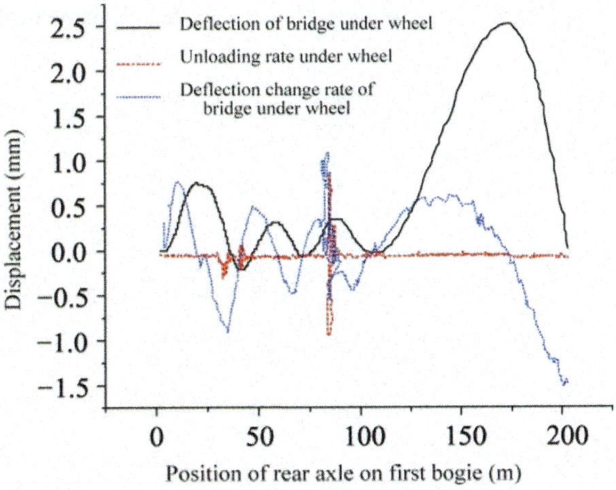

Fig. 9.101 Vertical displacement of bridge under wheel, deflection change rate of bridge, and unloading rate

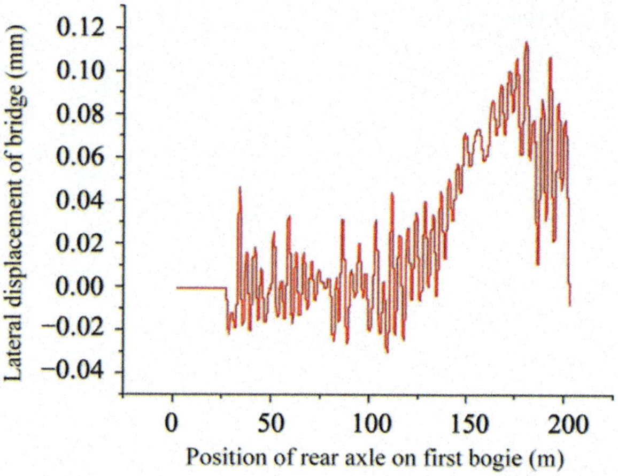

Fig. 9.102 Lateral displacement of bridge under wheel

Figure 9.103 illustrates the vertical rotation angle distribution at the bearings of the 102-m-long girder span, and the maximum dynamic rotation angle was recorded at approximately 0.14%. Figure 9.104 illustrates the horizontal rotation angle distribution at the bearings in middle spans, and the maximum dynamic rotation angle was approximately 0.0034%. All the displacement indexes fall within the permissible limits.

(10) Vibration acceleration of bridge

The vertical and lateral vibration accelerations of the main span of the continuous girder under the wheels are shown in Figs. 9.105 and 9.106. The maximum vertical vibration acceleration was about 0.44 m/s^2 and the maximum lateral vibration acceleration was about 0.11 m/s^2, both falling within the permissible limit.

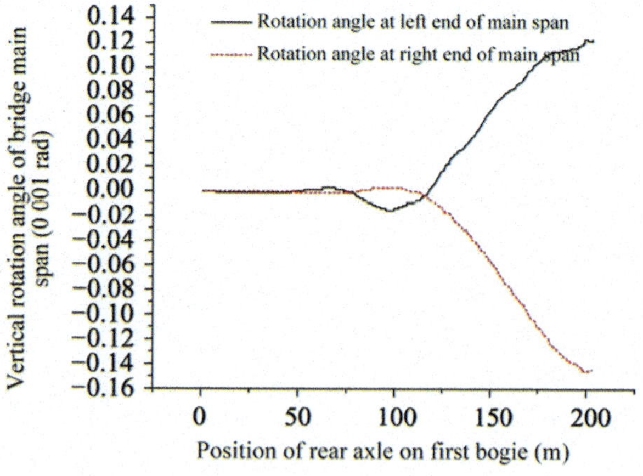

Fig. 9.103 Vertical rotation angle distribution at bearings in right side span

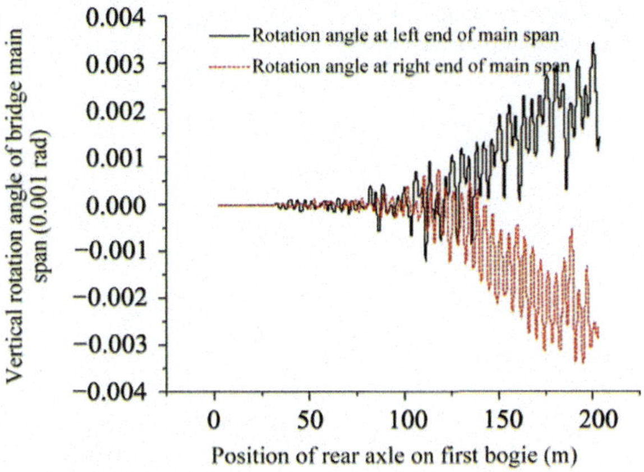

Fig. 9.104 Horizontal rotation angle distribution at bearings in right side span

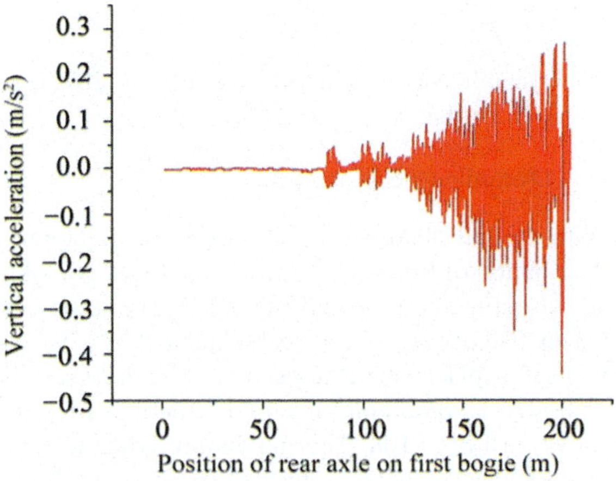

Fig. 9.105 Vertical vibration acceleration of main span

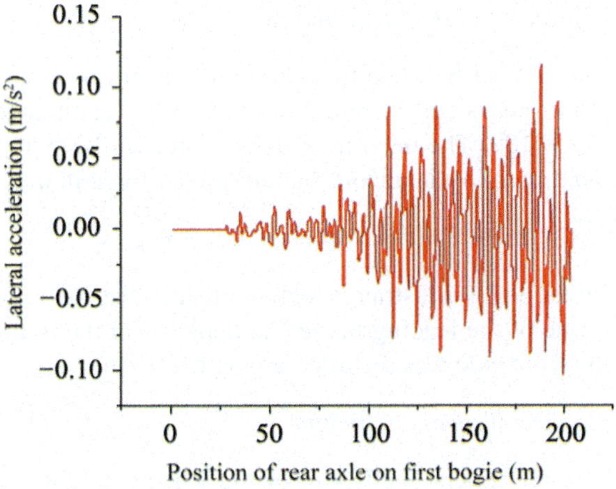

Fig. 9.106 Lateral vibration acceleration of main span

Fig. 9.107 Dynamic wheel load distribution under wheel

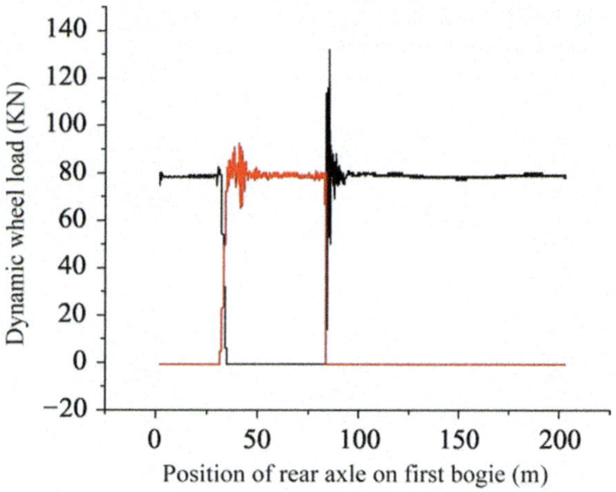

2. Dynamic Simulation Analysis of No. 207 Turnout on Continuous Girder and Arch Bridge

(1) Dynamic wheel load distribution

With the left abutment as the origin of the coordinates, the dynamic wheel load distribution of the inner wheel on the front axle of the first motor car's first bogie is shown in Fig. 9.107. As a result of the parametric vibration of the bridge, the dynamic wheel load undergoes noticeable fluctuations. This is primarily due to the intensified vertical displacement change of the wheelsets resulting from the combined influence of bridge vibration and inherent structural irregularities present in the turnout, as shown in Fig. 9.108. The maximum dynamic wheel load in the frog section was about 132.7 kN, falling within the permissible limit.

(2) Flange force distribution

The contact between the wheelset's inner wheel and the switch rail, as well as the point rail, occurs within a limited range. The flange force distribution is shown in Fig. 9.109. The maximum flange force in the switch section was 51.0 kN, and the maximum instantaneous lateral impact force in the frog section was about 44.0 kN.

(3) Lateral Shift of Wheelset and Bogie

Figure 9.110 illustrates the lateral shift distribution of the front axle, rear axle, and frame of the leading bogie. As depicted in the figure, the maximum lateral shift of the front axle was recorded at about 7.43 mm.

(4) Derailment coefficient

The derailment coefficient of the wheels is represented by the ratio of the sum of wheel-rail lateral forces to the dynamic wheel load, and its distribution is shown in

Fig. 9.108 Vertical displacement of wheelset

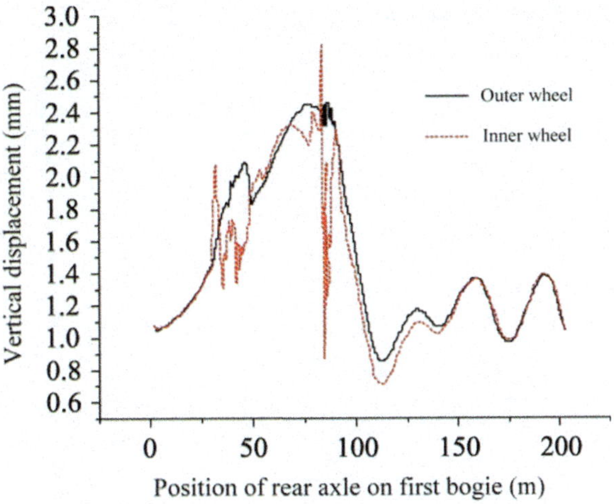

Fig. 9.109 Flange force distribution of inner wheel

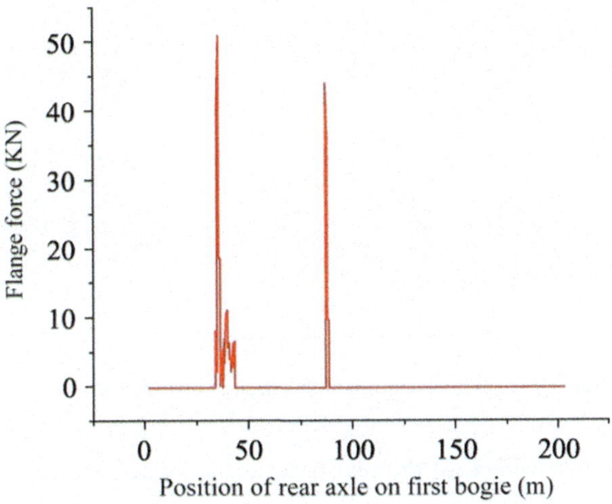

Fig. 9.111. As depicted in the figure, the maximum derailment coefficients of the front axle in the switch and frog sections were 0.20 and 0.22 respectively, while those of the rear axle in the switch and frog sections were 0.52 and 0.67, respectively.

(5) Wheel unloading rate

The unloading rate on wheels is represented by the ratio of the unloading amount of dynamic wheel load to static wheel load, and its distribution is shown in Fig. 9.112. As depicted in the figure, the unloading rates of the inner wheel on the front axle in the switch and frog sections were 0.24 and 0.78 respectively, while those of the

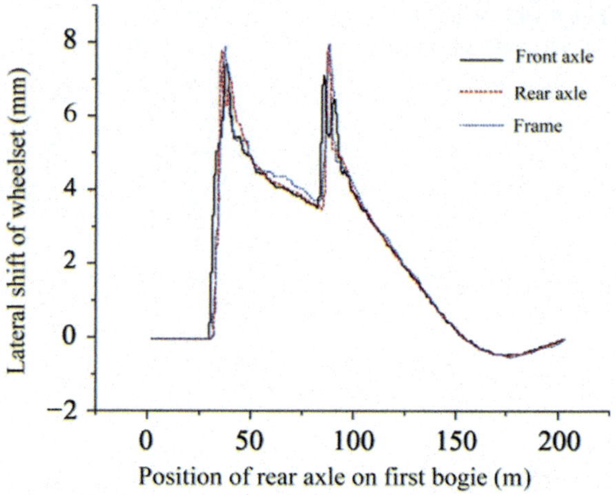

Fig. 9.110 Lateral shift of wheelset

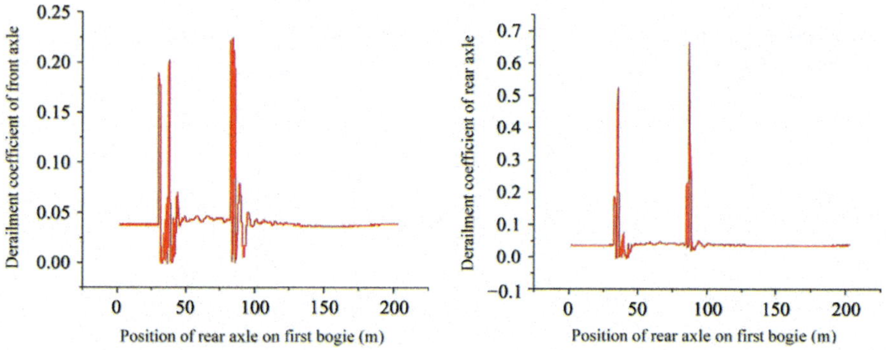

Fig. 9.111 Wheel derailment coefficient distribution

inner wheel on the rear axle in the switch and frog sections were 0.31 and 0.77, respectively.

(6) Riding quality of car body

The vertical and lateral vibration accelerations of the car body are shown in Fig. 9.113. The maximum lateral vibration acceleration in the switch section was about 0.35 m/ s^2, falling within the permissible limit.

(7) Dynamic Stress of Rail

Figure 9.114 illustrates the dynamic stress of the rails. The impact dynamic stress of the switch rail in the switch section was around 181.7 MPa, and that of the point rail in the frog section was approximately 83.8 MPa, both falling within the permissible limit.

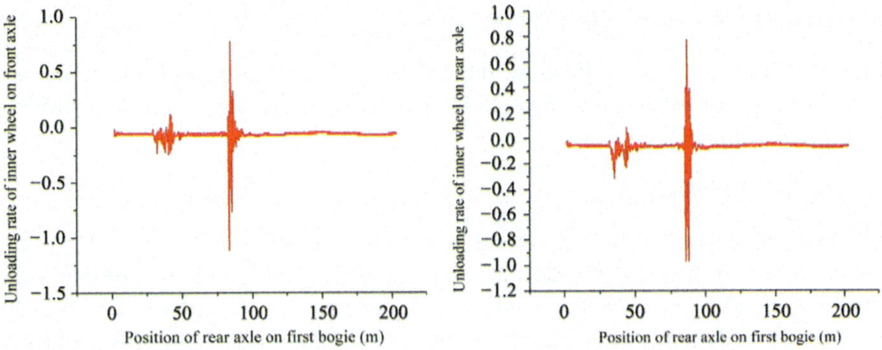

Fig. 9.112 Unloading rates of wheels

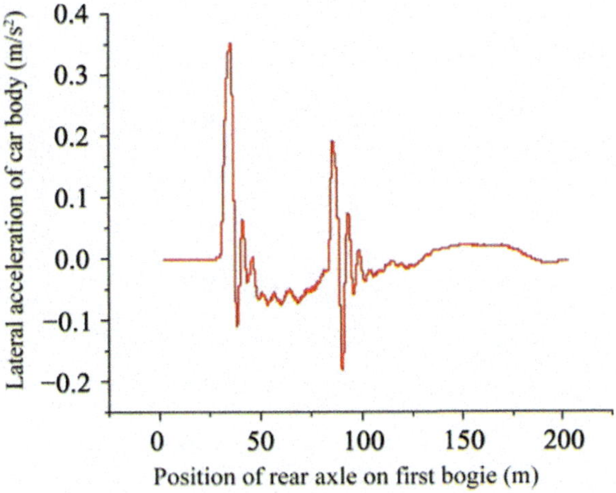

Fig. 9.113 Lateral vibration acceleration of car body

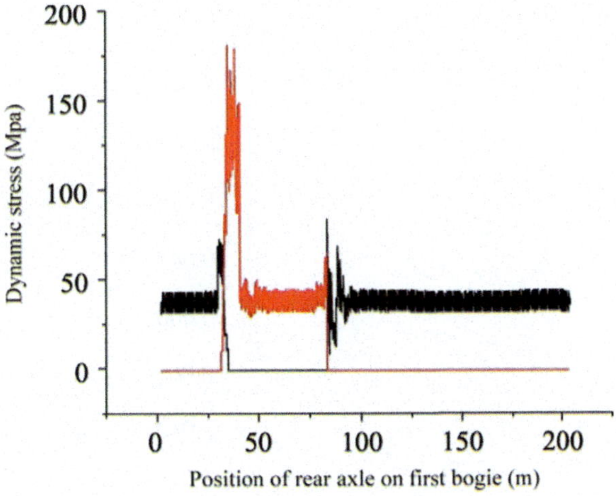

Fig. 9.114 Dynamic stress of rail

(8) Opening of switch rail and point rail

Figures 9.115 and 9.116 illustrate the opening of the switch rail and point rail, which were recorded around 0.53 and 0.019 mm respectively, both falling within the permissible limit.

(9) Bridge displacement

Figure 9.117 illustrates the corresponding relationship between the vertical dynamic displacement of the bridge under the front axle/deflection change rate of the bridge and the unloading rate. As depicted in the figure, under the action of EMU loads, the maximum dynamic displacement of the 102-m-long girder span was about 2.42 mm, while that of the 32-m-long span was about 0.58 mm. Figure 9.118 illustrates the lateral dynamic displacement of the bridge under the front axle. Under the sole

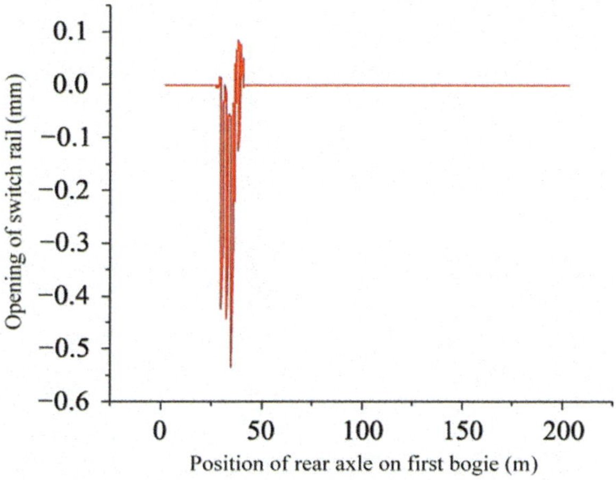

Fig. 9.115 Opening of switch rail

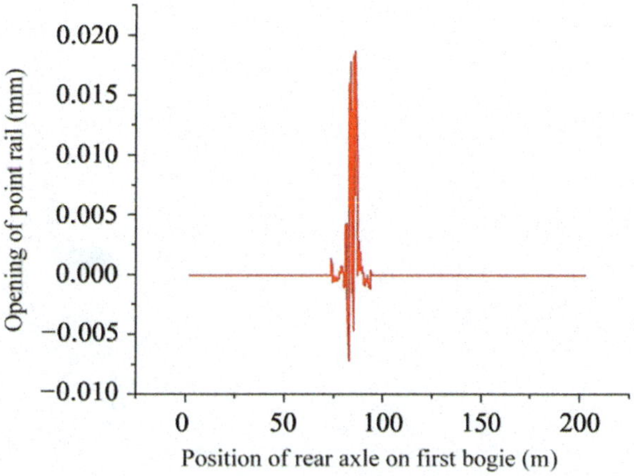

Fig. 9.116 Opening of point rail

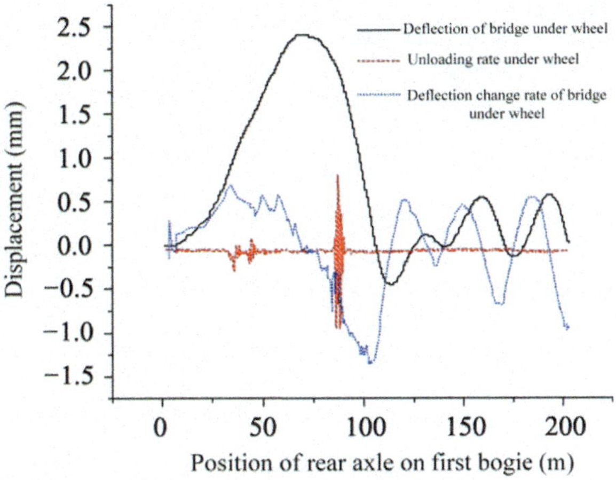

Fig. 9.117 Vertical displacement of bridge under wheel, deflection change rate of bridge, and unloading rate

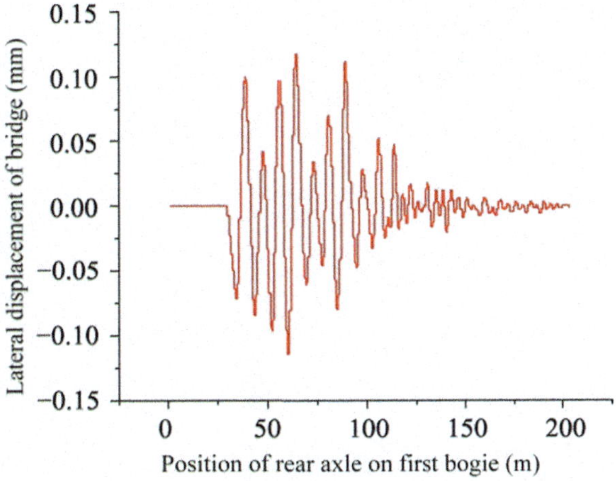

Fig. 9.118 Lateral displacement of bridge under wheel

action of turnout structural irregularities, the lateral dynamic displacement of the bridge did not exceed 0.12 mm, which was considered unnoticeable.

Figure 9.119 illustrates the vertical rotation angle distribution at the bearings of the 102-m-long girder span, and the maximum dynamic rotation angle was recorded at approximately 0.144‰. Figure 9.120 illustrates the horizontal rotation angle distribution at the bearings of the 102-m-long span, and the maximum dynamic rotation angle was approximately 0.0036‰. All the displacement indexes fall within the permissible limits.

(10) Vibration acceleration of bridge

The vertical and lateral vibration accelerations of the 102-m-long span of the continuous girder under the wheels are shown in Figs. 9.121 and 9.122. The maximum

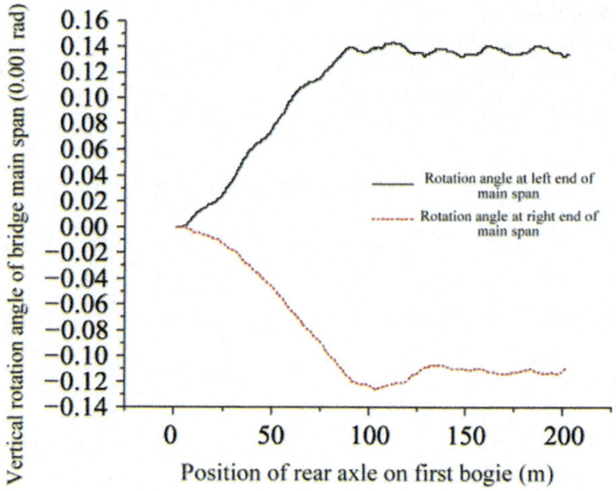

Fig. 9.119 Vertical rotation angle distribution at bearings in right side span

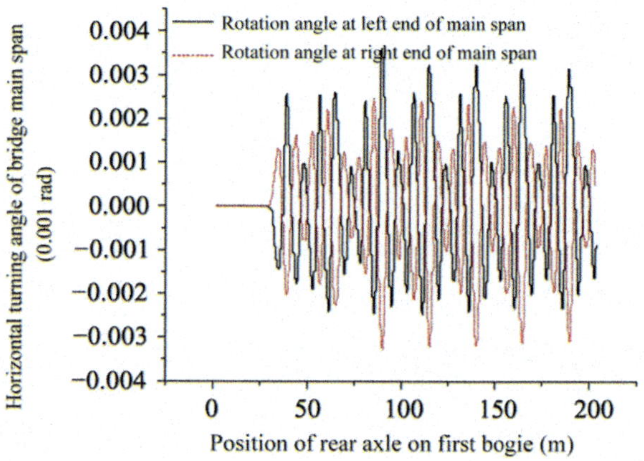

Fig. 9.120 Horizontal rotation angle distribution at bearings in right side span

vertical vibration acceleration was about 0.69 m/s^2 and the maximum lateral vibration acceleration was about 0.49 m/s^2, both falling within the permissible limit.

3. Summary of Train-Turnout-Bridge Dynamic Simulation Results

The above results are summarized in Table 9.23.

The above analysis leads to the following conclusions and recommendations:

(1) Dynamic responses between trains and turnouts are closely linked to the turnout passing speeds of trains on bridges. As turnout passing speeds escalate, the magnitude of various dynamic responses within the system also intensifies.
(2) When EMUs run at 250 km/h, all dynamic responses remain within acceptable limits, satisfying the requirements for driving safety, passenger comfort, and the structural safety of turnout bridges. The maximum wheel unloading rate within the No. 205 turnout section during EMU traversal was recorded at 0.79,

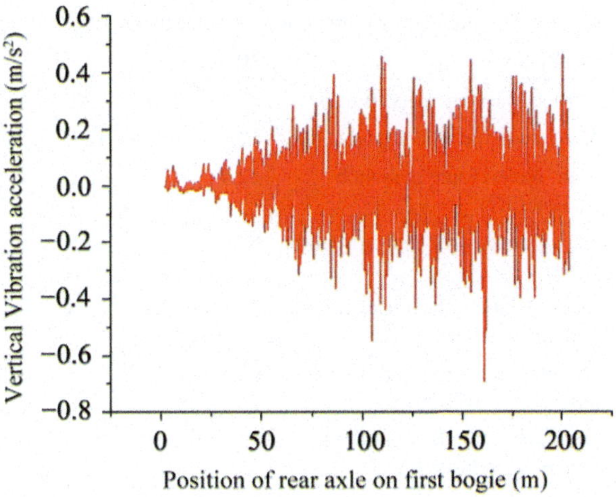

Fig. 9.121 Vertical vibration acceleration of main span

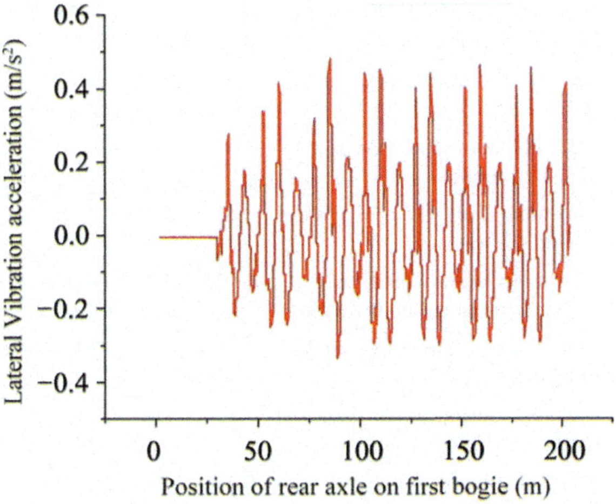

Fig. 9.122 Lateral vibration acceleration of main span

while in the No. 207 turnout section, this rate stood at 0.78, remaining within the specified limit for dynamic checking calculations of the train-turnout-bridge system.

(3) To address the stress and deformation issues associated with CWR turnouts laid on long-span continuous girder and arch combinations, it is necessary to carry out a comprehensive system analysis, taking into account the specific structures and arrangements of the turnouts and bridges.

(4) The simulation analysis results reveal that properly increasing the vertical rigidity of the bridge structures can effectively alleviate coupled vibration within the train-turnout-bridge system, creating beneficial conditions for maintaining

Table 9.23 Summary of maximum calculation results

Turnout No.	205#				207#			
Turnout passing speed (km/h)	200	220	230	250	200	220	230	250
Max. dynamic wheel load (kN)	123.5	126.4	128.1	128.3	130.9	132.6	133.4	132.7
Max. flange force in switch section (kN)	52.4	51.5	48.6	51.5	52.0	51.6	46.9	51.0
Max. flange force in frog section (kN)	53.6	51.3	45.9	37.4	49.9	49.6	47.6	44.0
Derailment coefficient of front axle in switch section	0.28	0.29	0.24	0.22	0.26	0.22	0.24	0.20
Derailment coefficient of front axle in frog section	0.23	0.23	0.23	0.24	0.23	0.24	0.22	0.22
Derailment coefficient of rear axle in switch section	0.53	0.53	0.53	0.52	0.53	0.53	0.51	0.52
Derailment coefficient of rear axle in frog section	0.65	0.70	0.74	0.76	0.57	0.64	0.65	0.67
Max. unloading rate of front axle in switch section	0.22	0.23	0.25	0.23	0.22	0.24	0.26	0.24
Max. unloading rate of front axle in frog section	0.72	0.75	0.77	0.78	0.71	0.75	0.78	0.78
Max. unloading rate of rear axle in switch section	0.30	0.27	0.28	0.31	0.31	0.27	0.28	0.31
Max. unloading rate of rear axle in frog section	0.67	0.75	0.78	0.79	0.60	0.68	0.76	0.77
Maximum lateral vibration acceleration of car body ($\times g$)	0.035	0.035	0.036	0.036	0.034	0.035	0.035	0.035
Max. dynamic stress of switch rail (MPa)	175.5	174.6	177.9	180.5	173.0	172.9	182.3	181.7
Max. dynamic stress of point rail (MPa)	79.7	81.0	81.5	82.5	80.4	80.9	81.9	83.8
Max. opening of switch rail (mm)	0.47	0.53	0.49	0.49	0.52	0.53	0.51	0.53
Max. opening of point rail (mm)	0.020	0.020	0.019	0.018	0.020	0.020	0.019	0.019
Max. vertical dynamic displacement of bridge (mm)	2.48	2.49	2.41	2.53	2.42	2.43	2.42	2.42
Max. lateral dynamic displacement of bridge (mm)	0.08	0.09	0.06	0.11	0.16	0.15	0.13	0.12
Vertical vibration acceleration in main span of bridge ($\times g$)	0.033	0.039	0.033	0.044	0.054	0.061	0.057	0.069
Lateral vibration acceleration in main span of bridge ($\times g$)	0.009	0.010	0.007	0.011	0.054	0.056	0.050	0.049

the service state of the turnouts on bridges and facilitating maintenance after putting into operation.

(5) The turnout passing speed on this bridge should be controlled below 250 km/h, and field comprehensive evaluations and tests are recommended for research on ballastless CWR turnouts on continuous girder and arch bridges.

9.6 Design of CWR Turnout on the Bridge at Zhangjiagang Station of Shanghai–Nanjing HSR

9.6.1 Design Overview

Shanghai–Nanjing High-Speed Railway along the Yangtze River is an important component of China's national high-speed railway network for passenger transport and the high-speed railway network across the Yangtze River Delta. It is the second passenger transport route for the Shanghai–Nanjing Corridor. This railway plays an important part in the comprehensive multimodal traffic corridor in the Yangtze River Economic Belt and reflects a move to improve and extend Shanghai–Wuhan–Chengdu High-Speed Railway Line. This railway extends for a total length of 281.12 km and is designed for train operation up to 350 km/h. It was put into operation on September 28, 2023, linking 8 stations: Nanjingnan, Jurong, Jintan, Wujin, Jiangyin, Zhangjiagang, Changshu, and Taicang. It is equipped with CRTS III slab ballastless tracks along its entire stretch. Within the turnout sections, sleeper-embedded ballastless tracks are employed, complemented by CRTS bi-block ballastless tracks for the transitional sections leading into and out of these turnout sections.

Zhangjiagang Station on this railway is an elevated station. The bridge span arrangement at the throat area at the small mileage end consists of 24 m simply-supported girder + (2 × 32) m simply-supported girder + (6 × 32) m turnout continuous girder + 32 m simply-supported girder + (28.6 + 34.7 + 34.7 + 32.7 + 32.6) m turnout continuous girder + (2 × 24) m simply-supported girder + 32 m simply-supported girder. The arrangement of the No. 18 single crossover and single turnout on the bridge are shown in Fig. 9.123. The distance between centers of tracks is 5.3 m.

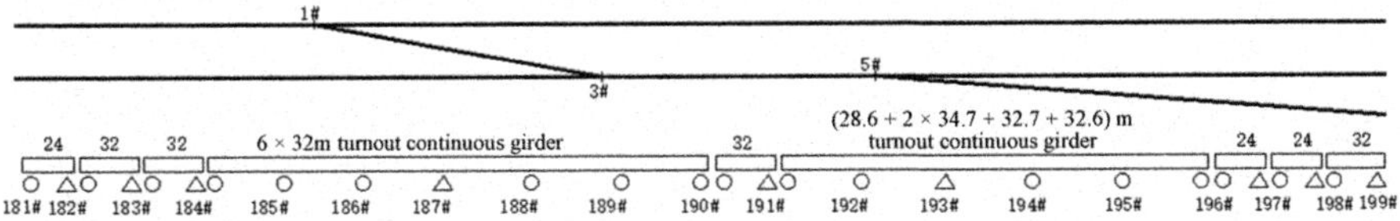

Fig. 9.123 Arrangement of CWR turnout on bridge at throat area at small mileage end on left track at Zhangjiagang station

9.6.2 Computation of CWR Turnout on Bridge

Checking calculations were conducted on rail strength, stress, and displacement within the CWR turnout at the throat area at the small mileage end of Zhangjiagang Station as follows.

1. Calculation Parameters

(1) Pier rigidity

See Table 9.24 for the rigidity of piers beneath the simply-supported girder and continuous girder for the turnout.

2. Calculation of Temperature Stress

The local maximum rail temperature is 59.1 °C, and the minimum rail temperature is −11.3 °C. The design stress-free rail temperature is 25 ± 3 °C.

Table 9.24 Rigidity of piers/abutments

Pier No.	Type of girder	Span length (m)	Longitudinal rigidity (kN/cm)	Remarks
181#				
182#	Simply-supported girder	24	551.3	
183#	Simply-supported girder	32	531.9	
184#	Simply-supported girder	32	905	
185#	(6 × 32) m continuous girder beneath turnout	32		
186#		32		
187#		32	1363.3	Fixed bearing
188#		32		
189#		32		
190#		32		
191#	Simply-supported girder	32	1629.2	
192#	(28.6 + 34.7 + 34.7 + 32.7 + 32.6) m continuous girder beneath turnout	28.6		
193#		34.7	3335.6	Fixed bearing
194#		34.7		
195#		32.7		
196#		32.6		
197#	Simply-supported girder	24	1130.9	
198#	Simply-supported girder	24	1041.7	
199#	Simply-supported girder	32	1326.2	

The maximum temperature drop range of rail is calculated as $\Delta Tj = 25 + 3 - (-11.3) = 39.3$ °C.

The maximum temperature rise range of rail is calculated as $\Delta Ts = 59.1 - 25 + 3 = 37.1$ °C.

The temperature stress of rail is calculated as $\sigma t = 2.1 \times 10^{11} \times 1.18 \times 10^{-5} \times 39.3 = 97.39$ MPa.

3. Calculation of Dynamic Flexural Stress

The maximum axle load of high-speed EMUs is taken as 170 kN, and the rigid wheelbase as 2.5 m. The design speed is 350 km/h and the speed coefficient is $\alpha = 1$. This bridge is located on a straight section with an eccentric load coefficient of $\beta = 0$ and lateral force coefficient of $f = 1.25$. The rail support rigidity D of the ballastless track is taken as 23 kN/mm. The sleeper spacing is taken as a = 600 mm. The rail section parameter is taken as 9 mm for wear:

$$\sigma_{\mathrm{bd}} = \frac{M_0(1 + \alpha + \beta)}{W_{\mathrm{b}}} \cdot f$$

where M_0—static bending moment of rail (N·mm);

α—velocity coefficient;

β—eccentric load coefficient.

See Table 9.25 for the calculation results of the dynamic flexural stress of rails.

Based on the calculation, the dynamic flexural tensile stress of the rail base is $\sigma_{\mathrm{d}} = 119.78$ MPa.

Table 9.25 Calculation of dynamic flexural stress of rail

Calculation of dynamic flexural stress			Calculation parameters	
Wheel position	I	II	E	210,000
P_0	85,000	85,000	I	26,900,000
x	0	2500	W_h	264,000
kx	0	2.8532	W_b	363,000
u_0	1.00	−0.0717	D	23
$P_0u_0u_0$	85,000.00	−6094.50	σ	600
ΣP_0	78,905.500000		α	1
M_j	17,284,734.00		β	0
M_d	34,569,468.00		f	1.25
σ_b	119.04		k	0.00114126
σ_h	163.68			

4. Analysis and Checking Calculation

1) Strength of Rails within Turnout Section

(1) Longitudinal force due to temperature variation of rail and girder-track relative displacement

The maximum longitudinal force due to temperature variation + temperature stress of the rail was 1500.3 kN, and the maximum girder-rail relative displacement was 18.18 mm, as shown in Figs. 9.124 and 9.125.

(2) Braking force on rail and girder-rail relative displacement

The maximum braking force on the rail was 241.7 kN, and the maximum girder-rail relative displacement was 1.635 mm, as shown in Figs. 9.126 and 9.127.

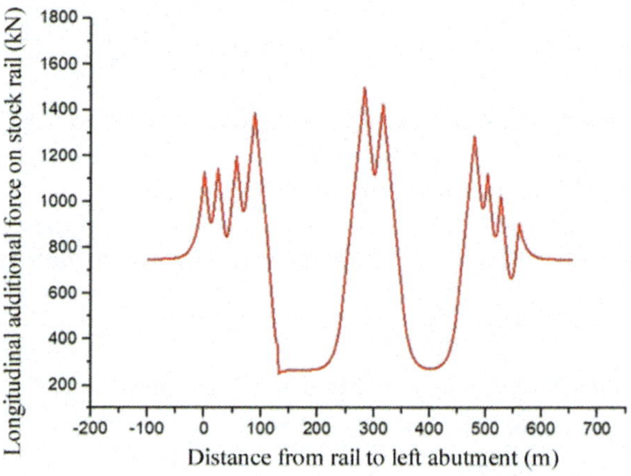

Fig. 9.124 Additional longitudinal forces due to temperature variation on rails

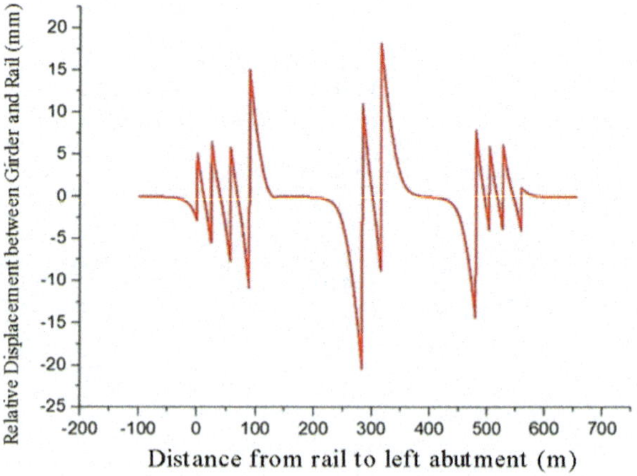

Fig. 9.125 Girder-rail relative expansion and contraction displacements

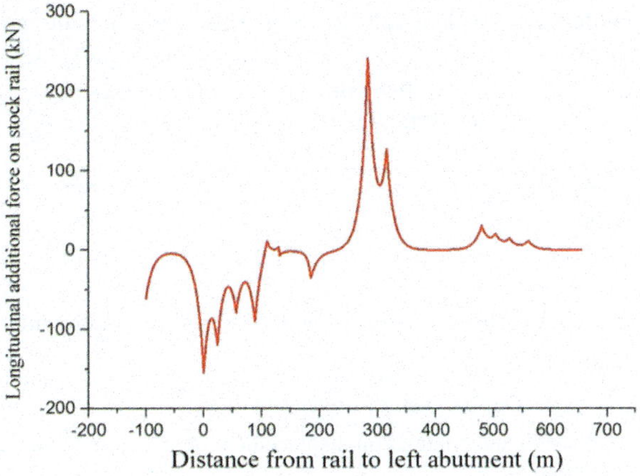

Fig. 9.126 Additional braking forces on rails

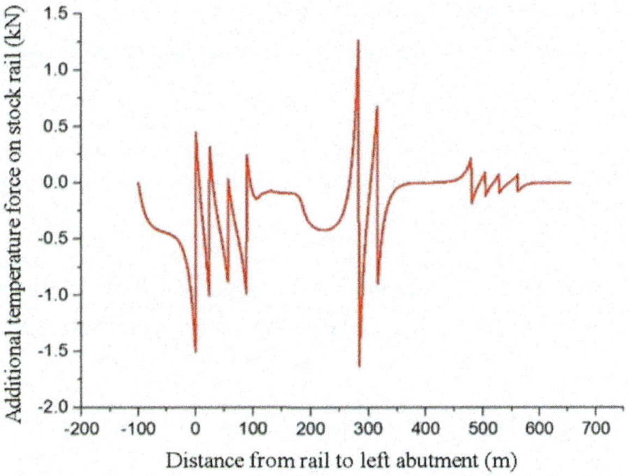

Fig. 9.127 Relative girder-rail braking displacements

(3) Calculation of rail strength

According to the checking calculation formula for rail strength, the calculation results of rail strength are shown in Table 9.26.

Table 9.26 Checking calculation of turnout rail strength (unit: MPa)

Item	Dynamic flexural stress of rail	Rail temperature stress + additional longitudinal force due to temperature variation	Braking force on rail	Total stress of rail	Permissible stress of rail	Conclusion
Calculated value	119.04	193.71	31.21	343.96	351.54	Satisfied

Table 9.27 Checking calculation of broken rail gap (unit: mm)

Item	Design stress-free rail temperature (°C)	Maximum temperature drop (°C)	Broken rail gap (mm)	Permissible broken rail gap (mm)	Conclusion
Calculated value	25	39.3	14.57	70	Satisfied

Table 9.28 Displacements of switch rail and point rail (unit: mm)

Turnout No.	1#	3#	5#
Absolute displacement of switch rail	−31.1	−21.01	15.16
Expansion and contraction displacement of switch rail relative to stock rail	7.38	−8.73	4.34
Displacement of stock rail relative to bridge at switch	−3.43	−5.0	0.35
Absolute displacement of point rail	12.32	−2.28	−3.44
Expansion and contraction displacement of point rail relative to wing rail	1.56	−1.56	1.76

Table 9.29 Checking calculation of stress of connecting bolts at heel end of switch rail and point rail (unit: mm)

Turnout No.	1#	3#	5#
Shear stress of connecting bolt of stopper at heel end of switch rail	70.1	75	64.6
Shear stress of connecting bolt at heel end of point rail	57.1	57.1	64.6

2) Checking calculation of broken rail gap

According to the calculation formula for broken rail gaps, the checking calculation results of broken rail gaps are shown in the following table (Tables 9.27 and 9.28).

3) Checking calculation of turnout displacement

4) Checking calculation on strength of turnout components

The maximum load on the switch rails and point rails within all the turnouts was taken for checking calculation. The permissible shear stress of Grade 10.9 high-tensile bolts was taken as 415 MPa, and the bolt diameter as 27 mm. The longitudinal stress limit of bolts was deduced inversely using the formula, which was 238 kN. The shear stress calculation results for the connecting bolt of the stopper at the heel end of the switch rail and that of at the heal end of the point rail are shown in Table 9.29.

According to the checking calculation, for the CWR turnout at the throat area at the small mileage end on the left track at Zhangjiagang Station, the rail strength, broken rail gaps, girder-rail relative displacement at the switch, turnout displacement, and strength of turnout components all meet the requirements.

References

1. Liu Y, Ren J, Wu Y et al (2014) Influence of void beneath turnout slab on stress performance of longitudinally coupled slab ballastless turnout on bridge. China Railw Sci 35(6):21–27
2. Wang Y (2014) Design and research of slab ballastless CWR turnout on bridges in Xuzhou-Shanghai section of Beijing-Shanghai high-speed railway. Railw Eng 5:120–123
3. Li Q (2017) Main technical innovations of CRTS III slab ballastless track on Zhengzhou-Xuzhou high-speed railway. J Railw Eng Soc 34(12):18–23
4. Zhang J (2010) Construction technology of long-sleeper embedded ballastless turnout for Wuhan-Guangzhou passenger dedicated railway. Railw Stand Des (1):90–93; 104
5. Cui G (2009) Research on structural design of domestic ballastless track for turnout section. J Entrep Sci Technol 22(8):152–153
6. Ma H (2015) Research on technologies of ballastless No. 18 single turnout with 60 kg/m rails for 350 km/h passenger dedicated line. Shijiazhuang Tiedao University, Shijiazhuang
7. Zhang D (2009) Structural design of No.42 simple turnout with 60 kg/m rails for 350 km/h passenger dedicated railway. Railw Stand Des (5):6–9
8. Yan B, Cheng R, Xie H et al (2021) Stress characteristics of CRTS II ballastless track on bridge under extreme temperature. J Railw Sci Eng 18(4):830–836
9. Zhang P, Tu J, Gui H et al (2021) Mechanical properties of CRTS II slab ballastless track on bridge under temperature gradient loads. J Southwest Jiaotong Univ 56(5):945–952
10. Lai Z, Jiang L (2022) Mapping relationship between longitudinal deformation and rail top deformation of ballastless track-bridge system of high-speed railway. J Central South Univ (Sci Technol) 53(2):535–546
11. Yapeng MI, Suixiao LIU (2010) Research on positioning method of shear notch and lateral stop on bridge deck with CRTS II slab ballastless track. Railway Stand Des 9:44–46
12. Dong C (2015) A discussion of Shaoxing north station transport hub design of Hangzhou-Ningbo passenger special line. China Municipal Eng (3):83–85; 119
13. Wenxiang SONG (2012) Study on route selection of Hangzhou-Ningbo passenger dedicated line. Railw Construct Technol 12:40–43
14. Xuezhong XIN (2012) Engineering technical characteristics and new construction technologies of Hangzhou-Ningbo passenger dedicated railway. Chin Railw 8:5–9
15. Zhou T, Liu Z, Wu H et al (2012) Static fine adjustment technology of Hangzhou-Ningbo passenger dedicated line. Railw Construct Technol (12):6–9; 19
16. Sun J (2017) Research on interactions between CWR turnout and bridge under temperature field of high-speed railway. China Railw Sci 38(1):43–48
17. Yan Z, Tang J, Tan Y et al (2014) Calculation of temperature force and displacement of CWR turnout on bridge with ANSYS. Sci Technol West China 13(3):41–42
18. Chu W (2008) Research on temperature force and displacement of CWR turnout with ballasted track on bridge based on ANSYS. Central South University, Changsha
19. Ye Q, Tang J, Yu L et al (2009) Analysis of longitudinal force due to deflection and displacement characteristics of CWR turnout with ballastless track on bridge. Sci Technol West China 8(15):7–9; 5
20. Ren J, Wang P, Wang X et al (2011) Braking forces and displacements of longitudinally coupled ballastless welded turnout on bridges. J Southwest Jiaotong Univ 46(1):49–55
21. Gao L, Jiang H (2022) Research on model updating of train-track-bridge spatial coupling system on high-speed railway based on multi-objective optimization. China Railw Sci 43(3):50–57
22. Yu L (2009) Study on spatial mechanics of CWR turnout with ballastless track on bridge. Central South University, Changsha
23. Ren J (2009) Research on stress characteristics and structural optimization of longitudinally coupled ballastless track within CWR turnout section on bridge. Southwest Jiaotong University
24. Chen R (2009) Research on theory and application of train-turnout-bridge coupling vibration of high-speed railway. Southwest Jiaotong University, Chengdu
25. Wu Z (2021) Research on resonance response of train-bridge coupling system based on pseudo excitation method. Beijing Jiaotong University, Beijing

26. Li C (2011) Dynamics research of ballastless turnout on bridge of high-speed railway based on rigid-flexible coupling co-simulation. Beijing Jiaotong University, Beijing
27. Guo F (2012) No.18 ballastless turnout with 60 kg/m rails for 350 km/h high-speed railway. Engineering Management Center of the Ministry of Railways, Beijing
28. Geng C (2021) Study on track service status monitoring of continuous girder and arch beneath turnout on Ganjiang super major bridge crossing Jiulong avenue. Chin Railw 6:85–91

The manufacturer's authorised representative in the EU is Springer Nature Customer Service Centre GmbH, Europaplatz 3, 69115 Heidelberg, Germany. If you have any concerns regarding our products, please contact ProductSafety@springernature.com

Printed and bound by CPI Group (UK) Ltd, Croydon, CR0 4YY
01/10/2026
02225907-0016